AA004331

Rudenko International Conference "Methodological Problems in Reliability Study of Large Energy Systems" (RSES 2024)

E3S Web of Conferences Volume 584 (2024)

Arkhangelsk, Russia
15-19 July 2024

Editors:

Valeriy Stennikov
Sergey Senderov
Alexander Michalevich
Husein Guliev

ISBN: 979-8-3313-0871-1

Printed from e-media with permission by:

Curran Associates, Inc.
57 Morehouse Lane
Red Hook, NY 12571

Some format issues inherent in the e-media version may also appear in this print version.

This work is licensed under a Creative Commons Attribution 4.0 International License. License details: http://creativecommons.org/licenses/by/4.0/.

No changes have been made to the content of these proceedings. There may be changes to pagination and minor adjustments for aesthetics.

Printed with permission by Curran Associates, Inc. (2025)

For additional information, please contact EDP Sciences – Web of Conferences
at the address below.

EDP Sciences – Web of Conferences
17, Avenue du Hoggar
Parc d'Activité de Courtabœuf
BP 112
F-91944 Les Ulis Cedex A
France

Phone: +33 (0) 1 69 18 75 75
Fax: +33 (0) 1 69 28 84 91

contact-edps@webofconferences.org

Additional copies of this publication are available from:

Curran Associates, Inc.
57 Morehouse Lane
Red Hook, NY 12571 USA
Phone: 845-758-0400
Fax: 845-758-2633
Email: curran@proceedings.com
Web: www.proceedings.com

TABLE OF CONTENTS

Taking into Account Balance Reliability in Planning of Power System Development 1
Chukreyev Iurii, Chukreyev Mikhail

Transformation of the Model of the Design Scheme of the Power System in the Problem of Ensuring Balance Reliability 5
Chukreyev Iurii

Impact of the Market Model in the Electric Power System on Its Reliability 9
Chukreyev Mikhail

Reproduction of the Mineral Resource Base of the Energy Sector as the Most Important Component of Ensuring Russia's Energy Security 14
Senderov Sergey

Finding Ways to Minimize Gas Shortages for Consumers in the Context of Large-Scale Emergencies in the Gas Industry 19
Senderov Sergey, Khamisov Oleg, Vorobev Sergey, Danilov Gleb

The Impact of Changes in Gas Consumption on the Selection of Critical Gas Industry Objects 24
Vorobev Sergey, Edelev Alexey

Assessment of the Level of Energy Security in the Regions of the Northwestern Federal District Under Emergency Conditions in the Gas Industry 28
Smirnova Elena

Use and Failure Cause Analysis of Gas-Turbine Units at Distributed Generation Facilities 36
Ilyushin Pavel, Shavlovskiy Sergey

Energy Storage Use Efficiency in the Context of Belorussian Power System Technological Limitations 41
Kashyn Mikhail, Molochko Andrey, Novikov Nikolay, Novikov Alexander

Battery Energy Storage System Placement and Sizing in Distribution Networks 45
Sorokin Dmitry

Markers of Cellulose Insulation Ageing: Calculation of Total Methanol Quantity in the Internal Insulation of Power Transformers 50
Darian L. A., Obraztsov R. M., Maksimchenko A. V., Verakso D. E., Polishchuk V. P., Logunov V. A.

Problems of Fuel and Energy Complex Management: Increasing Vulnerability of Intelligent Elements of Information and Control Systems to Electromagnetic Influences 55
Grabchak Evgeny, Grigoriev Vladimir, Loginov Evgeny

Digital Technologies for Managing the Turnover of Fuel and Energy Resources Within the Framework of an Expanded International Energy Association 59
Grigoriev Vladimir, Loginov Evgeny, Shkuta Alexander

Development of Effective Algorithms for Calculating Current Distribution Coefficients of an Electric Power System 64
Akhmetbaev Dauren, Dzhandigulov Abdygali, Akhmetbaev Arman, Bystrova Svetlana

Application of Static Thyristor Compensators to Increase the Reliability of Distribution Networks................ 69
Akhmetbaev Dauren, Akhmetbaev Arman

Modification of Control System Based on Virtual Synchronous Generator with Feedforward Control for Power Converters of Distributed Generation... 75
Askarov Alisher, Ufa Ruslan, Suvorov Aleksey, Kievets Anton

Control Algorithm of Energy Storage System Based on Virtual Synchronous Generator................................. 80
Suvorov Aleksey, Askarov Alisher, Rudnik Vladimir

On Reliability of 110-500 kV GIS: Operational Experience 85
Korotchenko Evgeniy, Galiaskarov Irek, Misrikhanov Misrikhan, Ryabchenko Vladimir, Shuntov Andrey

Placement of Controlled Shunt Reactors in 330 kV Electrical Networks and Study of Their Operating Modes 89
Hashimov Arif, Guliyev Huseyngulu, Babayeva Aytek

Determination of Optimal Dimensions of Hybrid AC/DC Distributed Generation System with Renewable Sources for Autonomous Power Supply of Remote Locations 95
Rahmanov Nariman, Guliyev Huseyngulu, Ýbtahimov Famil, Mammadov Zaur

Issues of Risk-Oriented Approach to Electrical Networks Reliability Management....................................... 101
Vanin Artem, Gabdushev Dmitry, Tulsky Vladimir

G Factor of the ESG Agenda in the Context of Sustainable Development of Russian Electric Power Companies 107
Sheveleva Galina I.

Methodology for Analysis of System Reliability When Planning States and Balances of Electric Power Systems 112
Krupenev Dmitry

Model for Minimizing Energy Consumption of Electric Power Systems Taking into Account the Balance of Active and Reactive Power...116
Iakubovskii Dmitrii, Krupenev Dmitry, Boyarkin Denis

Methodological Aspects of Reliability Assessment of Asynchronous Motors Under Current Operation Conditions at Mineral Resources Enterprises.. 121
Romanova Victoria, Batukhtin Andrey, Khromov Sergey, Alzakkar Ahmad

Improving the Reliability of the Power System in Postfault Condition 126
Solovyev Ivan, Motovilov Alexey, Hvijuzov Mikhail, Emelyanov Alexander

Methods for Calculating the Resistance of Grounding Devices with Backfilling Special Compositions.. 131
Drako Mikhail, Pavlovich Ivan, Baraishuk Siarhei

Measurement and Analysis of Electric Power Quality Indicators with Targeted Identification of the Perpetrators of Non-Compliances on the Example of the Belarusian Energy System 136
Drako Mikhail, Maiseyenkà Àleh, Kolik Viachaslau

Research on Impact of Solar Power Plants on 110 kV Line Current Differential Protection Operation.. 141
Navolochny Alexander, Onisova Olga, Salmin Alexander

Influence of the Temperature Schedule of Heating Network and Using Differential Pressure Regulators on the Reliability of Heat Supply Systems 146
Zvonareva Julia, Vankov Yuri, Sabirov Irek

Improving Protection of Medium Voltage Networks 151
Piskunov Sergey, Mokeev Alexey

Transformer Protection, Automation and Monitoring Systems 156
Piskunov Sergey, Mokeev Alexey, Ulyanov Dmitry

Methods for Ensuring the Efficiency and Safety of Energy System Facility Equipment Based on Digital Technologies 161
Sultanov Mahsud, Boldyrev Ilya, Ivanitckii Maxim, Menshikov Pavel

Prospective Power Systems Models Development Algorithms 166
Mashalov Eugene, Neuymin Vladimir, Snegirev Denis, Mihajlenko Andrej, Teplukhin Daniil, Klassen Viktor

Development of a Vector Control System Based on the DFSMC for the Study of Transients in the Conditions of Creating a Digital Twin of the Power System 171
Suprunov Igor, Novikov Nikolay, Novikov Alexander

Medium Voltage Digital Instrument Transformer 177
Ulyanov Dmitriy, Plakidin Roman

Low-Frequency Oscillations Analysis and Big Data 181
Popov Aleksandr, Butin Kirill, Rodionov Andrey

Practical Experience in Monitoring Low-Frequency Oscillations at Electric Power Station of United Power System of Russia 186
Govorun Mikhail, Zhukov Andrey, Satsuk Evgeniy, Dubinin Dmitriy, Karimov Nikita

Experience in Implementing the Algorithm for Emergency Control System of Unloading in Case of Overload in Terms of Power Based on Synchrophasor Angle 190
Govorun Mikhail, Satsuk Evgeniy, Luzhkovskiy Yuriy, Dubinin Dmitriy, Titov Sergey, Khokhrin Aleksandr

Fault Location with Increased Reliability for Overhead Transmission Lines 194
Lachugin Vladimir, Podshivalin Andrey, Ismukov Grigory, Lvov Alexandr

Monitoring of Medium Voltage Distribution Networks by Balance Smart Meters 198
Golub Irina, Voitov Oleg, Boloev Evgeny, Buchinsky Anatoly

Optimization of Data Processing Schemes for Locating Low-Frequency Oscillation Sources 204
Butin Kirill, Popov Aleksandr, Rodionov Andrey

Neural Networks and Damage Pattern Recognition in Power Transformer Diagnostics 209
Khrennikov Alexander, Aleksandrov Nikolay, Mikhailov Konstantin, Mikhailov Sergey

Principles of Controlling Intersystem Power Flows in the Conditions of the Electricity Market 213
Yusifbayli Nurali, Alizade Rana

Operation Modes of the Power System of Azerbaijan Under Conditions of Penetration of Large Volumes of RES 217
Alizade Rana, Nasibov Valeh

Method of Identification and Operational Control of Unauthorized Electricity Withdrawal in Distribution Networks 222
Omorov Turatbek, Zakiriaev Kubnychbek, Takyrbashev Beishenaly, Zholdoshov Tolkunbek

Automated Experimental Complex on the Platform of Modern ASCME 228
Omorov Turatbek, Asiev Abai, Osmonova Rima, Takyrbashev Beishenaly, Imanakunova Zgenishkul, Suerkulov Manas

The Use of Information and Computing Technologies in Solving Problems of Monitoring and Managing the Condition of Cable Equipment to Ensure Their Reliability and Efficiency........ 233
Ivanova V., Tsipkina V., Mamadalieva U.

Ensuring the Reliability of Energy Systems Through the Application of a New Method for Introducing Soil-Replacement Mixtures During the Installation of Deep Grounding Devices 238
Pavlovich Ivan, Baraishuk Siarhei, Murodov Muzaffar, Abdulkhayev Khurshed

Efficiency Evaluation of Evacuated Solar Tube Collectors in Sakha Republic (Yakutia) 242
Prokhorov Dmitry V., Vasilev Semen S.

Author Index

Statement of Peer review

In submitting conference proceedings to *Web of Conferences*, the editors of the proceedings certify to the Publisher that

1. They adhere to its **Policy on Publishing Integrity** in order to safeguard good scientific practice in publishing.
2. All articles have been subjected to peer review administered by the proceedings editors.
3. Reviews have been conducted by expert referees, who have been requested to provide unbiased and constructive comments aimed, whenever possible, at improving the work.
4. Proceedings editors have taken all reasonable steps to ensure the quality of the materials they publish and their decision to accept or reject a paper for publication has been based only on the merits of the work and the relevance to the journal.

Title, date and place of the conference

Methodological problems in reliability study of large energy systems (RSES) 2024
15-19 July 2024, Arkhangelsk, Russia

Proceedings editor(s):

1. Academician of Russian Academy of Science Valeriy Stennikov, Melentiev Energy Systems Institute, Russia.
2. Prof. Sergey Senderov, Melentiev Energy Systems Institute, Russia.
3. Prof. Alexander Michalevich, Energy Institute of Belarus NAS, Belarus.
4. Prof. Husein Guliev, Azerbaijan Research and Design and Exploration Institute of Power Engineering, Azerbaijan.

Date and editor's signature 15/09/2024

Taking into account balance reliability in Planning of power system development

Iurii Chukreyev[1,*], *Mikhail* Chukreyev[1]

[1] Institute for Socio-Economic & Energy Problems of the North, Federal Research Center Komi Science Centre Ural Branch of the Russian Academy of Sciences, Syktyvkar, Russia

Abstract. The article deals with the peculiarities of the methodology for determining balance reliability indicators and its synthesis given in the new edition of the methodological guidelines for designing the development of power systems of Russia. The analysis of power and electric power balances and the results of balance reliability indicators estimation given in the Scheme and Program of UPS of Russia for 2024-2029 is carried out. It is concluded that the decisions on substantiation of the required generating capacity to ensure balance reliability do not correspond to the new methodological guidelines.

1 Introduction

To ensure balance reliability (BR) of the power systems that form the UPS of Russia, the Ministry of Energy has been developing and approving methodological guidelines (hereinafter referred to as MI or MR) for designing the development of power systems since the 1980s. Until 2023, the MRs of 2003 were in effect. [1]. In them, the recommended values of the normative reserve of generating capacity (NRGC) were determined as a percentage maximum load of the united EPS of Russian UPS. They were obtained as a result of research and development work by design and scientific organizations.

New methodological guidelines for designing the development of power systems [2] were developed in the bowels of SO UPS (STC UPS JSC) and put into effect from January 2023 by the order of the Russian Ministry of Energy. Their difference is that in them the means of ensuring BR are found by comparing the calculated reliability indicators with the normative values directly when justifying decisions for the considered variants UPS of Russia development. For this purpose, a special Section VI on the methodology of assessment balance reliability indicators (BRI) was introduced in them.

2 Balance reliability indicators and information support required for their determination

According to recommendations [3], the BRI suitable for the purposes of making managerial decisions - justification of NRGC should have a simple physical meaning and be sensitive to perturbations. In domestic [3, 4, 6, etc.] and in foreign publications [5, etc.] the given recommendations are most fully satisfied by probabilistic indicators of BR:

- integral probabilities of occurrence of power deficit (J_d) or its absence (ρ) in the territorial zones of EPS;
- average number of days power deficit per year (Loss of Load Expectation - LOLE [5]);
- average number of hours capacity deficit per year (Loss of Load Hours - LOLH [5]).

Determination of probabilistic indicators of BR in the models developed both in our country and abroad is carried out by modeling random states of generating capacity caused by withdrawal generating equipment for emergency repair at a certain time interval of load change. In our country, in MI 2003, modeling was carried out for 24 steps of daily schedule the December day, with its extension to all working days of the year (250) [1, 4, 7]. In North America, modeling is performed for 365 values of daily load maximums to obtain the BR LOLE indicator. At present, when determining the BR J_d indicator, as well as when determining the European BR LOLH indicator, modeling is performed for all 8760 intervals of hourly load variation [2, 8].

Justification of NRGC in a multizone EPS is carried out by comparing the BRI obtained as a result of solving the problem of their assessment with the normative values. Domestic [3, 4, 6, etc.] and foreign publications [5, etc.] show that the following factors influence the BRI and the justification of NRGC:

- disposable capacities and the structure of generating capacities of individual territorial zones;

* Corresponding author: chukreev@iespn.komisc.ru

© The Authors, published by EDP Sciences. This is an open access article distributed under the terms of the Creative Commons Attribution License 4.0 (https://creativecommons.org/licenses/by/4.0/).

- reduction of generating capacity of territorial zones due to withdrawal of power plant units for emergency and scheduled repairs;
- schedules of load changes in the territorial zones by year and day;
- errors in forecasting consumer demand caused by temperature changes;
- model of the calculation scheme UPS of Russia;
- reserves of transmission capacity of links in normal and emergency modes between territorial zones.

The solution of the last two positions is partially reflected in [9, 10] and the publication of one of the authors in this collection. The first four positions are more or less studied and the methodological principles of their accounting in the models for assessing the BRI have been developed [6]. At the same time, it should be noted that the provisions included in the new MI concerning the issues of accounting for scheduled equipment repairs (item 102) and especially the representation of HPP and RES capacity (items 83, e, f, g, h) have not been practically tested and published. At the same time, their representation in the model for estimation of BRI can have a significant impact on the decisions on BR justification. These inconsistencies are explicitly present in the materials of the Scheme and Development Program UPS of Russia for 2024-2029 (hereinafter referred to as the UPS SPD). The results given in Appendices 3 and 5 of this work on power and electricity balances and BRI served as an impetus for the study below in terms of justification for the increase in NRGC. It is noteworthy that the values of BRI are extremely low in the territorial zones that have HPPs and RES as part of their generating sources. Specifically, these are the territories of Krasnodar Krai, Crimea, the East and Siberia regions.

3 Experimental studies

Let's consider the influence of the presented market models on the total reserve value. The value of the capacity reserve is primarily influenced by the forecasted demand and the adopted indicators of balance reliability. The latter is constant for one country, while the former parameter varies depending on the chosen market model in the energy system. Let us consider the presented models from this point of view in order.

It should be noted that in the work of the UPS SPD for 2024-2029, the issues of determining additional NRGC were solved without taking into account the requirement of 113p of the new MI concerning the provision of the required normative indicators of BR. In particular, for the UPS of East the approach based on determination of additional generating capacity on the basis of power balance deficit was used, for the South UPS - on the basis of power balance deficit according to the N-1 reliability criterion without taking into account the different timing of the annual maximum load. The recommendations for obtaining additional generating capacity based on the recommendations of the new MI 2023 on accounting of the BR are given below using the example UPS of East.

In accordance with the materials of Appendix 5 of the SPD UPS work, the model of the design scheme the East UPS for solving the problem of estimating the BRI included 10 territorial zones with external connections to China and the UPS of Siberia (Transbaikalia and the Republic of Buryatia). The results of the BRI presented in Appendix 5 show that all territorial zones, except for Yakutsk energy district, have the same and rather low BRI (ρ = 0.7435 ÷ 0.6744 for the period from 2026 to 2029). This indicates, firstly, the deficit of this part the system and, secondly, that the transmission capacity of the links connecting the territorial zones of this part EPS is sufficiently large (mainly 500 kV transmission lines) and does not affect the distribution of capacity deficits in the East UPS.

Let's consider the data on power and electric power surpluses and deficits of the East UPS from the point

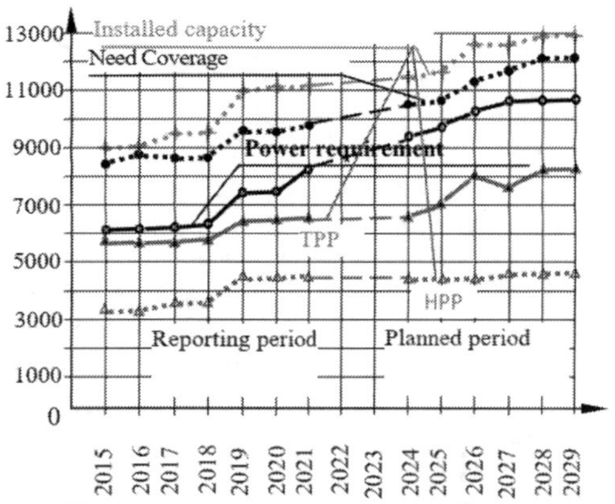

Fig. 1. Power balance parameters as applied to BR tasks

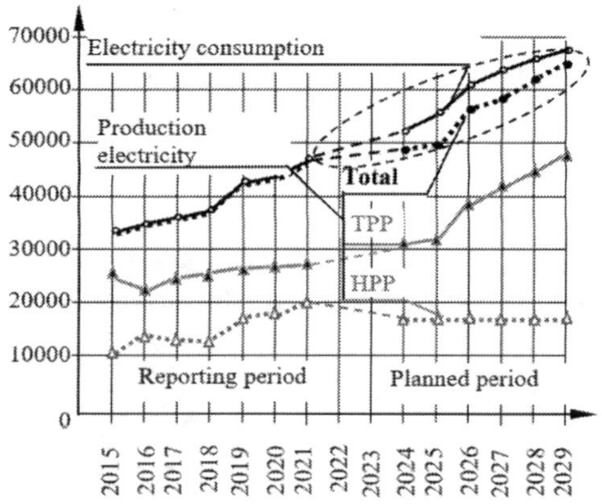

Fig. 2. Electricity balance parameters in relation to BR tasks

© The Authors, published by EDP Sciences. This is an open access article distributed under the terms of the Creative Commons Attribution License 4.0 (https://creativecommons.org/licenses/by/4.0/).

of view ensuring the BR (tab. 1). One should pay attention to a rather large (from 7 to 11%) surplus of capacity for all years of the forecast period (item 1) and at the same time a rather large (from 9.3 to 13.3%) deficit of electric power (item 2). Fig. 1 and Fig. 2 show the dependences of changes in the main parameters, which are the most important for the BR justification.

Table 1. Characteristics of the power and electricity balance required for the analysis BR East UPS (Appendix 3 of the SPD UPS)

Title	2024	2025	2026	2027	2028	2029
1. Capacity surplus, MW / %	1044 / 11,18	810 / 8,30	1007 / 9,77	761 / 7,06	1082/ 10,02	1056 / 9,76
2. Electricity deficit including transmission to adjacent energy systems MW·h / %	4921 / 9,27	7504 / 13,31	6875 / 11,10	8849 / 13,32	8387 / 12,42	8763 / 12,91
3. BR indicators (Annex 5)	0,9492	0,8162	0,7435	0,6744	0,7193	0,7455

For the majority of territorial zones of the East UPS, the BRI given in Appendix 5 and in item 3 of Table 1 in the form of probability deficit-free operation can be said to be critically low, for example, for 2027 the BRI ρ = 0.6744. Such values mean that consumers of most territorial zones of the East UPS will be limited in electric power during one third of the year. During the public discussion in November 2023 of the UPS SPD work project, SO UPS specialists, when responding to the comment about low reliability, limited themselves to standard explanations (dynamics of demand increase and high equipment failure rate). One should not have expected any other answer, since only these factors can affect such a significant decrease in the BRI.

The absence of all necessary information in the public domain to carry out calculations for the assessment of BRI allows us to make some quite adequate assumptions. Information on the composition of generating equipment, forms of the load schedule during peak hours and its reduction by days of the year was taken from the materials of the work performed in 2012 under the assignment of SO UPS as part of the justification the BR of the East UPS. Taking into account the opinion of SO UPS specialists about rather high accident rate of generating equipment in the East UPS, the version of their variation from 3.5% to 6% on average was adopted. In accordance with clause 103 of the new MI 2023, scheduled repairs at thermal and hydraulic power plants should be carried out during seasonal load reductions. Low BRI (ρ=0,6744) for territorial zones UPS of the East can be when the volume of scheduled repairs practically fills the area of seasonal load reductions. The results of BRI for different level of generating equipment accident rate are given in item 1, Table 2. As one could expect, the BRI significantly exceed those given in Appendix 5 of the UPS SPD work (item 3, Table 1).

Table 2. Estimated BR indicators and parameters of its provision for the East UPS

Indicator		2024	2025	2026	2027	2028	2029
1. Excluding electricity deficit caused by shortage of energy resource of HPPs							
in case of emergency of generating equipment	0,060	0,9598	0,8916	0,9313	0,8814	0,9292	0,9498
	0,035	0,9911	0,9584	0,9835	0,9568	0,9890	0,9875
2. Addition of required generating capacity							
2.1. According to the methodology SPD UPS		757	1154	1058	1361	1290	1348
2.2. According to the methodology new Guidelines 2023	from	450	810	1125	1310	1170	1110
	up to	500	930	1350	1650	1475	1340

Using the Orion-M software [5], the scenarios of HPP power limitation were identified that would correspond to the BRI given in item 3, Table 1 and Appendix 5 of the UPS SPD. The power additions (item 2, Table 2, item 2.2), at which the normative index of BR indicators ρ = 0.996 is provided (Table 2, item 2.2), were determined for the HPP power limitations found in this way. They correlate differently with those obtained in the UPS SPD for the values of power deficit (Table 2, item 2.1). For example, the years of the forecast period with relatively small values of power deficit (2024-2025) and not so low indicators of BR (Table 1, item 3), they turned out to be lower than those proposed in the UPS SPD (item 2.1). At the same time, for the years with significant power deficits (more than 8000 mln kW·h) and rather low BN indicators (2027-2029), the capacity surcharges proposed in the UPS SIPR fit into the scissors of the considered scenarios of HPP capacity limitations (p. 2.2, Table 2).

Under a regulated natural monopoly, a single governing body is responsible for the development and security of the energy system. It plans capacity inputs in its territory based on actual need and economic feasibility. Since the cost of capacity commissioning is included as an investment component in electricity

© The Authors, published by EDP Sciences. This is an open access article distributed under the terms of the Creative Commons Attribution License 4.0 (https://creativecommons.org/licenses/by/4.0/).

tariffs, the work on capacity commissioning will always be paid for, and therefore there are many people willing to invest in new generating facilities, which removes a number of uncertainties, primarily concerning the need to interest a potential investor. The possibility of spot commissioning capacities also contributes to a smaller volume of their commissioning, which in market conditions will significantly affect the tariff.

The model based on competition in the wholesale market implies control over the development of the energy system through an independent System Operator. The SO is responsible for making decisions on the amount of capacity required for the uninterrupted operation of the energy system. At the same time, these decisions are often dictated by the need to create an economically sound basis for investment. This is a very important point, because it should be understood that in this market model, market power is on the side of suppliers, which means that they dictate the conditions under which investments are interesting to them.

4 Conclusion

1 The concerns about the availability of baseline information when solving the problems of balance reliability for researchers, noted when discussing the new edition of the methodological guidelines for power system design, were fully manifested when developing and publicly discussing the work of the UPS SPD for 2024-2029.
2. The approaches to justification of the required additional generating capacity given in the UPS SPD for 2024-2029 do not comply with the methodological provisions set forth in Section VI "Ensuring Balance Reliability" of the new edition of the methodological guidelines for the design of power systems development. In these guidelines, the required additional generating capacity should be determined, among other things, on the basis of ensuring the normative indicator of balance reliability for all territorial zones of the model UPS of Russia design scheme.

The article was prepared within the framework of the R&D "Scientific bases of energy transition research at the regional level" (№124013000621-4).

References

1. Methodological recommendations for design development of energy systems. (Approved by order of the Ministry of Energy of Russia No. 281 of June 30, 2003).
2. Methodological instructions for the design development of energy systems. (Approved by order the Russian Ministry of Energy dated December 6, 2022, No. 1286).
3. Yu. N. Rudenko, I. A. Ushakov Reliability of energy systems. Science, Moscow, 1986, 252 p.
4. Yu. Ya. Chukreyev Models for ensuring the reliability of energy power systems. Komi SC UB RAS, Syktyvkar, 1995, 176 p.
5. Billinton R., Allan R.N. Reliability Evaluation of Power Systems. Second Edition. New York and London, Plenum Press, 1996. 509 p.
6. Chukreev Yu.Ya., Chukreev M.Yu. Models for assessing the indicators of balance reliability in managing the development of electric power systems. Syktyvkar: Komi NC URS RAS, 2014, 207 p.
7. Reference book on design of electric power systems. Edited by S.S. Rokotyan and I.M. Shapiro. Moscow: Energia, 1995, 352 p.
8. National Standard of the Russian Federation. Unified energy system and isolated power systems. Planning of power systems development. Calculations of balance reliability - norms and requirements. GOST P 58730-2019.
9. Chukreev M.Yu. Justification the model calculation scheme UPS of Russia for the task of ensuring balance reliability // Methodical issUPS of reliability research large power systems: Issue. 70, in 2 books. / Book 2 / ed. by N.I. Voropai. Irkutsk: ISEM SB RAS, 2019, 97-107 pp.
10. Chukreev Yu.Ya. Influence of state estimation models in the task of balance reliability provision on managerial decisions at EPS planning // Methodical issUPS of reliability research large power systems: Issue. 74. Reliability of Power Engineering Systems in the Conditions of Modern Challenges and Threats. Editor-in-Chief, Academician of the Russian Academy of Sciences V.A. Stennikov. Irkutsk: ISEM SB RAS. 2023, 166-175 pp.

© The Authors, published by EDP Sciences. This is an open access article distributed under the terms of the Creative Commons Attribution License 4.0 (https://creativecommons.org/licenses/by/4.0/).

Transformation of the model of the design scheme of the power system in the problem of ensuring balance reliability

Iurii Chukreyev[1,*]

[1] Institute for Socio-Economic & Energy Problems of the North, Federal Research Center Komi Science Centre Ural Branch of the Russian Academy of Sciences, Syktyvkar, Russia

Abstract. The paper deals with the issues reducing the number of territorial zones in the model calculation scheme UPS of Russia as applied to the problem of substantiating the required generating capacity (capacity reserve) under the conditions of ensuring normative indicators of balance reliability in accordance with the requirements of the new guidelines for design of power systems. On the example of the model UPS of Russia calculation scheme developed by SO UPS specialists, the possibilities of reducing the number of territorial zones to reasonable limits are shown.

1 Problems of forming a model calculation scheme UPS of Russia for the problem of balance reliability

The task justification of power reserves electric power systems when managing their development has always been given due attention. It is reflected in many monographic publications [1-5], reference books [6] and normative and technical documents [7-9]. Information both on forecasted parameters - load maximums, transmission capacity of links, cost characteristics of generating and grid equipment and many other things, and probabilistically determined - power equipment failure, power consumption forecasting errors, etc., had and has a certain error today. This is the basis for the application of various simplifications in the applied models calculation scheme, accounting of power consumption modes and other aspects, but, nevertheless, adequately reflecting the limitations of the real processes in the system. At the same time, it should be noted that in terms of the assumptions made, the studies conducted in our country differed little from those conducted abroad.

The model UPS design scheme under the above assumptions was considered in the form of aggregated territorial zones - united electric power systems (EPS) with unlimited possibilities of interconnection capacity within them. At the same time, it should be noted that the models calculation scheme UPS of Russia were also considered with the division of some EPS into two or more territorial zones, within which unlimited possibilities of interconnection capacity were also assumed. It should be noted that these simplifications were quite justified, allowing to find an adequate solution to the problem justifying the means of redundancy that meets the interests of both generating companies and consumers.

In the current conditions of planning the development EPS of the country, specialists of SOs and servicing research institutes (STC UPS) have formed the opinion that availability of opportunities in obtaining information and, to some extent, market relations require greater detailing model calculation scheme UPS of Russia. It is quite likely that the models for solving operational problems played a certain role here. It is not a secret that today, when solving a number of tasks UPS of Russia regime planning for the conditions of current and operational regime management, information availability and market relations have led to the possibility of more detailed representation the power system calculation scheme in form of hundreds and thousands of nodes and links between them. This is quite understandable due to sufficiently reliable information both on power consumption and generation modes and the time factor of planning for a day ahead or in terms of hours current day.

When planning the development, the increase UPS model to solve, to a greater extent, the problems of estimating the balance reliability indicators (BRI) is reflected in open publications [4, 10], UPS reports and regulatory materials [7-9]. It should be noted that there was no scientific and practical justification of increasing the model UPS design scheme for the tasks justifying the means of ensuring the balance reliability (BR) - the reserve of generating capacity or, as it is

* Corresponding author: chukreev@iespn.komisc.ru

© The Authors, published by EDP Sciences. This is an open access article distributed under the terms of the Creative Commons Attribution License 4.0 (https://creativecommons.org/licenses/by/4.0/).

customary today, the required generating capacity. It can be recognized that when solving the problem of identifying the most "bottlenecks" in the aspect of balance reliability of places for the reporting year and for one year in advance, the model UPS of Russia calculation scheme is quite appropriate to detail up to 50-150 territorial zones. At the same time, it is no secret that the scheme detailing is always associated with various simplifications [11, 12]. At the same time, the solution the problem of justifying means of ensuring balance reliability - the required generating capacities (capacity reserves), on the calculation schemes of large dimensions, causes significant difficulties, both from the standpoint of time costs and from the standpoint of analyzing the obtained results BRI.

2 Balance reliability indicators

According to the recommendations [1], the BRI suitable for the purposes of making managerial decisions - justification of capacity reserves - of the required generating capacity should have a simple physical meaning and be sensitive to perturbations. In domestic [1, 2, 4, etc.] and in foreign publications [3, etc.] probabilistic BRI satisfy the above recommendations most fully:

- integral probabilities of occurrence of power deficit (J_d) or its absence (ρ) in the territorial zones of EPS;

- average number of days of power deficit per year (Loss of Load Expectation - LOLE [3]);

- average number of hours of capacity deficit per year (Loss of Load Hours - LOLH [3]).

Determination of probabilistic BRI in the models developed in our country or abroad is carried out by modeling of random states generating capacity caused by withdrawal of generating equipment for emergency repair at a certain time interval of load change. In our country, until recently, modeling was carried out for 24 steps of the daily schedule December day, with its extension to all working days of the year (250) [2, 6, 7]. In North America, to obtain the BR LOLE index, modeling is performed for 365 values of load maxima, without taking into account their daily decreases. At present, when determining the J_d indicator, as well as when determining the European LOLH indicator, modeling is performed for all 8760 intervals of hourly load changes [8, 9]. At the same time, it should be noted that the normative value of LOLH indicators, when changing the modeling principles, remained at the same level (ρ = 0.996, Jd = 0.004), which is not quite correct, but this is not the issue of this article. For information, in European countries LOLH = 3 - 8 hours, which corresponds to ρ = 0.99909 - 0.99966 or Jd = 0.00034 - 0.00091.

3 Model design scheme, its influence on the BRI from the point of view justification of the required generating capacity

At all times, sufficient attention has been paid to the formation of models design scheme UPS of Russia for BR tasks. It should be noted that for the first time the approach of formation of the calculation scheme for the BR task based on the use of controlled sections of the real scheme electrical connections UPS of Russia was applied by the specialists of INEI RAS and Institute "Energosetproekt". The problematic point in this approach was and remains the issues of determining the capacity links included in a particular controlled section [11, 12]. At the same time, we leave behind brackets the influence of such important in the aspect of BR issues of their adequate application for conditions constantly changing composition of generating equipment under random modeling in the task of estimation of BRI.

It is quite obvious that the model calculation scheme UPS of Russia for BR tasks should be represented in the form of a graph, the vertices of which are zones of equal reliability, and the edges - intersystem links. Zones of equal reliability (territorial zones) are understood as a set of nodes included in it, network constraints between which, as in a concentrated system, do not affect the BRI in it. Taking this into account and analyzing the controlled cross-sections, schemes with 50 territorial zones were first proposed (2012, INEI RAS), then with 56 (taking into account Ukraine, Belarus and the Baltic states in 2014, STC UPS), but without taking into account the East UPS. In the last paper Scheme and Program of UPS Development for 2024-2029 (hereinafter referred to as UPS SPD without specifying the years). SO UPS and STC UPS propose a calculation scheme with 100 and 110 territorial zones with the East UPS.

The use of UPS calculation schemes with a large number of territorial zones in the task of estimating the BRI can be somehow explained by the possibility of identifying "bottlenecks". The BRI for territorial zones and the links connecting them will be low in such places UPS of Russia calculation scheme. The solution to the problem of justifying the capacity reserve, or, in today's interpretation the required generating capacity, is to bring these BRI up to the normative value. This is also stated in the new methodological guidelines for the design of power systems (hereinafter referred to as the new MI) [8]. Fig. 1, a shows the design scheme of UPS. It is based on the materials of Appendix 5 of the

© The Authors, published by EDP Sciences. This is an open access article distributed under the terms of the Creative Commons Attribution License 4.0 (https://creativecommons.org/licenses/by/4.0/).

UPS SPD work. Small circles indicate territorial zones allocated in each of the 7 UPS. The territorial zones included in the UPS Center, Volga and North-West have a high level of BRI ρ > 0.9999 (white circles). A small number of territorial zones of the Ural UPS (1), South (6), Siberia (6) and practically all UPS of the East (9) have BRI below the normative value (shaded circles on the scheme). Many connections between territorial zones within the UPS are not shown in the figure. At the same time, the links limiting the redistribution of capacity between deficit and surplus parts UPS are highlighted in bolder colors.

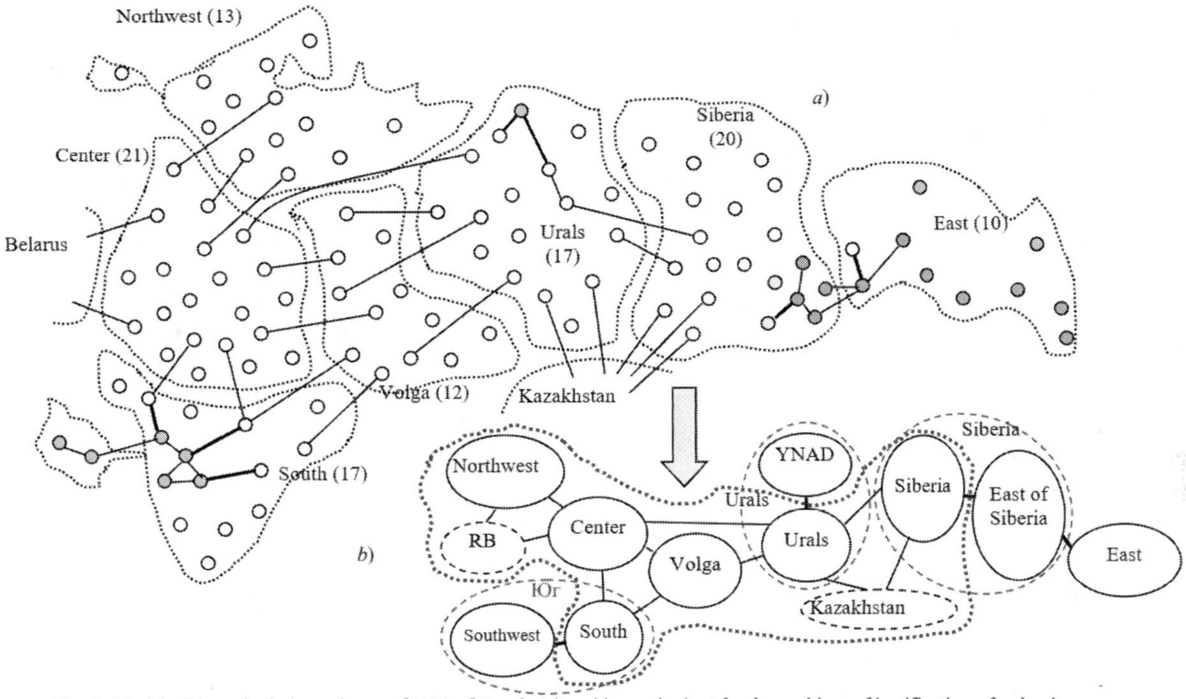

Fig. 1. Model of the calculation scheme of UES of Russia (a) and its equivalent for the problem of justification of redundancy means (b).

According to the values of probabilistic indicators of BR, in accordance with the theory of their obtaining, it is possible to distinguish territorial zones of practically the same reliability and thus to unite them into one zone. The capacity of the links within them practically does not affect the redistribution of capacity deficits. Such territorial zones, in accordance with the results of Appendix 5 of the UPS SPD work are shown in Fig. 1, b). In this case, the calculation scheme of 110 territorial zones is reduced to 10, not counting the energy systems of Belarus and Kazakhstan. In fact, according to the BRI it is still possible to combine the territorial zones UPS of the North-West, Center, Volga and part UPS of the South, Urals and Siberia into one zone, highlighted in Fig. 1, b by a small dotted line. Then the scheme will be reduced to 5 territorial zones. The latter is not quite correct due to the fact that the reporting materials of SOs and FGCs, especially those typical for the BR tasks, are prepared in the context of UPS.

As a matter of fact, this is the essence of the proposed hierarchical approach of transforming the model UPS of Russia calculation scheme for solving the problem of substantiating the required generating capacity. The scheme model used for solving the assessment of BRI on the basis analysis of the obtained results can be transformed into a simpler model for solving the problem of substantiating the requirements for the required generating capacity of power plants under the conditions of balance reliability. Based on Fig. 1, it is clear that in order to substantiate the required generating capacity in the problematic territorial zones UPS of the East. Siberia and the South, it is enough to identify these reliability zones and for them to carry out more detailed calculations to justify the required generating capacity from the standpoint of ensuring the normative indicator of BR.

© The Authors, published by EDP Sciences. This is an open access article distributed under the terms of the Creative Commons Attribution License 4.0 (https://creativecommons.org/licenses/by/4.0/).

4 Conclusion

The task of justification of the required generating capacity (capacity reserve) involves multiple estimation of the BRI under different variants of the generating capacity composition. For the model calculation scheme UPS of Russia (110 territorial zones) developed by SO UPS the solution of the problem is associated with significant difficulties, both of informational and computational nature, and of methodological nature related to the management decision-making process. As applied to the solution of the problem justification of the required capacity on the basis analysis BRI of the initial version the calculation model scheme, the possibilities of reducing the number of territorial zones UPS calculation model to reasonable limits are shown, which will lead to a significant reduction in the time of obtaining the result.

The article was prepared within the framework of the R&D "Scientific bases of energy transition research at the regional level" (№124013000621-4).

References

1. Yu. N. Rudenko, I. A. Ushakov Reliability of energy systems. Science, Moscow, 1986, 252 p.
2. Yu. Ya. Chukreyev Models for ensuring the reliability of energy power systems. Komi SC UB RAS, Syktyvkar, 1995, 176 p.
3. Billinton R., Allan R.N. Reliability Evaluation of Power Systems. Second Edition. New York and London, Plenum Press, 1996. 509 p.
4. Chukreev Yu.Ya., Chukreev M.Yu. Models for assessing the indicators of balance reliability in managing the development of electric power systems. Syktyvkar: Komi NC URS RAS, 2014, 207 p.
5. Reliability of Power Engineering Systems: Problems, Models and Methods of Their Solution / A.F. Dyakov, V.A. Stennikov, S.M. Senderov, et al. N.I. Voropai. Novosibirsk: Nauka, 2014, 284 p.
6. Reference book on design of electric power systems. Edited by S.S. Rokotyan and I.M. Shapiro. Moscow: Energia, 1995, 352 p.
7. Methodological recommendations for design development of energy systems. (Approved by order of the Ministry of Energy of Russia No. 281 of June 30, 2003).
8. Methodological instructions for the design development of energy systems. (Approved by order the Russian Ministry of Energy dated December 6, 2022, No. 1286).
9. National Standard of the Russian Federation. Unified energy system and isolated power systems. Planning of power systems development. Calculations of balance reliability - norms and requirements. GOST P 58730-2019.
10. Chukreev, Yu.Ya.; Chukreev, M.Yu. Balance reliability indicators for justification components of the normative power reserve in relation to the modern conditions UPS of Russia // Izvestiya RAN. Energetika. 2022, № 5, 22-35 pp.
11. Chukreev M.Yu. Justification the model calculation scheme UPS of Russia for the task of ensuring balance reliability // Methodical issues of reliability research large power systems: Issue. 70, in 2 books. / Book 2 / ed. by N.I. Voropai. Irkutsk: ISEM SB RAS, 2019, 97-107 pp.
12. Chukreev Yu.Ya. Influence of state estimation models in the task of balance reliability provision on managerial decisions at EPS planning // Methodical issues of reliability research large power systems: Issue. 74. Reliability of Power Engineering Systems in the Conditions of Modern Challenges and Threats. Editor-in-Chief, Academician of the Russian Academy of Sciences V.A. Stennikov. Irkutsk: ISEM SB RAS. 2023, 166-175 pp.

© The Authors, published by EDP Sciences. This is an open access article distributed under the terms of the Creative Commons Attribution License 4.0 (https://creativecommons.org/licenses/by/4.0/).

Impact of the market model in the electric power system on its reliability

Mikhail Chukreyev[1,*]

[1] Institute for Socio-Economic & Energy Problems of the North, Federal Research Center Komi Science Centre Ural Branch of the Russian Academy of Sciences, Syktyvkar, Russia

Abstract. In today's competitive international markets, the cost of electricity has a huge impact on a nation's ability to produce a competitive product. The percentage of electricity costs in manufactured goods can be up to 50% of the cost. Reducing the costs of this component undoubtedly affects the competitiveness, and consequently leads to an increase in the growth rate of Russia's GDP. Therefore, work on the efficiency of the energy sector is very important. The article considers the mechanisms of energy markets, their peculiarities are given. The influence of the chosen mechanism on the required full capacity reserve is shown, and the author's vision for further actions in these matters is given.

1 Introduction

For the last decades, the electric power industry can be considered as the most significant catalyst of progress. Its development directly affects all other sectors of our life. In the absolute majority of industries, electricity has a significant weight in the cost of the product. In everyday life and social life we cannot imagine ourselves without this resource.

In order to provide all areas of our life it is necessary to plan strictly the development of electric power industry. Many organizations work in this direction, forming the planned level of load, generation by zones of reliability, capacity of connections between them, etc. All this is consistent with the selected reliability criteria. This generally reflects the essence of the balance reliability problem (the property of the object to satisfy the consumers' requirements within the specified values and limitations on energy resource supply, taking into account planned and unplanned interruptions in the operation of its elements and operational limitations). And it is its solution that determines the ability and speed of progress. It turns out that the work on the assessment of balance reliability is a priority task for the formation of society, and the management subjects dealing with these issues are responsible for the realization of the development potential.

The development of the electric power industry follows the path of creating unified power systems (hereinafter - UPS). Examples are the UPS of Russia, unions of power systems the European Union, unions North America power systems. The reliability the power balance of such systems is always analyzed. In our country, within the UPS of Russia, zones in the form of united energy systems are allocated with their fragmentation into separate territories. The reliability of the power balance is ensured both by own sources and generation of adjacent zones without any restrictions on power supply from them.

On this basis, rather strict hierarchical systems are formed, within the framework of which effective methods have been developed, realized in software products, which allow solving the whole complex of tasks related to the management of the development of the power systems of the UPS of the country for a perspective of 5 to 20 years, including the task of ensuring balance reliability.

At the present stage the most acceptable indicators of balance reliability (IBR) from the point of view of justification of means of reliability provision are considered [1,2]:

- in domestic practice probabilities of deficit-free operation of territorial zones $\rho = 1 - J_д$ ($J_д$ - integral probability of power deficit appearance);
- in foreign practice - average probability values days of power shortage (Loss of Load Expectation, LOLE) in days/year and hours of power shortage per year (Loss of Load Hours, LOLH).

As applied to our country, the most acceptable in the practice of planning the development EPS remains the IBR in the form of integral probability of occurrence ($J_д$) or absence ($\rho = 1 - J_д$) of capacity deficit in the considered territorial zones of the UPS design scheme. The Methodological Recommendations [3] use values of 0.004 and 0.996 as normative parameters of this indicator, respectively. The countries of Western Europe and the USA have adopted somewhat different, but also probabilistic IBR and their normative values. In our country many works, for example [3-6], are devoted to the issues of development of models for estimation of IBR.

The goal in solving the problem of balance reliability is to obtain data on the means of its provision. The operational reserve can be considered as

* Corresponding author: chukreev@iespn.komisc.ru

© The Authors, published by EDP Sciences. This is an open access article distributed under the terms of the Creative Commons Attribution License 4.0 (https://creativecommons.org/licenses/by/4.0/).

the main one (its value is also affected by the operation of the "aggregator" and the presence of local power systems). The resulting value of the operational capacity reserve worries both consumers and suppliers. There are many factors influencing the reserve value, but one of the most important is the form of power system management and decision-making criteria. Let's try to estimate the influence of different structural features of the organization responsible for reliability on the level of operational reserve.

In humanity's striving for competition, the development of markets in the electric power industry is inevitable, but it cannot be uncontrolled due to its special nature and importance for the state. For many years of market relations development it is customary to distinguish four types of markets [7-9]: regulated natural monopoly, single buyer, competition in the wholesale market and retail markets. Let us consider two of their varieties: regulated natural monopoly and competition in the wholesale market. The other two are a development of the ideas of the former.

2 Market models

Let's consider two models from the position of ensuring their reliability and "interests" influencing the result, as well as predictability of power system development.

Regulated natural monopoly. This is a type of vertically integrated companies, as a rule, subject to state regulation. In its absence, there would be a price increase. This type of market is very natural for the power industry due to the specifics of the products produced and the need to maintain proper reliability. This option increases its positive qualities as the system becomes larger.

Monopoly companies plan the development of EPS (both power plants and power grids) on their territory based on the reliability and efficiency of power supply to consumers, environmental requirements, etc. The monopoly companies plan the construction of new power plants and power grids on their territories. Construction plans for new power plants and power transmission lines (transmission lines) are subject to approval by the regulator, after which construction costs are included as an investment component in electricity tariffs. In this way, the costs of EPS development are guaranteed to be paid by consumers.

A good example of such a structure was the former USSR. During that period, all information was collected within the walls of the Energosetproekt organization, which had research status, NIIPT and ORGRES. This made it possible to obtain information and a roadmap for independent researchers, scientific and scientific-educational organizations to carry out work on the assessment of balance reliability indicators. The state dictated the rules of the game, but at the same time supported third-party research in the field of balance reliability. At the same time, the interests of consumers represented by the state were taken into account first of all. This allowed to control the dynamics of development of the industry and, as a consequence, the society.

Competition in the wholesale market. This model differs significantly from the previous one. Instead of one Procurement Agency (monopoly supplier) there are several distribution and sales companies. These companies, as a rule, do not have their own generating capacities and monopolistically supply electricity to consumers on their territory. They are responsible for reliable electricity supply and remain subject to regulation by regional or municipal authorities. They own the low-voltage distribution networks, which they must develop if necessary.

In this market model, an independent System Operator (SO) is responsible for controlling the development of the power system and maintaining the proper level of balance reliability. At the same time, the existence of a wholesale market deprives consumers of a privileged position, which gives market power to suppliers and negatively affects the ability of the system to develop predictably and maintain the proper level of reliability. Now it is necessary to create economic opportunities for this purpose, which significantly affects the increase in the capacity tariff.

If we continue to compare with historical periods, this is the time of the electricity reform started in 2003. At this time, despite the declared competition in the sector, the SO started a consistent movement towards collecting and storing information in-house. All works related to the tasks of reserves determination and, accordingly, IBR assessment were performed by the SO or delegated as contractual works to specialized organizations. The information provided for these works was the intellectual property and could not be made public. But the works were carried out partly for the purpose of accumulation of certain competences within the SO, partly for solving tasks in the interests of the SO. Thus, the SO obliged to ensure a proper level of reliability is not interested in obtaining a minimum sufficient level of reserve, the ease with which it can ensure it is more important for it, and consumers interested in the lowest tariff have no leverage. At the same time, it should be understood that the SO covers all its needs by tariff and high tariff promotes the interest of generating companies, which creates a vicious circle - higher tariff - higher construction of new generation - easier to maintain the proper level of reliability of the SO and there is no consumer in this circle.

3 Influence of the market model on the formed value of the reserve

Let's consider the influence of the presented market models on the total reserve value. The value of the capacity reserve is primarily influenced by the forecasted demand and the adopted indicators of balance reliability. The latter is constant for one country, while the former parameter varies depending on the chosen market model in the energy system. Let us consider the presented models from this point of view in order.

© The Authors, published by EDP Sciences. This is an open access article distributed under the terms of the Creative Commons Attribution License 4.0 (https://creativecommons.org/licenses/by/4.0/).

Under a regulated natural monopoly, a single governing body is responsible for the development and security of the energy system. It plans capacity inputs in its territory based on actual need and economic feasibility. Since the cost of capacity commissioning is included as an investment component in electricity tariffs, the work on capacity commissioning will always be paid for, and therefore there are many people willing to invest in new generating facilities, which removes a number of uncertainties, primarily concerning the need to interest a potential investor. The possibility of spot commissioning of capacities also contributes to a smaller volume of their commissioning, which in market conditions will significantly affect the tariff.

The model based on competition in the wholesale market implies control over the development of the energy system through an independent System Operator. The System Operator is responsible for making decisions on the amount of capacity required for the uninterrupted operation of the energy system. At the same time, these decisions are often dictated by the need to create an economically sound basis for investment. This is a very important point, because it should be understood that in this market model, market power is on the side of suppliers, which means that they dictate the conditions under which investments are interesting to them.

4 Total capacity reserve for different market models

A seven-node scheme is used for the analysis (Figure 1). The initial information was obtained from the scheme and development program [10] and earlier studies. Then the data were finalized according to the

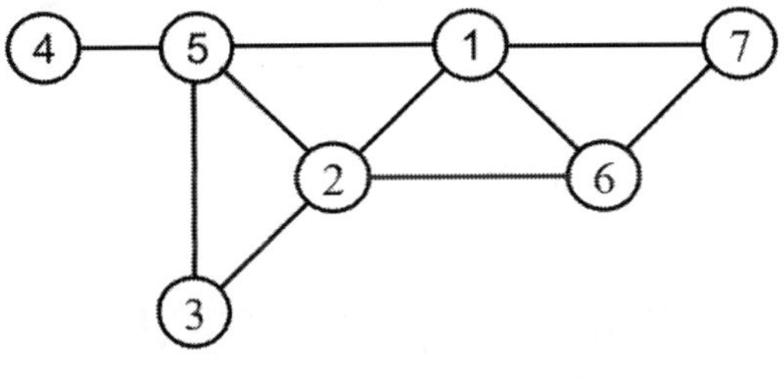

Fig. 1. Model of the computational scheme.

development strategy for each scenario (Table 1 and Table 2). In the case of the market model (competition on the wholesale market), the real statistics of the Russian energy system was used and everything was reduced only to modeling this behavior on the proposed calculation scheme. In the case of a regulated natural monopoly, the same starting point was taken as for the previous model, but its development took place in accordance with the principles fair for this type of market. In most countries the development of this model takes into account the retail price index or the deflator of the gross national product. Orientation to the retail price index is more favorable to the consumer, following the deflator of gross national product or producer price index is more favorable to producers. The factor of productivity increase is determined on the basis of estimates of prospective demand, estimates of capital investment, the value of profits, the probability of cost reduction, productivity growth, the need for investment, other unregulated activities [11-13]. In this model, average values for the world were taken, adjusted for our economic situation in the country. When constructing the actual values, we considered data on power systems with a close market model to a regulated natural monopoly in terms of the correlation between their forecasts and the resulting actual values [14]. Of course, such assumptions lead to errors in the result, but still create a tentative vision of an alternative path of development, which seems very interesting to us in this situation.

Let us analyze the information about the projected and actual demand and the required coverage. Table 1 and 2 show information for 10 years.

Table 1. Capacity balance for the model "competition in the wholesale market"

Year	Projected values			
	Consumption maximum, GW	Change year-on-year, %	Coverage, GW	Change year-on-year, %
Projected values				
1				
2	168809		236850,4	
3	173096	2,54	240725,2	1,64
4	177475	2,53	244637	1,62
5	181478	2,56	242291,3	-0,96
6	165151	-9	232516,4	-4,03
7	166262	0,67	232335,4	-0,08
8	160484	-3,47	232620,4	0,12
9	161311	0,51	225686,2	-2,98
10	162011	0,43	228545	1,27
Factual values				
1	158659		207700,5	
2	157219	-0,91	213875,1	2,97
3	158871	1,05	223733	4,61
4	154116	-2,99	215379,9	-3,73
5	157355	2,1	221166,2	2,69
6	156570	-0,5	215904,4	-2,38
7	159360	1,78	218713,4	1,3
8	155003	-2,73	217433,8	-0,58
9	157543	1,64	217082,6	-0,16
10	156946	-0,38	221925,1	2,23

© The Authors, published by EDP Sciences. This is an open access article distributed under the terms of the Creative Commons Attribution License 4.0 (https://creativecommons.org/licenses/by/4.0/).

Table 2. Capacity balance for the "regulated natural monopoly" model

Y ea r	Projected values			
	Consump-tion maximum, GW	Change year-on-year, %	Cover-age, GW	Change year-on-year, %
Projected values				
1				
2	166277		197720	
3	170952	2,81	204680	3,52
4	175285	2,53	208747	1,99
5	179476	2,39	214564	2,79
6	163499	-8,9	196035	-8,63
7	165706	1,35	196163	0,06
8	161062	-2,80	191583	-2,33
9	160897	-0,10	191162	-0,22
10	161661	0,47	191196	0,02
Factual values				
1	158659		189962	
2	157219	-0,91	185141	-2,54
3	158871	1,05	187515	1,28
4	154116	-2,99	182535	-2,65
5	157355	2,1	186103	1,95
6	156570	-0,5	186083	-0,01
7	159360	1,78	188012	1,04
8	155003	-2,73	182206	-3,09
9	157543	1,64	185601	1,86
10	156946	-0,38	184631	-0,52

*Factual values, obtained by calculation on the basis of statistics and operation of markets of this type in other countries.

Table 3. Full reserve for different market models

Year	Full reserve projected (model 2), %	Full reserve Factual (model 2), %	Full reserve projected (model 1), %	Full reserve Factual (model 1), %
1	-	30,91		19,73
2	40,31	36,03	18,91	17,76
3	39,07	40,83	19,73	18,03
4	37,84	39,75	19,09	18,44
5	33,51	40,55	19,55	18,27
6	40,79	37,9	19,90	18,85
7	39,74	37,24	18,38	17,98
8	44,95	40,28	18,95	17,55
9	39,91	37,79	18,81	17,81
10	41,07	41,4	18,27	17,64

Table 1 shows that the forecast for the 2nd model is always highly overestimated, despite the presence of huge capacity surpluses (Table 3). This is primarily due to the need to ensure attractiveness for potential investors, which requires a comfortable tariff level for them, i.e. the main focus is not on optimizing the system and reducing the tariff in favor of the consumer, but on working for the supplier. This is the main difference between models based on competition without direct state control and models that imply such control. For comparison, the use of model 1 and allows us to focus on the forecast demand in the perspective of 4-5 years (it is possible and more, but these values are optimal), which reduces the impact of forecast error and does not depend on economic attractiveness. This is possible due to the fact that in this model the most important indicator for the consumer - the price level - is in the center of attention. This is achieved through simplified decision-making, greater transparency, ease of regulation and incentives for efficiency gains. With this in mind, ORION-M has modeled the required level of full reserve with the same level of regulatory reliability. The results are presented in Table 3. The advantage of models with state control over purely market-based mechanisms is clearly visible.

5 Conclusion

Electric energy is a resource that has no alternative, so the creation of any type of market should be a reasonable decision. The adopted model of the energy market directly affects the way of controlling the reliability of the energy system, its ability to develop predictably, to be demanded by both the buyer and the seller.

The research shows that Model 1 can achieve predictability of the energy system, simplifying the task of maintaining proper reliability. In addition, the regulated market will reduce the problem of transparency in decision making for its participants. In the current realities, when one player in the market has all available information, it becomes even more important. With the current structure of the energy sector, the construction of new generation requires a significant price increase. Partly because of this, the price of selected capacities based on the results of competitive capacity selection tends to reach a value that allows financing new facilities. This disadvantage of the electricity market in the long term can be overcome only through state regulation.

The article was prepared within the framework of the R&D "Scientific bases of energy transition research at the regional level" (№124013000621-4).

References

1. Chukreev Yu.Ya., Chukreev M.Yu. Models for assessing the indicators of balance reliability in managing the development of electric power systems. Syktyvkar: Komi NC URS RAS, 2014, 207 p.
2. Billinton R., Allan R.N. Reliability Evaluation of Power Systems. Second Edition. New York and London, Plenum Press, 1996. 509 p.
3. Global Demand Response Capacity is Expected to Grow to 144 GW in 2025: https://www.navigantresearch.com/news-and-views/global-demand-response-capacity-is-expected-to-grow-to-144-gw-in-2025
4. Methodological recommendations for design development of energy systems. (Approved by order of the

© The Authors, published by EDP Sciences. This is an open access article distributed under the terms of the Creative Commons Attribution License 4.0 (https://creativecommons.org/licenses/by/4.0/).

https://doi.org/10.1051/e3sconf/202458401003

Ministry of Energy of Russia No. 281 of June 30, 2003).

5. Markovich I.M. Modes of power systems. M.: Energia, 1969, 351 p.
6. Yu. Ya. Chukreyev Models for ensuring the reliability of energy power systems. Komi SC UB RAS, Syktyvkar, 1995, 176 p.
7. Belyaev L.S., Podkovalnikov S.V. Market in the Electric Power Industry: Problems of Development Generating Capacities. - Novosibirsk: Nauka, 2004, 220 p.
8. Hunt S., Shuttleworth G. Competetion and Choice in Electricity. - Chichester: John Wiley, 1996.
9. Volkov E.P., Barinov V.A., Manevich A.S. Problems and Prospects of Development Electric Power Industry of Russia. - Moscow: Energoatomizdat, 2001, 432 p.
10. https://www.so-ups.ru/future-planning/dev-sch
11. Malinnikova E. V. Foreign experience in the regulation of natural monopolies // Economic Journal of the Higher School of Economics. V. Foreign experience in the regulation of natural monopolies // Economic Journal of the Higher School of Economics. 1998, №3, 342-357 pp.
12. Gunina E. N. Economic prerequisites of transformation of integrated corporate structures in the electric power industry of the Russian Federation // Space of economy. 2009, №3-3, 59-63 pp.
13. Davydovsky F.N. Monopoly and competition in the electric power industry: development alternatives and the problem of efficiency // Economics, entrepreneurship and law. 2011, №6, 30-44 pp.
14. https://www.mordorintelligence.com/ru/industry-reports/united-states-power-market

© The Authors, published by EDP Sciences. This is an open access article distributed under the terms of the Creative Commons Attribution License 4.0 (https://creativecommons.org/licenses/by/4.0/).

Reproduction of the mineral resource base of the energy sector as the most important component of ensuring Russia's energy security

Sergey Senderov[1, 2*]

[1] Melentiev Energy Systems Institute of Siberian Branch of the Russian Academy of Sciences, 130 Lermontov str., Irkutsk, Russia
[2] Federal State Budget Educational Institution of Higher Education «Irkutsk National Research Technical University», 83 Lermontov str., Irkutsk, Russia

Abstract. The situation with the production and consumption of primary fuel and energy resources in 2021-2023 is analyzed. The main threats to energy security and analysis groups declared in the Doctrine of Energy Security of the Russian Federation are characterized. The situation with the reproduction of the mineral resource base of the energy sector is analyzed based on the assessment of the state of the relevant indicators of energy security at the federal level. The main reasons are analyzed and conclusions are made on the emerging trends in the renewal of industrially recoverable reserves of the main primary fuel and energy resources.

1 Introduction

Currently, energy security in the Russian Federation is ensured in accordance with the main provisions of the Energy Security Doctrine approved by Decree of the President of the Russian Federation dated May 13, 2019 No. 216 [1]. The main threats to energy security (ES) and areas of indicative analysis in this Doctrine are generally consistent with those proposed in the first Energy Security Doctrine approved in 2012. In [2], the most important federal-level indicators, as well as strategic threats to ensuring the country's energy security, were proposed and discussed for all groups of indicative analysis.

Considerable attention was paid to foreign economic and foreign policy threats to Russia's energy security. This important component of the problem of ensuring ES has become significantly more acute since 2022 and has aggravated the problems with investing in the development of capital-intensive energy industries, primarily oil and gas. Investment problems directly affect the degree of realization of the threat of reducing the possibilities for developing new hydrocarbon deposits, especially in hard-to-reach expensive areas. In addition, it is necessary to talk about the problem of reducing the possibilities of importing modern equipment into these industries.

2 Consumption of primary fuel and energy resources.

Table 1 presents the actual indicators of consumption of primary fuel and energy resources (FER) within Russia for the last three years, their production volumes, as well as the volumes of export of oil, gas and oil products.

Table 1. Consumption of primary fuel and energy resources within Russia, their production and export, [3, 4]

Indicators	Year		
	2021	2022	2023
Domestic consumption of primary FER, total, mln toe, incl.	**1173**	**1229**	**1248**
- natural and associated gas*, bcm	470	484	500
- oil, mln t	274	293	296
- coal, mln t	158	182	189
- electricity hydroelectric power station, nuclear power plant, etc. FER, mln toe	136	135	127
Production of primary FER, total, mln toe, incl.	**2018**	**1936**	**1877**
- natural and associated gas*, bcm	764	674	638
- oil, mln t	524	535	530
- coal, mln t	432	444	438
electricity hydroelectric power station, nuclear power plant, etc. FER, mln toe	136	135	127
Export of Russian gas, bcm	*244*	*184*	*145*
Oil export, mln t	*230*	*242*	*234*
Export of oil products, mln t	*144*	*130*	*130*

* Does not include excess gas injection into UGS facilities and losses.

The values shown in Table 1 indicate a gradual increase in the volume of domestic consumption of primary FER. To a greater extent, this increase is influenced by the increase in the volume of natural gas

* Corresponding author: ssm@isem.irk.ru

© The Authors, published by EDP Sciences. This is an open access article distributed under the terms of the Creative Commons Attribution License 4.0 (https://creativecommons.org/licenses/by/4.0/).

consumption, which in turn is a consequence of the active policy of increasing the level of gasification of the regions. The consumption of oil and coal also grew. At the same time, it is clear that in these years the production of primary fuel and energy resources had the opposite trend, dictated by a decrease in gas production in volumes close to the volumes of its export reduction. Gas exports from Russia in 2023 amounted to 145 billion m3 (99.6 billion m3 were exported via main gas pipelines, 45.4 - in the form of LNG) [5].

The volumes of domestic oil consumption are calculated as follows: in 2023 530 million tons of oil were produced in Russia, of which 234 million tons were exported [6]. Accordingly, the difference, taking into account other and direct consumption, as well as oil losses, amounted to 296 million tons [7]. 275 million tons were processed. In the same year, 137.9 million tons of light oil products were produced, of which 49.8 million tons were gasolines, 88.1 million tons were diesel fuels.

Electricity generation at hydroelectric power plants amounted to 203 billion kWh, at nuclear power plants - 217 billion kWh, electricity generation using renewable energy sources (except hydroelectric power plants) - 7.8 billion kWh, which in total amounted to 427.8 billion kWh [8]. Using the value of the average consumption of equivalent fuel at thermal power plants in Russia in recent years of 296 g of oil equivalent/kWh [9], we can say that this volume of electricity produced (427.8 billion kWh) replaces approximately 127 million toe of fuel that would have to be spent for production of this electricity.

3 Main threats and energy security analysis groups

The Doctrine of Energy Security of Russia, which has been in force since 2019, describes the most important threats to energy security by type of implementation. From the standpoint of reliable fuel and energy supply to consumers within the country, the following can be considered:

– discrepancy between the capabilities of the fuel and energy complex and the needs of the socio-economic development of the Russian Federation;
– declining quality of the mineral resource base (MRB) of the energy sector (depletion of existing deposits, reduction in the size and decline in the quality of discovered deposits);
– insufficient provision of energy sector organizations with labor resources, especially highly qualified personnel.

Certainly, among the most important threats to the energy security, it is worth considering the significant depreciation of the main production assets of the energy sector.

In addition to these internal threats, the Doctrine also considers so-called transboundary threats. From the standpoint of reliability of fuel and energy supplies, the most relevant in their list are:

- terrorist and sabotage activities that damage the infrastructure and facilities of the energy sector;
- illegal use of information and communication technologies, including computer attacks on information infrastructure facilities and communication networks used to organize their interaction, which can lead to disruptions in the functioning of the infrastructure and facilities of the energy sector;
- adverse and dangerous natural phenomena, changes in the environment that lead to disruption of the normal functioning and destruction of the infrastructure and facilities of the energy sector.

The Doctrine states that, in general, the implementation of the specified threats, among other negative consequences, may lead to:

- disruption of the normal functioning of organizations, including energy sector organizations and sectors of the Russian Federation economy;
- an increase in prices (tariffs) for products of fuel and energy complex organizations and services in the energy sector;
- a decrease in the financial stability and termination of the activities of fuel and energy complex organizations.

In addition to the above, negative consequences may also be expressed in a slowdown in the rate of reduction of the specific energy intensity of the economy, as well as in the lag of the professional education system behind the existing and future needs of the energy sector for qualified personnel with a corresponding increase in the threat of imperfect management in the functioning and development of energy industries.

Based on the consideration of the possibilities of the implementation of the main threats and the potential consequences of this implementation, the Doctrine of ES declares that the normative level of ES should be ensured by achieving an acceptable situation for the objects of indicative analysis. In [10], in order to form an integrated qualitative assessment of the state of ES at the federal level, an attempt was made to form relative specific weights of each of the objects of analysis in the general system of their value. This was done using the principles of pairwise comparison of the conditional significance of individual objects of analysis and is given in Table 2. For specific indicators within the objects of analysis, in the same place in [10], the assumption was made that the weights of all indicators in each object of analysis are equal.

Table 2. Objects of indicative analysis and their specific weights in assessing the quality of the ES at the federal level

№	Object of indicative analysis	Specific weight
1	Ratio of growth in balance reserves of primary FER to their production volumes	0,073
2	Share of natural gas in the structure of the primary FER balance	0,078
3	Dynamics of the price level within the country for the main types of FER	0,066
4	Volume of non-payments in the retail and wholesale electricity markets	0,070
5	Implementation of investment programs by energy sectors	0,085
6	Relative change in the value of specific energy intensity of GDP	0,078

© The Authors, published by EDP Sciences. This is an open access article distributed under the terms of the Creative Commons Attribution License 4.0 (https://creativecommons.org/licenses/by/4.0/).

7	Share of fossil primary FER used in the domestic market in the total volume of their production	0,065
8	Terrorist attacks at energy sector facilities, main factors	0,064
9	Stability of providing consumers with various types of FER (incl. regional level)	0,047
10	Strikes at energy sector facilities	0,088
11	Physical deterioration of fixed production assets by energy sector and related industries	0,091
12	Emergencies at energy sector facilities	0,079
13	Share of Russia's FER exports in total exports, including the share of the Asia-Pacific region in total exports	0,058
14	Share of oil refining and petrochemical products in total exports of Russia's FER	0,059

The first of the listed objects of analysis that is quite significant is the ratio of the annual increase in the balance reserves of primary FER to the volumes of their production or the reproduction of the energy sector's primary reserves. Due to the structure of the Russian energy sector, it is one of the most important objects of indicative analysis for the conditions of Russia. Below, we will dwell in more detail on this object.

4 Problems of energy sector's mineral resource base reproduction

From the data in Table 1 it is evident that in recent years the share of gas (in terms of equivalent fuel) in the domestic consumption of primary FER was 47%, coal – 9%, oil – 34%. Electricity from nuclear power plants, renewable energy sources and other fuel and energy resources replaced about 10% of equivalent fuel that could be spent on generating the corresponding volume of electricity. Thus, the domestic consumption of primary FER in the country is 90% provided by gas, coal and oil. At the same time, if we talk about the consumption of boiler and furnace fuel (BF), realizing that only heating oil mainly participates in its consumption from oil, we should keep in mind the share of gas in the BF balance exceeding 54% and coal – about 14%, 75% of Russian electricity is generated using natural gas.

We can assess the current situation and the conditions influencing it using the example of natural gas, which dominates the structure of the Russian fuel and energy balance, Fig. 1.

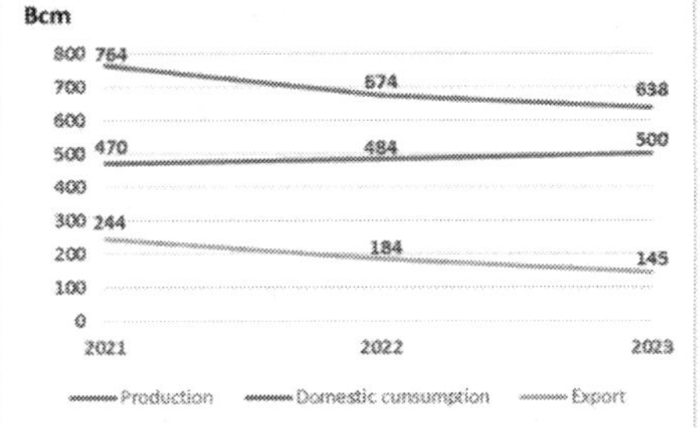

Fig. 1. Production, export and domestic consumption of natural gas in 2021-2023.

From the data in Table 1 and Fig. 1 it is evident that gas exports have decreased by 99 bcm (41%) over three years, while domestic consumption has increased by only 30 bcm (6%) with gas production decreasing by 126 bcm (16%) (the figures for the imbalance between production, export and consumption in the corresponding years characterize the excess of gas injection into UGS facilities over extraction from there and vice versa). Considering that the prices of the domestic and foreign gas markets differ significantly, and domestic prices are strictly regulated by the state, such a decrease in exports could not but affect the volumes and rates of investment in the development of the industry, including the corresponding geological exploration work.

As for gas exports, significant hopes for its development in the future were associated with the production and transportation of LNG. However, the main operator of the Arctic LNG 2 project, PJSC Novatek, suspended LNG production due to sanctions and a shortage of tankers. Thus, according to [11], the company cannot receive three gas carriers from the Korean shipyard Hanwha Ocean, since their customers are Sovcomflot structures that fell under US sanctions. This led to the fact that at the beginning of 2024, gas production in the company sharply decreased [12].

Taking into account the data [13, 14], we will analyze the information on the growth and depletion of industrially recoverable gas reserves over the past 5 years, taking into account exploration, production and revaluation of reserves. Over the past 5-year period, provided for the analysis of this indicator on the basis of the corresponding methodology [10], the growth rate of industrially recoverable gas reserves lagged behind the value of their disposal by 30%, which amounted to almost 6 trillion m^3. Only the revaluation for this period reduced the reserves by slightly more than 7 trillion m^3. This minus was compensated by the results of geological exploration by only 4.4 trillion m^3. Thus, the negative situation here is due to the correction of significant errors in geological exploration in previous periods. A striking example is the Kovyktinskoye gas condensate field in the Irkutsk region, which is classified as one of the largest. The field was discovered in 1987 with confirmed gas reserves of categories A+B1+C1 of 2.7 trillion m^3 [15]. As a result of the revaluations [14], the gas reserves of the field in these categories as of 01.01.2023 decreased to 1.4 trillion m^3. Thus, the recoverable reserves for the Kovyktinskoye gas condensate field alone decreased by almost 2 times.

It has already been noted above that the process of replenishing the mineral resource base of the energy sector largely depends on the level of investment. The situation with underinvestment in the oil and gas industry was highlighted by Deputy Prime Minister of the Russian Federation A. Novak on 12.10.2023 at the Russian Energy Week forum. From [16] it follows that over the past few years the industry has received $1 trillion in underinvestment. Today you can see the reporting picture for 2023. According to the results of the year, PJSC Gazprom received a loss of 629 billion rubles [17, 18], while a year earlier, according to the results of 2022, it received a profit of 1.23 trillion rubles.

© The Authors, published by EDP Sciences. This is an open access article distributed under the terms of the Creative Commons Attribution License 4.0 (https://creativecommons.org/licenses/by/4.0/).

 https://doi.org/10.1051/e3sconf/202458401004

Gazprom's revenue decreased by 26.8%, to 8.54 trillion rubles, operating expenses decreased by 7.8%, to 8.58 billion rubles. At the same time, the revenue of the gas business fell by 43%, to 4.41 trillion rubles. This situation, coupled with the chronic, in recent years, reduction of industrially extractable gas reserves, makes one doubt the possibility of improving the situation with investments to support small and medium-sized businesses in gas in the coming years.

The situation with oil is slightly better. At the same time, [20] notes that investments by the state and the largest oil producing companies in exploration from 2020 to 2023 decreased by 7% to 324 billion rubles. This led to a decrease in exploratory drilling by 18% over the specified period, analysts emphasize. At the same time, the share of company investments in exploration in the total costs of field development usually amounts to 5-15%, depending on the degree of exploration. This is also explained by the fact that investments in exploration are high-risk, do not bring income to producing companies and give a positive effect only in the long term. As a result, companies in difficult modern conditions are the first to reduce investments in exploration.

Taking into account [13, 14, 19], we will consider information on the growth and depletion of industrially recoverable oil reserves, taking into account exploration, production and revaluation of reserves. Over the past 5-year period provided for the analysis of this indicator, the growth in industrially recoverable oil reserves exceeded their depletion by 7% (492 million tons). At the same time, in 2023, the growth in industrially recoverable oil reserves was 2% less than the amount of their depletion. It is necessary to note the problem of the growth of the share of hard-to-recover reserves (HRR). Currently, their share in the structure of oil production and reserves is more than 22 and 55%, respectively [20]. Exploration of HRR and their development are characterized by unfavorable geological and climatic conditions, therefore they require more complex technologies. In addition, HRR fields have a shorter service life than traditional ones, which makes it necessary to increase investments in exploration and development. In addition to the above, Russian geological exploration is on average 40% dependent on imports, for some positions this dependence is significantly higher. The same problem is relevant for gas.

Taking into account the data [13, 14, 21], we will examine information on the growth and depletion of industrially extractable coal reserves, taking into account exploration, production and revaluation of reserves. Over the 5-year period from 2018 to 2022, the growth rate of industrially extractable coal reserves lagged behind the amount of their disposal by 30% or 1.2 billion tons. Only revaluation, write-off of reserves for other reasons and losses of reserves during this period reduced reserves by almost 2 billion tons. This minus was offset by the difference in geological exploration and production by only 545 million tons over the entire period.

5 Ensuring energy security for the "Ratio of the increase in balance reserves of primary FER to the volumes of their production"

Based on the conducted analysis of the level of reproduction of the mineral resource base of the energy sector, it is possible to calculate the values of the corresponding indicators and get an idea of the situation in this object of analysis as a whole. In [23], the apparatus for calculating the values included in this object, the indicators "Ratio of the increase in the extractable balance reserves of the corresponding energy resource for the five-year period preceding the analysis to the total volume of its extraction (production) in the country for the same period" is presented. The value of the indicator for a specific (i-th) primary energy resource ($K_{res\_5}^{ER_i}$) is determined as follows:

$$K_{res\_5}^{ER_i} = \frac{\Delta Q_{res\_5}^{ER_i}}{Q_{extr\_5}^{ER_i}}$$

where $\Delta Q_{res\_5}^{ER_i}$ is the increase in recoverable balance reserves of the i-th primary energy resource for the five-year period preceding the analysis, units; $Q_{extr\_5}^{ER_i}$ is the total volume of disposal of balance reserves of the i-th primary energy resource for the five-year period preceding the analysis, units.

The threshold values of this indicator are as follows: crisis (C) – 0.95, pre-crisis (PC) – 1.0. The analysis of the considered data allowed us to form the values of this indicator for each of the analyzed primary FER. This is shown in Table 3. In addition, the table provides a qualitative assessment of the state of the indicators according to the FER data and shows their specific weight in the general value system of this object of analysis. The specific weights of the indicators for each of the given FER are calculated according to the share of participation of a specific energy resource in their total consumption, taking into account energy production from renewable energy sources (5% of the total consumption of primary FER).

Table 3. Values of the indicators of renewal of the mineral resource base of the energy sector for specific primary FER and the final assessment of the general indicator

FER	Value	Qualitative assessment*	Specific weight
Natural gas	0,69	C	0,47
Oil	1,07	A	0,34
Coal	0,70	C	0,09
Uranium	0,13	C	0,05
Final assessment		**C**	

* C – crisis: A - acceptable

From the information in Table 3 it is evident that 61% of the total weight of the indicators is in the crisis value zone, which indicates a crisis state of this general indicator as a whole.

© The Authors, published by EDP Sciences. This is an open access article distributed under the terms of the Creative Commons Attribution License 4.0 (https://creativecommons.org/licenses/by/4.0/).

6 Conclusion

Taking into account the situation with the replenishment of the main primary FER it can be argued that from the standpoint of energy security, the situation should be characterized as a crisis. First of all this can lead to a significant reduction in the possibilities of ensuring the prospective production of Russian natural gas and uranium and forms significant additional requirements for the effectiveness of geological exploration work in the relevant industries.

This situation is aggravated by the decrease in profitability from the extraction of primary FER due to the decrease in their export volumes, the decrease in sales prices on world markets (especially oil and oil products) in the absence of opportunities to compensate for the decrease in these revenues through sales on the domestic market. All this leads to a decrease in the volume of investment in Russia's energy sector with a corresponding decrease in its prospective production capabilities. It would seem that with restrictions on hydrocarbon exports, the existing FER in the subsoil are spent more slowly. At the same time, the decrease in investment in the energy sectors postpones the development of new areas of their extraction, thereby increasing the share of hard-to-recover resources in the existing balance reserves.

The research was carried out under State Assignment Project (no. FWEU-2021-0003) of the Fundamental Research Program of Russian Federation 2021-2030.

References

1. Doctrine of Energy Security of the Russian Federation. http://publication.pravo.gov.ru/Document/View/0001201905140010?index=0&rangeSize=1
2. S.M. Senderov, V.I. Rabchuk. Energeticheskaya Politika. 3 (141) (2019)
3. Rosstat estimated gas production in Russia in 2023 at 638 bcm. https://www.interfax.ru/business/943942
4. The Russian coal industry in 2023. https://delprof.ru/press-center/open-analytics/ugolnaya-otrasl-rossii-v-2023-godu/
5. Russia reduced pipeline gas exports by 29.9% in 2023. https://www.vedomosti.ru/economics/news/2024/02/06/1018782-rossiya-snizila-eksport
6. Russia reduced oil exports by 3.3% in 2023 to 234 million tons. https://www.interfax.ru/business/944802
7. Oil refining in Russia in 2023 increased by 1.2% to 275 million tons. https://www.interfax.ru/business/944803
8. Electricity generation in Russia in 2023 increased by 0.7%. https://www.bigpowernews.ru/markets/document112771.phtm
9. Rosstat. Form 4-TER. Balances, receipt and consumption of fuel and thermal energy, waste oil products, secondary combustible and thermal resources in the Russian Federation in 2023.
10. S. M. Senderov, V. I. Rabchuk. Bulletin of the Russian Academy of Sciences. Energy. 2 (2018)
11. Arctic LNG-2 falls under US sanctions. https://www.kommersant.ru/doc/6612003
12. The start of gas supplies from Arctic LNG 2 will be postponed due to US sanctions. https://www.kommersant.ru/doc/6534921
13. State report on the state and use of mineral resources of the Russian Federation for 2019-2022. https://gd2021.data-geo.ru/fuel/gas/
14. State balance of mineral reserves of the Russian Federation as of 01 Jan. 2023. Issue 82: combustible gases, consolidated data on the Russian Federation. - Moscow, FGBU "Rosgeolfond", 2023, 786 p.
15. The stage of production drilling has begun at the Kovyktinskoye field. https://irkutsk-dobycha.gazprom.ru/press/about-company/2019/08/39/
16. Alexander Novak October 12 at the forum "Russian Energy Week". https://rusenergyweek.com/news/aleksandr-novak-prinjal-uchastie-v-rabote-vtorogo-dnja-rossijskoj-energeticheskoj-nedeli//
17. Gazprom reported a net loss of 629 billion rubles for 2023. https://www.forbes.ru/investicii/511630-gazprom-otcitalsa-o-cistom-ubytke-v-razmere-629-mlrd-rublej-po-itogam-2023-goda
18. Gazprom posted a loss for the year for the first time in 25 years. https://www.rbc.ru/business/02/05/2024/663392319a79474c1bacd13a
19. Russia's oil reserves growth in 2023 was the lowest in six years. https://www.vedomosti.ru/business/articles/2024/04/12/1031371-prirost-zapasov-nefti-v-rossii-stal-minimalnim
20. By 2030, we need to start producing another 230 million tons of oil per year. https://www.vedomosti.ru/business/articles/2023/10/31/1003416-do-2030-goda-nuzhno-nachat-dobivat-esche-do-230-mln-tonn-nefti-v-god
21. State Reserves Committee: 43 hydrocarbon deposits discovered in Russia by 2023. https://neftegaz.ru/news/gas/814172-gkz-za-2023-g-v-rossii-otkryto-43-mestorozhdeniya-uglevodorodnogo-syrya/
22. JSC Atomredmetzoloto. https://www.armz.ru/o-kompanii/o-nas
23. Senderov S., Rabchuk V., Slavin G., Pyatkova N. and oth. Methodological recommendations for assessing the state of energy security of the Russian Federation at the federal level. ISEM SB RAS. – Irkutsk, 2014. – 35 p.

© The Authors, published by EDP Sciences. This is an open access article distributed under the terms of the Creative Commons Attribution License 4.0 (https://creativecommons.org/licenses/by/4.0/).

Finding ways to minimize gas shortages for consumers in the context of large-scale emergencies in the gas industry

Sergey Senderov, *Oleg* Khamisov, *Sergey* Vorobev[1*], and *Gleb* Danilov[1]

[1]Melentiev Energy Systems Institute, 130 Lermontov str., Irkutsk, Russia

Abstract. The existing structure of the gas transportation network of Russia is characterized by the presence of critically important objects, the loss of operability of which cannot be compensated by any of the previously considered measures. Therefore, it is necessary to justify the development of the gas transportation network with the creation of additional gas transportation capacities outside the arcs of the existing graph. This problem can be solved by segmenting the existing network into simple polygons with the possibility of creating arcs connecting previously unconnected nodes, i.e. polygons starting with a quadrangle are of interest. The specific cost of creating such arcs, as a first approximation, is an order of magnitude higher than the specific cost of creating additional arcs in the existing corridors of main gas pipelines. The authors of the article propose a mathematical model for solving this problem. The results of the study on the aggregated calculation scheme of the gas transportation network of Russia are presented, conclusions are made on the operability of the proposed approach.

1 Introduction

Large geographically distributed energy systems are subject to a significant number of external threats. The main ones include both the aging and wear of the linear part of the main production assets, and the threat of deliberate impacts of various negative nature. Large-scale accidents in energy systems [1, 2], caused by the failure of various critically important objects of these systems, entail damage to consumers in the form of undersupply of fuel and energy resources. For the gas industry, critically important objects are hub, head compressor stations, intersections of main gas pipelines, or individual sections of main gas pipelines, which, in the event of loss of their operability, are capable of causing noticeable restrictions in gas supply to consumers. Therefore, the search for and determination of critical elements and their combinations in energy systems are important current tasks.

2 Current state of the problem under study

A number of studies have been conducted at the Energy Systems Institute to identify critical facilities in the gas transportation network. A list of main gas pipeline intersections in the Unified Gas Supply System of Russia has been determined, the disruption of which will lead to a relative deficit of daily gas supplies across the system as a whole in the amount of 5% or more [3, 4]. Studies have been conducted to find and determine combinations of individual sections of main gas pipelines, the simultaneous disruption of which can lead to a significant deficit of daily gas supplies across the system (5% or more) [5, 6]. The issues of analyzing critical facilities in electric power systems are the subject of works [7 - 9]. Studies have also been conducted to determine critical combinations of gas industry facilities. Such facilities have been found, the failure of which in a pair with one of the critical facilities will lead to a much greater gas deficit for consumers than if these facilities fail separately [10]. After identifying possible critical combinations of objects in each specific case, the possibilities of bypassing bottlenecks by temporarily increasing the capacity of individual sections of main gas pipelines were analyzed. This technological measure was used to minimize gas shortages for consumers. As a result of such measures, the significance of a number of potential critical combinations of objects was reduced. Thus, all possible critical combinations of gas industry objects were identified and a list of them ranked by importance was formed.

Research to solve the problem of minimizing gas shortages to consumers in the event of large-scale emergency situations in the gas transportation network developed over time according to the following logic:

- for isolated cases of loss of operability of gas transportation network objects, sections were determined on which the capacity could be temporarily increased (within technically permissible limits) until the emergency situation was eliminated;

- if there are results of multi-iteration studies that make it possible to identify possible shortfalls in gas

* Corresponding author: seregavorobev@isem.irk.ru

© The Authors, published by EDP Sciences. This is an open access article distributed under the terms of the Creative Commons Attribution License 4.0 (https://creativecommons.org/licenses/by/4.0/).

supplies to consumers in the conditions of a representative number of abnormal losses in the operability of various critically important objects of the gas transportation network with the identification of invariant measures expressed in an increase in the throughput of the same arcs of the gas transportation network within 10% of the initial throughput of these arcs, we can talk about implementing measures to constantly increase the throughput of a section in an amount of up to 10% of its initial throughput;

- if there are results of multi-iteration studies that indicate the preservation of a stable mathematical expectation of significant residual shortfalls in gas supplies to consumers, a problem can be solved that, in the conditions of a representative number of abnormal losses in the operability of various critically important objects of the gas transportation network, will make it possible to identify sections of the gas transportation network where we can talk about implementing measures to create additional gas transportation capacities in the volumes justified by the solution.

Due to the fact that the existing structure of the gas transportation network of Russia allows us to talk about the presence of critically important objects, the loss of operability of which cannot be compensated by any of the above measures, there is a need to justify the development of the gas transportation network with the creation of additional gas transportation capacities outside the arcs of the existing graph. The problem can be solved by segmenting the existing network into simple polygons with the possibility of creating arcs connecting previously unconnected nodes, i.e. polygons starting with a quadrangle are of interest. The specific cost of creating such arcs, as a first approximation, should be an order of magnitude higher than the specific cost of creating additional arcs in existing corridors. To estimate the cost of creating such arcs at this stage of the study, data are used that allow us to estimate their approximate length.

3 Creating new lines in the graph

A directed graph consisting of m nodes and n arcs (lines) is given. The matrix A with elements a_{ij}, i = 1, ... , m, j = 1, ... , n is the incidence matrix of the graph,

$$a_{ij} = \begin{cases} 1; \\ -1; \\ 0 \end{cases} \tag{1}$$

The set $I_s \in \{1,\ldots,m\}$ is the set of supplier nodes in which production si > 0, $i \in I_s$ is specified; the set $I_D \in \{1,\ldots,m\}$ is the set of consumer nodes in which demand $d_i > 0$, $i \in I_D$ is specified; the set $I_0 \in \{1,\ldots,m\}$ is the set of branch nodes in which there are neither suppliers nor consumers. The constants $c_j > 0$, j = 1, ..., n determine the specific transportation costs along the corresponding arcs. The optimal flow distribution is found by solving the following problem

$$\sum_{j=1}^{n} c_j |x_j| \to \min, \tag{2}$$

$$\sum_{j=1}^{n} a_{ij} x_j = \begin{cases} s_i, & i \in I_S; \\ -d_i, & i \in I_D; \\ 0, & i \in I_0, \end{cases} \tag{3}$$

$$\underline{x}_j \leqslant x_j \leqslant \overline{x}_j, \; j = 1, \ldots, n. \tag{4}$$

The variables x_j can take negative values if the direction of flows along some lines does not coincide with the initial orientation of these lines. This is why the modulus has to be used in the objective function (2). The objective function itself is a convex piecewise linear coercive function. Since A is the incidence matrix of a directed graph, then by virtue of Fredholm's theorem the system of linear equations (3) is compatible if and only if

$$\sum_{i=1}^{m} S_i = \sum_{i=1}^{m} d_i \, , \tag{5}$$

i.e. when the total demand is equal to the total supply. When condition (5) is satisfied, due to the coercivity of the objective function, a solution to problem (2)-(3) exists. Consequently, only capacity constraints (4) can lead to the fact that a solution to problem (2)-(4) may not exist. In this case, the network cannot transport the required volume of gas from the fields to consumers.

The solvability of problem (2)-(4) can be checked by simply trying to solve it. To do this, we will make the following transformation to get rid of the modulus in (2). We introduce auxiliary variables y_i, i = 1, ..., n and rewrite problem (2)-(4) in the following form

$$\sum_{j=1}^{n} c_i y_i \to min, \tag{6}$$

$$\sum_{j=1}^{n} a_{ij} x_j = \begin{cases} s_i, & i \in I_S; \\ -d_i, & i \in I_D; \\ 0, & i \in I_0, \end{cases} \tag{7}$$

$$y_j \geqslant x_j, \; y_j \geqslant -x_j, \; j = 1, \ldots, n, \tag{8}$$

$$\underline{x}_j \leqslant x_j \leqslant \overline{x}_j, \; j = 1, \ldots, n. \tag{9}$$

Conditions (8) are equivalent to conditions $y_j \geq |x_j|$, j = 1, ..., n, and the latter conditions guarantee the equivalence of problems (2)–(4) and (6)–(9). Both problems are solvable or unsolvable simultaneously. If (x∗, y∗) is a solution to problem (1.6)–(1.9), then x∗ is a solution to problem (2)–(4). If x∗ is a solution to problem (2)–(4), then there exists a vector y∗ such that the pair (x∗, y∗) is a solution to problem (6)–(9). The advantage of problem (6)–(9) is that it is a linear programming problem.

The basic model assumes that some line (or lines) stop functioning and it is necessary to determine between which nodes it is necessary to build new lines so that the

© The Authors, published by EDP Sciences. This is an open access article distributed under the terms of the Creative Commons Attribution License 4.0 (https://creativecommons.org/licenses/by/4.0/).

demand is covered with minimal total costs. In this case, the basis is the problem (6)-(9). New parameters: $\tilde{c}_j$ - specific cost of transportation along new line j, $\tilde{p}_j$ - specific cost of construction of new line j. At the initial stage, it is determined by experts between which nodes new lines can be built. Let the number of new lines be q.

New lines correspond to their own m X q incidence matrix $\tilde{A}$ with elements $\tilde{a}_{ij}$. The corresponding optimization problem has the form:

$$\sum_{j=1}^{n} c_j y_j + \sum_{j=1}^{q} \tilde{p}_j \tilde{y}_j + \sum_{j=1}^{q} \tilde{c}_j \tilde{y}_j \to \min, \tag{10}$$

$$\sum_{j=1}^{n} a_{ij} x_j + \sum_{j=1}^{q} \tilde{a}_{ij} \tilde{x}_j = \begin{cases} s_i, & i \in I_S; \\ -d_i, & i \in I_D; \\ 0, & i \in I_0, \end{cases} \tag{11}$$

$$y_j \geqslant x_j,\ y_j \geqslant -x_j,\ j = 1, \ldots, n, \tag{12}$$

$$\tilde{y}_j \geqslant \tilde{x}_j,\ \tilde{y}_j \geqslant -\tilde{x}_j,\ j = 1, \ldots, n, \tag{13}$$

$$\underline{x}_j \leqslant x_j \leqslant \overline{x}_j,\ j = 1, \ldots, n. \tag{14}$$

$$z_j \underline{\xi}_j \leqslant \tilde{x}_j \leqslant z_j \overline{\xi}_j,\ j = 1, \ldots, q, \tag{15}$$

$$z_j = 0 \vee 1,\ j = 1, \ldots, q. \tag{16}$$

In this problem $\underline{\xi}_j < 0$, $\overline{\xi}_j > 0$ - the maximum throughput capacities in the reverse and forward directions of the new lines are considered to be given. If the integer variable z_j in the solution of the problem is equal to zero, then from (15) it follows that $\tilde{x}_j = 0$, i.e. line j does not need to be built. If $z_j = 1$, then the line is built, and its throughput capacity coincides with the found flow. The advantage of this formulation is that the number of integer variables coincides with the number of "reasonable" new lines, as a result of which redundant integer variables do not appear. The disadvantage is the need for experts to select these "reasonable" lines, which for networks with a large number of nodes can be a very labor-intensive task.

4 Calculation results

To test the model (10)-(16), an aggregated calculation scheme of the Unified Gas Supply System was assembled, including 23 nodes and 26 arcs. Table 1 presents a description of the nodes.

Table 2 contains: description of lines (arcs). The direction of lines is indicated, if the minimum throughput is given with a minus sign, then flow is allowed in the direction opposite to the specified one, if the minimum throughput is zero, then the flow along this line is one-way.

Table 1. Nodes of the calculation scheme

№	Gas consumption, mln m³/day	Gas production, mln m³/day
1	11.56	0.0
2	186.27	0.0
3	0.0	0.0
4	49.86	0.0
5	109.42	0.0
6	0.55	0.0
7	0.0	0.0
8	0.0	0.0
9	0.0	0.0
10	35.89	0.0
11	56.19	0.0
12	0.0	13.3
13	0.0	25.1
14	112.76	0.0
15	0.0	2.9
16	0.0	0.0
17	0.0	6.4
18	0.0	30.0
19	0.91	0.0
20	0.0	0.0
21	73.93	0.0
22	16.06	0.0
23	0.0	575.5

Figure 1 shows the given calculation scheme. Solid lines correspond to existing ones (L1-L26), dotted lines correspond to possible ones (L27-L30). Numbers with a + sign near the node correspond to production, with a - sign correspond to consumption.

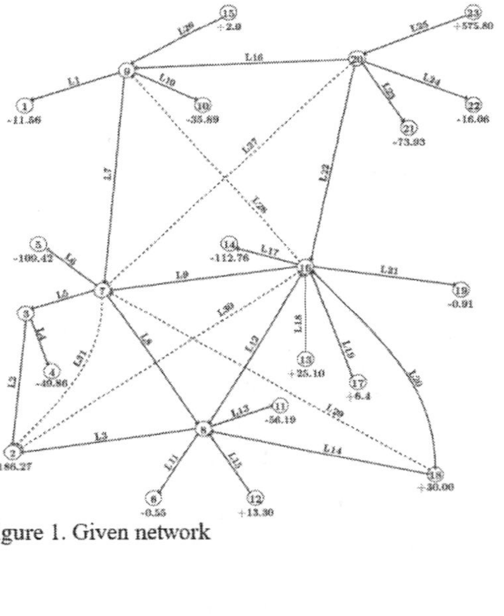

Figure 1. Given network

© The Authors, published by EDP Sciences. This is an open access article distributed under the terms of the Creative Commons Attribution License 4.0 (https://creativecommons.org/licenses/by/4.0/).

Table 2. Description of the arcs of the calculation scheme

Arc	Starting node	End node	Min flow, mln m^3/day.	Max flow, mln m^3/day.	Cost of gas transportation along the arc, USD.
L1	9	1	-21.40	21.40	810.0
L2	3	2	-205.20	205.20	472.5
L3	8	2	-4.00	4.00	877.5
L4	3	4	-85.00	85.00	202.5
L5	7	3	-258.20	258.20	405.0
L6	7	5	-110.00	110.00	202.5
L7	9	7	-85.00	85.00	472.5
L8	7	8	-27.40	27.40	405.0
L9	16	7	-347.50	347.50	1012.5
L10	9	10	-85.00	85.00	202.5
L11	8	6	-0.55	0.55	405.0
L12	16	8	-73.65	73.65	1012.5
L13	8	11	-85.00	85.00	202.5
L14	18	8	-0.55	0.55	1485.0
L15	12	8	-13.50	13.50	270.0
L16	20	9	-210.60	210.60	1485.0
L17	16	14	-115.00	115.00	202.5
L18	13	16	-25.10	25.10	337.5
L19	17	16	-10.00	10.00	540.0
L20	18	16	-71.10	71.10	1012.5
L21	16	19	-7.65	7.65	675.0
L22	20	16	-522.00	522.00	675.0
L23	20	21	-85.00	85.00	202.5
L24	20	22	-23.00	23.00	945.0
L25	23	20	-592.50	592.50	945.0
L26	15	9	-44.50	44.50	810.0

In the test example, arc L9 was disconnected, then the model built a pre-set arc L27 with a capacity corresponding to L9. Thus, ensuring the supply of gas to consumers in nodes No. 2, 4, 5.

5 Conclusion

This study considers the issues of creating additional gas transportation capacities outside the arcs of the existing graph. To solve this problem, a mathematical model is proposed for segmenting the existing network into simple polygons with the ability to create arcs connecting previously unconnected nodes. The studies conducted on the aggregated calculation scheme showed the adequacy and applicability of the proposed approach and the corresponding mathematical model.

Further development of this study will be aimed at optimizing the model for working with a detailed calculation scheme of the gas transportation network, in order to form a list of new arcs that minimize gas shortages for consumers in the event of failure of various critical gas industry facilities.

The research was carried out within the framework of the state assignment project FWEU-2021-0003 (registration number AAAA-A21-121012090014-5) for fundamental research of the Siberian Branch of the Russian Academy of Sciences, using the resources of the Shared Use Center Siberian Supercomputer Center ICM&MG SB RAS.

References

1. Accidents on main gas pipelines in Russia in 2018-2019. URL: https://ria.ru/20190728/1556953028.html.
2. Top 5 largest and most destructive accidents on gas pipelines. URL: https://sila-sibiri-rabota.ru/avarii-na-gazoprovodax/.
3. Senderov S.M., Rabchuk V.I., Edelev A.V. (2016) News of the Russian Academy of Sciences. Energetika, **No. 1**, pp. 70-78. (in Russian)
4. Senderov S., Edelev A. (2017) Energy, DOI:10.1016/J.ENERGY.2017.11.063.
5. Vorobev S., Edelev A. Analysis of the importance of critical objects of the gas industry with the method of determining critical elements in networks of technical

© The Authors, published by EDP Sciences. This is an open access article distributed under the terms of the Creative Commons Attribution License 4.0 (https://creativecommons.org/licenses/by/4.0/).

infrastructures / Management of Large-Scale System Development (MLSD), 2017 Tenth International Conference. IEEE, 2017. DOI: 10.1109/MLSD.2017.8109707.
6. Vorobev S., Edelev A., Smirnova E. Search of critically important objects of the gas industry with the method of determining critical elements in networks of technical infrastructures / Methodological Problems in Reliability Study of Large Energy Systems (RSES 2017). E3S Web Conf. Volume 25, 2017. DOI: 10.1051/e3sconf/20172501004.
7. Senderov S., Krupenev D. Energy security and critical facilities of energy systems: methodology and practice of their identification on the example of Russia's gas and electric power industries, Energy Systems Research. 2019. T. 2. № 2 (6). p. 41-50.
8. Krupenev D.S. Principles of Determining Critical Objects of Electric Power Systems / Methodological Issues of Studying the Reliability of Large Power Systems. International Scientific Seminar named after Yu.N. Rudenko: In 2 books. Responsible editor Voropai N.I., 2018. P 329-337.
9. Krupenev D., Boyarkin D., Iakubovskii D. Improvement in the computational efficiency of a technique for assessing the reliability of electric power systems based on the Monte Carlo method, Reliability Engineering & System Safety. 2020. T. 204.
10. Senderov S., Vorobev S., Edelev A. (2018). E3S Web Conf. **Vol. 58**, DOI: 10.1051/e3sconf/ 20185803002.

© The Authors, published by EDP Sciences. This is an open access article distributed under the terms of the Creative Commons Attribution License 4.0 (https://creativecommons.org/licenses/by/4.0/).

The impact of changes in gas consumption on the selection of critical gas industry objects

Sergey Vorobev[1*] and *Alexey* Edelev[1]

[1]Melentiev Energy Systems Institute, 130 Lermontov str., Irkutsk, Russia

Abstract. The article is devoted to the problem of finding critically important objects of the gas industry from the standpoint of ensuring energy security. A critically important object is a partial or complete failure of which can cause significant damage to the country from the fuel and energy complex as a whole or within a separate energy system. The current state of the gas industry is characterized, on the one hand, by a significant decrease in gas exports to Europe, and on the other hand, by a significantly increased gas consumption within the country. The article presents the results of calculations in which the authors analyze how the change in the structure of gas consumption affected the previously found critically important objects of the gas industry. The change in the significance of these objects on the operability of the Russian gas industry is shown. Conclusions are made on the need to adjust the list of critically important objects of the gas industry.

1 Introduction

Over time, old fields are depleted, new gas production centers come to take their place, the structure and directions of gas exports change. Thus, the gas transportation system is a rather complex and detailed transport graph that changes over time. The Russian gas transportation network is characterized by a significant number of critically important objects, the operability of which critically affects the operability of the entire gas transportation network and the corresponding gas supply capabilities for consumers. According to [1], a critically important object is one whose partial or complete failure can cause significant social and economic damage to the country from the fuel and energy complex.

2 Current state of the problem under study

Research is being conducted worldwide on various aspects of analyzing the most important objects and their combinations in energy systems. Among the most relevant are the following points.

The study [2] proposes a methodology for analyzing the reliability of gas supply to consumers in the gas transmission network. Demand volumes and the possibility of supplying gas volumes are modeled using the Monte Carlo method. For each scenario, an indicator is calculated that is proposed for a quantitative assessment of the reliability of gas supply. Thus, based on Monte Carlo tests, the average reliability of gas supply was obtained, and measures to improve the reliability of gas supplies were proposed.

In [3], a systematic basis for analyzing the reliability of energy supply in integrated energy systems is proposed, taking into account the uncertainty factor. Based on the linear block of the integrated energy system functioning model, a quasi-stationary model is created, which is used to model the dynamic behavior of the integrated energy system. This model takes into account practical engineering and operational solutions. Then, considering the cross-correlations between different uncertainty sources and the time-dependent relationships of each variable, a model that combines statistical framework with machine learning method is used. Monte Carlo simulation combined with failure statistics is used to evaluate the reliability of energy supply to consumers. Case studies are conducted on a realistic integrated power system that combines an IEEE-15 power grid with an 18-node gas pipeline network. The effectiveness and accuracy of the quasi-stationary model are verified. The reliability assessment results show that the cross-correlations between variables and the time-dependent relationships of each variable have a great impact on the reliability assessment of the system.

In [4], a systematic framework for dynamic real-time reliability analysis of integrated power system is proposed by integrating various machine learning and statistical methods. The dynamic behavior of the integrated power system is modeled based on a multi-level autoencoder model instead of using traditional simulation models. In addition, variables reflecting the transient properties of gas pipeline networks, such as gas

* Corresponding author: seregavorobev@isem.irk.ru

© The Authors, published by EDP Sciences. This is an open access article distributed under the terms of the Creative Commons Attribution License 4.0 (https://creativecommons.org/licenses/by/4.0/).

pressure and flow rate, are taken as indicators for quantitatively assessing the reliability of power supply. The final reliability assessment is performed by estimating the probability distribution of each functional state of the target integrated energy system.

In this study [5], an integrated method based on demand analysis is proposed for assessing the reliability of gas supply of a large-scale and complex gas pipeline network. The method consists of establishing a system of indicators, analyzing demand, assessing the probability of unit failure, and calculating gas supply volumes. Demand analysis focuses on forecasting market demand and investigating the importance of gas consumers. The influence of consumer importance, hydraulic and pressure limitations, and failures of network elements are taken into account. The proposed method is practically applied to a model of a real gas pipeline network located in China. Based on the results of gas supply reliability assessment, the weakest nodes and key links of the gas pipeline network are identified, and proposals are made to improve the reliability of gas supply.

Previously, a number of studies were conducted on identifying critical facilities in the gas transportation network. A list of intersections of main gas pipelines in the Unified Gas Supply System of Russia was determined, the disruption of which would lead to a relative deficit of daily gas supplies across the system as a whole in the amount of 5% or more [6]. Studies were conducted to find and determine combinations of individual sections of main gas pipelines, the simultaneous disruption of which could lead to a significant deficit of daily gas supplies across the system (5% or more) [7, 8]. Taking into account previously accumulated experience and based on an analysis of studies currently being conducted in the world, a methodology for forming lists of critical facilities of energy systems from the standpoint of ensuring the operability of these systems was formulated in [9, 10] using the example of the Russian gas industry.

The works reviewed identified critical facilities and significant facilities of various levels, all calculations in these studies were carried out on the calculation schemes of the gas transportation network for 2016-2019. In this study, the authors propose to look at how the importance of critical objects may change under new design conditions.

3 Determination of critical elements in the gas industry

In the ongoing study, both to determine the critical important objects themselves and to search for critically important combinations of objects, a flow model was used, which is the core of the software-computing complex "Oil and Gas of Russia" [11, 12]. The use of this internal control system makes it possible to determine the degree to which gas needs are met within the country and export supplies are ensured. In addition, the software-computing complex "Oil and Gas of Russia" makes it possible to identify bottlenecks - sections of the gas transportation network that, in some cases, limit the production capabilities of the system.

The flow distribution model in the Unified Gas Supply System of Russia, laid down in the software-computing complex "Oil and Gas of Russia", is intended to assess the production capabilities of the Unified Gas Supply System of Russia under conditions of various types of disturbances. The purpose of such studies is to minimize gas shortages at consumption points. The Unified Gas Supply System of Russia is presented in the model as a combination of three subsystems: gas sources, the main transport network and consumers.

When solving the problem of assessing the state of the system after a disturbance, the criterion for the optimal distribution of flows is the minimum gas deficit for the consumer at the minimum cost of delivering gas to consumers. This problem can be solved by finding the maximum flow through the network, followed by minimizing the cost of delivering gas to consumers [13]. The mathematical formulation of this problem is described in [14].

In the Unified Gas Supply System of Russia flow distribution model, as already mentioned, the Basaker-Gowen algorithm is used to calculate the maximum flow of minimum cost, which as a result allows us to determine the possible level of consumer satisfaction with gas. As a result of various emergency situations, gas shortages may occur among consumers, caused by a lack of throughput capacity of certain sections of gas pipelines. Bypassing such bottlenecks or areas of the system that limit production capabilities, in acceptable quantities, will reduce the gas shortage among consumers that arose in the situation under consideration.

If there is a gas shortage among consumers caused by a lack of capacity of the relevant gas pipelines, other branches of the main gas pipelines not affected by the violation in question take on increased volumes of gas. In such a situation, the congestion of the network changes and a lack of capacity may appear in other sections of the main gas pipelines.

Subsequent debottlenecking in the gas transportation network will minimize gas shortages among consumers and makes the assessment and identification of possible critical combinations of gas industry facilities as adequate as possible.

To solve the problem of bypassing the identified bottlenecks, the flow distribution model included the possibility of increasing the gas flow through the arcs within 10% of their throughput. Such a short-term increase in the throughput of a section of the main gas pipeline is possible with an increase in the operating power of compressors at large main compressor stations [15]. As a result of increasing the operating pressure in the gas pipeline, an increase in the throughput of the main gas pipeline section by up to 10% is achieved. As a result (by using the technical capabilities of the gas transportation network), the problem of minimizing gas shortages among consumers is solved.

© The Authors, published by EDP Sciences. This is an open access article distributed under the terms of the Creative Commons Attribution License 4.0 (https://creativecommons.org/licenses/by/4.0/).

4 Calculation results

The Unified Gas Supply System of Russia calculation scheme used to model the operation of the Unified Gas Supply System of Russia in this work takes into account all the main features of the functioning of the Unified Gas Supply System of Russia and contains:

- 378 nodes, including: 28 gas sources; 64 gas consumers (subjects of the Russian Federation); 24 underground gas storage facilities; 266 hub compressor stations;
- 486 arcs representing main gas pipelines and branches to gas distribution networks.

Initial data, such as daily volumes of gas consumption, production, export and import, and the capacity of existing gas pipelines, were adopted in accordance with official statistics [16-18] for 2021.

The study [19] conducted a corresponding analysis on the 2017 calculation scheme, as a result of which 61 gas industry facilities were classified as critical gas industry facilities. Among these facilities: 25 arcs between nodal compressor stations and 36 nodes, including 30 nodal compressor stations, 5 main compressor stations at the exits of large gas fields and a compressor station at one underground gas storage facility. The largest gas deficit among consumers in the system as a whole reached 21%.

Table 1 presents the results of identifying critical facilities on the 2021 calculation scheme.

Table 1. Results of identifying critical objects in the 2021 calculation scheme.

№	Object type	Total gas deficit, %
1	Node	11
2	Node	10
3	Node	10
4	Arc	10
5	Arc	9
6	Node	8
7	Node	8
8	Arc	8
9	Node	7
10	Node	6
11	Node	6
12	Node	6
13	Node	6
14	Node	6
15	Node	6
16	Node	6
17	Arc	6
18	Arc	6
19	Arc	6
20	Arc	6
21	Arc	6
22	Arc	6
23	Node	6
24	Node	6
25	Node	5
26	Node	5
27	Arc	5

In Table 1, the object name "node" refers to nodal compressor stations, the object name "arc" refers to sections of main gas pipelines located between compressor stations. The results obtained (Table 1) clearly demonstrate how the importance of critical objects has decreased over 4 years of development of the gas transportation network.

Thus, out of 61 objects, 27 remained, the maximum possible gas deficit in the system as a whole also decreased from 21% to 11%. Such a significant decrease in the importance of the objects in question was influenced by the development of gas production in Yamal, reducing the load from the Northern regions of the Tyumen region.

A certain role was played by the decrease in Russian gas exports to Europe. The reorientation of gas supplies to the domestic market led to a greater balance in the demand for gas within the country.

However, even taking into account the development of new gas production centers and the development of the gas transportation network itself, there are still facilities in the system whose failure could lead to a gas deficit of 10-11% in the system as a whole.

5 Conclusion

The article is devoted to the problem of finding critical gas industry facilities from the standpoint of ensuring energy security. The article shows how the change in the structure of gas consumption affected the previously found critical gas industry facilities. The change in the significance of these facilities and their impact on the performance of the Russian gas industry is shown. It is noted that under the considered scenarios, a number of critical facilities remain in the system, and if they fail to operate, consumers will continue to experience a gas shortage. Conclusions are made about the need to adjust the list of critical gas industry facilities.

The research was funded by the Russian Science Foundation, grant no. 23-29-00460, using the resources of the High-Temperature Circuit Multi-Access Research Center (Ministry of Science and Higher Education of the Russian Federation, project no 13.CKP.21.0038).

References

1. S. Senderov, V. Rabchuk, A. Edelev. Features of the formation of the list of critical facilities of the gas transportation network of Russia taking into account the requirements of energy security and possible measures to minimize the negative consequences of emergency situations at such facilities // Bulletin of the Russian Academy of Sciences. Power Engineering, 2016, No. 1, pp. 70-78.
2. Yu W., Song S., Li Y., Min Y., Huang W., Wen K., Gong J. Gas supply reliability assessment of natural gas

© The Authors, published by EDP Sciences. This is an open access article distributed under the terms of the Creative Commons Attribution License 4.0 (https://creativecommons.org/licenses/by/4.0/).

transmission pipeline systems. Energy 2018, 162, p. 853–870
3. L. Chi, H. Su, E. Zio, M. Qadrdan, J. Zhou, L. Zhang, L. Fan, Z. Yang, F. Xie, L. Zuo, J. Zhang, A systematic framework for the assessment of the reliability of energy supply in Integrated Energy Systems based on a quasi-steady-state model, Energy, Volume 263, Part B, 2023, 125740, https://doi.org /10.1016/j.energy.2022.125740
4. L. Chi, H. Su, E. Zio, M. Qadrdan, X. Li, L. Zhang, L. Fan, J. Zhou, Z. Yang, J. Zhang, Data-driven reliability assessment method of Integrated Energy Systems based on probabilistic deep learning and Gauss-ian mixture Model-Hidden Markov Model, Renewable Energy, Volume 174, 2021, p. 952-970, https://doi.org/10.1016/j.renene.2021.04.102.
5. W. Yu, W. Huang, Y. Wen, Y. Li, H. Liu, K. Wen, J. Gong, Y. Lu, An integrated gas supply reliability evaluation method of the large-scale and complex natural gas pipeline network based on demand-side analysis, Reliability Engineering & System Safety, Volume 212, 2021, 107651, https://doi.org/10.1016/j.ress.2021.107651
6. Senderov S., Edelev A. Formation of a List of Critical Facilities in the Gas Transportation System of Russia in Terms of Energy Security // 2017. Energy. doi:10.1016/J.ENERGY.2017.11.063.
7. Vorobev S., Edelev A. Analysis of the importance of critical objects of the gas industry with the method of determining critical elements in networks of technical infrastructures // Management of Large-Scale System Development (MLSD), 2017 Tenth International Conference. IEEE, 2017. DOI 10.1109/MLSD.2017.8109707.
8. Vorobev S., Edelev A., Smirnova E. Search of critically im-portant objects of the gas industry with the method of determining critical elements in networks of technical infrastructures // Methodological Prob-lems in Reliability Study of Large Energy Systems (RSES 2017). E3S Web Conf. Volume 25, 2017. DOI 10.1051/e3sconf/ 20172501004.
9. Senderov S., Edelev A. Formation of a list of critical facilities in the gas transportation system of Russia in terms of energy security. Energy, 2019, doi:10.1016/j.energy.2017.11.063.
10. S.M. Senderov, S.V. Vorobev, Approaches to the identification of critical facilities and critical combinations of facilities in the gas industry in terms of its operability. Reliability Engineering & System Safety, Volume 203, 107046, 2020, doi: 10.1016/j.ress.2020.107046.
11. Edelev A.V., Enikeeva S.M., Senderov S.M. Information support in the study of issues of functioning of large pipeline systems / Computational technologies, 1999, Vol. 4, No. 5, pp. 30 – 35.
12. Vorobyov S.V., Edelev A.V. Methodology for determining bottlenecks in the operation of large pipeline systems / Software products and systems, 2014, No. 3, pp. 174 – 177.
13. Ford L.R., Fulkerson D.R. Flows in networks. Princeton university press, Princeton, New Jersey, 1962. 276 p.
14. Vorobyov S.V., Edelev A.V. Features of mathematical modeling in the distribution of excess gas in the Unified Gas Supply System of Russia / Scientific Bulletin of NSTU Vol. 62, No. 1, 2016. – pp. 181-194.
15. Production, preparation and transportation of natural gas and condensate. Reference manual in 2 volumes. Volume II / Ed. Yu.P. Korotaev, R.D. Margulov. – Moscow: Nedra, 1984. 288 p.
16. Export of the most important goods of the Russian Federation in 2012-2023 (according to the Federal Customs Service of Russia) http://customs.ru/index.php?option =com_ newsfts&view=category&id=52&Item id=1978&limitstart=60.
17. InfoTEK Monthly oil and gas journal. №1, 2022, p. 150.
18. Ministry of Energy of the Russian Federation. Statistics. http://minenergo.gov.ru/activity/statistic.
19. Senderov S.M., Rabchuk V.I., Vorobyov S.V. Formation of a list of critical gas industry facilities from the standpoint of the country's energy security / Materials of the collection of reports Methodological issues of studying the reliability of large energy systems. 90th meeting "Reliability of developing energy systems". July 1-7, 2018, Irkutsk.

© The Authors, published by EDP Sciences. This is an open access article distributed under the terms of the Creative Commons Attribution License 4.0 (https://creativecommons.org/licenses/by/4.0/).

Assessment of the level of energy security in the regions of the northwestern federal district under emergency conditions in the gas industry

Elena Smirnova [1*]

[1] Melentiev Energy Systems Institute ESI SB RAS, Irkutsk, Russia

Abstract. The article presents an assessment of the level of energy security using the example of the subjects of the Russian Federation located in the territory of the Northwestern Federal District. The approach to the assessment is based on the use of the apparatus for monitoring and indicative analysis of energy security (ES), developed in the ISEM SB RAS. An assessment of the level of ES in the regions is carried out based on modeling the situation with the cessation of operation of the most important facilities of the gas industry, and an analysis of the dynamics, scale and trends in the change in the state of the level of ES when modeling the situation using peak UGS facilities provided for by the General Scheme for the Development of the Gas Industry for the Period up to 2030, for the analyzed subjects of the Russian Federation.

Introduction

Energy security is a state of protection of citizens, society, the state, and the economy from threats of a deficit in meeting their energy needs with economically accessible energy resources of acceptable quality, from threats of disruption of uninterruptible energy supply. In other words, any level of energy security is a balance between energy supply and demand or a deficit-free energy balance. The reasons for the imbalance may lie both in the state of fuel and energy complex facilities and in unfavorable changes in natural, political, and economic conditions.

One of the most important areas of activity to achieve and maintain the required level of energy security in Russia is monitoring and indicative analysis of ES.

It is proposed to carry out an indicative assessment of the level of energy security of a specific region of the country using three, largely interrelated, blocks of indicators (table 1).

The objectives of monitoring the energy security of the Russian Federation and its regions are to identify the observed and expected processes, phenomena and parameters that determine the level and threats to energy security. Thus, the meaning and essence of monitoring and indicative analysis are to display information on the degree of implementation of threats to energy security, using a system of indicators, when comparing the numerical values of these indicators with their threshold values. In accordance with this, the ISEM SB RAS has developed an information base for substantiating and making decisions to ensure the energy security of the Russian Federation and its regions.

Table 1. The composition of the most important indicators of regional energy security.

1. The block of industrial and resource security of the fuel- and energy supply of the region
1.1. The ratio of the total available capacity of the region's power plants to the maximum electric load of consumers on its territory. 1.2. The ratio of the amount of available capacity of power plants and the capacity of inter-system connections of the region with neighboring consumers to the maximum electric load on its territory. 1.3. Opportunities to meet the needs of boiler heating oil (BHO) from the region's own sources.
2. The block of reliability of fuel and energy supply of the region
2.1. The share of the dominant resource in total consumption of BHO in the region. 2.2. Share of the largest power plant in the installed electric capacity of the region. 2.3. The level of potential supply of demand for fuel in the conditions of a sharp cooling (10% consumption of consumption) in the region.
3. Block of the state of the basic production assets (BPA) of energy systems in the territory of the region
3.1. Degree of depreciation of the BPA in the energy sector of the region. 3.2. The ratio of the average annual input of installed capacity and reconstruction of power plants in the region over the previous 5-year period to the established capacity of the region.

This article presents the process of assessing the level of energy security of the subjects of the Russian Federation located on the territory of the Northwestern Federal District (NWFD).

* Corresponding author: smirnova.e.m@isem.irk.ru

© The Authors, published by EDP Sciences. This is an open access article distributed under the terms of the Creative Commons Attribution License 4.0 (https://creativecommons.org/licenses/by/4.0/).

Results of the conducted indicative analysis

When conducting the study, the values of the indicators by region were quantitatively correlated with their threshold values adopted in [1] (tables 2-4).

Along with a qualitative assessment of the level of energy security of the regions for 2022, in order to show some trends, information on the qualitative assessment of the relevant indicators in 2018 is provided for reference. As a result of the analysis of the data of the relevant tables for the subjects of the Northwestern Federal District, it is possible to briefly characterize the trends inherent in the energy sector of the territories from the standpoint of energy security requirements.

Table 2. Characteristics of the state of indicators on the territory of the subjects of the Northwestern Federal District for the block of production and resource security of the fuel and energy supply system.

Region, area	Indicator	Dimension	The threshold values of the indicator		The meaning and status of the indicator, year			
			N	C	2018		2022	
Republic of Karelia	1.1	un.	0,5	0,3	0,9	N	0,8	N
	1.2	un.	1,5	1,2	5,9	N	3,2	N
	1.3	%	60	40	21,1	C	17,8	C
Komi Republic	1.1	un.	0,7	0,5	2,0	N	1,9	N
	1.2	un.	1,5	1,2	2,4	N	2,3	N
	1.3	%	60	40	133	N	108	N
Arkhangelsk region	1.1	un.	0,7	0,5	1,9	N	1,2	N
	1.2	un.	1,5	1,2	2,7	N	1,9	N
	1.3	%	40	20	21,8	C	22,8	C
Vologda region	1.1	un.	0,5	0,3	0,8	N	0,9	N
	1.2	un.	1,5	1,2	4,1	N	4,3	N
	1.3	%	40	20	1,7	C	1,3	C
Kaliningrad region	1.1	un.	0,7	0,5	1,5	N	2,4	N
	1.2	un.	1,5	1,2	2,5	N	3,4	N
	1.3	%	60	40	0,2	C	0,3	C
Leningrad region	1.1	un.	0,5	0,3	1,78	N	1,59	N
	1.2	un.	1,5	1,2	2,4	N	2,8	N
	1.3	%	40	20	46,7	N	36,7	PC
Murmansk region	1.1	un.	0,5	0,3	1,9	N	1,9	N
	1.2	un.	1,5	1,2	3,4	N	3,3	N
	1.3	%	60	40	4,6	C	5,3	C
Novgorod region	1.1	un.	0,5	0,3	0,6	N	0,6	N
	1.2	un.	1,5	1,2	3,1	N	3,1	N
	1.3	%	60	40	1,9	C	0,6	C
Pskov region	1.1	un.	0,5	0,3	1,1	N	1,0	N
	1.2	un.	1,5	1,2	6,6	N	6,3	N
	1.3	%	40	20	6,4	C	2,2	C

According to the first block of indicators, an acceptable situation from the standpoint of economic security is observed only in the Komi Republic. In other regions, it is important to note the crisis state for indicator 1.3. "Possibilities of meeting the needs for BHO from the region's own sources" (table 2). According to the second block of indicators, an acceptable situation is observed only in the Vologda Region. In particular, according to indicator 2.1 "Share of the dominant resource in the total consumption of BHO in the region" (table 3), crisis and pre-crisis values prevail, due to the high share of natural gas in the fuel consumption balance. In turn, this is also reflected in the values of indicator 2.3 "Level of potential supply of fuel demand in conditions of a sharp cold snap (10% increase in consumption) in the region". Since the Northwestern Federal District belongs to regions with a cold climate (the temperature of the coldest five-day period is below minus 30 ° C) [2], the crisis threshold value for such regions is 100%. The studies were conducted under a simultaneous hypothetical decrease in the average January temperature throughout the entire Northwestern Federal District. The results of calculations on the level of provision of the specified demand for fuel and energy resources are presented in table 3.

© The Authors, published by EDP Sciences. This is an open access article distributed under the terms of the Creative Commons Attribution License 4.0 (https://creativecommons.org/licenses/by/4.0/).

https://doi.org/10.1051/e3sconf/202458401007

Table 3. Characteristics of the state of indicators in the territory of the subjects of the Northwestern Federal District for the block of reliability of fuel and energy supply.

Region, area	**Indica-tor**	**Dimen-sion**	**The threshold values of the indicator**		**The meaning and status of the indicator, year**			
			N	**C**	**2018**		**2022**	
Republic of Karelia	2.1	%	40	70	54,6	PC	58,2	PC
	2.2	%	50	70	25,4	N	25,6	N
	2.3		100	<100	90	C	90	C
Komi Republic	2.1	%	90	>90	84,5	N	89,5	N
	2.2	%	50	70	38,7	N	42,3	PC
	2.3		100	<100	94	C	94	C
Arkhangelsk region	2.1	%	40	70	64,8	PC	64,5	PC
	2.2	%	50	70	18,5	N	28	N
	2.3		100	<100	97	C	97	C
Vologda region	2.1	%	40	70	74,5	C	83,4	N
	2.2	%	50	70	22,4	N	21,8	N
	2.3		>100		100	N	100	N
Kaliningrad region	2.1	%	90	90	96,4	C	94,8	C
	2.2	%	40	50	69,4	C	46,9	PC
	2.3		100	80	89	PC	89	PC
Leningrad region	2.1	%	40	70	97	C	97,6	C
	2.2	%	50	70	31,8	N	33,4	N
	2.3		100	90	92	PC	92	PC
Murmansk region	2.1	%	40	70	81,9	C	79	C
	2.2	%	50	70	48,7	N	48,6	N
	2.3		100	90	94	PC	94	PC
Novgorod region	2.1	%	40	70	96,8	C	98,5	C
	2.2	%	50	70	81,4	C	80,3	C
	2.3		100	<100	94	C	94	C
Pskov region	2.1	%	40	70	86,2	C	92	C
	2.2	%	50	70	98,5	C	98,7	C
	2.3		100	90	97	PC	97	PC

The situation in any of the subjects also cannot be called acceptable for the third block of indicators (table 4) due to exceeding the crisis threshold for the wear and tear of the energy sector's fixed assets, as well as the insufficiently active introduction of new generating capacities over the past five-year period.

Table 4. Characteristics of the state of indicators in the territory of the subjects of the Northwestern Federal District for the block of the state of the fixed assets of energy systems.

Region, area	**Indica-tor**	**Dimen-sion**	**The threshold values of the indicator**		**The meaning and status of the indicator, year**			
			N	**C**	**2018**		**2022**	
Republic of Karelia	3.1	%	40	60	71,6	C	74,8	C
	3.2	%	2	1	0	C	0	C
Komi Republic	3.1	%	40	60	55	PC	60	C
	3.2	%	2	1	1,4	PC	0	C
Arkhangelsk region	3.1	%	40	60	42,1	PC	53,0	PC
	3.2	%	2	1	0	C	0	C
Vologda region	3.1	%	40	60	54,2	PC	57,9	PC
	3.2	%	2	1	5,4	N	0	C
Kaliningrad region	3.1	%	40	60	50,7	PC	5,7	N
	3.2	%	2	1	51	PC	8,5	N
Leningrad region	3.1	%	40	60	44,8	PC	47,4	PC
	3.2	%	2	1	1,0	N	4,3	N
Murmansk region	3.1	%	40	60	69,5	C	73,2	C
	3.2	%	2	1	0	C	0	C

© The Authors, published by EDP Sciences. This is an open access article distributed under the terms of the Creative Commons Attribution License 4.0 (https://creativecommons.org/licenses/by/4.0/).

Novgorod region	3.1	%	40	60	58,8	PC	64,4	C
	3.2	%	2	1	1,0	C	0,5	C
Pskov region	3.1	%	40	60	70,5	C	72,9	C
	3.2	%	2	1	0	C	0	C

The obtained and analyzed values of the main indicators, in turn, formed the basis for an integrated assessment of the level of energy security in the territory of the subjects of the Northwestern Federal District. To obtain such an assessment, an approach based on the convolution of indicator values was used, taking into account their specific weights. Qualitative characteristics of the state of all discussed indicators from tables 2-4 were collected for the relevant territories and processed according to [1].

As a result, a qualitative final assessment of the state of energy security of the territories was obtained, presented in table 5. According to its results, an acceptable state from the standpoint of ensuring ES cannot be stated in any of the subjects of the NWFD.

Table 5. Qualitative assessment of the state of energy security in the territory of the subjects of the Northwestern Federal District.

Years	**The order numbers of the estimated ES indicators**								**The sum of the specific weights by state**			**Quality condition ES**
	1.1	**1.2**	**1.3**	**2.1**	**2.2**	**2.3**	**3.1**	**3.2**	**Boundaries of states**			
	Specific weights of indicators								**C**[†]	**PC**	**N**[‡]	
	0,104	**0,138**	**0,133**	**0,120**	**0,079**	**0,170**	**0,127**	**0,129**				
Republic of Karelia												
2018	N	N	C	PC	N	C	C	C	0,55	0,12	0,321	C
2022	N	N	C	PC	N	C	C	C	0,55	0,12	0,321	C
Komi Republic												
2018	N	N	N	N	N	C	PC	PC	0,17	0,256	0,574	PC
2022	N	N	N	N	PC	C	C	C	0,42	0,079	0,495	C
Arkhangelsk region												
2018	N	N	C	PC	N	C	PC	C	0,43	0,247	0,321	C
2022	N	N	C	PC	N	C	PC	C	0,43	0,247	0,321	C
Vologda region												
2018	N	N	C	C	N	N	PC	N	0,25	0,127	0,62	PC
2022	N	N	C	C	N	N	PC	C	0,38	0,127	0,491	PC
Kaliningrad region												
2018	N	N	C	C	C	PC	PC	N	0,33	0,297	0,371	PC
2022	N	N	C	C	PC	PC	PC	N	0,25	0,376	0,371	PC
Leningrad region												
2018	N	N	N	C	N	PC	PC	C	0,24	0,297	0,454	PC
2022	N	N	PC	C	N	PC	PC	N	0,12	0,43	0,45	PC
Murmansk region												
2018	N	N	C	C	N	PC	C	C	0,50	0,17	0,321	C
2022	N	N	C	C	N	PC	C	PC	0,38	0,299	0,321	PC
Novgorod region												
2018	N	N	C	C	C	C	PC	C	0,63	0,127	0,242	C
2022	N	N	C	C	C	C	C	C	0,75	0	0,242	C
Pskov region												
2018	N	N	C	C	C	PC	C	C	0,58	0,17	0,242	C
2022	N	N	C	C	C	PC	C	C	0,58	0,17	0,242	C

[†] The state of ES in the region is recognized as a crisis if the sum of the shares of indicators in the state "C" exceeds 0,4
[‡] The state of ES in the region is recognized as normal if the sum of the specific weights of the indicators in the "N" state exceeds 0,7

© The Authors, published by EDP Sciences. This is an open access article distributed under the terms of the Creative Commons Attribution License 4.0 (https://creativecommons.org/licenses/by/4.0/).

Modeling the situation with the termination of the operation of the most important gas industry facilities

Earlier in [3], the results of the work on identifying and analyzing critical gas facilities (gas industry facilities, the loss of which can lead to a significant gas deficit for consumers of 5% or more) affecting the UGSS as a whole and directly the gas supply systems of specific regions were presented. In addition, in [4], an assessment was made of the vulnerability of the fuel supply systems of all regions of the Russian Federation in the unified gas supply system coverage area in the event of the termination of operation of individual especially significant gas industry facilities. A list of regions was compiled, ranked by the degree of vulnerability of their fuel supply system under the conditions of the specified shutdowns.

In [5], the results of model studies were presented, reflecting the degree of possible gas shortages to consumers in the event of an emergency on the unified gas supply system. The study revealed that, when the conditions are implemented for all subjects of the Russian Federation, gas shortages have a greater negative impact on the regions of the Northwestern Federal District. Also, the largest deficit in electric and thermal energy during the peak increase in demand for fuel, taking into account possible shutdowns of the unified gas supply system, was identified among the regions of the Northwestern Federal District (table 6).

Table 6. Gas shortages in regions of the Northwestern Federal District.

Region, area	Gas shortage, %	Number of influencing critical facilities UGSS, pcs.
Republic of Karelia	100	10
Komi Republic	100	4
Arkhangelsk region	100	8
Vologda region	100	10
Kaliningrad region	100	10
Leningrad region	100	10
Murmansk region	100	10
Novgorod region	100	10
Pskov region	100	10

Taking into account this calculated information on gas shortages in the regions of the Northwestern Federal District, as well as on the basis of using the method of normalizing the values of indicators [6], a new integral assessment was obtained (fig. 1), reflecting the qualitative and quantitative state of the indicators from the standpoint of ensuring ES in the event of an emergency in the gas transportation system.

The conversion of indicator values expressed in different units of measurement into normalized ones is performed according to the following expression:

$$X_i^N = \frac{X_{PCi} - X_i^t}{X_{C,i} - X_{PC,i}}, \quad (1)$$

where X_i^N – is the normalized value of indicator i in the analyzed period, rel. units; X_i^t – is the actual value of the indicator in the system of initial units; $X_{PC,i}$, $X_{C,i}$ - respectively, the threshold values of the pre-crisis and crisis states of the indicator i in the system of initial units.

In accordance with the calculation algorithm, the normalized threshold value $X_{PC,i}^N$ is always equal to zero, since this value is the starting point for unfavorable conditions, and the value of $X_{C,i}^N$ is always equal to -1.

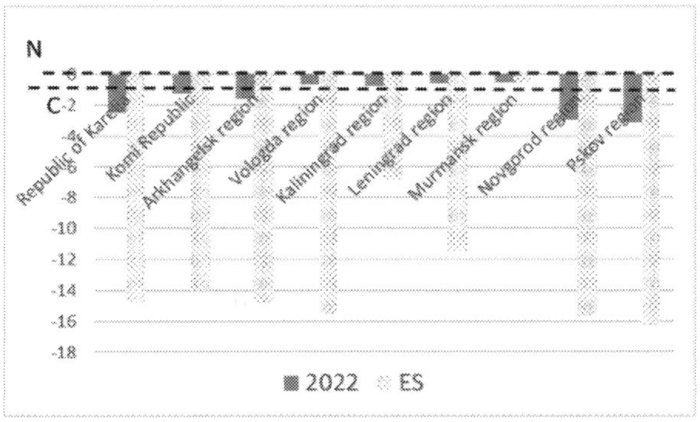

Fig. 1. Integrated qualitative assessment of the state of the ES of the regions of the Northwestern Federal District in the event of an emergency.

© The Authors, published by EDP Sciences. This is an open access article distributed under the terms of the Creative Commons Attribution License 4.0 (https://creativecommons.org/licenses/by/4.0/).

Peak underground gas storage

One of the possible measures to reduce the vulnerability of regional gas supply systems may be a change in the configuration of the gas transmission network, including the connection of an additional gas source in the event of an emergency. Peak-type underground gas storage facilities (UGS), which can be created in the Russian gas transmission system, can be considered as such a source. In general, gas reserves in UGS are intended to regulate unevenness (seasonal and peak) of gas consumption, as well as to provide gas in emergency and force majeure situations [7].

According to [8,9], for the period up to 2030, it is planned to commission 14 new UGS facilities to level out peak unevenness of gas consumption, the indicators of which are presented in table. 7

Table 7. Characteristics of peak UGS facilities in the Russian Federation according to projects for commissioning by 2030.

№	Peak underground gas storage	Region, area	Active volume of gas	Maximum daily withdrawal volume
			million m^3	
1	Udmurt Reservation Complex	Udmurt Republic	640	7,5
2	Kaliningrad	Kaliningrad Region	174	12
3	Bednodemyanovskoye	Penza Region	2000	30
4	Novomoskovskoye	Tula Region	340	30
5	Arbuzovskoye	Republic of Tatarstan	700	12,8
6	Shatrovskoye	Kurgan Region	1000	14
7	Volgogradskoye	Volgograd Region	300	0,25
8	Skalinskoye	Volgograd Region	2292	30
9	Bereznyakovskoye	Perm Region	324	17
10	Tulskoye	Tula Region	300	30
11	Smolenskoye	Smolensk Region	600	50
12	Seregovskoye	Komi Republic	1000	50
13	Serpukhovskoye	Moscow Region	300	30
14	Shedokskoye	Krasnodar Region	300	30

Table 7 clearly demonstrates that by 2030, it is planned to create several large peak UGS facilities with a daily withdrawal capacity of 30 million m^3 or more.

In this paper, the task was to assess how the vulnerability of regional gas supply systems would be affected by the creation of peak UGS facilities for the Northwestern Federal District, in order to change the situation with the gas shortfall discussed above. The results obtained in modeling the Kaliningrad and Seregovsky UGS facilities included in the UGSS system are presented in Table 8. In this case, a 100% gas deficit remained only for consumers in the Republic of Karelia, while in other regions, the reliability of gas supplies was almost completely restored.

A qualitative integral assessment taking into account the UGS facilities is presented in Table 9 and Fig. 2, in order to assess the changed situation in the regions of the Northwestern Federal District from the standpoint of ensuring the required level of energy security.

Table 8. Relative estimated gas shortfalls for direct consumption by region of the Northwestern Federal District, taking into account the planned UGS facilities.

Region, area	Relative gas deficit in emergency situations on the Unified Gas Supply System, %	Relative gas deficit taking into account UGS, %
Republic of Karelia	100	100
Komi Republic	100	0
Arkhangelsk Region	100	0
Kaliningrad Region	100	0
Leningrad Region	100	27
Novgorod Region	100	16.5
Pskov Region	100	0
Vologda Region	100	0

© The Authors, published by EDP Sciences. This is an open access article distributed under the terms of the Creative Commons Attribution License 4.0 (https://creativecommons.org/licenses/by/4.0/).

 https://doi.org/10.1051/e3sconf/202458401007

Table 9. Qualitative assessment of the state of energy security in the territory of the subjects of the Northwestern Federal District taking into account peak gas storage.

Years	The order numbers of the estimated ES indicators								The sum of the specific weights by state			Quality condition ES
	1.1	1.2	1.3	2.1	2.2	2.3	3.1	3.2	Boundaries of states			
	Specific weights of indicators								C[§]	PC	N[**]	
	0,104	0,138	0,133	0,120	0,079	0,170	0,127	0,129				
1	2	3	4	5	6	7	8	9	10	11	12	13
Republic of Karelia												
2022	N	N	C	PC	N	C	C	C	0,559	0,12	0,321	C
PUGS	N	N	C	PC	N	C	C	C	0,559	0,12	0,321	C
Komi Republic												
2022	N	N	N	N	PC	C	C	C	0,426	0,079	0,495	C
PUGS	N	N	N	N	PC	N	C	C	0,256	0,079	0,665	PC
Arkhangelsk region												
2022	N	N	C	PC	N	C	PC	C	0,432	0,247	0,321	C
PUGS	N	N	C	PC	N	N	PC	C	0,262	0,247	0,491	PC
Vologda region												
2022	N	N	C	C	N	N	PC	C	0,382	0,127	0,491	PC
PUGS	N	N	C	C	N	N	PC	C	0,382	0,127	0,491	PC
Kaliningrad region												
2022	N	N	C	C	PC	PC	PC	N	0,253	0,376	0,371	PC
PUGS	N	N	C	C	PC	N	PC	N	0,253	0,206	0,541	PC
Leningrad region												
2022	N	N	PC	C	N	C	PC	N	0,12	0,43	0,45	PC
PUGS	N	N	PC	C	N	PC	PC	N	0,29	0,26	0,45	PC
Murmansk region												
2022	N	N	C	C	N	PC	C	PC	0,38	0,299	0,321	PC
PUGS	N	N	C	C	N	PC	C	PC	0,38	0,299	0,321	PC
Novgorod region												
2022	N	N	C	C	C	C	C	C	0,758	0	0,242	C
PUGS	N	N	C	C	C	C	C	C	0,758	0	0,242	C
Pskov region												
2022	N	N	C	C	C	PC	C	C	0,588	0	0,412	C
PUGS	N	N	C	C	C	N	C	C	0,588	0,17	0,242	C

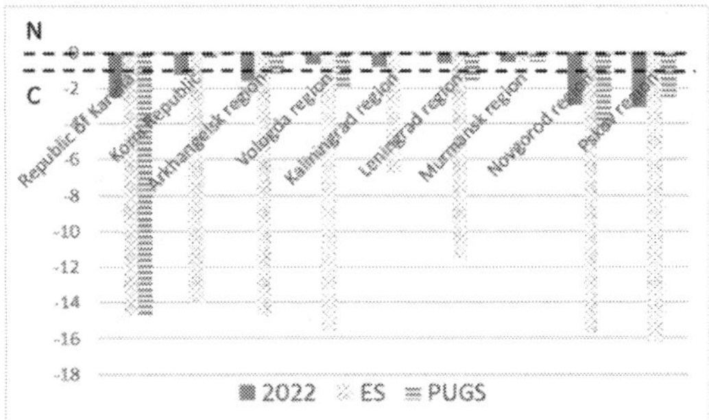

Fig. 2. Integrated qualitative assessment of the state of the ES of the regions of the Northwestern Federal District taking into account peak UGS.

Conclusion

Table 9. Relative estimated gas shortfalls for direct consumption by region of the Northwestern Federal District, taking into account the planned UGS facilities The results of the assessment of the level of ES of the regions of the Northwestern Federal District for three scenarios (under normal conditions, in case of an emergency in the UGSS, in case of an emergency in the UGSS taking into account the introduction of peak UGS) presented in the article allow us to conclude and confirm the need to create peak UGS in the UGSS for the analyzed territories. In addition, the work substantiates the effectiveness of using simulation models of the gas industry to analyze the vulnerability

[§] The state of ES in the region is recognized as a crisis if the sum of the shares of indicators in the state "C" exceeds 0,4

[**] The state of ES in the region is recognized as normal if the sum of the specific weights of the indicators in the "N" state exceeds 0,7

© The Authors, published by EDP Sciences. This is an open access article distributed under the terms of the Creative Commons Attribution License 4.0 (https://creativecommons.org/licenses/by/4.0/).

of regional gas supply systems. Peak UGS planned for creation by 2030 reduce the vulnerability of the gas supply system of the regions, increasing the reliability of their fuel supply and their level of ES as a whole.

When disconnecting the critical infrastructure facilities affecting the gas supply of the NWFD, both of the planned peak UGS were able to reduce the growing gas deficit in all regions, with the exception of the Republic of Karelia.

The study was carried out within the framework of the state assignment project FWEU-2021-0003 (registration number AAAA-A21-121012090014-5) of fundamental research of the SB RAS.

References

1. S.M. Senderov, N.I. Pyatkova, V.I. Rabchuk, G.B. Slavin, S.V. Vorobyov, E.M. Smirnova *Methodology for monitoring the state of energy security of Russia at the regional level*, Irkutsk: ISEM SB RAS (2014)
2. SP131.13330.2020. Construction climatology (2021)
3. S.M. Senderov, S.V. Vorobev, *Approaches to the identification of critical facilities and critical combinations of facilities in the gas industry in terms of its operability. Reliability Engineering & System Safety*, Volume **203**, 107046 (2020), doi: 10.1016/j.ress.2020.107046
4. Senderov S.M., Smirnova E.M., Vorobyov S.V. *Approaches to Assessing the Vulnerability of Fuel Supply Systems in Gas-Consuming Regions of Russia in the Context of the Shutdown of Particularly Significant Facilities in the Gas Industry* Bulletin of the Russian Academy of Sciences. Power Engineering, **1** (2020)
5. Beresneva N.M., Senderov S.M. *Analysis of the impact of large-scale emergencies in the gas industry on the capabilities of fuel and energy supply to consumers in conditions of significant cold snaps.* Methodological issues of studying the reliability of large energy systems: Issue. **74**. Reliability of energy systems in the context of energy transition (2023)
6. Senderov S.M., Smirnova E.M. *Use of normalized indicator values to determine the level of energy security using the example of the Central and Southern Federal Districts*. Methodological issues of studying the reliability of large energy systems: Issue. **73**. Reliability of energy systems in the context of energy transition (2022)
7. Kazaryan V.A. Underground *Storage of Gases and Liquids*. Vol. **2**. *Operation. Repair. Conservation. Liquidation. Use of Gas and Liquid Storage Facilities in Various Sectors of the Economy*, Izhevsk: Institute of Computer Research (2019)
8. *General development plan for the Russian gas industry until 2030*, Moscow (2008)
9. https://www.gazprom.ru/about/production/underground-storage/*[Electronic resource]*

© The Authors, published by EDP Sciences. This is an open access article distributed under the terms of the Creative Commons Attribution License 4.0 (https://creativecommons.org/licenses/by/4.0/).

Use and Failure Cause Analysis of Gas-Turbine Units at Distributed Generation Facilities

Pavel Ilyushin[1]*, and *Sergey* Shavlovskiy[1]

[1]Energy Research Institute of the Russian Academy of Sciences, 117186, Nagornaya st., 31, building 2, Moscow, Russia

Abstract. Distributed generation (DG) facilities commonly use gas-turbine units (GTU) varying in design and power. High manufacturing quality, reliability, efficiency, and low environmental impact are among the factors driving their popularity. This paper analyzes GTU reliability metrics and overviews the factors that may affect reliability of the units in use. It discusses the GTU design considerations that shall be borne in mind while engineering DG facilities and emergency automation devices. The paper further analyzes the recorded use GTUs at existing DG facilities from the perspective of the main causes and consequences of their failure.

1 Introduction

The commissioning of ever more renewable energy facilities fuels the advancement and expansion of distributed generation, a consistent trend that changes the way grids operate and guides the economic development all over the world [1, 2].

Newer technology for the generation, transport, and use of energy, distributed generation, smart grids, and coordinated control of DG facilities is the driving force behind the transition to "energy of the future" [3, 4]. It also calls for a wider adoption of energy-saving and energy-efficient technologies in transport, housing, and industry.

The complexity and cost of building and operating large-scale power plants, substations, and transmission lines, as well as the all-time strong consumer demand for high-quality electrical service is what creates a fertile ground for DG advancement [5]. Focus is made on using distributed energy more efficiently in order to integrate all possible energy sources [6].

DG facilities commonly use GTUs from different manufacturers offering different power levels. GTUs offer high manufacturing quality; they are reliable (as measured by the availability factor and the mean time between outages), efficient (measured as efficiency per se and/or as the heat utilization factor), and environmentally friendly thanks to minimum emission [7, 8]. Other advantages include general longevity, long time to repair, and extended maintenance-free operation.

Large and medium-sized industrial and commercial facilities keep commissioning their own GTU-based DG facilities for in-house heat and power generation [9]. Converting gas-fired boiler rooms into GTU-based DG facilities allows owners to reduce manufacturing costs and ensure uninterrupted power supply to their critical loads [10].

This research sought to analyze the reliability metrics and designs of GTUs as well as their recorded use and failure causes, for use as reference in DG site engineering.

2 GTUs Reliability Metrics

We analyzed reliability statistics on GTUs of different designs and power levels operated under different conditions at different DG facilities; the analysis revealed:

GTU availability factor (% of time a GTU can generate electricity) to be:

- 97.1% for 0.5–2.99 MW units,
- 95.0% for 3.0–19.99 MW units,
- 93.5% for 20–25 MW units.

Mean time between outages, i.e., the time between unscheduled GTU shutdowns due to any internal (failure; unscheduled repair; shutdown by the process protection system) or external (shutdown by the relay protection devices) cause to be:

- 2,220 hours for 0.5–2.99 MW units,
- 1,956 hours for 3.0–19.99 MW units,
- 3,605 hours for 20–25 MW units [11].

Thus:

- Large steam turbines and GTUs are 5 to 8 times less prone to shut down and have 2% to 7% lower downtime than less powerful GTUs at DG facilities. This is due to their design, longer electrical distance from the source of accident-associated disturbance, the presence of modern relay protection devices at power plants and grids, better trained personnel, and a more consistent load schedule,
- Relatively frequent but short GTU shutdowns at DG facilities are attributable to accident-associated disturbances in the outer grid, relay protection devices

* Corresponding author: ilyushin.pv@mail.ru

© The Authors, published by EDP Sciences. This is an open access article distributed under the terms of the Creative Commons Attribution License 4.0 (https://creativecommons.org/licenses/by/4.0/).

https://doi.org/10.1051/e3sconf/202458401008

flaws, as well as malfunctions in the secondary GTU hardware,

• GTU availability factor decreases by 0.2% for each MW in the 0.5 to 25 MW range since immediate on-site repair becomes less and less feasible for more powerful GTUs,

• Russian DG facilities have a 3% to 7% lower GTU availability factor for a number of reasons including remote placement from a service center, lack of spare parts inventory, and long spare parts delivery time from the manufacturers,

• Mean time between outages increases with per-unit GTU power. In the analyzed power range, it is about 100 h/MW thanks to more powerful GTUs having better design and being operated under better conditions,

• When analyzing GTUs in isolation, their reliability metrics vary greatly with operating environment, which could be classified as light, moderate, or harsh. Under harsh conditions, the availability factor may go as far down as > 40% below average,

• Internal causes account for $\alpha \approx 32.5\%$, while external causes account for $\beta \approx 67.5\%$ of outages due to the electrical proximity of accident-related disturbances.

When analyzing GTU reliability, statistics needs to be adjusted for GTU generation and for the effect of the internal and external causes of outage.

GTU reliability metrics should be borne in mind when engineering DG facilities; the concurrent factors to consider include the distance to a service center, the specifics of the outer grid in the area, and the operating conditions [12]. The thorough consideration of these factors could help provide uninterrupted supply of electricity to the loads throughout the GTU-based DG facility's lifecycle.

3 GTUs Design Considerations

DG facilities use two GTUs types:

• Jet engine-based GTUs;

• Power plant-grade GTUs.

The former are smaller-sized forced turbines with a lower specific weight; the jet engine could be replaced entirely or part by part when repaired at the manufacturer's facility. They are the more maneuverable alternative that could be used at peak loads while being easier to maintain, with less stringent infrastructural requirements; however, they lose in longevity. These can have two or three independent shafts with a variable speed of 6-14 thous [13]. rpm. Fig. 1 shows a simplified three-shaft GTU diagram: 1. Combustion chamber. 2. Low-pressure compressor. 3. High-pressure compressor. 4. High-pressure gas turbine. 5. Low-pressure gas turbine. 6. Power turbine. 7. Synchronous generator.

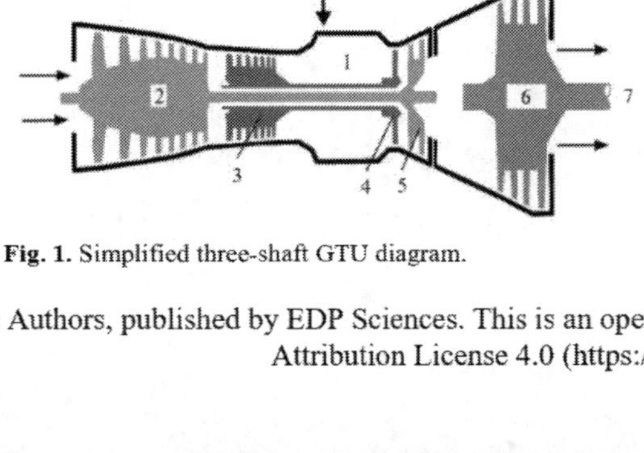

Fig. 1. Simplified three-shaft GTU diagram.

Jet engine-based GTUs are often connected to a synchronous generator via a low-maintenance reduction drive, sometimes via a power inverter.

They lack, however, dynamic stability. GTU power turbine has a mechanical inertia constant (T_j) in range of 2 to 4 seconds, which is 3-4 times less than that of a single-shaft GTU and 2-3 times less than that of a steam-turbine unit. Short-term load shedding will cause the GTU power turbine and the synchronous generator to accelerate, causing pole slipping, which results in the power system or process protections shutting the GTU down. For some GTUs, pole slipping is not acceptable at all for mechanical strength reasons that are a consequence of the unit's design [14].

Fig. 2 shows the results of calculations of the dynamic stability of GTUs of the first and second types with three-phase short circuit (SC), in comparison with steam turbine units (STU).

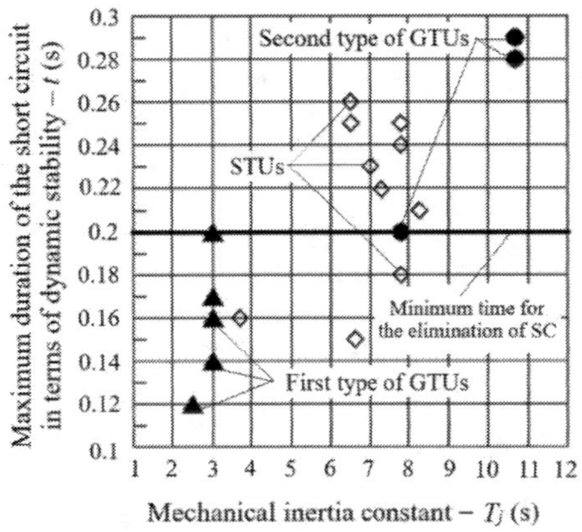

Fig. 2. Graph of changes in dynamic stability limits for different types of generating units.

The calculation results show that the limits of the dynamic stability of the first type of GTU are in many cases less than required for localization of close three-phase short circuits by relay protection devices [15, 16].

Fig. 3 shows an example of a transient process with a long asynchronous mode at a combined-cycle power plant with three first-type GTUs units and one steam turbine unit.

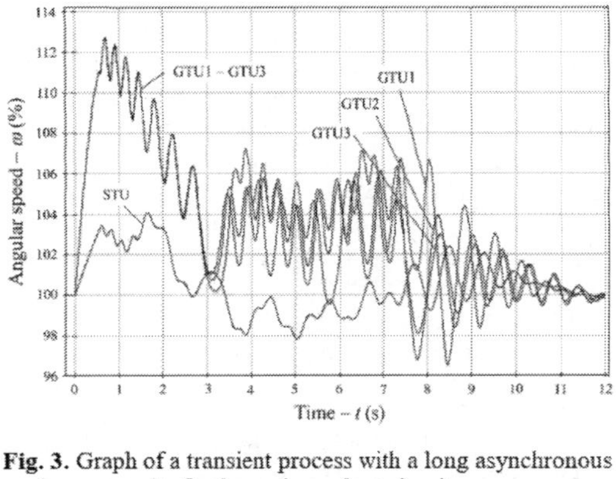

Fig. 3. Graph of a transient process with a long asynchronous mode as a result of a three-phase short circuit.

© The Authors, published by EDP Sciences. This is an open access article distributed under the terms of the Creative Commons Attribution License 4.0 (https://creativecommons.org/licenses/by/4.0/).

The analysis of Fig. 3 shows that as a result of the three-phase short circuit, all GTUs and the steam turbine unit switch to asynchronous mode. An attempt to resynchronize them with each other and with the power system (at $t \approx 3$ s) turns out to be unsuccessful and the asynchronous mode continues. After $t \approx 8.5$ s, resynchronization of all generators occurs [17, 18].

GTUs have peculiar pole-slipping behavior that must be taken into account when designing emergency automation devices for the DG facility and the outer grid:

• When pole-slipping, the power turbine and the synchronous generator will have a higher slip; however, fast response from the automatic speed controller will facilitate self-induced resynchronisation of the GTU. Since the slip fluctuations have a relatively high amplitude, given the substantial differential frequency pole slip will cede within seconds (4-8 revolutions max.),

• Pole slipping in the outer grid may cause significant voltage sag at the end-user loads' buses. This is critical for electric motors since their tripping will cause further voltage sag in the grid nodes, compromising the GTU's ability to self-synchronize [19, 20],

• Pole slipping at a DG facility is likely to propagate to machines in the outer grid due to the large difference in the mechanical constants of inertia between different generator types. Multimachine pole slip is more likely to occur in grids carrying industrial loads with multiple synchronous motors. In that case, the swing center moves substantially, making locating and addressing the slip difficult [21].

When choosing emergency automation devices, the engineers shall decide on the how fast pole slip needs to be handled. It can be addressed immediately or with an offset of up to a few seconds, giving the GTU time to self-synchronize. In case of the former, GTU shutdown may cause grid hardware overload, leading to further system collapse. In case of the latter, there is a high chance of secondary destabilization of motors and other generators in the outer grid. Optimizing the pole slip procedure timing requires calculating the electromechanical transients in various operating situations.

Power plant-grade GTUs generally have a single shaft and are heavier, constant-rpm units locked to the synchronous generator's rpm. They have a simpler design for better reliability and cost-effectiveness. Such a GTU has a heavy rigid shaft, plain bearings, and uniform profile blades in the blading section (except the first stages of the compressor and the last stages of the power turbine). All blades are mainly cooled by the compressor-fed air. This GTU type is associated with stricter structural and infrastructural requirements to the site. However, they also have a longer service life.

GTU design considerations must be taken into account when engineering DG facilities and power evacuation, placing the relay protection devices, emergency automation devices and choosing their algorithms, and calculating the preliminary setpoints [22–24]. This helps minimize unnecessary GTU downtime of any nature, wear, maintenance and repair costs.

4 Analysis of the operating experience and causes of damage to GTUs

GTUs ≤ 25 MW are dominated by the jet engine type; their reliability is a function of many factors:

• Quality of engineering – were steady states and electromechanical transients calculated, with adjustments for the effects of the load? Is the GTU in use suitable, type and power-wise? Were the GTUs reviewed in terms of their rated operating conditions? Was the emergency action plan drafted for the DG facility and the outer grid?

• Were the calculations and operating conditions analysis used to draft appropriate technical requirements to the GTUs before purchasing?

• Was the GTU acceptance from the manufacturer done appropriately with all the necessary steps and precautions? Were the customer's representatives involved in the acceptance testing at the factory?

• Were the GTUs transported to the site in an appropriate manner?

• Were the GTUs stored appropriately before installation at the DG facility?

• Were the GTUs conserved and de-conserved appropriately if long-term on-site storage was necessary?

• Were the construction, installation, startup and commissioning done to the appropriate standards? Did the works involve specialized firms?

• Had the facility run the full spectrum of launch tests and addressed all the identified shortfalls before operating the GTUs?

• Has GTU monitoring and diagnosis system been set up and verified to operate normally, in order to prevent their outage and/or unnecessary change in setpoints?

• Does the facility only use manufacturer-recommended oils and other process fluids and lubricants for GTU maintenance?

• Is the gas prepared to the standards?

• Is the personnel trained to operate the GTUs appropriately?

• Had a full set of user manuals for the GTUs and their secondary hardware been developed and presented to the staff before startup?

• Have the GTU and secondary hardware operations been arranged in full accordance with all the applicable standards and manuals?

• Is maintenance done in a timely manner and to a high quality?

• Is there a stock of spare parts? Can spare parts be delivered quickly for maintenance and repair?

• What are the operating conditions on the site?

• Are there manufacturing defects in the GTUs and the secondary hardware that went undetected during factory and on-site testing?

GTU failures and damage are to a lesser extent attributable to undetected flaws in the GTUs and secondary hardware, or poorly done installation, startup, and commissioning [25, 26]. Investigation into the causes of accidents revealed the main causes to be: design oversights; human error on the personnel's part;

© The Authors, published by EDP Sciences. This is an open access article distributed under the terms of the Creative Commons Attribution License 4.0 (https://creativecommons.org/licenses/by/4.0/).

inappropriate preparation of gas; delayed maintenance and repair.

Consider the consequences of design oversights:

• DG facility difficult to connect to the outer grid, grid auditors unwilling to approve the design,

• GTU damage caused by accident-associated disturbances in the outer grid or the on-site power system,

• The facility's inability to deliver power reliably to the critical loads when islanded,

• Faster wear and tear necessitating unscheduled repair or maintenance, e.g., due to frequent startups and shutdowns, etc.,

• Inability to reach the efficiency stated in the data sheet when testing the GTU on site,

• Poor efficiency due to inability to evacuate excessive generated power to the outer grid, operational issues while at minimum process load;

• Longer payback due to inflated project costs, higher per-unit fuel consumption, etc.

Thus, designing a GTU-based DG facility is a process that should involve specialized agencies with a track record of such projects.

Some DG facility owners have set up contracts to get their GTUs serviced by the manufacturer with payment per machine hour. In such cases, the manufacturer is motivated to maximize the GTU uptime, and thus to provide quality service and ensure GTU reliability. The manufacturer's staff receives telemetry data from the GTU monitoring and diagnosis system to decide whether and when to shut the unit down, run the maintenance, or organize repairs [27, 28].

Below are examples of GTU outages and causes thereof, including cases where unscheduled repairs were necessary; these GTUs were delivered by different manufacturers and operated at different DG facilities.

1. GTU and secondary hardware shutdown:

• Control automation failure,

• False positives on the part of the relay protection devices,

• Partial SCADA failure,

• Failure of the lubricator of the standalone booster compressor when running on associated petroleum gas,

• Booster compressor cylinder valve failure.

2. GTU shutdown and damage to the GTE:

• Blade root cracks, blade fracture in the high-pressure compressor,

• Breakage of the low-pressure compressor blades,

• Cracks in the gas generator housing,

• Fracture of the GTE shaft train due to multiple close tri-phase short circuits and asynchronous automatic reclosing in the outer grid,

• Defective operation of the input guide,

• Heat pipe burnout at < 6,000 hours in service,

• Breakage of the pads under the rear stand cone,

• GTE regulator channel failure,

• Jammed gas dispenser needle.

3. Reduction drive:

• Fracture of the driving impeller teeth.

4. Turbo generator:

• Damage, including stator winding collapse,

• Misalignment, shaft neck in contact with the labyrinth seals and the diffuser,

• Burnout of the "star" bus of the stator winding, with cross-sectional fracture,

• Short circuit in the exciter winding, incl. cases with resulting winding bandage rupture,

• Failure of the excitation control cabinet hardware.

When the oils in use are not quality-assured or the caloric content of the gaseous fuel (which is critical when using associated petroleum gas) or the contamination of the air fed to the GTU compressor is not kept track of, primary GTU parts are at risk.

Lack of sensors for continuous bearing support monitoring and processor protection may cause the GTUs to operate at unnecessarily high vibration amplitude, resulting in faster wear and tear, and causing the blading section to deteriorate as the defect propagates uncontrollably. On-site GTU manuals must contain clear instructions for the personnel to address increased GTU vibrations.

During startup and commissioning, load-shedding tests are often omitted; load shedding is associated with a drastic rpm decline. This may cause GTU outage in case of load shedding, e.g., when islanding the DG facility's area due to excessive generation. Significant load shedding on the GTU may also occur due to short-circuiting and disconnection of the relay protection devices of the transmission line, as well as due to synchronous swings. In this case, GTU outage is a result of the combustion chamber flare being extinguished. Load shedding causes gas feed to the GTE to drop significantly to reduce rpm and power output. In case of reduced gas flow, the compressor air-cooled flare may go out; in that case, the process protection system will shut the GTU down. If gas continues to be fed to the combustion chamber afterwards, the ignition of the fuel mix may cause the GTE to explode.

5 Conclusions

Good manufacturing quality, reliability, efficiency, and environmental friendliness are the factors behind the popularity of GTUs at DG facilities.

Relatively frequent but short GTU shutdowns are attributable to accident-associated disturbances in the grid, relay protection devices flaws, as well as malfunctions in the secondary hardware. The reliability of each particular GTU depends on its operating conditions, the availability and quality of service, and the qualifications of the operating personnel.

GTU design is a factor to be considered when engineering a DG facility and its power output diagram, in order to minimize GTU outages, damage sustained in operation, maintenance and repair costs.

Review of GTU use records and damage causes at the existing DG facilities could help their owners to take timely action to prevent similar accidents and potential damage to the GTUs.

Acknowledgment

This work is supported by the Russian Science Foundation under grant 21-79-30013 in the Energy Research Institute of the Russian Academy of Sciences.

© The Authors, published by EDP Sciences. This is an open access article distributed under the terms of the Creative Commons Attribution License 4.0 (https://creativecommons.org/licenses/by/4.0/).

References

1. S.R. Sinsel, R.L. Riemke, V.H. Hoffmann, Renewable Energy, **145**, 2271–2285 (2020)
2. B. Zeng, J. Wen, J. Shi, J. Zhang, Y. Zhang, Energy, **96**, 614–624 (2016)
3. B. Buchholz, Z. Styczynski, *Smart Grids – fundamentals and technologies in electricity networks* (Springer Heidelberg, 2014)
4. M. Ghadi, A. Rajabi, S. Ghavidel, A. Azizivahed, L. Li, J. Zhang, Applied Energy, **253**, 113543 (2019)
5. P.V. Ilyushin, O.V. Shepovalova, S.P. Filippov, A.A. Nekrasov, Energy Reports, **7**, 5, 308–327 (2021)
6. A.L. Kulikov, P.V. Ilyushin, A.A. Loskutov, Russian Electrical Engineering, **92**, 2, 90–96 (2021)
7. G.G. Ol'khovskii, Thermal Engineering, **60**, 79–88 (2013)
8. P. Ilyushin, S. Filippov, A. Kulikov, K. Suslov, D. Karamov, Machines, **10**, 693 (2022)
9. D.M. López González, J. Garcia Rendon, Renewable and Sustainable Energy Reviews, **157**, 112018 (2022)
10. A. Kavousi-Fard, A. Khodaei, S. Bahramirad, The Electricity Journal, **29**, 10, 22–27 (2016)
11. V.O. Samoylenko, P.V. Ilyushin, A.V. Pazderin, Renewable Energy and Power Quality Journal, **18**, 70–75 (2020)
12. M. Memari, A. Karimi, H. Hashemi-Dezaki, International Transactions on Electrical Energy Systems, **31**, 6 (2021)
13. Z. Zhang, S. Zang, B. Ge Aircraft Engineering and Aerospace Technology, **91**, 7, 1002-1010 (2019)
14. P. Ilyushin, E3S Web of Conferences **25**, 02002 (2017)
15. P.V. Ilyushin, Power Technology and Engineering, **51**, 6, 713–718 (2018)
16. D. Zhou, J. Hao, D. Huang, et al. Frontiers in Energy, **14**, 817–835 (2020)
17. P. Ilyushin, E3S Web of Conferences **58**, 02001 (2018)
18. B. Sahoo, S.R. Samantaray, P.K. Rout, IEEE Transactions on Industry Applications, 1–13 (2024)
19. P.V. Ilyushin, A.L. Kulikov, S.P. Filippov, *Adaptive algorithm for automated undervoltage protection of industrial power districts with distributed generation facilities,* in Proceedings of the 2019 International Russian Automation Conference, RusAutoCon, 08-14 September 2019, Sochi, Russia (2019)
20. M. Song, L. Li, Q. Wu, C. Li IEEE Transactions on Applied Superconductivity, 1–5 (2024)
21. P.V. Ilyushin, A.V. Pazderin, *Requirements for power stations islanding automation*, in Proceedings of the 2018 International Conference on Industrial Engineering, Applications and Manufacturing, ICIEAM, 15-18 May 2018, Moscow, Russia (2018)
22. J.O.C.P. Pinto, M. Moreto, Electric Power Systems Research, **190**, 106572 (2021)
23. D. Zheng, W. Zhang, S. Netsanet Alemu, P. Wang, G.T. Bitew, D. Wei, J. Yue, Microgrid Protection and Control, 45–56 (2021)
24. S.A. Eroshenko, P.V. Ilyushin, *Features of implementing multi-parameter islanding protection in power districts with distributed generation units,* in Proceedings of the 2018 IEEE 59th International Scientific Conference on Power and Electrical Engineering of Riga Technical University, RTUCON, 12-14 November 2018, Riga, Latvia (2018)
25. E.I. Velichko, A.V. Bunyakin, A.V. Muzykantova, IOP Conf. Ser.: Earth Environ. Sci., **666**, 022066 (2021)
26. N. Shamarova, K. Suslov, P. Ilyushin, I. Shushpanov, Energies, **15**, 6967 (2022)
27. U. Sarwar, M. Muhammad, A.A. Mokhtar, R. Khan, P. Behrani, S. Kaka, Results in Engineering, **21**, 101841 (2024)
28. F. Nadir, B. Messaoud, H. Elias, Journal of Failure Analysis and Prevention, **24**, 97–107 (2024)

© The Authors, published by EDP Sciences. This is an open access article distributed under the terms of the Creative Commons Attribution License 4.0 (https://creativecommons.org/licenses/by/4.0/).

Energy storage use efficiency in the context of Belorussian power system technological limitations

Mikhail Kashyn[1,*], *Andrey* Molochko[2], *Nikolay* Novikov[3] and *Alexander* Novikov[4]

[1] RUE Belenergosetprojekt, Republic of Belarus
[2] RUE BelTEI, Republic of Belarus
[3] JSC STC FGC UES, JIHT RAS, National Research University MPEI, Russia
[4] STC Energobezopasnost, Russia

Abstract. The paper provides an efficiency assessment of lithium-ion energy storage unit installation in the Belarusian power system at thermal power plants, in power supply and distribution networks, together with renewable energy sources, at electric charging stations for electric vehicles.

Introduction

Currently, the Belarusian power system faces several technological and circuit challenges that require system solutions to ensure the operational reliability and efficiency of the entire energy system and its individual components. In the near future, circuit limitations include the disconnection from intersystem connections with Ukraine's energy system and the planned separation from parallel operation with Lithuania's energy system in 2025. This will result in a significant reduction in external connections to the 750-330 kV backbone network.

One of the significant technological challenges at present is the difficulty in maintaining 400 MW capacity CCGT units of Western origin used in the Belarusian power system, which has a total capacity of 1,200 MW. This situation necessitates extending the lifespan of other generating equipment, which has a more limited range of active power adjustment and higher specific fuel consumption.

At the same time, increasing the design capacity of the Belarusian Nuclear Power Plant (NPP) to 2,400 MW exacerbates the challenge of optimizing the use of generating capacity at other power plants in order to meet the significantly uneven daily and weekly energy demands of both individual energy nodes and the entire Belarusian power system (the coefficient of load profile unevenness is about 0.7).

The power units at the Belarusian NPP operate based on the daily load profile, and their combined installed capacity accounts for about 35–40% of the peak load of the energy system during winter months.

One way to address these challenges is through the use of energy storage systems (ESS) in the Belarusian power system. This would allow for the separation of electricity production and consumption over time, thereby smoothing out the load profile. With sufficient installed capacity, ESS could become a key component of the "smart grid" concept in the future.

The ESS can perform several functions: providing a backup power source, regulating voltage and frequency, providing emergency power to prevent system failures (during power system partitioning) and restoring the power system after a failure, smoothing load spikes, and long-term use in conjunction with renewable energy sources. The reliability of isolated energy systems is ensured by ESS, as they can perform these functions simultaneously [1].

1 Analysis of ESS usage experience in the world

In recent years, there has been a significant increase in the use of ESS in large-scale energy industry around the world.

The well-known energy storage technologies can be divided according to the type of stored energy into:

- mechanical (pumped storages, flywheels),
- electrochemical (accumulator batteries, flow batteries),
- chemical (fuel cells),
- electrical (capacitors, supercapacitors, superconducting magnetic accumulators),
- thermal (use of molten salts and hot water).

To determine ESS use effect, it is important to consider:

- the capacity of ESS (the amount of electricity the device can accumulate and store over a certain period),
- power output per unit of time during discharge,
- duration of discharge (ability to output required power over a finite time interval),
- efficiency of drive.

* Corresponding author: m.kashin@besp.by

© The Authors, published by EDP Sciences. This is an open access article distributed under the terms of the Creative Commons Attribution License 4.0 (https://creativecommons.org/licenses/by/4.0/).

https://doi.org/10.1051/e3sconf/202458401009

The ESS cost is currently quite high, and using them for one task alone does not usually lead to return on investment. But with multifunctional use, the economic effect is significantly improved. According to the VYGON Consulting study, key areas of ESS application include [2] (fig. 1):

- electric transport,
- rechargeable batteries for internal combustion engines,
- industrial sector, primarily special forklifts and UPS for communication infrastructure.
- portable electronics;
- storage systems used in the electric power industry, etc.

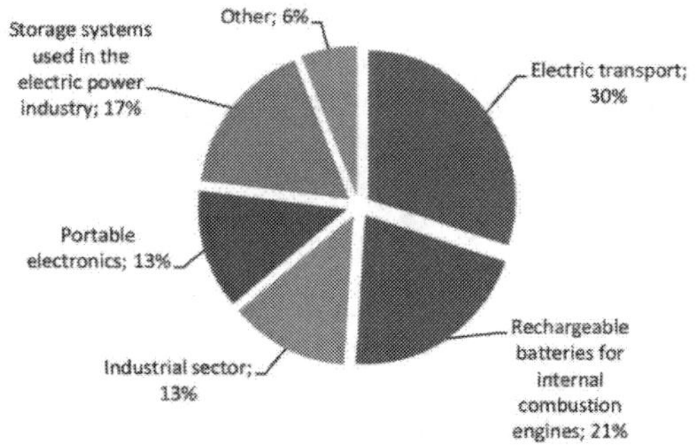

Fig. 1. ESSs usage area diagram

According to BNEF's forecast, the power of ESS, excluding PSPP, will grow to 358 GW worldwide and capacity to 1,028 GWh by the end of 2030, which is more than 20 times more than at the end of 2020 (17 GW and 34 GWh, respectively) [3].

The United States and China will be the largest markets, accounting for more than half of all ESS by 2030. Other significant markets include India, Australia, Germany, UK and Japan.

Currently, the global experience with ESS in energy industry shows the effectiveness of lithium-ion batteries, as they have better characteristics than other technologies, such as a long service life, good performance at low temperatures and high cell voltage.

The volume of use of lithium-ion batteries in the energy sector and other areas is growing due to their technical characteristics:

- energy density 90-250 Wh/kg,
- efficiency 85-95%,
- service life 10-20 years/4,500 cycles,
- element voltage 1.6-4.2V,
- operating temperature 0°C-40°C.

2 ESS technologies possible to use in the Belarusian Energy System

The ESS technology development level and the industrial designs use in CIS countries allows us to draw conclusions about the possibility of using a part of storage device types described above in Belarus' power system. Traditional means of regulating load schedules on the power system scale in the presence of nuclear power plants are pumped storage stations (PSPP), which are flexible sources of peak power and also act as load regulators. The single capacity of PSPP in Belarus is limited due to flat terrain and associated need for flooding significant areas. To ensure the operation of Belarus' NPP, several hydropower plants must be built. In addition, this solution requires significant investment and a long period of time [5].

In 2021-2023, the leading design institutes of the Belorussian power system carried out a number of research projects to determine the feasibility and expediency of using ESS in the Belarusian power system. As a result, a project entitled "Concepts for Using Energy Storage Systems Based on Lithium-Ion Batteries in the Belarus Power System" was developed. This project provides a feasibility analysis for the integrated use of ESS at generating sources, electric networks, and electric power facilities of industrial enterprises and transportation. According to the document, the technical potential of ESS installations in the Belarus energy system is approximately:

- at thermal power plants - 1,200 MWh and 150 MW,
- distribution networks of industrial consumers - 500 MWh and 100 MW,
- renewable energy sources - 500 MWh and 70MW,
- charging infrastructure for electric transport - 2,800 MWh and 300 MW.

3 Usage of ESS on Thermal Power Plants

ESS installed on generating sources with the ability to supply power to electric networks with voltages of 110 kV or higher can:

- participate in primary and secondary frequency control,
- smooth out peaks and regulate transmission line loads, contribute to ensuring static and dynamic network stability.

In some cases, these ESS can act as an alternative to peak generation. Their economic feasibility is determined by their ability to compete with other energy sources. Unlike a thermal power plant (TPP), ESS have time limitations for power output, but they also have advantages such as high discharge rates, the ability to output only the required amount of power, and the ability to meet consumer needs at constant voltages.

In the R&D of RUE BelTEI, the technically available potential for installing ESS at TPP in the Belarusian power system is determined variably:

- Option 1 - from the condition of covering own needs within 3 hours,
- Option 2 - for the 12 largest plants, from the condition ensuring the start-up mechanisms for own needs,
- Option 3 - from the possibility of eliminating power imbalances in the system during heating and interheating periods, considering the operation of two nuclear power units and peak reserve energy sources.

© The Authors, published by EDP Sciences. This is an open access article distributed under the terms of the Creative Commons Attribution License 4.0 (https://creativecommons.org/licenses/by/4.0/).

 https://doi.org/10.1051/e3sconf/202458401009

Table 1 presents the results of the analysis of technical and economic calculations for options under various scenarios of prices for 1 kWh of capacity per ESS. Scenario 1 assumes a price of $900 per kWh, while scenario 2 assumes a price of only $320 per kWh. To smooth out the load schedule of the power system and to provide limited ranges of generation regulation, option 3, with a power of ESS 150 megawatts and a total battery capacity of 1,200 MWh, is considered the main option for the Belarusian power system.

Table 1 – Technical and economic calculations for options for installing ESS at TPP

№	Attribute		Option		
			1	2	3
1	ESS capacity, MWh		730	400	1 200
2	ESS power, MW		240	60	150
3	Invest-ment, $ millions	scenario 1	940	500	1 490
		scenario 2	230	150	160
4	payback period, years	scenario 1	165	160	385
		scenario 2	40	40	40

The possibility of using ESS to provide household components of a 0.4 kV rural settlement network with electricity for heating, hot water and food preparation was evaluated in several research projects by RUE Belenergosetproject. The operability of power supply circuit options was assessed by the criterion of permissible voltage deviation at the most distant point of 0.4 kV and non-excess of permissible wire capacity in loading sections of lines. Aggregated estimates show that the cost of installing 10 kV ESS exceeds the cost of reconstructing grid facilities to achieve the same objectives by more than ten times [5]. This does not allow us to draw a clear conclusion about the feasibility of widespread use of SNS in these networks.

4 Using ESS in combination with Renewable Energy Sources

Currently, the installed capacity of power plants using renewable energy sources in the Belarusian power system is approximately 270 MW, which is about 2% of the total installed capacity of the system. The use of small-scale networks (ESS) is a key solution to effectively integrate large renewable energy sources with the energy system on the level of 10-110 kV networks. These ESS, installed as part of power plants based on renewable energy, are designed to:

- smooth out peaks in the load schedule of the grid,
- provide additional renewable energy generation capacity,
- solve problems with electricity quality supplied to the grid,
- minimize or eliminate overflows into the external grid and optimize operating modes of own renewable energy production, extending its lifespan.

The technically available potential for installing ESS at power plants based on renewable energy sources, determined from the possibility of ensuring full alignment of the daily schedule for generating electricity from renewable energy capacities installed in Belarus, is about 75 MW of installed power with storage capacity of about 490 MWh. Taking into account the possibility of reducing specific cost indicators in the future, the payback period for these projects may be 40 years. The calculation of the payback time for the renewable energy potential was done according to the approach of substituting electric energy generation from low-cost sources during peak periods with fuel-free renewable generation.

5 Usage of ESS at Electric Vehicles Charging Stations

The operation of electric charging stations is characterized by a sharply variable load schedule, which can be smoothed out by ESS to eliminate excessive power grid construction. RUE BelTEI estimates the potential for installing ESS in charging complexes within the Belarusian power system by 2030 at 320 MW and 2,800 MWh. Depending on the unit cost of ESS, the payback period for these projects can range from 7 to 28 years.

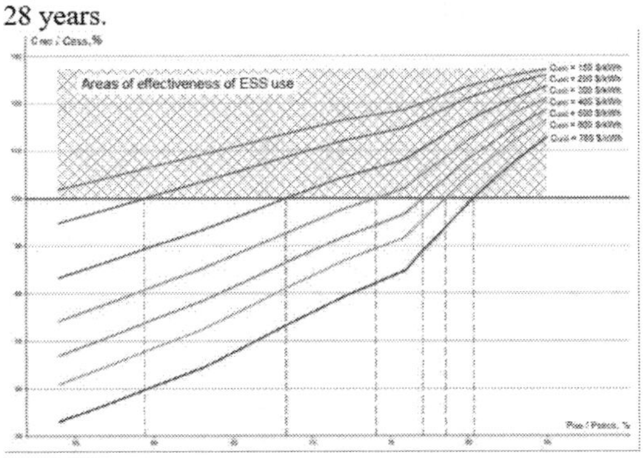

Fig. 2 - Assessment of the Effectiveness of the Integration of ESS into SECS.

A preliminary assessment of the effectiveness of integrating ESS into superfast electric charging stations (SECS) can be carried out based on a graphical dependence developed by the RUE BelTEI (fig. 2). The following parameters are used:

- On the X-axis, the ratio of the available reserve capacity of the 110/10 kV substation (P_{res}) to the capacity of SECS (P_{SECS}), in %.
- On Y-axis, a comparison of the increased capital costs for reconstructing and building electric networks to connect SECS without integrating it with ESS (C_{rec}), and with integration (C_{ESS}), in %.

Low load SECS at night allows for the installation of a ESS with 4,200 MWh and 600 MW.

For the share of electric transport at 10%, the annual fuel and energy savings due to the use of ESS could amount to 64-317 thousand ton of conventional fuel (t.c.f.) per year, depending on the comparable options for covering the load of SECS, capacity and power of the ESS.

© The Authors, published by EDP Sciences. This is an open access article distributed under the terms of the Creative Commons Attribution License 4.0 (https://creativecommons.org/licenses/by/4.0/).

In 2024 the RUE BelTEI conducted an assessment of the economic viability of involving ESS in regulating the daily load schedule for the Belarusian power system, with the economic impact of providing ESS with a total power of 350 MW and useful capacity of 2500 MWh as part of SECS storage services for electricity needs estimated at $4.5-$49.7 million annually.

5 Hydrogen technologies and iron air ESS

From a technical point of view, Belarus has prospects for the production of hydrogen using electrolysis to regulate its power system modes. In the future, hydrogen produced can be used as fuel, including for backup power, maintaining frequency and voltage in networks, eliminating network overloads, and as a raw material in industry. Taking into account the operation of Belarusian NPP units, it is technically feasible to produce climate-friendly "yellow" hydrogen.

This, in turn, leads to an increase in export potential due to global trends towards decarbonization. The consumable material for hydrogen fuel cells are only fuel tanks, while the main reaction product is ordinary water, so using hydrogen fuel cells instead of batteries significantly reduces the volume of hazardous waste containing toxic substances.

It should be noted that, with the current cost of electrolyzers ranging from $1,000 to $1,750 per kW, and, accordingly, the cost of "yellow" hydrogen at about $6-$8 per kg [7], the use of this method to manage the capacity of the Belarusian power system is not economically feasible. By 2050, there is projected to be a significant reduction in the cost of electrolyzer systems to under $200-$300 per kW for systems with power exceeding 10 MW. At this point, it would be advisable to consider hydrogen generation as a means of managing capacity in the Belarus power system.

Iron-air batteries utilizing redox reactions involving iron are also promising for ESS, as their cost is significantly lower than that of lithium-ion batteries. However, they also have drawbacks, such as a limited number of charge/discharge cycles. Russia is in the final stages of developing iron-air battery technology.

Conclusion

1. The world experience ESS use analysis has shown that there has been a significant increase in its use in large-scale energy, industrial and household sectors in recent years.

2. It is advisable to conduct feasibility studies on using ESS at thermal power plants after the actual market appearance of ESS costing no more than $300 per kWh in the Republic of Belarus.

3. Using ESS for power supply and distribution networks with voltages ranging from 0.4 to 10 to 110 kV to solve system problems is currently not economically feasible, while the economic feasibility threshold is $200 per kW.

4. ESS use is recommended for renewable energy sources with capacity exceeding 1 MW.

5. Use of ESS is recommended on all superfast electric charging stations (SECS), both to avoid excessive grid construction by leveling the peaks of SECS load and to provide system services.

6. The use of hydrogen technologies in the ESS in the Belarusian power system is possible in the future, with a reduction in the cost of the main technological processes. The use of iron-air batteries in the power system could be considered if their production technology is sufficiently developed.

References

1. Bushuev V.V., Novikov N.L. Infrastructure storage units in the electric power industry. Energy Policy, 2020, No. 10, p. 74-89.
2. Cumulative energy: "green" innovations for energy conservation. delprof.ru . [Online] November 22, 2021 https://delprof.ru/press-center/open-analytics/nakopitelnaya-energetika-zelenaya-innovatsiya-dlya-sokhraneniya-energii/.
3. The global capacity of energy storage systems will exceed 1 TWh in 2030. Rennen. [Online] 2021 https://renen.ru/globalnaya-yomkost-sistem-nakopleniya-energii-prevysit-1-tvt-ch-v-2030-godu/.
4. Operation and storage of lithium-ion batteries. [Online] January 27, 2019 https://electrodrive.by/news/expluatatia_i_hranenie_Li-ion_akkumulatorov.
5. Kashin M.A., Novikov N.L., Novikov A.N. The use of electric energy storage devices to increase the controllability and operational reliability of the Belarusian energy system // Collection: Methodological issues of reliability research of large energy systems: Issue 74. Reliability of energy systems in the context of modern challenges and threats. Ed. Academician of the Russian Academy of Sciences V.A. Stennikov. Irkutsk: ISEM SB RAS, 2023. pp. 125-134.
6. D.I. Mendeleev, D.A. Rossikhin, L.A. Galimzyanov, A.Y. Fedotov. Analysis of the use of SNES in thermal power plants.
7. S. Tikhonov. Novak: The most promising way to produce hydrogen is from atomic energy. Rossiyskaya gazeta. [Online] 14 01 2022 [Quoted: 31 08 2023] https://rg.ru/2022/01/14/novak-samyj-perspektivnyj-sposob-polucheniia-vodoroda-iz-atomnoj-energii.html.
8. Green Hydrogen Cost Reduction: scaling up electrolysers to meet the 1.5 C climate goal'. Abu Dhabi : IRENA, 2020.

© The Authors, published by EDP Sciences. This is an open access article distributed under the terms of the Creative Commons Attribution License 4.0 (https://creativecommons.org/licenses/by/4.0/).

 https://doi.org/10.1051/e3sconf/202458401010

Battery Energy Storage System Placement And Sizing In Distribution Networks

Dmitry Sorokin[1*]

[1]JSC Rosseti R&D Center, Russia

Abstract. The article discusses the methodology for selecting installation locations and parameters of battery energy storage systems (BESS) in electrical distribution networks. The methodology is applicable to BESS which implement the functions of ensuring the reliability of power supply to consumers (use of BESS as a backup or emergency source of power supply), as well as the function of regulating voltage levels in the electrical network (regulating reactive power by the BESS inverter).

1 Introduction

Trends in the development of distribution electric networks, caused, among other things, by the energy transition, are an increase in the capacity of renewable energy sources and distributed generation, the development of electric vehicles (EV), including the use of fast EV charging stations, the emergence of a new type of consumers (prosumers), a characteristic feature of which is not only the consumption of electric energy, but also the possibility of supplying electrical energy to the distribution network. These trends lead to a decrease in the predictability and complexity of control electric networks saturated with such devices, and, as a result, to an increase in the probability of power system failures. One of the possible solutions to these problems is the use of battery energy storage systems (BESS) equipped with intelligent automatic control systems.

BESS technologies are one of the priority directions for the development of the Russian power systems. This is reflected in many regulatory legal acts and documents, including the Strategy for Scientific and Technological Development of the Russian Federation [1], the Roadmap for the Development of the High-tech directions until 2030 [2], the Unified Technical Policy of PJSC Rosseti [3] and others. On the basis of the Rosseti R&D Center, the Competence Center for the use of BESS in the electric grids has been established. The tasks of the Competence Center are to supervise projects for the development and research of BESS technologies, as well as to cooperate in the development of technologies and the implementation of projects for the use of BESS. The development of BESS technologies can be carried out on the basis of the Innovation Laboratory of Digital Networks, opened by PJSC Rosseti in 2024 in St. Petersburg, Russia [4].

BESS based on the use of lithium-ion batteries have become the most widespread in both world and Russian practice in electrical networks. Promising areas of application of lithium-ion BESS in the Russian distribution networks are currently: improving the quality of power supply to consumers (using BESS as an emergency or backup power source, including the use of storage devices with renewable energy generation as part of autonomous or hybrid electrical installations), improving the electric power quality (using BESS for compensate the voltage sags), the use of BESS in the EV charging infrastructure to reduce the requirements for the connection of fast and ultra-fast EV charging stations (with a capacity of 150-250 kW or more) to electrical networks. Many studies have noted that in order to ensure not only the technical, but also the economic feasibility of using battery energy storage systems, it is necessary to implement various functions in one storage device.

Examples of projects implemented by PJSC Rosseti for the use of BESS are: mobile BESS (Tver region), BESS installed in the electrical networks of Smolensk region, BESS in Pushkin (a part of Saint-Petersburg) and others. Currently, the PJSC Rosseti has 36 battery energy storage systems in operation (35 stationary installations and one mobile installation). All BESS are installed in 0.4 kV distribution electric networks and implemented using lithium-ion batteries based on NMC or LFP technologies. The fields of application of most installed BESS are: an autonomous or backup power supply source for consumers and voltage regulation.

One of the main tasks to be solved in order to ensure the technical and economic feasibility of BESS projects is to determine the optimal places for installing storage devices in electric networks and optimal BESS parameters.

* Corresponding author: sorokin_dv@ntc-power.ru

© The Authors, published by EDP Sciences. This is an open access article distributed under the terms of the Creative Commons Attribution License 4.0 (https://creativecommons.org/licenses/by/4.0/).

2 Overview of BESS placement and sizing methods in distribution networks

There are quite many publications devoted to the methods of selecting installation sites and parameters of BESS in distribution electric networks [5]. In most cases, BESS with a bidirectional inverter based on lithium-ion batteries are used. The place of installation of the BESS is an electrical network node (transformer substation 10(6)/0.4 kV or 0.4 kV connection node of consumer). The main parameters of the BESS are: rated active power (Pn), rated reactive power (Qn), rated energy (En), rated voltage at the connection point, charge/discharge time. Algorithms for selecting installation sites and calculating the parameters of the BESS are primarily determined by the functionality implemented in the BESS and in some cases are quite complex for accurate (quantitative) formalization. For example, when choosing the installation location of a mobile BESS, the following criteria are accepted: the capacity of the central transformer substation 10(6)/0.4 kV is up to 25 kVA; the greatest length of 6-10 kV transmission line to the central transformer substation, which can be replaced with a BESS; the complexity of maintenance of an overhead line (wooded and swampy terrain); insignificant daily and hourly electricity consumption; absence of planned growth in electricity consumption [6].

Many methods consider the use of BESS in order to reduce active power losses and normalize voltage levels in electrical networks. This problem can be formulated as a multi-criteria optimization problem with constraints [7, 8, 9]. To solve this problem, both traditional optimization methods (linear and quadratic programming methods, simplex method, etc.) and heuristic algorithms (genetic algorithm, particle swarm optimisation, etc.) are used. In a number of techniques, it is proposed to use various sensitivity coefficients [10]. A number of techniques are based on the application of consumer load diagrams [11].

Optimization criteria can be: minimization of consumer payments for electric energy, minimization of capital and operating costs, minimization of undersupply of electric energy, ensuring of required power and energy capacity to compensate for peaks of power, reduction of fuel consumption of diesel generator, maximization of reliability levels (reduction of SAIDI, SAIFI), normalization of voltage levels in the electrical network, minimization of electrical energy losses, etc. (depending on the considered BESS functionalities).

It should be noted that due to the possibility of implementing various functionalities in the BESS and the actual absence of standardized storage devices, there is currently no universal, generally accepted method for selecting installation locations and parameters of battery energy storage systems.

3 Practical method for selecting installation locations and parameters of BESS in distribution networks

This study examines a practical method for selecting installation locations and parameters of battery energy storage systems that implement the functions of increasing the reliability of power supply to consumers and regulating voltages in electrical distribution networks. The technique consists of several stages and is based on the use of mathematical models of electric power systems.

At the first stage, the selection of nodes of the electrical network for connecting the BESS is performed. For radial electric networks, the choice of BESS installation sites is based on the following practical considerations. On the one hand, in order to minimize the number of storage devices and ensure the reliable supply of the maximum number of consumers, it is recommended to install BESS in nodes as close as possible to the power supply centers (transformer substations). On the other hand, it is possible to provide an acceptable (determined by the users of the method) ratio between the BESS parameters (rated powers, etc.) and the risks of loss of consumer power supply. In this case, it is possible to install BESS in different network nodes. A quantitative assessment of the risk of loss of consumer power supply is the value of the coefficient K_{res}:

$$K_{res} = \left(1 - \frac{P_{max}}{P_{load}}\right) \cdot 100\% \qquad (1)$$

where P_{max} is the maximum active power of consumer loads that can be switched off in the considered section of the electrical network in the event of an emergency disturbance (outage of a power line or power center), kW; P_{load} – total active power of consumer loads in the considered section of the electrical network, kW.

The coefficient K_{res} is a relative value of the maximum power of consumers that can be obtained in case of emergency situations in the considered section of the electric network. K_{res} =0% - indicates that switching-off of any power transmission line (power supply center) in the considered section of electrical networks leads to termination of power supply to the consumer (group of consumers). K_{res}= 100% indicates that when any power transmission line (power supply center) is switched off in the considered section of the electric networks, there is no termination of the power supply to consumers (in the case of the use of BESS – for the duration of the discharge of the battery energy storage system).

As already noted, minimizing the number and power of BESS in the considered section of electric networks can be performed by determining the required value of the coefficient K_{res}. The possible values of K_{res} are determined based on the steady-state power flow calculations with simulation of various emergency situations leading to termination of consumer power supply.

An additional criterion for assessing the feasibility of technical solutions is the cost of the installation and maintenance of BESS. To estimate these costs, it is

© The Authors, published by EDP Sciences. This is an open access article distributed under the terms of the Creative Commons Attribution License 4.0 (https://creativecommons.org/licenses/by/4.0/).

recommended to use the method of net present value (NPV) calculation. However, due to the fact that currently there is no sufficient and reliable information in Russian practice about the cost of installation and maintenance of BESS, a simplified approach is used in the study. It is assumed that the cost of installation and maintenance of the BESS is proportional to the rated power of the BESS, and minimizing the number and total capacity of the BESS leads to lower costs for their installation and maintenance.

At the second stage, the value of the reactive power of the BESS inverter is determined, which is required to minimize the level of active power losses in the section of the electric network and taking into account the need to maintain voltage levels in an acceptable range. The optimization criterion F in this case is:

$$F = \Delta P_{losses} + \gamma \cdot \sum_{i=1}^{N} \left(\frac{U_i - U_{rated}}{U_{rated}}\right)^2 \rightarrow min \qquad (2)$$

where ΔP_{losses} is the loss of active power, pu; γ is the coefficient determining the influence of the components of the objective function; U_i is the voltage at the ith node of the electrical network, kV; U_{rated} is the rated voltage of the ith node of the electrical network, kV.

By analogy with estimating the cost of installation and maintenance of BESS, instead of the cost of power losses in the objective function (2), the value of active power losses in the considered section of the electrical network is used.

At the third stage, the energy E is determined from the condition of the need to ensure power supply to consumers from the BESS for the required time:

$$E = P_{load}^{max} \cdot \Delta T \qquad (3)$$

where P_{load}^{max} is the maximum value of the active power of consumer loads, kW; ΔT – required time for backing up consumer loads, hours.

The value of energy E can be determined by the maximum value of consumer loads P_{load}^{max} based on the maximum electrical mode (for example, the winter maximum loads), in the absence of the required data – by the rated (maximum) power of the consumer loads. If there is information about daily consumer load schedules, it is advisable to use this information to estimate necessary BESS energy value more precisely [11].

4 Power system model

To test and evaluate the effectiveness of the proposed method for selecting BESS installation sites and parameters, a mathematical model of an electric power system has been developed. The model is based on the equivalent model of the IEEE-33 radial distribution network [12]. The model contains 13 nodes and 12 branches, as well as one equivalent generator simulating the power supply center for consumer loads. Consumer loads are set by a symmetrical three-phase loads. The model was developed in the DIgSILENT PowerFactory. Figure 1 shows the developed model. The model corresponds to a network in which BESS are installed in each node of the electrical network.

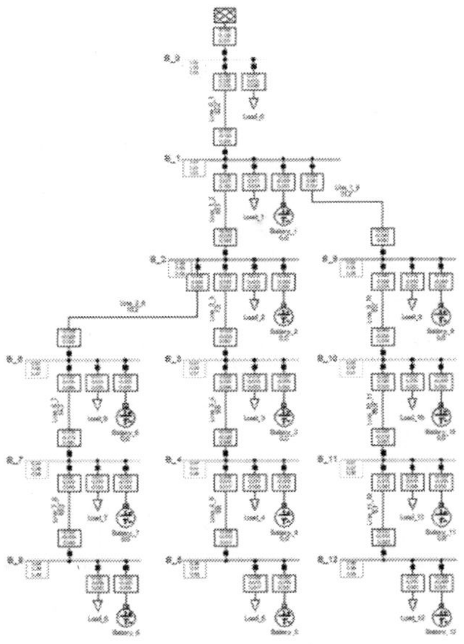

Fig. 1. Test power system model.

5 Practical BESS placement and sizing

Based on the results of the calculations performed, four options for installing the BESS are considered (see Fig. 2):

Option No. 1. There are no BESS in the electrical network (K_{res}= 0%).

Option No. 2. BESS are installed in all nodes of the electrical network (K_{res}=100%). Each BESS provides backup power supply only for the loads of consumers connected to the BESS installation node.

Option No. 3. Optimal (quasi-optimal) placement of the BESS: in nodes 3, 6 and 10 of the electrical network. The option provides a sufficiently high value of the K_{res} (75%). However, in case of an emergency switching-off of the head power lines (1-2, 2-3, 2-6, 1-9, 9-10) it requires the implementation of automation (with the installation of switching devices) to disconnect redundant consumers. The rated power of the installed BESS is determined by the power of the consumer loads required to obtain backup power supply of these consumers.

Option No. 4. Installing one BESS at the power supply center (node 1). In this case, the storage capacity is selected based on the need to supply power to consumer loads of the considered electrical network. However, due to the fact that the BESS does not provide backup power supply of consumer loads in full when the head power lines 1-2, 1-9 are switched off, the value of the coefficient K_{res} in this case is reduced to 65%. The results of the calculations of the first stage are shown in Table 1.

© The Authors, published by EDP Sciences. This is an open access article distributed under the terms of the Creative Commons Attribution License 4.0 (https://creativecommons.org/licenses/by/4.0/).

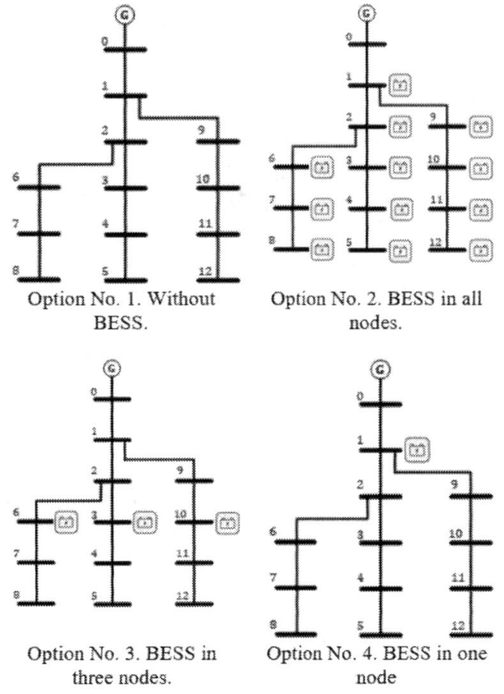

Fig. 2. BESS placement options.

At the second stage, optimization was performed according to criterion (2). Optimization was performed using a genetic algorithm implemented in the Matlab. Based on the results of optimization, the optimal values of the reactive power loads were obtained, ensuring a minimum of active power losses in the considered section of the electrical network and normalization of voltage levels. The voltage values in the nodes of the electric network is shown in Fig. 3. It can be seen from the figure that the use of BESS allows reducing the voltage deviations from the nominal value for all nodes of the network. At the same time, Option No. 2 (12 BESS) and Option No. 3 (3 BESS) are comparable options in terms of active power loss levels and the degree of voltage normalization. The results of the calculations of this stage are shown in Table 1.

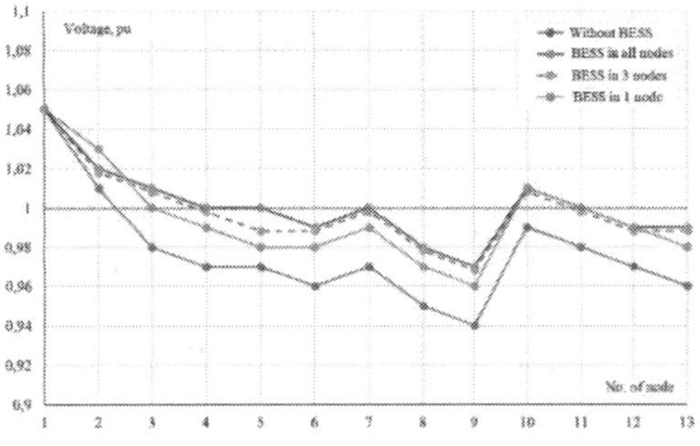

Fig. 3. Voltage values in the nodes of the test electrical network.

6 Results

The obtained results of the selection of installation locations and parameters of the BESS in the model of the electrical network show that Option No. 3 is the optimal for the considered network in terms of technical indicators. This option corresponds to the minimum total power of the installed BESS, ensures the achievement of a sufficiently high value of the coefficient K_{res} (75%), and also leads to a significant reduction of active power losses in the considered section of the electric network. However, automation needs to be implemented with the installation of switching devices aimed at disconnecting consumers for which the backup power supply is not provided. The logic of the operation of such automation is quite simple and is feasible in practice. It should be noted that the choice of installation sites and parameters of the BESS should be confirmed by an assessment of the economic performance indicators of the project for the use of battery energy storage systems, which will be possible in the future when developing BESS technologies and when reliable data on the cost of the life cycle of typical BESS for use in the Russian power grid appear.

Table 1. Summary of the calculation results.

Option	BESS parameters	BESS total rated power, kVA	Number of BESS	K_{res}, %	U_{min}, pu	P_{losses}, %
No. 1	–	–	–	0	0,94	7,0
No. 2	12*30 kVA	360	12	100	0,97	5,5
No. 3	1*125 kVA, 2*60 kVA	245	3	75	0,97	5,7
No. 4	1*300 kVA	300	1	65	0,96	6,5

where K_{res} is the coefficient calculated according to (1), %; U_{min} is the minimum voltage in the nodes of the electric network, pu; P_{losses} is the loss of active power in the electric (relative to the total value of consumer loads), %.

7 Conclusions

The article considers a practical technique for selecting installation sites and parameters of the BESS in electric networks. The technique is quite universal and can be adapted for application to BESS with various functionalities. The combination of various functions within a single device (BESS) will contribute to improving the technical and economic feasibility of projects for the use of battery energy storage systems. The application of the developed methodology with subsequent installation and operation of BESS in the electric networks will lead to an increase in the reliability of electric power systems, including those saturated with different active adaptive devices (equipped with inverter), and will also contribute to the development of BESS technologies in Russia.

© The Authors, published by EDP Sciences. This is an open access article distributed under the terms of the Creative Commons Attribution License 4.0 (https://creativecommons.org/licenses/by/4.0/).

References

1. Strategy of scientific and Technological development of the Russian Federation (approved by Decree of the Government of the Russian Federation dated June 9, 2020 No. 1523-r).
2. Roadmap for the Development of the High-tech directions until 2030 (approved by Decree of the Government of the Russian Federation dated February 28, 2022 No. 356).
3. Unified Technical Policy in the Electric grids (approved by the Board of Directors of the Company, Protocol No. 592 dated 10/20/2022).
4. Intellectual laboratory of digital networks of PJSC Rosseti. URL: https://ntc-power.ru/news/novosti-otrasli/rosseti-otkryli-intellektualnuyu-laboratoriyu-tsifrovykh-setey-v-sankt-peterburge / (Date of reference: 06/12/2024).
5. N. Y. Dahlan, S. Nordin. *Review of optimization approaches for determining the optimal size and placement of battery energy storage systems (BESS),* in 3rd IEEE International Conference on Energy Applications (ICPEA), 2023.
6. V.O. Akulichev and others. *The use of mobile electric energy storage systems for power supply to remote consumers of low and medium power,* in Electric Power. Transmission And Distribution, **6(75)**, 2022.
7. B. Cortez-Caicedo. *Optimization of BESS placement, technology selection and operation in micro grids to minimize energy losses and CO2 emissions: A Hybrid approach,* in Journal of Energy Storage, **73**, 2023.
8. A. Alzahrani. *Minimization of electricity losses due to the optimal placement of batteries in a distributed network with a high level of penetration of photovoltaic cells,* in Energia, **13 (1)**, 2020.
9. J. M. Rasol. *Optimal placement, dimensions and daily charging/discharge of batteries for energy storage in a low-voltage distribution network with a high degree of photovoltaic energy penetration,* in Applied Energy, 2018. URL: https://findresearcher.sdu.dk/ws/portalfiles/portal/161247175/Applied_Jannesar.pdf (Date of application: 06/12/2024).
10. S. Shafi. *Optimal placement of batteries in a distribution network using the voltage sensitivity method,* in IEEE Conference on Energy in Illinois (PECI), 2019.
11. P.V. Subbotin. *The method of selecting electric energy storage devices and their locations in electric distribution networks*. Dissertation for the academic degree of Candidate of Technical Sciences, Moscow, 2019.
12. N. Pompern. *Optimal placement and capacity of a battery energy storage system in distribution networks integrated with photovoltaic and electric vehicle installations using metaheuristic algorithms,* in IEEE Access, **11**, 2023.

© The Authors, published by EDP Sciences. This is an open access article distributed under the terms of the Creative Commons Attribution License 4.0 (https://creativecommons.org/licenses/by/4.0/).

Markers of cellulose insulation ageing: calculation of total methanol quantity in the internal insulation of power transformers

L.A. Darian [1,3*], *R.M.* Obraztsov [1], *A.V.* Maksimchenko [1], *D.E.* Verakso [1], *V.P.* Polishchuk [2], *V.A.* Logunov [3,4]

[1]JSC "Technical Inspection of UES", Moscow, Russia
[2] United Institute of High Temperatures, Russian Academy of Sciences, Moscow, Russia
[3] Federal State Budgetary Educational Institution of Higher Education "NIU MPEI", Moscow, Russia
[4] FGUP RFNC-VNIITF, Snezhinsk, Russia

Abstract. The methods of calculation of methanol quantity in the internal insulation of power transformers (PT) based on the values of its concentration in transformer oil (TO) are considered. The total methanol quantity formed during the operation of PTs allows estimating the degree of degradation of cellulose insulation. We are considering the models for calculation of the total methanol quantity in PT on the base of:

- literature data characterizing the methanol balance in the cellulose-oil system;
- data on the reduction of methanol concentration after TO replacement during total overhaul.

The errors analysis on the base of the results of calculations using the models showed that their value can reach 30%. Reduction of errors in the calculation results should be based on a deeper study of the kinetics of methanol sorption processes by cellulose and taking into account the structural features of different types of cellulose insulation in the particular PT.

The considered models can be recommended for estimation of total quantity of any dissolved decomposition products of cellulose insulation in TO, which can become an important indicator of technical condition of PT.

1 Introduction

The ageing of the PT internal insulation resulting from various factors, becomes apparent in a gradual deterioration of its characteristics, which can eventually cause an internal short circuit, often leading to a large-scale accident. Analyzing the quantity of the TO ageing markers, i.e. specific chemical compounds formed as a result of the internal insulation ageing, obtain a clear picture of the degree of its degradation.

This paper discusses the measuring results of the methanol concentration in TO conducted over a five-year period. Methanol is formed mainly at lower temperatures as a result of cellulose hydrolysis. The total methanol quantity in the PT internal insulation Q_t is proportional to the change of cellulose polymerization degree [1-3]. It is considered that when the degree of polymerization is less than 250, the service life of PT cellulose insulation has completely expired [3]. The quantity of Q_t allows us to estimate directly the degree of cellulose insulation ageing and, consequently, to estimate PT residual life. During the cellulose insulation aging process, the generated methanol migrates between cellulose and TO. The methanol quantity in the TO Q_o is found from measured methanol concentrations C_o , but the methanol quantity Q_c contained in the cellulose can be a significant part of its total quantity Q_t :

$$Q_t = Q_c + Q_o \tag{1}$$

Currently, there are no available standardized procedure to determine the methanol quantity in cellulose insulation. Therefore, the methanol quantity Q_c can only be determined by calculation methods.

The aim of this work was to develop and analyze non-equilibrium and equilibrium models for calculation of the total methanol quantity in PT and to assess the errors of these methods.

2 Object and methods of research

Measurements of the internal insulation ageing markers concentration were carried out on a three-phase PT of 110 kV voltage class with 80 MVA power, TO mass M_o = 41 tones, its volume V_o =47.8 m^3 . The transformer was put into operation in 1986. The PT was equipped with a diagnostic monitoring system to control the concentration of nine gases dissolved in the TO (carbon oxides, hydrocarbons, hydrogen, oxygen and nitrogen). The monitoring system also recorded hourly the ambient temperature, the temperature of TO upper and lower layers and the PT load current. Based on these data, the temperature of the hottest point of the insulation was calculated in accordance with the methodology [4]. Approximately once a month TO

* Corresponding author: daryan-la@ti-ees.ru

© The Authors, published by EDP Sciences. This is an open access article distributed under the terms of the Creative Commons Attribution License 4.0 (https://creativecommons.org/licenses/by/4.0/).

samples were taken to determine the concentration of methanol.

During the observation period, PT was taken out for repair twice. At the end of spring 2019, an overhaul was carried out which included replacement of the TO, drying of cellulose insulation and revision of the active part. Inspection of the PT inner surface did not reveal any signs of cellulose insulation deterioration and winding clamping. In late spring 2022, the sealing gasket between the PT cover and tank was replaced and the sealing of the welded joints was improved. During this overhaul, methanol concentrations were measured daily.

3 Measurement results

Figure 1 shows the temperature values of the upper and lower layers of the TO; the characteristic difference between these temperatures was ≈10 °C. The ambient temperature differed from the temperature of the lower oil layers by no more than ≈5 °C. During the observation period, the PT load was maintained mainly at the same level (≈25% of the nominal one), so the periodic type of TO temperature dependence was determined by the seasonal change of the ambient temperature. Further, the arithmetic mean value of the temperature of the upper and lower layers of TO is taken as the representative temperature of internal insulation T_i.

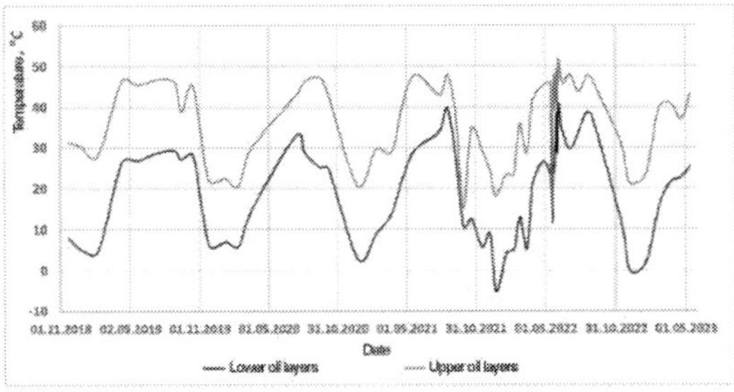

Fig. 1. Temperature of lower and upper oil layers in the PT

Figure 2 shows the values of methanol concentration C_o over the observation time and its average concentration $C_{av}(t)$ after TO replacement, which was calculated using the formula:

$$C_{av}(t) = \frac{1}{t} \times \int_o^t C_o(t')dt' \qquad (2)$$

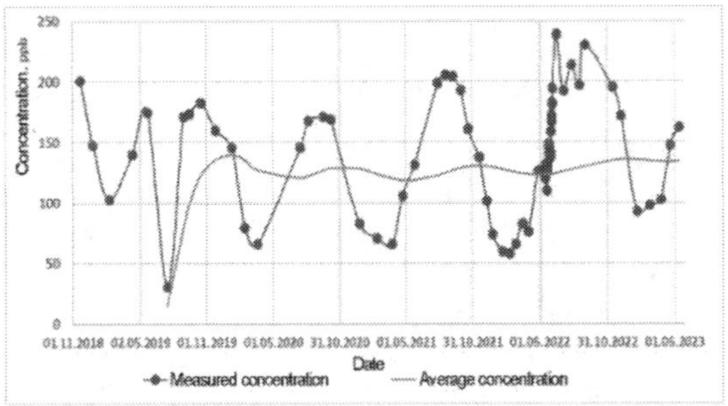

Fig. 2. Methanol concentration over the observation period

The sharp decrease in concentration in summer 2019 is explained by TO replacement during the overhaul. The minimum value of methanol concentration was ≈30 ppb and measured 20 days after TO replacement. "Fresh" TO does not contain methanol, therefore its appearance in samples after TO replacement is explained by desorption from cellulose insulation. In Figure 2 the concentration of points in early summer 2022 is associated with the increased frequency of measurements during the overhaul. The change in methanol concentration during this period is shown in Figure 3 in more detail. This figure also shows the internal insulation temperature T_i .

If we disregard a sharp decrease in methanol concentration due to TO replacement, the periodic time dependence of methanol concentration (Fig. 2) and TO temperature (Fig. 1) are similar at the qualitative level. The value of methanol concentration is established as a result of the competing processes of adsorption and desorption from the cellulose surface. The methanol sorption rate depends exponentially on the cellulosic surface temperature. Thus, it is possible to speak about correlation between extremes of methanol concentration and insulation temperature.

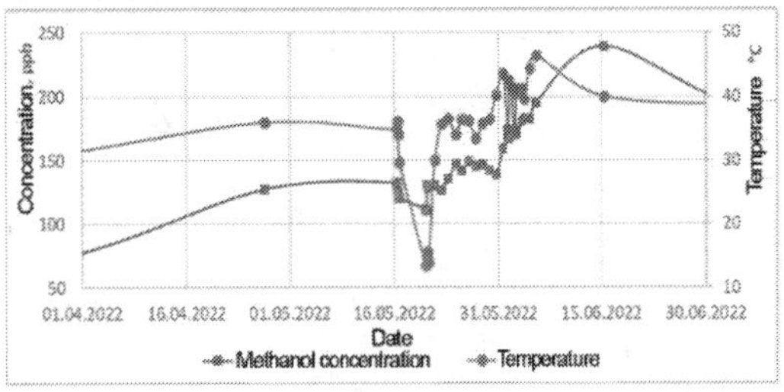

Fig. 3. Methanol concentration and internal insulation temperature T_i in the second quarter of 2022

Figure 4 shows the values of the extremes of methanol concentration and internal insulation temperature during the observation period.

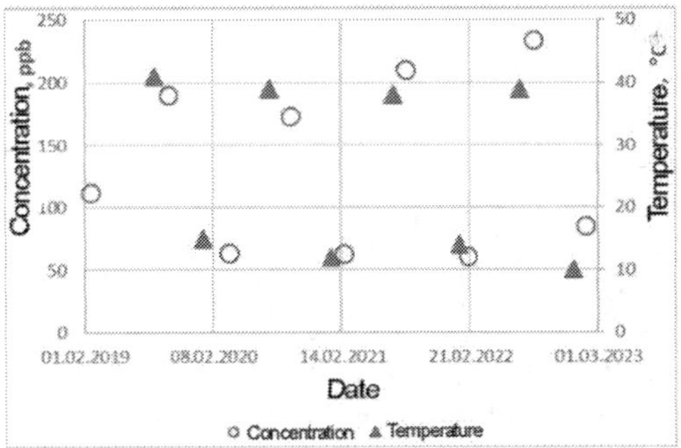

Fig. 4. Extremes of methanol concentration and internal insulation temperature

It can be seen that concentration extremes were preceded by temperature extremes, the time lag between these extremes ranged from 25 to 75 days [5]. Concentration maxima were recorded in late summer or early autumn, and concentration minima were recorded in the second half of winter or early spring. The maximum concentration of $C_{o.r.}$ =175 ppb (22.05.2019) and the minimum concentration of 30 ppb (15.07.2019) are not shown in Figure 4 because these extremes are

© The Authors, published by EDP Sciences. This is an open access article distributed under the terms of the Creative Commons Attribution License 4.0 (https://creativecommons.org/licenses/by/4.0/).

explained by TO replacement.

Table 1 summarizes the values of methanol concentration extremes $C_{o.i}$, where i is the extremum number. The estimate of the maximum concentration and representative values of methanol concentration $C_o(90)$ at temperature 90 °C in 2018 have been added to Table 1, and the calculation methods are explained below.

Table 1. Extremes of methanol concentration

№	Extremum type	Extreme date	$C_{o.i}$, ppb	$C_o(90)$, ppb
1	Max	*≈01.10.2018*	*≈240*	*≈1334*
2	Min	17.02.2019	112	1995
3	Max	01.10.2019	190	1056
4	Min	27.03.2020	64	1082
5	Max	17.09.2020	173	913
6	Min	21.02.2021	63	1134
7	Max	16.08.2021	210	1098
8	Min	17.02.2022	61	1090
9	Max	29.08.2022	234	1203
10	Min	25.01.2023	85	1366

The data in Figure 2, Figure 4, and Table 1 provides an estimate of the maximum methanol concentration in 2018. In Figure 2, the earliest methanol concentration value measured on 21.11.2018 was 201 ppb. This value was obtained shortly after the autumn maximum. Figure 2 shows that by 21.11.2018 the methanol concentration should have decreased by ≈20%, meaning that the maximum concentration in 2018 could presumably have been ≈240 ppb. Approximately the same maximum methanol concentration could have been recorded in 2019 if the TO had not been replaced. This follows from the extrapolation of the concentration value of 175 ppb measured on 22.05.2019, three to four months before the autumn maximum. Thus, the maximum methanol concentration of $C_{o.3}$ =190 ppb recorded on 01.10.2019 turned out to be ≈50 ppb less than it would have been without the TO replacement. The methanol concentration in the second minimum (64 ppb) decreased by about the same amount compared to the first (112 ppb).

From the data in Table 1 it can be concluded that in 2021-2023 the methanol concentration in the extremes owing to cellulose aging increased at a rate of ≈20 ppb/yr. Figure 2 shows that from 2021-2023 the average methanol concentration in the TO increased at a rate of ≈5 ppb/yr. Note that over the 32 years of PT operation the average rate of increase of methanol concentration in TO was ≈7 ppb/year. Taking into account the obtained spread of values, the most reasonable value for the average rate of methanol concentration increase is ≈10 ppb/year.

4 Models for calculating total methanol quantity

When calculating the volume of methanol contained in TO, it can apparently be assumed that due to diffusion and convective currents it is distributed uniformly over the volume of TO, therefore:

$$Q_o = C_o \times V_o \quad (3)$$

To analyze the methanol quantity in the cellulose, we can formally introduce an integral distribution coefficient K_{in}, determined from the correlation:

$$K_{in} = Q_c/Q_o \quad (4)$$

The total methanol quantity in the internal insulation is:

$$Q_t = Q_o \times (1 + K_{in}) \quad (5)$$

Equilibrium and non-equilibrium models of methanol distribution in the oil-cellulose system can be used to calculate the total quantity of methanol Q_t. Note that at equilibrium distribution extremes of methanol concentration should coincide with extremes of insulation temperature T_i, but the data in Figure 4 clearly show that this condition of existence of equilibrium is not fulfilled.

4.1 Non-equilibrium model

In the absence of equilibrium, some additional information is needed to estimate the methanol quantity of cellulose insulation Q_c. Such information can be the change in the methanol concentration in the inner insulation during TO replacement. Since the generation rate of methanol in the inner insulation is relatively low, the values of its total quantity Q_t during the first and second concentration maxima (≈01.10.2018 and 01.10.2019) differ by the methanol quantity $Q_{o.r.} = V_o \times C_{o.r.}$ ($C_{o.r.}$ =175 ppb) that was removed during the overhaul:

$$Q_{o.1} \times (1 + K_{in.1}) \cong Q_{o.3} \times (1 + K_{in.3}) + Q_{o.r.}, \quad (6)$$

where $K_{in.1}$ and $K_{in.3}$ are integral distribution coefficients during the first and second maxima.

The above data show that $Q_{o.1}$ ≈*11.*5 mL (≈*01.*10.2018, $C_{o.1}$ ≈240 ppb) and $Q_{o.3}$ =9.1 mL (01.10.2019, $C_{o.3}$ =190 ppb). The thermograms of the internal insulation of the PT in autumn 2018 and 2019 were similar, so there is reason to believe that:

$$K_{in.1} \cong K_{in.3}. \quad (7)$$

From formulas (6) and (7), we can derive a correlation for the integral distribution coefficient ($K_{in} = K_{in.1}$):

$$K_{in} \cong (C_{o.3} - C_{o.1} + C_{o.r})/(C_{o.1} - C_{o.3}) \quad (8)$$

Substituting the above concentration values into this correlation, we obtain K_{in} =2.5. Accordingly, the value of the total methanol concentration in the internal insulation of the PT was ≈40.3 ml before overhaul, and ≈31.8 ml after. The values of the integral coefficient K_{in} and the volume of methanol concentration, are probably slightly underestimated due to incomplete accounting of the volume of methanol released from the internal insulation during the overhaul, because the calculation does not take into account the methanol which was removed during the drying of cellulose insulation.

4.2 Equilibrium models

If the distribution of methanol in the cellulose-oil system is close to equilibrium, the Q_t value can be estimated by two different methods based on the

© The Authors, published by EDP Sciences. This is an open access article distributed under the terms of the Creative Commons Attribution License 4.0 (https://creativecommons.org/licenses/by/4.0/).

available literature data.

The first method is based on the use of the temperature coefficient $K_{c.f}$, which is the ratio of methanol concentrations in TO at the reference temperature of 20 °C and at the insulation temperature T_i [3, 6]:

$$K_{c.f}(T_i) = C_o(20)/C_o(T_i) \quad (9)$$

In [6], the temperature coefficient for methanol was measured on the PT model up to a temperature of ≈*100* °C. At high temperature intensive desorption of methanol occurs, so that at temperature ≈90 °C it practically ceases to be absorbed by cellulose [6]. Accordingly, at 90 °C it can be assumed that $Q_t = Q_o$. The experimental results [6] are well described by the correlation:

$$K_{c.f}(T_i) = a \times exp(b/(T_i + 273)), \quad (10)$$

where a and b are empirical constants (b = 3720 K).

From formulas (10) and (11) it follows that the concentration of methanol in the TO at a temperature of 90 °C is calculated by the formula:

$$C_o(90) = C_0(T_i) \times \exp(b \times (1/(T_i + 273) - 1/363)) \quad (11)$$

The total volume of methanol in the inner insulation is found from correlation (3):

$$Q_t \cong C_o(90) \times V_o \quad (12)$$

Calculations according to formula (12) are reasonable for concentration extremes, since in these modes the rate of change of methanol quantity in TO is minimal, so the error of using the equilibrium approximation should be the smallest. Calculated values of concentration $C_o(90)$ are given in Table 1. To simplify the calculation, the correlation of the internal insulation temperature in the calculation of $C_o(90)$ was described (interpolated) by a harmonic function.

It is interesting to note that when the time lag is taken into account, the ratio of $C_o(90)$ concentrations in the neighboring extremes (maximum-minimum) differs by no more than 25%, which indicates that the accuracy of methanol quantity determination in the internal insulation of PTs is high enough for practical calculations.

From the correlation (12) follows, that the total volume of methanol during the first maximum concentration after overhaul (01.10.2019) is 44.9 ml. It should be noted that the calculation takes into account the time lag between the temperature of the internal insulation and the concentration of methanol [5], i.e. as a representative temperature of the internal insulation was taken the temperature of its previous maximum (41 °C, 24.08.2019) Under this condition, the concentration $C_o(90)$ = 940 ppb.

The second method to estimate the total methanol quantity is based on use in formula (5) to calculate the integral factor K_{in} the equilibrium distribution coefficient K_e (T_i), which is the ratio of methanol concentrations in cellulosic insulation and TO. The integral coefficient K_{in} is calculated by the formula:

$$K_{in} = K_e(T_i) \times \rho_o \times M_c/(\rho_c \times M_o), \quad (13)$$

where M_c is the mass of cellulosic pulp; ρ_c is its density; ρ_o is the density of TO.

The error in calculation of the distribution coefficient K_{in} with formula (13) can be large, and it is not easy to estimate it correctly. Firstly, it is explained by the fact that the equilibrium distribution of methanol in the internal insulation of PTs in operation is apparently unattainable. Secondly, quite often, as, for example, in the researched PT, the mass of cellulose insulation M_c remains unknown.

The time it takes to achieve the equilibrium concentration of methanol at temperature change T_i is determined mainly by the time of equalization of its concentration in cellulose owing to diffusion. The diffusion coefficient in cellulose materials depends on their porosity and increases with the temperature rise. Two types of cellulosic materials are used in PT: cable paper and electric-grade cardboard with densities of ≈*750* and ≈*950* kg/m³ respectively [7]. The temperature of cable paper is higher than that of electric-grade cardboard since it is used as turn insulation. The typical thickness of turn insulation is ≈*0.5* mm; electric-grade cardboard has a thickness of ≈1-3 mm and mainly used in barrier insulation. Therefore, the methanol quantity of cable paper may be closer to the equilibrium methanol quantity than in electric-grade cardboard, however, it makes up only ≈*10%* of the total cellulosic pulp mass [3]. This may mean that the amount of adsorbed and desorbed methanol depends not only on the mass of cellulose insulation, but also on paper grades and their operational conditions in PT.

Taking into account this uncertainty, the calculation using formula (13) is mainly of methodological interest. In accordance with the survey data [3], the calculation was carried out for two ratios M_c/M_o: ≈0.07 and 0.1. For the calculation of the distribution coefficient K_{in} we used values of the equilibrium coefficient K_e (T_i), which was measured in experiments with 0.5 mm thick kraft paper [8]. The calculation was performed for the second maximum of methanol concentration (C_{o3} =190 ppb), which was recorded on 01.10.2019. The temperature of its previous maximum (41 °C, 24.08.2019) was taken as the representative temperature of the internal insulation. At this temperature, the coefficient K_e (T_i) is 59.4 [8], consequently, the integral distribution coefficient K_{in} from (13) is 3.8, and the total methanol quantity Q_t, determined from (5), with M_c/M_o =0.07 and M_c/M_o =0.1, is 43.9 mL and 58.5 mL respectively. Thus, the ratio of the cellulose insulation mass to the oil mass has a significant effect on the result of calculating the total methanol quantity in the internal insulation of the PT.

Table 2 shows the values of the total methanol quantity Q_t, which were calculated above on the base of different models. As can be seen, the "non-equilibrium" value of methanol volume (31.9 ml) was noticeably smaller than that obtained by "equilibrium" calculations. This was to be expected, since at equilibrium there is no methanol concentration gradient inside the cellulose insulation, which in reality always exists. At the same time, it has already been noted that the "non-equilibrium" calculation seems to have slightly underestimated value of Q_t.

In view of the above, it can be concluded that after the overhaul the value of the total methanol quantity Q_t was in the range from 35 to 40 ml. Accordingly, it can be assumed that the error of calculation of the total quantity of methanol on the base of equilibrium models for the researched PT was ≈*30%*. The value of the error of the "equilibrium" calculation can be specified after

© The Authors, published by EDP Sciences. This is an open access article distributed under the terms of the Creative Commons Attribution License 4.0 (https://creativecommons.org/licenses/by/4.0/).

analyzing the kinetics of the processes of methanol sorption-desorption by the cellulose surface taking into account the diffusion transfer of absorbed methanol. At present, apparently, there is no necessary set of experimental and theoretical data to carry out such an analysis.

Table 2. Total methanol quantity in PT as of 01.10.2019.

№	Calculation model	Q_t, ml
1	Non-equilibrium model, data on methanol quantity decrease with TO replacement are used, integral distribution coefficient K_{in} is calculated by formula (8)	≈32
2	Equilibrium model, calculation by formula (12) using data on temperature coefficient from [6]	≈45
3	Equilibrium model, integral distribution coefficient K_{in} calculated by formula (13) using data on the equilibrium distribution coefficient from [8]	≈44 (M_c/M_o = 0.07) ≈58 (M_c/M_o = 0.10)

The error of equilibrium models for the calculation of methanol distribution in the PT can be relatively small at low temperature change rate for the internal insulation on account of changes in ambient temperature and PT load.

5 Conclusion

1. We considered non-equilibrium and equilibrium models for calculating the total methanol quantity in PT internal insulation on the base of methanol reduction data obtained during overhaul and literature data on the effect of temperature on methanol distribution in the oil-cellulose system.
2. Data on the change in methanol concentration in transformer oil after replacement may allow estimation of the total methanol quantity of PT internal insulation. However, these data may be underestimated due to the difficulty in accounting for methanol losses from cellulose insulation during the insulation drying process.
3. A significant influence of "oil-to-paper" mass ratios on the results of calculation of the total methanol quantity is shown. To reduce the error of methanol quantity calculation in the internal insulation of the PT, it is necessary to take into account structural features.

References

1. E.M. Rodriguez-Celis, S. Duchesne, J. Jalbert and M. Ryadi. Cellulose, 22, pp. 3225–3236, (2015)
2. E.M. Rodriguez-Celis, J. Jalbert, S. Duchesne, B. Noirhomme, and M-C. Lessard. CIGRE-083, CIGRE Canada Conference (2012)
3. CIGRE TB 779. WG A2/D1-46. (2019)
4. IEC 60076-7:2018 Power transformers. Part 7. Loading guide for the loading of oil-filled power transformers (2018)
5. L. Darian, R. Obraztsov, A. Maksimchenko, V. Polistchook, S. Asoskov. Seasonal changes in the concentrations of cellulose aging markers in transformer oil, RSES 2023, E3S Web of Conferences 461, 01017 (2023)
6. J. Jalbert, M.-C. Lessard, and M. Ryadi. IEEE Trans. Dielectr. Electr. Insul. 20(6), pp. 2287-2291 (2013)
7. Reference book on electrical insulating materials, Part I. M.: Energia (1986)
8. H. Zheng, K. Shi, Y. Zhang, C. Zhang, E. Zhang, C. Zhang, Effect of Temperature on Methanol Equilibrium in Oil-paper Insulation System of Power Transformers (Inter. Conf. on High Voltage Engineering and Application, ICHVE, 6-10 Sept., 2020)

© The Authors, published by EDP Sciences. This is an open access article distributed under the terms of the Creative Commons Attribution License 4.0 (https://creativecommons.org/licenses/by/4.0/).

Problems of fuel and energy complex management: increasing vulnerability of intelligent elements of information and control systems to electromagnetic influences

Evgeny Grabchak [1], *Vladimir* Grigoriev [2], and *Evgeny* Loginov [3*]

[1] Joint Institute for High Temperatures, Russian Academy of Sciences, 125412, Moscow, Izhorskaya Street, Building 13, Building 2
[2] Moscow State Institute of International Relations (University) of the Ministry of Foreign Affairs of the Russian Federation, 119454, Moscow, Vernadsky Ave., 76
[3] Central Economics and Mathematics Institute of the Russian Academy of Sciences, 117418, Moscow, Nakhimovsky Prospekt, 47

Abstract. The problems of counteracting the risks of initiated loss of control in the fuel and energy supersystem are considered. The development of tools for monitoring and forecasting the state of technical elements of energy facilities as part of the fuel and energy supersystem, as well as systems for remote monitoring and control of the state of equipment is proposed.

Introduction

The fuel and energy complex of the Russian Federation is a significant part of the unified national economic system. The development and security of the state depend on its compliance with today's challenges. At present, the risks and threats of destructive impacts on electric power systems have sharply increased [1-2]. Moreover, the largest accidents of recent decades have confirmed the vulnerability of the energy systems of economically developed countries. The results of the analysis of the causes of these accidents did not give a definitive answer to whether the causes were a random coincidence of various unrelated factors or attacks disguised as natural (natural) impacts.

It is necessary to develop approaches to counteracting the risks of initiated loss of control in the fuel and energy supersystem under conditions of impacts based on new physical principles. This topic has become even more relevant given the trend of digitalization and intellectualization of complex technological systems.

In the modern world, where global problems such as climate change and depletion of natural resources are becoming increasingly pronounced, the fuel and energy complex of the Russian Federation must adapt to new realities. To ensure the sustainability of energy systems, it is necessary to implement innovative technologies, including digital solutions and artificial intelligence systems. These technologies can help in monitoring the state of infrastructure, forecasting energy consumption and optimizing resource allocation. In addition, it is important to develop flexible mechanisms for managing energy systems that will allow an effective response to sudden changes in supply and demand. This includes the use of renewable energy sources and the development of energy storage systems, which also contributes to increasing the environmental sustainability of the sector.

An equally important aspect is the creation of a cybersecurity system capable of protecting the fuel and energy complex from potential threats associated with cyber attacks. The use of multi-level defense mechanisms and regular software updates will help minimize the risks associated with vulnerabilities in information systems.

In addition to this, it is worth paying attention to the need to form interdepartmental and interregional cooperation. Effective interaction between different sectors and levels of governance will allow for more rapid identification and response to potential threats, as well as the exchange of best practices and experiences.

In general, the development of the fuel and energy complex in modern conditions requires a comprehensive approach that combines technological advances, protection mechanisms and the ability to adapt to rapidly changing conditions. This will ensure not only energy security, but also sustainable economic growth in the long term.

* Corresponding author: loginovel@mail.ru

© The Authors, published by EDP Sciences. This is an open access article distributed under the terms of the Creative Commons Attribution License 4.0 (https://creativecommons.org/licenses/by/4.0/).

The need for new approaches to solving the issue of preventing and counteracting the risks of initiated loss of control in the fuel and energy supersystem

In these conditions, it is necessary to search for methodological approaches to solving the issue of preventing and counteracting the risks of initiated loss of control in the fuel and energy supersystem both in the context of traditional attacks on fundamentally new digital information and control technologies being implemented in the supersystem, and in view of new methods of influence, for example, of an electromagnetic nature [3-7]. First of all, this can be implemented through structural changes in the technological and organizational layers of the industry aimed at increasing the stability and survivability of the supersystem [8-10].

The following measures are necessary:

- development, simulation modeling and practical confirmation of the effectiveness of tools for monitoring and forecasting the state of technical elements of energy facilities as part of the fuel and energy supersystem, as well as systems for remote monitoring and control of equipment condition;

- identification of factors of destructive impact on elements of the fuel and energy supersystem, further modeling of the consequences of these impacts using digital twin technologies;

- modernization of the mechanism for maintaining stable operation of energy facilities in conditions of uncertainty in the development of an emergency situation using controlled clustering and subsequent restoration of functionality;

- development and implementation of a decision support system in conditions with a significant uncertainty component to solve the problems of maintaining the stability and survivability of the fuel and energy supersystem and its elements.

In the context of digitalization of the industry, it is necessary to comprehensively improve the methods of analyzing risks and threats arising from the development of the key trend of increasing the vulnerability of intelligent elements of information and control systems to physical and network influences, analyzing current situations and forecasting problems in the operation of elements of a super-large fuel and energy supersystem as the basis of critical infrastructures of energy-dependent industries in the Russian economy.

In addition, an important area is the integration of innovative technologies, such as blockchain, to increase the transparency of processes and increase trust in information systems. It is also necessary to consider the possibilities of using artificial intelligence to automate monitoring and diagnostic processes, which will allow for a faster response to emerging threats.

Improving the legislative framework and mechanisms for interaction between various industry entities can also significantly improve the protection of the fuel and energy supersystem. The formation of interdepartmental and intersectoral working groups for the prompt exchange of information and the development of joint solutions will help in creating a unified approach to risk management.

In conclusion, strategic planning and the implementation of comprehensive programs aimed at preventing and countering risks are becoming necessary to ensure the effective functioning of the fuel and energy supersystem and its resilience to various impacts. The transition to new approaches requires active cooperation between both government agencies and private companies, as well as constant monitoring of the development of technologies and protection methods.

Development of a package of scientific and technical solutions and conducting comprehensive tests of electrical power equipment

It is necessary to substantiate technical measures for a group of complex mutually stimulating increasing effects of a technical, economic and organizational nature, including those that form the systemic stability of the energy infrastructure in the conditions of:

(a) known, modeled and monitored threats of physical and network impacts,

(b) threats of new attack technologies that are only moving to the prototype testing stage abroad.

Developing a package of scientific and technical solutions and conducting comprehensive tests of electric power equipment are key factors in ensuring the reliability and efficiency of energy systems. In the context of constantly increasing threats, both physical and cyber, it is necessary to systemically integrate various technical solutions aimed at strengthening the resilience of the energy infrastructure.

© The Authors, published by EDP Sciences. This is an open access article distributed under the terms of the Creative Commons Attribution License 4.0 (https://creativecommons.org/licenses/by/4.0/).

To achieve this goal, it is necessary to thoroughly justify technical measures, which can be grouped into the following areas:

Development of protection and monitoring mechanisms: One of the main components of a resilient energy infrastructure is the creation of a monitoring system that allows real-time tracking of any deviations in the operation of equipment, as well as identifying potential threats. This includes the use of modern sensors and data analysis systems with a high degree of sensitivity.

Integration of new technologies: Given the danger posed by new attack technologies, it is necessary to actively research and implement secure architectures and components that are able to withstand potential threats. This may include data encryption, the use of blockchain technologies to ensure the integrity and reliability of information, as well as machine learning algorithms to predict and prevent attacks.

Cost-effectiveness of measures: It is important not only to implement new technologies, but also to evaluate their cost-effectiveness. Measures must be justified in terms of costs and benefits, taking into account not only direct costs, but also possible losses from system inefficiency in the event of threats being realized.

Organizational changes: Successful implementation of new solutions requires a review of the organizational structure and management processes. It is necessary to create teams responsible for security and organize training for employees to work with new technical means of protection.

Modeling of threats and scenarios: Models must be developed that allow predicting various attack scenarios and consequences for the energy system. This will help in timely preparation for potential threats and will allow prompt decision-making in the event of their implementation.

Thus, an integrated approach to solving issues of increasing the systemic stability of electric power equipment includes not only technical measures, but also economic and organizational justification, which generally contributes to the creation of a reliable and safe energy infrastructure capable of withstanding both known and new threats.

Conclusion

To counteract the vulnerability of digital information and control systems to EMI, it is necessary to develop new means of recording geo-induced currents, install them on the most vulnerable networks, collect statistics on current pulses directly related to solar and geomagnetic activity and compare these data with accidents. The information obtained will allow a more substantiated determination of the scale of the threat to the electric grid economy from "space weather" and related geomagnetic processes of geo-induced current impacts, and in general from any situation leading to accidents in the fuel and energy supersystem. The proposed scientific and technical measures will significantly reduce losses in accidents caused by magnetic storms.

An important aspect is also the creation of a forecasting and analysis system that will allow predicting possible changes in geomagnetic activity and informing grid operators in advance of potential risks. This can be achieved by integrating meteorological and space data into a single analytical platform.

Equally important is the training of personnel responsible for operational management and technical support of energy facilities. Conducting specialized training and advanced training courses on the impact of EMI and geo-induced currents on equipment will improve the readiness of personnel to act in crisis situations.

The proposed scientific and technical activities may also include the development and implementation of new methods for protecting equipment from adverse effects, such as shielding, the use of high-quality transformers with increased stability, and the use of technologies for dynamic response to changes in electrical networks.

Ultimately, a systematic approach to solving the problem of vulnerability of energy systems to geomagnetic processes will not only reduce losses during accidents caused by magnetic storms, but also increase the overall stability of the energy infrastructure, which is extremely important in the context of global climate change and the increasing frequency of extreme atmospheric phenomena.

The work was carried out with the support of the Ministry of Science and Higher Education of the Russian Federation within the framework of project No. 075-15-2024-525 dated 04/23/2024.

References

1. V.V. Grigoriev, E.L. Loginov, V.E. Loginova *Agent-based human interaction with the digital educational environment, based on a set of digital intelligent agents* // Proceedings of SPIE - The International Society for Optical Engineering. Ser. "Computer Applications for Management and Sustainable Development of Production and Industry, CMSD 2021" (2022) C. 122510L.

© The Authors, published by EDP Sciences. This is an open access article distributed under the terms of the Creative Commons Attribution License 4.0 (https://creativecommons.org/licenses/by/4.0/).

2. A. Bugaev *Ensuring the security of cyber-physical systems in the energy sector in the context of the expansion of digital management services* // CEUR Workshop Proceedings. Proceedings of the II International Scientific and Practical Conference *"Information Technologies and Intelligent Decision Making Systems"* (ITIDMS-II-2021). Aachen, Germany (2021) C. 164-173.
3. E.P. Grabchak *Ensuring observability and controllability of complex technical systems in difficult and irregular situations when commands with a large distortion component are received* // Lecture Notes in Electrical Engineering (2021) T. **729** LNEE. C. 624-631.
4. V.Y. Afanasiev, E.P. Grabchak, S.V. Mischeryakov *Digital transformation of TPP control systems: transition to an intelligent equipment life cycle management model* // Socio-economic Systems: Paradigms for the Future. Springer International Publishing (**2021**) C. 1765-1772.
5. .E.P Grabchak. *Ensuring observability and controllability of complex technical systems in difficult and irregular situations when commands with a large distortion component are received* // Advances in Automation II. Proceedings of the International Russian Automation Conference. Ser. *"Lecture Notes in Electrical Engineering"* (2020) C. 624-631.
6. V.S. Balandin, A.A. Shkuta, P.A. Boyko *The use of electronic semantization of the cognitive activity manifestations with the aim of detection of intentions of the group of people leading to the destabilization of the digital super system* // IOP Conference Series: Materials Science and Engineering. 1. Ser. *"1st International Conference on Innovative Informational and Engineering Technologies, IIET 2020"* (2020) C. 012002.
7. P.A. Bojko, V.S. Balandin, G.S. Ivanovskij *Convergent use of neural network information technologies for monitoring and evaluating the performance of digital control operators of situational centers at critical infrastructure facilities* // IOP Conference Series: Materials Science and Engineering. 3. Ser. *"III International Scientific and Practical Conference on Innovations in Engineering and Technology"* (2020) C. 012044.
8. S.I Bortalevich, V.U. Chinaliev *Strategic planning of integrated development in the field of science and technology of the enterprises of the machine-building production* // Journal of Physics: Conference Series (2019) C. 12041.
9. V Bahtizin, V.Y. Bortalevich, A.I. Soldatov *Using artificial intelligence to optimize intermodal networking of organizational agents within the digital economy* // Journal of Physics: Conference Series (2019) C. 12042.
10. A.A. Shkuta, V.Y. Bortalevich, D.D. Sorokin *Intelligent monitoring, modeling and regulation of information traffic to specify the trajectories of the behavior of organizational agents in the context of receipt of difficult-interpreted information* // IOP Conference Series: Materials Science and Engineering (2019) C. 012015.

© The Authors, published by EDP Sciences. This is an open access article distributed under the terms of the Creative Commons Attribution License 4.0 (https://creativecommons.org/licenses/by/4.0/).

E3S Web of Conferences **584**, 01013 (2024) https://doi.org/10.1051/e3sconf/202458401013
RSES 2024

Digital technologies for managing the turnover of fuel and energy resources within the framework of an expanded international energy association

Vladimir Grigoriev [1], *Evgeny* Loginov [2*], and *Alexander* Shkuta [3]

[1] Moscow State Institute of International Relations (University) of the Ministry of Foreign Affairs of the Russian Federation, 119454, Moscow, Vernadsky Ave., 76
[2] Central Economics and Mathematics Institute of the Russian Academy of Sciences, 117418, Moscow, Nakhimovsky Prospekt, 47
[3] Financial University under the Government of the Russian Federation, 125167, Moscow, Leningradsky Prospekt, 49

Abstract. The paper proposes to form a coordinated energy supply system by creating and developing an expanded international energy association (Russia+). Digital approaches to organizing, on the principles of the Unified Energy System of Russia, synchronous parallel operation of territorially distributed facilities for generating electricity, extracting and supplying fuel and energy resources based on a digital information platform are considered.

Introduction

It is necessary to create a toolkit that ensures fast and convenient use of information on the turnover of fuel and energy resources so that the operator [and regulator] of the expanded international energy association can perform a set of functional tasks that can be quickly adapted to it.

It is proposed that the considered information and analytical services for monitoring trade turnover in distributed electronic markets of fuel and energy resources be placed in the information systems that service electronic trading platforms. Such services should be available to the operator [and regulator] of the expanded international energy association, as well as to the bodies that manage fuel and energy activities. Some of the services may be available to companies participating in trades.

On this basis, it will be possible to optimize relations between interconnected energy companies and buyers of fuel and energy resources in managing the processes of extraction, transportation, processing and marketing (supply) of fuel and energy resources with an exit to the cumulative effect of socio-economic development.

Solution to the problem

The proposed policy should allow for a sharp expansion of the use of modern automation tools for management processes and the ability to coordinate the work of production (mining, etc.) and transport capacities in the industries of the countries participating in the expanded international energy association.

Here, strategic segmentation is required within the expanded international energy association as a group of national infrastructure segments servicing the turnover of certain volumes of fuel and energy resources and the added value associated with them, tied to specific fuel and energy resource markets [1].

The economic and technological parameters of these infrastructure segments are the basis for choosing options for state support for companies (technical and economic agents) with the aim of developing the aggregate structure of production (mining, etc.) and transport capacities [2].

A similar principle of identifying the system-forming aggregate structure of production (extractive, etc.) and transport capacities interconnected by specializations and resources as a group of infrastructure segments tied to specific markets (including consumer groups) is applicable to both territorially isolated and interconnected national economic systems and facilitates the possibility of a coordinated energy supply system. The fundamental feature of the proposed

* Corresponding author: loginovel@mail.ru

© The Authors, published by EDP Sciences. This is an open access article distributed under the terms of the Creative Commons Attribution License 4.0 (https://creativecommons.org/licenses/by/4.0/).

mechanism is the ability to optimize the process of compiling fuel and energy and economic balances taking into account the flows of input information from available sources in the corporate-industry and territorial context [3-4].

The set of functional tasks for information and analytical services for monitoring trade turnover in distributed electronic markets of fuel and energy resources can be expanded by the following points:

1. Data collection and analysis: development of mechanisms for automatic collection of data on prices, trading volumes and the number of participants on electronic platforms. Integration with external sources of information, such as the gas, electricity and other energy resource markets, is also necessary.

2. Monitoring market dynamics: creation of tools for tracking changes in the market situation, including price fluctuations, changes in supply and demand, as well as identifying patterns and trends.

3. Forecasting and modeling: implementation of machine learning and artificial intelligence algorithms to predict market events, which will allow market participants to plan their actions in advance.

4. Risk assessment: development of models for assessing risks associated with trading operations and potential market fluctuations, which will help participants minimize losses.

5. Creation of analytical reports: generation of regular reports for operators and regulators, which will contain key market indicators, analysis of changes and recommendations compared to historical data.

6. Interaction management: providing a platform for information exchange and coordination between market participants, including the ability to quickly respond to changing market conditions.

7. User training: developing educational programs for service users so that they can effectively use the tools and technologies provided.

8. Regulatory functions: integrating modules that allow regulators and supervisory authorities to monitor and analyze compliance with standards and legal requirements, as well as identify unfair practices.

9. Sustainable development: taking into account environmental aspects and implementing mechanisms to assess the impact of trading operations on the environment, which is an important aspect of modern energy resource management.

On this basis, it will be possible to optimize relations between interconnected energy companies and buyers of fuel and energy resources in managing the processes of extraction, transportation, processing and marketing (supply) of fuel and energy resources with an exit to the cumulative effect of socio-economic development. These integrated approaches will create a more transparent and efficient market, which in turn will strengthen trust between all market participants and improve the security and sustainability of the energy system as a whole.

10. Sustainable development: taking into account environmental aspects and principles of sustainable resource management. This includes analyzing the impact of trading operations on the environment, as well as recommendations for reducing the carbon footprint and optimizing resource use.

11. User interfaces: developing intuitive interfaces that will allow users to easily navigate the system, access the necessary data and quickly respond to changes in the market situation.

12. Data security: ensuring a reliable information protection system, including data encryption, as well as user authentication and authorization mechanisms to prevent unauthorized access.

13. Competition analysis: implementing tools for studying the competitive environment in distributed markets, which will allow participants to better understand each other's positions and develop strategies to improve their competitiveness.

14. Blockchain Integration: Explore possible integration with blockchain technologies to increase transparency and trust in trade data, as well as to protect against fraud.

15. Cross-Marketing Opportunities: Explore opportunities for cross-selling and sharing experiences between different sectors of the economy, which can lead to new forms of cooperation and accelerate innovation processes.

16. User Analytics: Develop tools to analyze the behavior and preferences of service users. This will improve the quality of service and tailor products and services to specific customer needs.

17. Mobile Applications: Create mobile solutions for accessing monitoring services, which will allow users to stay informed about trade operations at any time and from anywhere.

These functional tasks will help to create a multifaceted approach to monitoring and analyzing trade turnover, ensuring reliability, a high level of service and promoting the development of distribution electronic markets in the fuel and energy sector.

On this basis, it is possible to more accurately identify the national and cross-border movement of arrays of added value from the export-import of fuel and energy resources, formed within the framework of the work of distributed infrastructure systems [5].

It is possible to implement the establishment of stable operational and communication links between a complex of information systems of government agencies and socially significant commercial structures within the framework of an expanded international

© The Authors, published by EDP Sciences. This is an open access article distributed under the terms of the Creative Commons Attribution License 4.0 (https://creativecommons.org/licenses/by/4.0/).

energy association (Russia+) with a common organizational and information platform and a system of coordinated management of information exchange [6]. The development of the system must be carried out on the principles of the Unified Energy System of Russia, which have proven their effectiveness, with synchronous parallel operation of territorially distributed facilities for generating electricity, extracting and supplying fuel and energy resources [7].

This will not only optimize production and distribution processes, but will also strengthen the competitiveness of participating countries in the global energy market. It is important to take into account that during the implementation of this policy, it is necessary to ensure active cooperation between government agencies, the private sector and scientific institutions in order to accumulate and process experience, produce analytical data and develop innovative solutions that meet modern challenges.

Adaptation to changing market conditions and technological trends should become a continuous process. To this end, platforms for the exchange of information and best practices should be created, as well as regular seminars and conferences should be held where participants can share knowledge and experience. An important aspect is also the integration of new technologies, such as digitalization, automation and the use of artificial intelligence, which can significantly improve the efficiency of energy resource management.

This policy should include elements of forecasting and risk analysis, which will allow participating countries to respond more flexibly to changes in the global economic situation and the needs of domestic and foreign markets. A systematic approach to designing energy infrastructures will ensure sustainable development and avoid duplication of resources, which in turn will facilitate more efficient use of all available capacities and resources within the association.

For the exchange of information and technologies that would facilitate rapid response to changes in supply and demand in global markets. These platforms should unite producers, consumers, scientific and research laboratories, as well as government agencies, which will ensure an integrated approach to solving emerging issues.

One of the key aspects of such a policy is the need to create a unified system for monitoring and analyzing data on the state of the energy market. This will allow for changes in needs to be anticipated and production and transport processes to be adapted accordingly. With the use of modern technologies such as artificial intelligence and big data, it will be possible not only to promptly process information, but also to predict trends in the development of the industry.

It is also important to develop infrastructure to support the transportation and storage of energy resources. In this context, it is necessary to consider the possibility of introducing environmentally friendly and modern technologies, which corresponds to global trends in reducing the carbon footprint. Support for such initiatives can become part of government policy aimed at sustainable development.

Another goal should be to create incentives for companies engaged in energy innovation. This could include financial support for start-ups, research projects and companies developing alternative and efficient energy sources. Improving legislation on tax incentives and subsidies would also be an important step in supporting these initiatives.

Education and training are also critical to the successful implementation of new technologies and practices. It is necessary to create educational programs and courses aimed at training specialists with relevant knowledge and skills in the field of new technologies, project management and sustainable development. In the long term, it is worth considering the possibility of forming an energy strategy of the participating countries based on cooperation and mutual interests. This will not only optimize internal processes, but also create effective external relations between countries, which will ultimately contribute to greater energy security and stability.Thus, the implementation of the proposed policy implies a comprehensive approach that includes innovative technologies, educational initiatives and active cooperation at all levels, which in the long term will lead to the creation of a sustainable and competitive energy system within the framework of an expanded international energy union.

This integration will create a single field for the exchange of data and information between various market participants, which will increase the transparency and efficiency of decision-making. An important aspect is the use of advanced information technologies, such as blockchain, to ensure the security and integrity of data, which is especially critical in the face of the risk of cyberattacks and unauthorized interference. In addition, the creation of the Russia+ platform involves not only coordination of actions at the domestic level, but also active international cooperation, which can be achieved through the exchange of experience and best practices with other countries. This will also take into account regional characteristics and potential threats, as well as create mechanisms for rapid response in situations requiring coordinated actions by participating countries.

For the successful implementation of this system, it is also necessary to pay attention to the training of qualified personnel capable of working in the conditions of new technologies and actively participating in the

© The Authors, published by EDP Sciences. This is an open access article distributed under the terms of the Creative Commons Attribution License 4.0 (https://creativecommons.org/licenses/by/4.0/).

digitalization processes of the energy sector. Retraining and advanced training programs for specialists in the field of information technology and energy resource management will help ensure the sustainability and efficiency of the unified system.

Thus, the creation and development of an international energy association not only contributes to strengthening energy security, but also promotes sustainable development, innovation and the implementation of environmentally friendly technologies in the energy sector, which are key points in the context of global climate change and the transition to a green economy.

Thus, the creation of an integrated information platform and a coordinated data exchange management system will form the basis for more efficient functioning of the energy sector in the "Russia+" format. This will not only affect internal processes, but will also expand opportunities for international cooperation related to energy security and sustainable development.

An important point in this strategy should be the development of a system for monitoring and analyzing data coming from various sources. The use of modern analytical tools and machine learning algorithms will allow not only to quickly respond to changes in the market situation, but also to predict possible crisis scenarios, which will significantly increase the level of preparation and risk management.

To ensure the sustainability and security of the system, it is also necessary to pay attention to information protection and cybersecurity. The development of comprehensive data protection solutions, as well as the implementation of multi-level security systems, will help minimize the risks of cyber attacks and leaks of confidential information. This is especially important in the context of globalization, when potential threats can come from outside. In addition, one of the key success factors will be interaction with regulatory authorities and key stakeholders. It is important to create conditions for an open dialogue between government agencies, businesses and the scientific community. This will allow taking into account the interests of all participants in the process and more effectively address emerging issues.
Harmonization of standards and regulations regarding data exchange will also have a positive impact on the creation of a single energy market. The introduction of uniform standards will simplify the process of integrating various systems and reduce the costs of their support and maintenance. This, in turn, will increase the competitiveness of market participants and attract new investors interested in a stable and transparently functioning segment.

As a result, the implementation of the integrated system "Russia+" will become a strategic step towards the creation of a sustainable, efficient and secure energy infrastructure that will be able not only to satisfy the country's domestic needs, but also to become an important player in the international arena, contributing to the development of global energy security.

Conclusion

Thus, the overall result of the implementation of the project under consideration is the possibility of synchronous parallel operation of territorially distributed facilities for generating electricity, extracting and supplying fuel and energy resources. That is, based on the principles of the Unified Energy System of Russia, it is possible to coordinate the work of the system-forming aggregate structure of national production (extracting, etc.) and transport capacities within the framework of an expanded international energy association (Russia+) interconnected by specialization and resources as a group of infrastructure segments generating certain volumes of fuel and energy resources and the added value associated with them, tied to specific markets (including consumer groups).

In addition, the implementation of this project will improve the efficiency of operation of both existing and new capacities, thanks to the introduction of modern monitoring and control technologies. The control system will be based on the principles of digitalization, which will allow real-time monitoring of production processes and prompt response to changes in operating conditions.

Another important component of the project will be the development of a single platform for the exchange of information between all participants in the process - from manufacturers to end consumers. This will create a transparent and predictable market, where each decision will be made on the basis of relevant and reliable data, which will significantly reduce risks and increase the safety of the energy infrastructure.

Moreover, the project involves active interaction with neighboring countries and international partners, which will create additional opportunities for cooperation in the field of energy exchange, the development of joint infrastructure projects and the exchange of technologies. This approach will not only strengthen Russia's position in the international arena, but will also contribute to the sustainable development of the energy sector as a whole, including environmental aspects such as reducing greenhouse gas emissions and switching to cleaner energy sources.
Thus, the Russia+ project will become an important milestone in the evolution of the country's economy and will open up new horizons for the development of both domestic and international energy systems.

© The Authors, published by EDP Sciences. This is an open access article distributed under the terms of the Creative Commons Attribution License 4.0 (https://creativecommons.org/licenses/by/4.0/).

The introduction of renewable energy sources, such as solar and wind power plants, will also be an important part of this project. This will allow for the diversification of generation sources and reduce dependence on traditional fossil resources.

One of the key aspects will be the development of programs for training and advanced training of personnel in the field of new technologies and management methods. The training of qualified specialists will provide support for the implemented solutions and will help to integrate them into existing systems as effectively as possible. It is important to create conditions for the exchange of knowledge and experience not only within the country, but also abroad.

Thus, the project is not limited to technical aspects. It involves the integration of social, economic and environmental components, which will contribute to the creation of a more sustainable energy ecosystem. This will not only increase the reliability of energy supplies, but will also contribute to improving the quality of life of citizens, which is ultimately of strategic importance for the country. In addition, sustainable development of this project involves the development of measures to ensure energy security, including capacity backup and the creation of rapid response systems for possible emergencies. Given global challenges such as climate change and geopolitical instability, the development of such aspects will become a necessity.

Successful implementation of the project will require active cooperation with government agencies, businesses and scientific institutions. This will ensure an approach based on multidisciplinary approach, which in turn will open new horizons for innovation and further development of the industry.

In conclusion, the implementation of this project provides a unique opportunity to transform the energy system. The synergy of all the above components will create a powerful foundation for sustainable growth, allowing Russia to take a leading position in the global energy market. Thus, work on the implementation of this program will contribute not only to energy independence and security, but also to the socio-economic development of the entire country.

The work was supported by the Ministry of Science and Higher Education of the Russian Federation within the framework of project No. 075-15-2024-525 dated 23.04.2024.

References

1. V.V. Grigoriev, E.L. Loginov, V.E. Loginova *Agent-based human interaction with the digital educational environment, based on a set of digital intelligent agents* // Proceedings of SPIE - The International Society for Optical Engineering. Ser. "Computer Applications for Management and Sustainable Development of Production and Industry, CMSD **2021**" (2022) C. 122510L
2. E.P. Grabchak *Ensuring observability and controllability of complex technical systems in difficult and irregular situations when commands with a large distortion component are received* // Advances in Automation II. Proceedings of the International Russian Automation Conference. Ser. *"Lecture Notes in Electrical Engineering"* (2020).
3. V.S. Balandin, A.A. Shkuta, P.A. Boyko *The use of electronic semantization of the cognitive activity manifestations with the aim of detection of intentions of the group of people leading to the destabilization of the digital super system* // IOP Conference Series: Materials Science and Engineering. **1**. Ser. *"1st International Conference on Innovative Informational and Engineering Technologies, IIET 2020"* **2020**. P. 012002.
4. P.A. Bojko, V.S. Balandin, G.S. Ivanovskij *Convergent use of neural network information technologies for monitoring and evaluating the performance of digital control operators of situational centers at critical infrastructure facilities* // IOP Conference Series: Materials Science and Engineering. **3**. Ser. *"III International Scientific and Practical Conference on Innovations in Engineering and Technology"* **2020**. P. 012044.
5. S.I. Bortalevich, V.U. Chinaliev *Strategic planning of integrated development in the field of science and technology of the enterprises of the machine-building production* // Journal of Physics: Conference Series. **2019**. P. 12041.
6. A.R. Bahtizin, V.Y. Bortalevich, A.I. Soldatov *Using artificial intelligence to optimize intermodal networking of organizational agents within the digital economy* // Journal of Physics: Conference Series. **2019**. P. 12042.
7. A.A. Shkuta, V.Y. Bortalevich, D.D. Sorokin *Intelligent monitoring, modeling and regulation of information traffic to specify the trajectories of the behavior of organizational agents in the context of receipt of difficult-interpreted information* // IOP Conference Series: Materials Science and Engineering. **2019**. C. 012015.

© The Authors, published by EDP Sciences. This is an open access article distributed under the terms of the Creative Commons Attribution License 4.0 (https://creativecommons.org/licenses/by/4.0/).

Development of effective algorithms for calculating current distribution coefficients of an electric power system

Dauren Akhmetbaev [1], *Abdygali* Dzhandigulov [1*], *Arman* Akhmetbaev [2], *Svetlana* Bystrova [3]

[1] L. Gumilyov Eurasian National University, 010000, Astana, Republic of Kazakhstan
[2] Nizhny Novgorod State Technical University named after R.E. Alekseev, Russia
[3] Ekibastuz Engineering and Technical Institute named after Academician K. Satpayev, Ekibastuz, Republic of Kazakhstan

Abstract. Calculations of steady-state modes of a complex electric power system are significantly simplified if we use the coefficients of distribution of the setting currents. The use of the theory of directed graphs for the formation of the matrix of the coefficients of distribution of the setting currents is directly related to the determination of the values of all possible and specific trees of the graph of a complex electric network. The authors have developed and implemented an improved algorithm. calculation of current distribution coefficients of the electric power system. The concepts of the "base" of the graph and the "hanging" contour are introduced. In this case, to calculate the current distribution coefficients, it is necessary to take into account all possible and specific trees of only the graph base. A method is indicated for supplementing the matrix of current distribution coefficients of the graph base to the current distribution matrix of the original graph. The paper proposes methods for simplifying the network topology based on the development of diacoptic principles.
Key words: electrical network, topology, graph, graph tree, diakoptics, distribution coefficients.

Introduction

The study of electrical networks is always a pressing and complex task. The complexity lies primarily in the large dimension and nonlinearity of the resulting systems of equations that are solved by iterative methods, with all the consequences that arise from this, such as an increase in the number of operations, weak convergence of methods, finding additional restrictions to select the desired solution from a variety of solutions, etc. Calculations of electrical networks are significantly simplified if the distribution coefficients of the setting currents are known [1]. The matrix of distribution coefficients can be formalized both by scheme transformation methods and by analytical methods.

Topological methods for determining the matrix of distribution coefficients were developed in [2]. These methods were associated with the difficulties of searching and determining the values of all possible trees, especially specific 2-trees of the graph of a complex electrical network. In [3], an analytical approach to determining the topological content of the distribution coefficients of the setting currents is presented, which significantly simplified the technology of searching for specific trees of a directed graph of a complex network. This work is a continuation of a series of works devoted to optimizing algorithms for finding all possible and specific spanning trees of a directed graph, with the aim of reducing the number of operations and the amount of RAM on a computer.

Graph theory and topological content of current distribution coefficients

The foundations of the topology of the equivalent circuit of electrical networks were laid in the classical works of Kirchhoff and Maxwell [4,5]. They were the first to introduce the concept of trees and to obtain the topological properties of the determinants of the matrix of nodal conductivities $\underline{\overline{Y}}$and the matrix of circuit resistances $\underline{Z_{K}}$.

The current distribution coefficients can be found by solving a system of nodal equations:

$$\underline{C_{ij}} = \frac{\underline{Y_{rs}}(\Delta_{ir}-\Delta_{js})}{\Delta} \quad (1)$$

where $Y_{rs}$is the conductivity of branch i between nodes r and s.

Based on the study of determinants (1) of nodal equations of a complex electrical network, in [6], the topological content of distribution coefficients was established, which, due to the complexity of search algorithms and the definition of all possible, especially 2-trees of the graph, has not found practical application. In [7], an analytical approach to determining the topological content of current distribution coefficients is proposed. The essence of the approach is that an analytical proof of the formation of specific trees of a directed graph is obtained, without determining 2-trees of the graph, based on all its possible trees [8]. Then, the value of the distribution coefficient in ia branch from the node current $j$is determined by the topological expression:

* Corresponding author: abeked@mail.ru

© The Authors, published by EDP Sciences. This is an open access article distributed under the terms of the Creative Commons Attribution License 4.0 (https://creativecommons.org/licenses/by/4.0/).

$$\underline{C}_{ij} = \frac{\underline{F}_{ij}}{\sum \underline{F}}, \quad (2)$$

where is $\sum \underline{F}$the total value of all possible trees in the graph; $\underline{F}_{ij}$is the algebraic sum of the values of specific trees in the graph of the i -th branch.

Specific trees of the i -th branch relative to the j -th node are formed among possible trees as follows: if the direction of the graph of the i -th branch coincides with the direction of the traversal from the j -th node to the base node, then the tree size is multiplied by +1, otherwise by -1, if the i -th branch is absent, then by 0.

Ways to improve algorithms for finding possible graph trees

In the work [9], an effective algorithm for searching and determining the weights of possible graph trees without using previously defined trees is implemented. The proposed algorithm works successfully, especially in the case of a poorly filled vertex adjacency matrix [10]. With a more filled vertex adjacency matrix, the program execution time increases many times.

For example, for a fairly simple 14-node scheme (http://energy.komisc.ru/dev/test_cases) the number of spanning trees is 3909, and for a 30-node scheme the number of spanning trees is 7,824,000. At the same time, for a 39- node scheme the number of spanning trees is 421,380. That is, each network has its own characteristics. This work is devoted to identifying and taking into account these characteristics.
Let us illustrate it with the following example.

Table 1. Node data. Unom = 220 kV

No	Voltage		Load power		Generation power	
	Phase, deg	module, kV	P, MW	Q, MVar	P, MW	Q, MVar
1	201,9	2,50	0	0	-128	-126
2	214,7	0.99	18	10	0	0
3	198,0	-1,59	20	15	60	35
4	193,5	0,19	0	0	-160	-105
5	199,5	-1,05	13	9	0	0
6	204,8	3,88	94	49,5	0	0
7	209,1	5,87	80	43	0	0
8	207,0	3,57	35	21	0	0
9	212,8	3,77	21	12	0	0
10	211,6	3,08	37	25	0	0
11	205,5	0,36	16	9	0	0
12	112	0	28	14	0	0

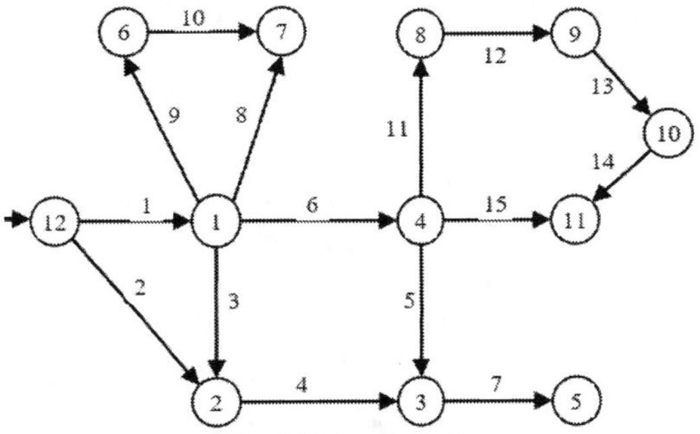

Fig. 1. Network graph.

Table 2. Branch data

No.	Start	End	R, Ohm	X, Ohm	B, mSm (capacity +, ind -)
1	12	1	87,0	88,0	-
2	12	2	14,4	13,0	-
3	1	2	40,0	86,0	-
4	2	3	6,10	71,0	0.8540
5	4	3	39.50	24,0	0,2640
6	1	4	19,00	35,0	0,1372
7	3	5	2,20	30,0	0,3740
8	1	7	3,30	44,0	2,1920
9	1	6	0,60	9,0	0,4320
10	6	7	5,20	73,5	3,6200
11	4	8	6,28	59,4	4,3800
12	8	9	13,00	47,0	0,1490
13	9	10	35,00	13,0	0,3460
14	10	11	14,70	47,0	0,5760
15	4	11	35,00	13,0	0,5760

Using the Kirchhoff matrix, we can calculate that the number of spanning trees is 165.

Recall that the indices of the element of the matrix of current distribution coefficients $\underline{C}_{ij}$denote $i$the -th branch relative to $j$the -th node. Formula (2) can be rewritten as

$$\underline{C}_{ij} = \left(\sum_k G_k \cdot \alpha_{i,j}^k\right)\left(\sum_k G_k\right)^{-1}, \quad (3)$$

here $G_k$is the weight of the tree, calculated as the product of the complex resistances of all the chords $k$of the tree, and the coefficient $\alpha_{i,j}^k$is calculated using the rule:

- $\alpha_{i,j}^k = 0$,if branch j is a chord of the k -th tree.
- $\alpha_{i,j}^k = 1$, if the direction of branch j coincides with the direction in the path from the i-th vertex to the basis vertex in the k -th tree.
- $\alpha_{i,j}^k = -1$, if the direction of branch j is opposite to the direction in the path from the i-th vertex to the basis vertex in the k -th tree.

During implementation, the following stages should be highlighted:

Stage 1. – Identification of all hanging branches. These branches are present in all trees, therefore in

© The Authors, published by EDP Sciences. This is an open access article distributed under the terms of the Creative Commons Attribution License 4.0 (https://creativecommons.org/licenses/by/4.0/).

formula (3) the complex resistances are reduced and the values of these resistances do not affect the values of $\underline{C}_{ij}$. The coefficients $\underline{C}_{ij}$will be equal to 1 if j-th node is the beginning of $i$the -th branch, are equal to -1, if $j$the -th node is the end of $i$the -th branch and zero, if $j$the -th node is not incident $i$to the -th branch. And in formula (4) the values $\alpha_{i,j}^{k}$for the terminal j-th node for other branches are equal to the values of such coefficients of the neighboring node q: $\alpha_{i,j}^{k} = \alpha_{i,q}^{k}, \forall i$. Thus, it is sufficient to consider a network without hanging branches. In the network under consideration (Fig. 1) the hanging branch is 7 and the terminal node is 5.

With such "removal" of a hanging branch, other hanging branches may arise. Then the first stage is repeated. In the considered scheme, stage 1 does not lead to the appearance of new hanging branches.

Thus, we need to consider a network with 11 nodes and 14 branches. The number of spanning trees of the network is preserved.

Stage 2. Selection of all "hanging" contours. By a hanging contour we mean a contour that has a single common connecting node with the "base" network, i.e. the remainder of the network after removing the "hanging" contours. In the scheme under consideration, there are two "hanging" contours (Fig. 2).

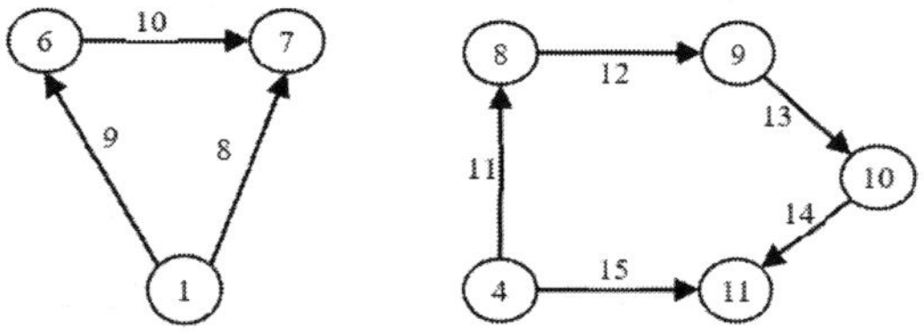

Fig. 2. Two "hanging" contours.

The basis of the graph will be as follows (Fig. 3).

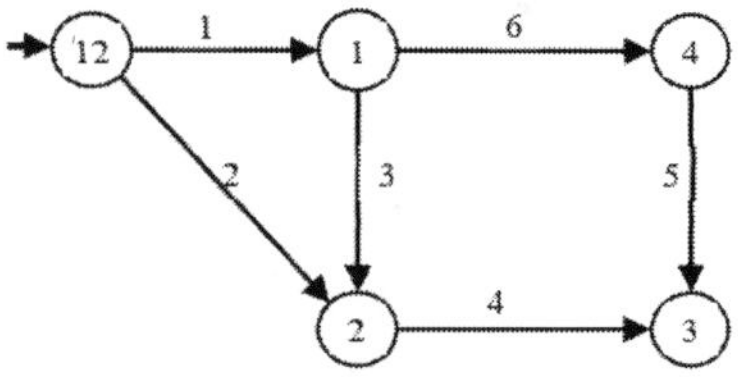

Fig. 3. The basis of the graph.

Table 3. Data on hanging contours

Contour number	Branches of the contour	Contour nodes	The connecting knot
1	8, 9, 10	1, 6, 7	1
2	11,12,13,14,15	4,8,9,10,11	4

Each spanning tree of the network base generates three trees from the first contour and five trees from the second contour. That is, each tree of the network base generates 15 trees of the original graph. So, the network base will be $165 : 15 = 11$spanning trees.

Let $\tilde{G}_k$the weight of the tree be the network base, and q_1, q_2let be the connecting nodes of two hanging contours. In our case $q_1 = 1, q_2 = 4$.Then for each tree of the base in the original graph there will correspond 15 trees with weights $G_k$obtained by multiplying $\tilde{G}_k$by a pair of edges, one from each contour, which are respectively a chord of the tree of the original graph. In formula (4), the values $\alpha_{i,j}^{k}$for each j-th node of the hanging contour for the branches of the contour are calculated in the same way as in the graph base, only instead of the base node, a connecting node is taken, and for the branches of the network base in the same way as for the connecting node. In particular, for the graph under consideration, 15 trees will be determined by adding pairs of chords (Table 4).

Table 4. Table of added pairs of chords

Tree number	1	2	3	4	5	6	7
Chord from the first circuit	8	8	8	8	8	9	9
Chord from the second circuit	11	12	13	14	15	11	12

Tree number	8	9	10	11	12	13	14	15
Chord from the first circuit	9	9	9	10	10	10	10	10
Chord from the second circuit	13	14	15	11	12	13	14	15

The values $\alpha_{i,j}^{k}$for all nodes of one circuit and branches of another circuit are equal to zero. The values $\alpha_{i,j}^{k}$for all nodes of each circuit and branches of the network base are equal to the values of $\alpha_{i,j}^{k}$the connecting node of the corresponding circuit. The latter will be known to us, since the connecting nodes are also nodes of the network base. Thus, the current distribution coefficients $\underline{C}_{ij}$for the branches $i$of the base and nodes $j$of the hanging circuit are equal to the values of the current distribution coefficients of the connecting node of the corresponding circuit.

The values $\alpha_{i,j}^{k}$for all nodes of the base and branches of any hanging branches and contours are zero. After the above "cuts" we need to find the weights of the spanning trees and calculate the values $\alpha_{i,j}^{k}$for the base of the graph, which already has 5 nodes and 6 branches, and the number of spanning trees 15, instead of 165 the trees of the original graph. That is, the volume of basic calculations has been reduced by 93.3%.

Let us describe the construction of the matrix of current distribution coefficients of the original network based on the current distribution matrix of the network base (Table 5).

Table 5. Current distribution matrix of the network basis

i \ j	1	2	3	4
1	-0.378-0.056i	-0.097+0.014i	-0.213-0.068i	-0.3-0.039i
2	-0.622+0.056i	-0.903-0.014i	-0.787+0.068i	-0.7+0.039i
3	0.376-0.038i	-0,059+0,009i	0,148+0,066i	0,257-0,021i
4	-0,246+0,018i	0,038-0,005i	-0,639+0,134i	-0,443+0,019i
5	0,246-0,018i	-0,038+0,005i	-0,361-0,134i	0,443-0,019i
6	0,246-0,018i	-0,038+0,005i	-0,361-0,134i	-0,557-0,019i

© The Authors, published by EDP Sciences. This is an open access article distributed under the terms of the Creative Commons Attribution License 4.0 (https://creativecommons.org/licenses/by/4.0/).

The current distribution matrix of the complete network will have a dimension of 11 × 15. It can be constructed in blocks (Table 6). The upper left block will be the same as that of the network base. The remaining elements of the first six rows will be as follows: at column 7, the same as that of column 5, since node 7 is connected to node 5. At columns 6 and 7, the same as that of column 1, and the remaining columns 8,9,10,11 are the same as that of column 4, since node 4 is a connecting node of the second "hanging" circuit with nodes 8,9,10,11. The seventh row, corresponding to the seventh branch, will be zero, with the exception of the 5th element, where the value -1 will be located.

Rows 8, 9, 10 also contain zero elements, with the exception of columns 6 and 7. The indicated 6 elements can be found as the coefficients of the current distribution matrix of the three-node network corresponding to contour 1 (Fig. 2). In this case, node 1, which connects this contour with the base, should be taken as the base vertex . Similarly, rows 11, 12, 13, 14, 15 will have nonzero elements in the columns corresponding to vertices 8, 9, 10 and 11. This block of 20 elements can be found as the current distribution matrix of the five-node network corresponding to contour 2 (Fig. 2). In this case, node 4, which connects this contour with the base, should be taken as the base vertex.

Table 6. Current distribution matrix of the original network

i \ j	1	2	3	4	5	6	7	8	9	10	11
1	-0,38-0,06i	-0,10 +0,01i	-0,21 - 0,07i	-0,3 - 0,04i	-0,21 - 0,07i	-0,38 - 0,06i	-0,38-0,06i	-0,3 - 0,04i	-0,3 - 0,04i	-0,3 - 0,04i	-0,3 - 0,04i
2	-0,62 +0,06i	-0,90-0,01i	-0,79 +0,07i	-0,7 +0,04i	-0,79 +0,07i	-0,62 +0,06i	-0,62 +0,06i	-0,7 +0,04i	-0,7 +0,04i	-0,7 +0,04i	-0,7 +0,04i
3	0,38 - 0,038i	-0,06 +0,01i	0,15 +0,07i	0,26 - 0,02i	0,15 +0,07i	0,38-0,04i	0,38 - 0,04i	0,26 - 0,02i	0,26 - 0,02i	0,26 - 0,02i	0,26 - 0,02i
4	-0,25 +0,02i	0,04 - 0,05i	-0,64 +0,13i	-0,44 +0,02i	-0,64 +0,13i	-0,25 +0,02i	-0,25 +0,02i	-0,44 +0,02i	-0,44 +0,02i	-0,44 +0,02i	-0,44 +0,02i
5	0,27 - 0,02i	-0,04 +0,01i	-0,36 - 0,13i	0,44 - 0,019i	-0,36 - 0,13i	0,25 - 0,02i	0,27 - 0,02i	0,44 - 0,02i	0,44-0,02i	0,44 - 0,02i	0,44 - 0,02i
6	0,25 - 0,02i	-0,04 +0,01i	-0,36 - 0,13i	-0,56 - 0,02i	-0,36 - 0,13i	0,25 - 0,08i	0,25 - 0,02i	-0,56 - 0,02i	-0,56-0,02i	-0,56 - 0,02i	-0,56 - 0,02i
7	0	0	0	0	-1	0	0	0	0	0	0
8	0	0	0	0	0	-0,07-0,00i	-0,65-0,00i	0	0	0	0
9	0	0	0	0	0	-0,93 +0,00i	-0,35 +0,00i	0	0	0	0
10	0	0	0	0	0	0,07 +0,00i	-0,35 +0,00i	0	0	0	0
11	0	0	0	0	0	0	0	-0,74 +0,12i	-0,51 +0,18i	-0,37 +0,06i	-0,14 +0,12i
12	0	0	0	0	0	0	0	0,26 +0,12i	-0,51 +0,18i	-0,37 +0,06i	-0,14 +0,12i
13	0	0	0	0	0	0	0	0,26 +0,12i	0,49 +0,18i	-0,37 +0,06i	-0,14 +0,12i
14	0	0	0	0	0	0	0	0,26 +0,12i	0,49 +0,18i	0,63 +0,06i	-0,14 +0,12i
15	0	0	0	0	0	0	0	-0.26-0.12i	-0.49-0.18i	-0.63 - 0.06i	-0.86 - 0.12i

Of course, for the considered example, the reduction of operations is not very significant for calculations using a computer. However, if we consider a 39- node test scheme, then the number of spanning trees is 421,380, while the graph base contains 22 nodes and 27 branches, and the number of spanning trees is 28,092, instead of 421,380 trees of the original graph. That is, the volume of basic calculations has also decreased by 93.3%, and this is already 393,288 spanning trees less.

Conclusions

The proposed method for identifying the graph base, hanging branches and contours significantly reduces the number of calculations when calculating the current distribution coefficients of the electric power system.

References

1. Christofides N. Graph Theory. Algorithmic Approach. Moscow: Mir, 1978, - 432 p.

© The Authors, published by EDP Sciences. This is an open access article distributed under the terms of the Creative Commons Attribution License 4.0 (https://creativecommons.org/licenses/by/4.0/).

2. Geraskin O.T. Topological analysis of current distribution coefficients in electrical networks. News of universities. Power engineering, 1967, No. 5, pp. 20–28.
3. David Avis, Komei Fukuda. Reverse search for enumeration. Discrete applied mathematics 65, pp . 21-46, 1996.
4. Oboskalov V. P. Application of probabilistic-statistical methods and graph theory in electric power engineering: a tutorial / V. P. Oboskalov , S. E. Kokin, I. L. Kirpikova ; Ministry of Education and Science of the Russian Federation, Ural Federal University named after the first President of Russia B. N. Yeltsin. – Ekaterinburg: Publishing House of Ural University, 2016. 271,[1] p. ISBN 978-5-321-02507-9.
5. Papkov B.V. Elements of graph theory in electric power engineering problems: a tutorial / B.V. Papkov, A.L. Kulikov. - Nizhny Novgorod: RANEPA, 2019. 175 p.: ill ., tab .: 20 cm .; ISBN 978-5-00036-323-7.
6. E.K. Tillaboev , Z.S. Makhmudov Possibilities of graph theory in electrical systems // Online Electrician: Electrical Power Engineering. New technologies , 2016.–URL: / articles.php?id =166
7. Barkan Ya.D. Automation of voltage regulation in distribution networks. Moscow: Energy, 1972. 120 With.
8. Akhmetbaev DS, Jandigulov AR, AkhmetbaevAD. Improving Algorithms for Searching for Trees of Directed Graphs of Electrical Power Systems. Power Technology and Engineering. V58, No2, July 2024, pp 349-355.
9. Akhmetbaev DS, Jandigulov AR, AkhmetbaevAD. New Approaches to the Topological Method of Analysis of Electrical Networks. International Multi-Conference on Industrial Engineering and Modern Technologies, FarEastCon 89933870, 156113 (2019).
10. . Akhmetbaev DS, Jandigulov AR, Bystrova SV Topological system method of formation of transformer transformation coefficients. E3S Web of Conferences 216, 01087 (2020).
11. Jandigulov AR, Rauan K, Bystrova SV. Simplified topological algorithm for forming a steady mode of electric power system. . E3S Web of Conferences 461, 01025 (2023).

© The Authors, published by EDP Sciences. This is an open access article distributed under the terms of the Creative Commons Attribution License 4.0 (https://creativecommons.org/licenses/by/4.0/).

Application of static thyristor compensators to increase the reliability of distribution networks

Dauren Akhmetbaev [1], *Arman* Akhmetbaev [2*]

[1] L. Gumilyov Eurasian National University, 010000, Astana, Republic of Kazakhstan
[2] Nizhny Novgorod State Technical University named after R.E. Alekseev, Russia

Abstract. Currently, the use of FACTS devices for voltage regulation and optimal control of electric power system modes is widely used. Controlled FACTS devices, based on power electronics, allow obtaining the desired characteristic of not only power transmission, but also a dedicated part of the electrical system. The paper presents the results of a study of the operating modes of a real distribution network, taking into account the installation of a static thyristor compensator (STC), in order to increase its operating reliability. The efficiency of distribution networks in and the area of existence of modes significantly depend on the installation location and the law of power control of compensating devices. To solve this problem, a systems approach based on graph theory methods is proposed. The conditions for stabilizing nodal voltages to reduce total active power losses in the electrical network are studied. The ranges of change in the required powers with the corresponding STC control laws are determined to ensure a minimum of total power loss in the network

Key words: distribution electric network, reactive power, voltage, topology, graph theory.

Introduction

Changes occurring in the electric networks of power systems show that the increase in reactive power consumption significantly exceeds the increase in active power consumption [1,2]. Increasing the output of reactive power by generators for the purpose of delivering it to the consumer is impractical. Therefore, the most appropriate is reactive power compensation (RPC) at the places of energy conversion, including at the places of connection of objects of electricity consumption (EE). Considering that in high-voltage electric networks, active power losses are the greatest compared to losses in lower-voltage networks (in percentage terms, significantly lower!). Therefore, the placement of RPC technical means in higher-voltage networks increases the energy efficiency of electric networks [3,4].

In recent years, static compensators (SCC) of reactive power, which are part of FACTS (Flexible Alternative Current Transmission Systems - flexible AC transmission systems) of technology, which includes a set of technical and information means of automatic control of parameters of power transmission lines. Shunt reactors and capacitors (hereinafter referred to as static capacitor banks - BSC or capacitor units - KU) are the main reactive elements of the STC. These elements can be used both individually and jointly. Shunt reactors can be regulated using power electronics (thyristor valves, etc.) or by changing the saturation of the magnetic core of the reactor, and capacitor banks can be either permanently connected or step-regulated using switching equipment or power electronics [5,6].

The choice of one or another type of STC and the locations of their installation are determined by the specific configuration of the electrical network, the balance of electrical load capacities, the equipment installed in the network and its capabilities for voltage regulation, as well as the technical and economic indicators of specific types of STC [7,8].

This study is devoted to improving the operational reliability of a 110 kV distribution electrical network using a STC.

Brief information about reactive power compensation

In industrially developed countries, as well as in the CIS countries, much attention is paid to reactive power compensation using SC. In particular, in France, Sweden, Germany, the capacity of the BSC is 35 percent of the active peak power, in the USA and Japan - about 70 percent. In some energy companies in the USA, the capacity of the installed compensating devices is 100 percent of the generator capacity [6-8].

The global practice of solving problems of the power factor control system shows the effective use for these purposes of the normative method, in accordance with which the values of the power factor (PF) are standardized on buses of certain voltage classes.

The optimal value of the reactive power factor $tg\varphi$depending on the nominal network voltage in the

* Corresponding author: abeked@mail.ru

© The Authors, published by EDP Sciences. This is an open access article distributed under the terms of the Creative Commons Attribution License 4.0 (https://creativecommons.org/licenses/by/4.0/).

maximum load mode in the USA, Japan and most European countries is maintained at the level of 0.2-0.4, which corresponds to $cos\varphi = 0{,}98 - 0{,}9$.In recent years, in many US power systems, distribution electric networks in the maximum load mode operate with $tg\varphi = 0$ [9,10].

If the voltage regulation is carried out by a static compensator, then the required reactive power in the node under consideration is determined by using the standard value of the power factor. In this case, the value $Q = P \cdot tg\varphi$determines the upper limit of the network voltage regulation range provided by the static compensator. If we assume that in power supply networks with a voltage of up to 220 kV inclusive, the determining factors are load power losses and EE, which are mainly inversely proportional to the square of the voltage level, then in such networks it is advantageous to maintain a higher voltage level (up to the technical limit). During hours of minimum loads in a node with a static compensator, reactive power can be generated in the network, which leads to a significant increase in voltage. Regulation of the static compensator power in the specified ranges leads to the provision of permissible voltage modes at the point of its connection [11].

It should be noted that the traditional method can determine the approximate ranges of the power of the control unit in the network nodes using the standard values of the power factor, which is insufficient under the conditions of using FACTS devices .

In this work, a new technology for determining the power of a compensating device is implemented based on a topological method from the standpoint of synthesis of voltage modes.

Calculations of the power of compensating devices.

Calculations of the powers of compensating devices can be obtained on the basis of a topological model of the formation of steady-state modes. After replacing the setting currents $\underline{\dot{J}}$with powers, the equation of the steady-state mode is written as follows [12,13]:

$$\dot{U} = U_0 + C^T Z_B C \overline{U_D^{-1}} \bar{S} \qquad (1)$$

where $C$is a rectangular complex matrix of current distribution coefficients; $Z_B$is a diagonal matrix of branch resistances; U_D – diagonal matrix of nodal conjugate voltages; $\bar{S}$– column vector of conjugate powers of nodal loads and generators, T –transposition sign

The complex power $i$of the -th node , taking into account the powers of the transverse branches, is determined as:

$$\dot{S}_i = P_i + jQ_i + (g_i + j \cdot (b_L - b_C))_i |U_i|^2 \qquad (2)$$

Where$j$ – imaginary unit , P_i, Q_i –active and reactive power $i$of the -th node; g_i, b_L, b_C –active, inductive, capacitive conductivity $i$of the -th node; U_i –modulus of complex voltage $i$of the -th node.

All components of the right-hand side of equation (1), except for the sought node voltages, are constant values for a given electrical network circuit. It is known that modern compensation devices equipped with controlled power electronics allow stabilizing the voltage of an arbitrary node regardless of the electrical system mode. Consequently, the desired voltage values on the buses of an arbitrary network node can be considered a given value. Then, the problem is reduced to determining the power of compensating devices, for example, STC, installed to provide the desired node voltages.

The values of the required powers of the compensating devices are determined by the equation [14,15] :

$$\dot{U}_ж = U_0 + C_d^T Z_B C_d \overline{U_D^{-1}} (\bar{S} - j\overline{Q_k}) \qquad (3)$$

where is $\dot{U}_ж$ –the complex value of voltage, and at certain nodes the voltage modulus must take the desired value; Q_k –reactive power of compensating devices at the same nodes (3).

$$\begin{cases} |U_{жk}| \cos\delta_k = U_0 - \sum |Z_{kj}| \dfrac{|S_j| cos(\delta_j - \phi_j + \psi_{kj}) + Q_{Kj} sin(\delta_j + \psi_{kj})}{|U_{жj}|} \\ |U_{жk}| \sin\delta_k - \sum |Z_{kj}| \dfrac{|S_j| sin(\delta_j - \phi_j + \psi_{kj}) + Q_{Kj} cos(\delta_j + \psi_{kj})}{|U_{жj}|} \end{cases}$$

where $Z = C^T Z_B C$ – symmetric matrix of nodal resistances, δ_k –phase of nodal voltage,
φ_k –phase of conjugate complex of nodal load,
$\psi_{k,j} = \arg(Z_{k,j})$ –phase of element of matrix of nodal resistances.

Improving the operational reliability and energy efficiency of a real 110 kV distribution network using the FACTS device

The initial data of the system under study are presented in Table 1, and the calculation scheme is shown in Fig. 1.

In the maximum load mode, all consumers have a standard power factor $cos\varphi = 0{,}92 \div 0{,}93$. Calculations were carried out based on the topological model of the formation of steady-state mode parameters (1). The study of the maximum mode without the use of the control unit showed that at the voltage of the power supply center (PC) $U_0 = 121$кВ, in all distribution substations, standard values are provided on the low-voltage buses, taking into account the action of the OLTC installed on the transformers. The total losses of active power are equal to $\Delta P_\Sigma = 6{,}431$ МВт,which is 9.9% of the active power entering the network, $P_{act} = 64{,}978$ МВт,which are presented in Table 2.

Calculations were made with a decrease in the voltage of the central control unit in order to determine the permissible modes for the voltage of the distribution network. The calculations established that the permissible mode for the voltage in the network is ensured with a decrease in the voltage of the base node to 117 kV inclusive.

© The Authors, published by EDP Sciences. This is an open access article distributed under the terms of the Creative Commons Attribution License 4.0 (https://creativecommons.org/licenses/by/4.0/).

Table 1. Initial data of the network calculation scheme

Node data					Branch data					
No.	Load powers				No.	start	con	R, Ohm	X, Ohm	b , mkSm (capacity+, ind-)
	Min. mode		Max mode							
	P, MW	Q,MVar	P, MW	Q,MVar						
1	0.02	0,1	0.02	0,1	1	B	1	4.95	10.5	134.9
2	0.02	0,1	0.02	0,1	2	1	2	7.92	16.8	215.9
3	0.042	0.224	0.042	0.224	3	2	3	1,485	3.15	40.48
4	0.02	0,1	0.02	0,1	4	3	4	2.59	4.44	13,785
5	0.02	0,1	0.02	0,1	5	4	5	11,118	24,211	75.15
6	0.028	0.14	0.028	0.14	6	5	6	6,225	10,675	33,135
7	0.038	0.22	0.038	0.22	7	6	7	7.47	12.81	39,765
8	0.028	0.151	0.028	0.151	8	7	8	14.94	25.62	79.55
9	0,011	0.076	0,011	0.076	9	8	9	18.36	26.04	78.35
10	2,419	1,814	6,048	4,536	10	3	9	18.36	26.04	78.35
11	2,419	1,814	6,048	4,536	11	1	10	8	110.2	0
12	3,072	2,304	7.68	11.52	12	2	11	8	110.2	0
13	2,419	1,814	6,048	4,536	13	3	12	2,2	43.24	0
14	2,419	1,814	6,048	4,536	14	4	13	8	110.2	0
15	3.84	2.88	9.6	7.2	15	5	14	8	110.2	0
16	0	0	0	0	16	6	15	3.95	69.45	0
17	1.92	1.44	4.8	3.6	17	7	16	2,515	71,085	0
18	1.92	1.44	4.8	3.6	18	16	17	2,515	0	0
19	0	0	0	0	19	16	18	2.515	41.33	0
20	1.21	0.907	3,024	2,268	20	8	19	4.83	112.83	0
21	1.21	0.907	3,024	2,268	21	19	20	4.83	0	0
22	0.48	0.36	1,2	0.9	22	19	21	4.83	65.6	0
B	0	0	0	0	23	9	22	42.6	508.2	0

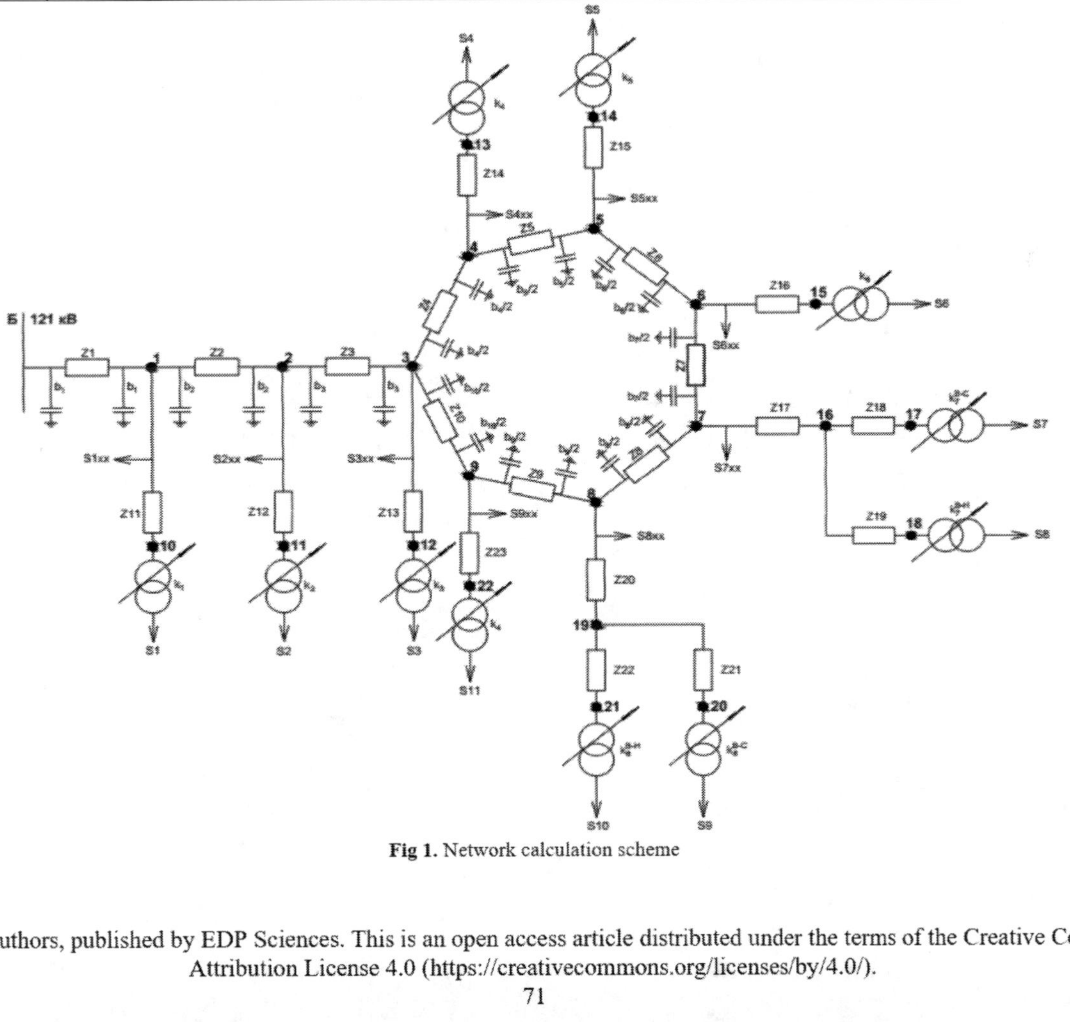

Fig 1. Network calculation scheme

© The Authors, published by EDP Sciences. This is an open access article distributed under the terms of the Creative Commons Attribution License 4.0 (https://creativecommons.org/licenses/by/4.0/).

Table 2. Changes in power loss depending on power centre voltage

U_0	P_{act}	P_{load}	ΔP_{Σ}	ΔP_{Σ}
kV	MW	MW	MW	%
121	64,98	58,55	6,43	9,90
120	65,73	58,55	7,18	10,90
119	66,51	58,55	7,96	11,90
118	67,32	58,55	8,77	13,00
117	68,16	58,55	9,61	14,10
116	69,03	58,55	10,48	15,20
115	69,94	58,55	11,4	16,30

Reducing the voltage of the CPU leads to an increase in the total losses of active power, which is not advisable. Increased voltage in the network creates a condition for the occurrence of additional losses of active power due to the transfer of reactive power in the network. Therefore, to reduce the losses of active power in the network while ensuring operational reliability, it is envisaged to use FACTS devices, which is of great importance both theoretically and practically.

For the experimental analysis during the installation of the STC, the third and seventh nodes of the closed network were selected.

When installing the control unit in the third node with stabilization of the voltage level in the region of permissible voltage modes, the reactive power of the control unit and the total losses of active power in the network were determined.

Table 3. Changes in power losses during compensation in the third node

U_3	P_{act}	P_{load}	ΔP_{Σ}	ΔP_{Σ}	Q_c
kV	MW	MW	MW	%	MVar
120	70,376	58,547	5,911	10,10	67,78
119	70,245	58,547	5,818	9,94	63,99
118	70,148	58,547	5,758	9,83	60,29
117	**70,086**	**58,547**	**5,732**	**9,79**	**56,66**
116	70,059	58,547	5,741	9,81	53,12
115	70,067	58,547	5,784	9,88	49,65
114	70,11	58,547	5,86	10,01	46,27
113	70,188	58,547	5,971	10,20	42,97

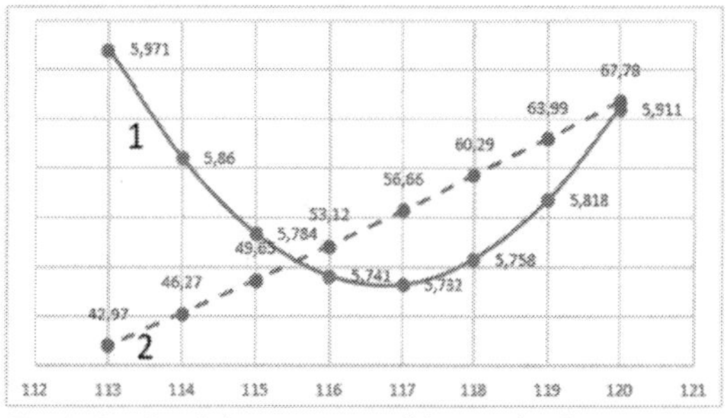

Fig. 2. Graphs of changes: 1 – total losses of active power, 2 – power of the compensating device.

The study showed that the appropriate mode for minimum active power losses in the network is ensured by stabilizing the voltage of the third node at 117 kV, with the power of the control unit equal to 56.661 MVAr . In this case, the total active power losses in the network are reduced and become equal to 5.732 MW, which is 8.92% of the active power entering the network (64.279 MW).

Table 4. Changes in power losses during compensation in the seventh node

U_7	P_{act}	P_{load}	ΔP_{Σ}	ΔP_{Σ}	Q_c
kV	MW	MW	MW	%	MVar
114	69,834	58,547	5,659	8,10	40,41
113	69,772	58,547	5,611	8,04	38,56
112	69,729	58,547	5,581	8,00	36,75
111	**69,704**	**58,547**	**5,57**	**7,99**	**34,98**
110	69,699	58,547	5,578	8,00	33,25
109	69,714	58,547	5,605	8,04	31,55
108	69,746	58,547	5,65	8,10	29,9

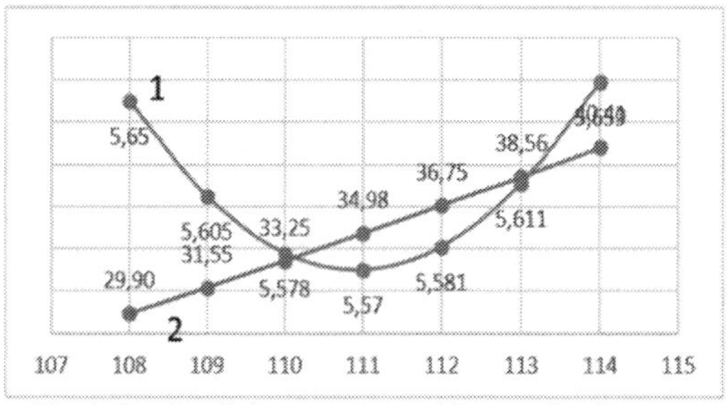

Fig. 3. Graphs of changes: 1 – total losses of active power, 2 – power of the compensating device.

When installing the control unit in the seventh node, the optimal mode occurs when the voltage stabilizes at 111 kV, with the control unit power of 34.978 MVAr . The total losses of active power in the network become equal to 5.57 MW, which will be 8.68% of the active power entering the network (64.117 MW).

Of the two options considered, preference is given to the second option, when the control unit is installed in node 7, since the power of the control unit is significantly lower and provides the greatest reduction in total active power losses in the network.

To further reduce active power losses in the distribution network, the option of installing the control unit simultaneously in the third and seventh nodes was considered.

When the voltage is stabilized at 116 kV, simultaneously in two nodes (3 and 5?), the active power losses in the network are reduced to 5.235 MW, which is 8.21% of the active power of the CPU supplied to the network (63.782 MW).

© The Authors, published by EDP Sciences. This is an open access article distributed under the terms of the Creative Commons Attribution License 4.0 (https://creativecommons.org/licenses/by/4.0/).

Table 5. Changes in power losses with compensation in the third and seventh nodes

U_7	P_{act}	P_{load}	ΔP_{Σ}	ΔP_{Σ}	$Q_{c,3}$	$Q_{c,7}$
kV	MW	MW	MW	%	MVar	MVar
119	63,92	58,55	5,37	8,41	40,41	35,01
118	63,84	58,55	5,29	8.29	38,56	35.13
117	63,79	58,55	5,25	8,23	36,75	35,25
116	**63,78**	**58,55**	**5,23**	**8,21**	**34,98**	35,37
115	63,80	58,55	5,25	8,23	33,25	35,49
114	63,85	58,55	5,31	8,31	31,55	35.62
113	63,94	58,55	5,39	8,43	29,9	35,75
112	64,05	58,55	5,51	8,60	1,53	35,88

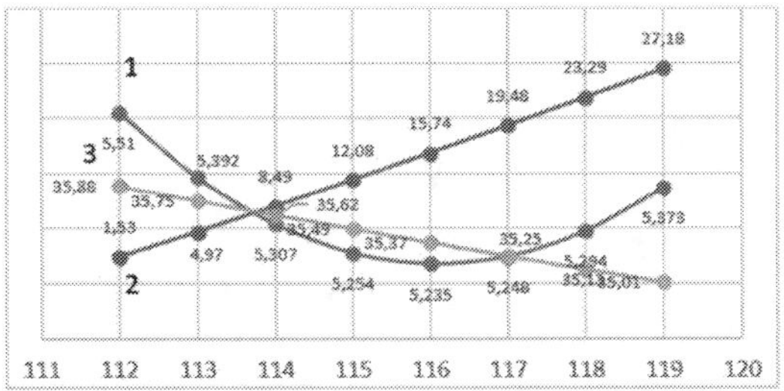

Fig. 4. Graphs of changes: 1 – total losses of active power, 2 – power of the compensating device in node 3, 3 – power of the compensating device in node 7.

From the analysis of the calculations performed for the maximum mode of the network under consideration, it can be stated that the use of a static compensator as a device for stabilizing the desired voltage level of the selected node significantly increases the operating reliability and reduces the total losses of active power while ensuring the quality of voltage on the busbars of distribution substations.

For the minimum load mode, an increase in the total active power losses in the network is also typical with a decrease in the CPU voltage. In terms of reducing the total active power losses, it is necessary to stabilize the increased (technically permissible) voltage on the CPU buses. The minimum value of the total active power losses, equal to 0.902 MW, corresponds to a CPU voltage of 121 kV. When the CPU voltage decreases, for example, to 117 kV, the total active power losses increase to 1.001 MW.

Table 6. Changes in power losses during compensation in the seventh node

U_7	P_{act}	P_{load}	ΔP_{Σ}	ΔP_{Σ}	Q_c
kV	MW	MW	MW	%	MVar
120	24,35	23,55	0,8	3,28	14,06
119,5	24,33	23,55	0,77	3,18	13,09
119	24,31	23,55	0,75	3,10	12,12
118,5	24,29	23,55	0,74	3,03	11,17
118	24,28	23,55	0,73	2,99	10,23
117,5	24,27	23,55	0,72	2,95	9,3
117	24,27	23,55	0,71	2,95	8,37
116,5	**24,27**	**23,55**	**0,72**	**2,95**	**7,46**
116	24,28	23,55	0,723	2,97	6,55

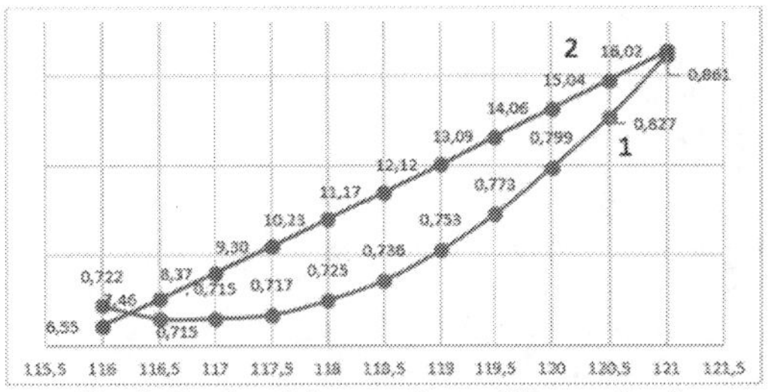

Fig. 5. Graphs of changes: 1 – total losses of active power, 2 – power of the compensating device.

The conducted analysis of the distribution network modes using the STC showed that the lowest value of active power losses is 0.715 MW, which corresponds to 2.946%, and is achieved by installing the control unit in node 7. At the same time, the voltage of this same node is maintained at 117 kV , with the voltage of the power center being 121 kV. When setting the voltage of the power center to 115 kV, the minimum value of power losses in the network is 0.833 MW. Therefore, in the minimum load mode, preference is also given to a higher CPU voltage level.

Conclusions

The use of the FACTS device for mode control significantly reduces technological losses of active power with an increase in the reliability of the distribution networks.

References

1. Hingorani Narain G. Understanding FACTS: concepts and technology of flexible AC transmission systems / Narain G. Hingorani, Laszlo Gyugyi. - IEEE Press - Wiley, 2000. - 428 p.
2. Arkhipov N.K. Voltage regime in electrical distribution networks. Moscow: VZEI Publishing House, 1964.-131 p.
3. Barkan Ya.D. Automation of voltage regulation in distribution networks. Moscow: Energy, 1972.-120 p.
4. Guidelines for reactive power compensation in distribution networks. VNIIE.M.: Energy, 1974.-73 p.
5. Zhelezko Yu.S. Determination of power and installation locations of local voltage regulation // Electric stations. Moscow. 1972.- No. 11.- P. 34-36.
6. Tukhvatullin M.M., Ivekeev V.S., Lozhkin I.A., Urmanova F.F. Analysis of modern FACTS devices used to improve the efficiency of electric power systems in Russia // Electrical Engineering. 2015. No. 3(28). P. 41-46.
7. Zhelezko Yu.S. Reactive power compensation and improvement of electric power quality. - M.: Energoatomizdat, 1985.
8. Neshtaev V.B. Optimal selection of reactive power sources in electric power distribution systems: author's abstract of dissertation ... candidate of technical sciences . Krasnoyarsk. 2012. 21 p.
9. Bryancev A.M., Bazylev B.I., Lurye A.I., Smolovik S.V. Regulation and stabilization of high-voltage electric network voltage by controlled inductive-

© The Authors, published by EDP Sciences. This is an open access article distributed under the terms of the Creative Commons Attribution License 4.0 (https://creativecommons.org/licenses/by/4.0/).

capacitive reactive power sources / "Electricity", No. 10, 2012. P.15-21.
10. Bryantsev A.M., Bryantsev M.A., Bazylev B.I., Dyagileva S.V., Karymov R. R. , Lurye A.I., Makletsova E.E., Negryshev A.A., Smolovik S.V. Status and prospects of application in electrical systems of the Russian Federation and the CIS of adjustable reactive power sources (IRP) with magnetization-controlled shunt reactors (USHR) and capacitor banks (BSC). Energy Expert No. 2 (19) - 2010. 88-93.
11. Bryantsev M.A., Bazylev B.I. Dyagileva S.V., Negryshev A.A., Karymov P . P . Automatic systems for reactive power compensation and voltage stabilization of electric networks based on USHR and BSC. Moscow. TRAVEK, 2010.
12. Akhmetbaev DS, Jandigulov AR, AkhmetbaevAD. Improving Algorithms for Searching for Trees of Directed Graphs of Electrical Power Systems. Power Technology and Engineering. V58, No2, July 2024, pp 349-355.
13. Akhmetbaev DS, Jandigulov AR, AkhmetbaevAD Topological algorithm for forming nodal stresses of complex networks energy systems. E3S Web of Conferences 139, 01066 (2019).
14. Akhmetbaev DS, Jandigulov AR, Bystrova SV Topological system method of formation of transformer transformation coefficients. E3S Web of Conferences 216, 01087 (2020).
15. Jandigulov AR, Rauan K, Bystrova SV. Simplified topological algorithm for forming a steady mode of electric power system. . E3S Web of Conferences 461, 01025 (2023).
16. Akhmetbaev DS, Jandigulov AR, AkhmetbaevAD. New Approaches to the Topological Method of Analysis of Electrical Networks. International Multi-Conference on Industrial Engineering and Modern Technologies, FarEastCon 89933870, 156113 (2019).

© The Authors, published by EDP Sciences. This is an open access article distributed under the terms of the Creative Commons Attribution License 4.0 (https://creativecommons.org/licenses/by/4.0/).

Modification of Control System Based on Virtual Synchronous Generator with Feedforward Control for Power Converters of Distributed Generation

Alisher Askarov[1*], *Ruslan* Ufa[1], *Aleksey* Suvorov[1], and *Anton* Kievets[2]

[1]National Research Tomsk Polytechnic University, School of Energy & Power Engineering, 634050 Tomsk, Russian Federation
[2]Gazprom Transgaz Yugorsk LLC, Department of Communications, 628260 Yugorsk, Russian Federation

Abstract. An important development trend for modern power systems is the application of control algorithms for power converters based on a virtual synchronous generator (VSG), which can mimic the characteristics of the conventional synchronous generation and support the grid-forming operation. This paper proposes a modified current-controlled VSG structure (CC-VSG) with a feedforward controller consisting of a simple proportional loop, which is added to the VSG active power control loop, to improve the damping properties of CC-VSG, reduce the model order of this control system and simplify the approach to its tuning. In order to qualitatively and quantitatively analyse the effectiveness of the proposed solution, the paper presents a comparison of dynamic responses when using control algorithms based on CC-VSG or the traditional VSG structures, including several of their modifications. For this purpose, time-domain simulation of the considered system was performed, demonstrating the absence of impact of the feedforward control on the inertial response of CC-VSG, which can occur when using other VSG structures or local negative feedbacks within the CC-VSG itself.

1 Introduction

Currently, one of the promising directions in the development of approaches to controlling power converters of distributed generation [1–3] is the development of control systems to ensure the grid-forming operation [4,5]. The latter is fundamentally different from the conventional approach, in which the grid inverter is a grid-following [6], i.e., often does not participate in any way in the regulation of the power system. Within the framework of this direction, the development of control algorithms based on the virtual synchronous generator (VSG) [7,8] is of the greatest interest. This concept allows imitating the main properties of conventional synchronous generation, taking into account the possibility of flexible change of control loops and their tuning to achieve the necessary regulation quality and stability margins of the power system [9–11].

Nevertheless, attempts to reproduce the behavior of a synchronous generator using VSG algorithms lead to the fact that along with the useful qualities of conventional generation there are also its disadvantages. In particular, one of the most difficult problems is the appearance of low-frequency oscillations during VSG operation [12–14]. Several approaches to improve the quality of oscillation damping using VSG have been proposed in the literature. For example, one way is to increase the value of the damping coefficient in the swing equation of the virtual rotor [15]. However, this leads to a regular change in frequency droop and non-uniform distribution of active power between inverters in steady state, which is undesirable and requires additional coordination of the tuning of the control loops of several inverters. Another approach is to use a phase-locked loop controller (PLL) within the swing equation of the virtual rotor to create an additional damping effect [16], where this controller is used to measure the frequency of the grid voltage. In order to reduce the negative effect of PLL on the VSG properties in such a control loop, a lead-lag loop can be used [17]. Instead of PLL, it is also possible to use the frequency of the virtual rotor obtained via a low-pass filter (LPF) [18], but this may cause undesirable overshoot and requires adequate tuning of the LPF itself. Taking into account the above mentioned, the improvement of the damping properties of the VSG is usually associated with the introduction of additional control loops, which lead to a corresponding complication of the algorithm structure, as well as an increase in its order and the appearance of additional roots of the characteristic equation [19]. Thus, despite the positive effects from modernization of the VSG structure, there is a problem of providing the necessary degree of freedom of the considered system due to the introduction of an additional tuning parameter.

Previously, the authors proposed an original structure of the control system based on the current-controlled VSG (CC-VSG) [20], which has a number of advantages in comparison with the conventional

* Corresponding author: aba7@tpu.ru

© The Authors, published by EDP Sciences. This is an open access article distributed under the terms of the Creative Commons Attribution License 4.0 (https://creativecommons.org/licenses/by/4.0/).

voltage-controlled VSG. At the same time, the considered CC-VSG model is a complex dynamic system of the third order, which causes certain difficulties with the study of this model, as well as, most importantly, with its adequate tuning. At the same time, the issue of tuning can be solved, for example, by using the Vyshnegradskii's stability diagram [21], but it remains difficult to realize the necessary level of damping at the expense of the initial CC-VSG structure in various operating conditions of the power converter.

In this connection, this paper proposes the use of the feedforward method to improve the damping properties of CC-VSG. This feedforward controller is a proportional loop, which is added to the VSG active power control loop. Consequently, the proposed modification of CC-VSG does not imply the introduction of additional differential components, which eliminates the use of virtual damper winding and reduces the order of the CC-VSG model from the original third to the second. The proposed controller has one tuning parameter – the gain of the proportional loop, the selection of which can be made on the basis of the known properties of second-order systems. As a result, due to the tuning of the feedforward controller it becomes possible to achieve the desired dynamic response of CC-VSG, as will be shown below, without adversely affecting any characteristics of the transient process with preserving the advantages of CC-VSG.

2 Analysis of structure of control system based on CC-VSG and its considered modifications

Fig. 1 shows a diagram of an electrical network with a power converter, which has a control system based on the CC-VSG algorithm considered in this paper.
In order to exclude a detailed description of the specific principles of practical realization of the primary energy source, which can be, for example, an energy storage device, and the features of functioning introduced in this case, the paper assumes that an ideal DC voltage source U_{DC} is connected in the DC link of the converter. Such an approach is widely used in the development of control algorithms for power converters and allows to obtain generalized results without reference to a specific generation source. The external power system is represented as an ideal AC voltage source U_{grid}.

The presented structure of the converter with CC-VSG consists of a power part and a control part. The power part includes a DC link, an inverter and an output filter. The control part is based on the CC-VSG algorithm, which reproduces the dynamics of a virtual synchronous machine (block "VSG Dynamics" in Fig. 1). The equations used in the considered CC-VSG algorithm are presented and described in detail in [22,23]. Fig. 1 also separately shows the VSG active power control loop taking into account the virtual governor with the droop coefficient k_{ω}.

The feature of this control system based on CC-VSG is the use of two control loops to form the necessary power reference values for the power converter. The first loop determines the value of the main output active and reactive power references (P_{ref} and Q_{ref}), which can be set either from the local level control system or from the plant level control system. The second loop takes into account the electromagnetic and electromechanical equations of VSG, resulting in the formation of the values of P_{VSG} and Q_{VSG}. The resulting P^*_{ref} and Q^*_{ref} values are formed by summing the power values from the two loops and are then used to determine the current reference values for the inner current control loop (ICCL).

When considering the initial CC-VSG structure within the active power control loop, all the elements shown by solid lines in Fig. 1 are taken into account. At the same time, the damping properties for the initial structure are determined by the presence of the VSG active power feedback P_{VSG} (highlighted separately by green solid lines in Fig. 1). When calculating the P_{VSG}, it is taken into account that the initial structure of the CC-VSG contains a virtual damper winding with the parameters R_{1q} and L_{1q}, which in turn forms the damping component of the VSG.

This paper considers several options for providing the damping properties of the CC-VSG while eliminating the virtual damper winding loop in order to reduce the resulting model order and reduce the number

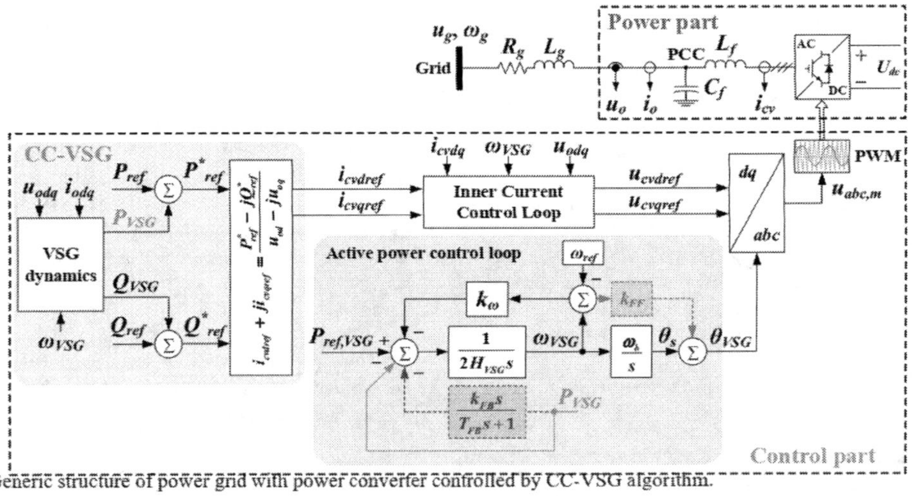

Fig. 1. Generic structure of power grid with power converter controlled by CC-VSG algorithm.

© The Authors, published by EDP Sciences. This is an open access article distributed under the terms of the Creative Commons Attribution License 4.0 (https://creativecommons.org/licenses/by/4.0/).

of tunable parameters. The first option is to use additional negative feedback, which is shown by the blue dashed line in Fig. 1. The second is to consider the proposed feedforward controller, which is highlighted by the red dashed line in Fig. 1. Separately, it is worth noting that the basis of the CC-VSG active power control loop, highlighted by the black and green solid lines, remains unchanged. There is only a change in the definition of the active power P_{VSG}, in the equations of which the components of the virtual damper winding are excluded.

3 Application of feedforward control to improve damping properties of CC-VSG

This paper proposes a new method of effective damping of low-frequency oscillations, which allows to preserve the advantages of CC-VSG, presented in detail in [22], and does not lead to the appearance of undesirable negative properties of the model. At the same time, due to this modification, the order of the CC-VSG model is reduced and it becomes possible to form the necessary dynamic response under various kinds of disturbances in the power grid. The proposed control loop is based on the feedforward proportional loop with a gain k_{FF}, the input of which is fed by the VSG frequency deviation, and the output signal is added to the VSG output phase angle. At the same time, the major changes affect only the VSG active power control loop, in which an additional controller is integrated. The structure of CC-VSG with the new controller is shown in Fig. 1 (the CC-VSG structure is highlighted in black, and the controller itself is highlighted in red).

To provide an opportunity for further theoretical analysis of the modified CC-VSG structure, an appropriate state-space model was formed, taking into account the change of equations describing the dynamics of the virtual synchronous machine operation (1)–(4):

$$\frac{d\omega_{VSG}}{dt} = \frac{1}{2H_{VSG}}\left[P_{ref,VSG} - P_{VSG} - k_{\omega}\left(\omega_{VSG} - 1\right)\right] \qquad (1)$$

$$\frac{d\theta_s}{dt} = \omega_b \omega_{VSG} \qquad (2)$$

$$\theta_{VSG} = \theta_s + k_{FF}\left(\omega_{VSG} - 1\right) \qquad (3)$$

$$i_{q,VSG} = -\frac{\psi_q}{L_v} \qquad (4)$$

The description of the state vector **x** also changes (5):

$$\Delta\mathbf{x} = \begin{bmatrix} \Delta u_{1d} & \Delta u_{1q} & \Delta i_{1d} & \Delta i_{1q} & \Delta\xi_d & \Delta\xi_q & \Delta i_{2d} & \Delta i_{2q} & \cdots \\ & & & \Delta\psi_d & \Delta\psi_q & \Delta\psi_{fd} & \Delta\theta_s & \Delta\omega_{VSG} \end{bmatrix}^{T} \qquad (5)$$

To prove the effectiveness and correctness of the proposed controller implementation based on the feedforward control within the CC-VSG structure, the participation factors were analyzed using a state-space model. From the obtained analysis, the conjugate low-frequency roots $\lambda_{12,13}$ (Table 1) are associated with five state variables: $\Delta\gamma_q$, $\Delta\psi_d$, $\Delta\psi_{fd}$, $\Delta\theta_s$, and $\Delta\omega_{VSG}$. Two state variables $\Delta\psi_d$ and $\Delta\psi_{fd}$ are related to the reactive power control loop of the VSG, the other two $\Delta\theta_s$ and $\Delta\omega_{VSG}$ are related directly to the active power control loop of the VSG, as shown in Fig. 1, and the state variable $\Delta\gamma_q$ reflects the operation of the ICCL on the transverse component q, which is responsible for the active power generation of the converter under the axis direction adopted in the control system. Thus, the increase in the damping ratio for $\lambda_{12,13}$ can be achieved by affecting the above control loops of the VSG. At the same time, as it will be shown further, the adjustment only on the active power control loop is enough to provide the necessary damping properties in the model of the modified CC-VSG, which does not lead to a significant complication of the control system structure and its further tuning.

Table 1. Analysis of eigenvalues of the modified CC-VSG structure.

Eigenvalue		Sensitivity of λ to k_{FF}	Mainly related state variable
λ_{12}	−10.6+*j*10.6	−0.321−*j*1.037	$\Delta\gamma_q$, $\Delta\psi_d$, $\Delta\psi_{fd}$, $\Delta\theta_s$, $\Delta\omega_{VSG}$
λ_{13}	−10.6−*j*10.6	−0.321+*j*1.037	

4 Simulation results and comparison of dynamic responses of different VSG structures

In this section, the results of time-domain simulation of the developed modified structure of CC-VSG with feedforward controller in the PSCAD software are presented. The inertial response of different VSG algorithms was evaluated under different damping factors ζ: (1) conventional voltage-controlled VSG structure (VC-VSG) [18], (2) VC-VSG with active power reference feedforward controller (RFF-VSG) [24], (3) VC-VSG with phase feedforward damping controller (PFD-VSG) [25], (4) CC-VSG with negative feedback (CC-VSG with FB) [26], (5) CC-VSG with proposed feedforward controller (CC-VSG with FF). To form the same values of the damping factor ζ and natural frequency ω_n in the considered VSG structures, the corresponding tuning parameters were set. As a disturbance, the grid frequency decrease from 60 to 59.9 Hz was reproduced. The resulting waveforms of frequency and output active power are presented in Fig. 2. The results of the rate of change of frequency (RoCoF) estimation are summarized in Table 2. The RoCoF values were calculated according to (6) [27]:

$$\text{RoCoF} = \frac{f_0 \cdot \Delta f_{3T}}{3T} \qquad (6)$$

where f_0 is the nominal grid frequency, Hz, T is the frequency period, s, Δf_{3T} is the value of Δf three periods after the disturbance, Hz.

© The Authors, published by EDP Sciences. This is an open access article distributed under the terms of the Creative Commons Attribution License 4.0 (https://creativecommons.org/licenses/by/4.0/).

Table 2. Estimation of the RoCoF with changing ζ for different VSG structures.

Model	RoCoF, Hz/s			
	$\zeta = 0.25$	$\zeta = 0.5$	$\zeta = 0.707$	$\zeta = 1$
VC-VSG	0.32	0.3	0.27	0.25
RFF-VSG	0.25	0.24	0.24	0.24
PFD-VSG	0.25	0.25	0.25	0.25
CC-VSG with FB	0.45	0.77	0.98	1.22
CC-VSG with FF	0.24	0.24	0.24	0.24

For all VSG models (Figs. 2a–2e), the maximum frequency deviation and overshoot of the converter output active power simultaneously decrease with increasing damping factor ζ, and the damping properties of VSG increase. At the same time, the change of ζ has a significant effect on the frequency droop for the conventional VC-VSG structure (Fig. 2a). The latter can be clearly seen from the steady-state value of active power after the disturbance (P_{out} = 0.08 pu at ζ = 0.25 and P_{out} = 0.32 pu at ζ = 1). A similar situation occurs in the case of RFF-VSG due to the use of the conventional damping coefficient in the swing equation (Fig. 2b). For the PFD-VSG (Fig. 2c) and both CC-VSG structures (Figs. 2d and 2e), the mentioned problem of the impact of the damping factor on the frequency droop is absent. However, the obtained RoCoF values (Table 2) show that the application of negative feedback within the CC-VSG structure leads to a significant increase in the RoCoF by almost 2.5 times with an increase in the damping factor ζ (Fig. 2d).

The addition of the feedforward controller to any VSG structure (RFF-VSG, PFD-VSG and CC-VSG with FF) does not affect the RoCoF, which remains equal to 0.24-0.25 Hz/s for all considered cases (Fig. 2b, 2c and 2e). The latter allows to conclude that there is no impact of the damping effect on the inertial response of the VSG with the feedforward controller, including the proposed modified CC-VSG structure, which proves the wide flexibility in model tuning to obtain any desired dynamic response.

It should also be noted that for the RFF-VSG, a larger value of active power overshoot is observed to provide the same frequency change profile compared to the developed CC-VSG with FF (Fig. 2c and 2e, respectively), which results in a larger value of energy consumption from the primary source.

5 Conclusion

In this paper, the authors propose a modification of the CC-VSG model by using the feedforward control method within the VSG active power control loop. This control loop is a simple proportional loop with a specified gain, included in parallel with the elements of the direct circuit of the active power control loop, and has a direct effect on the internal angle of the virtual synchronous machine. At the same time, the introduction of the feedforward controller into the CC-VSG model was carried out instead of the virtual damper winding, the equations of which were excluded from the virtual synchronous machine. The presented simulation results show that the developed modified CC-VSG structure allows to ensure the stability of the power converter operation under different grid conditions, which is necessary for the reliable operation of power systems. Also, as follows from the comparison of the dynamic response of different VSG structures, feedforward control in any form does not worsen the inertial response, but depending on the specific implementation of the feedforward controller and the VSG structure, the impact on the dynamic response in general will be different. The developed CC-VSG structure with FF simultaneously has all the advantages inherent in different VSG structures with differing damping approaches.

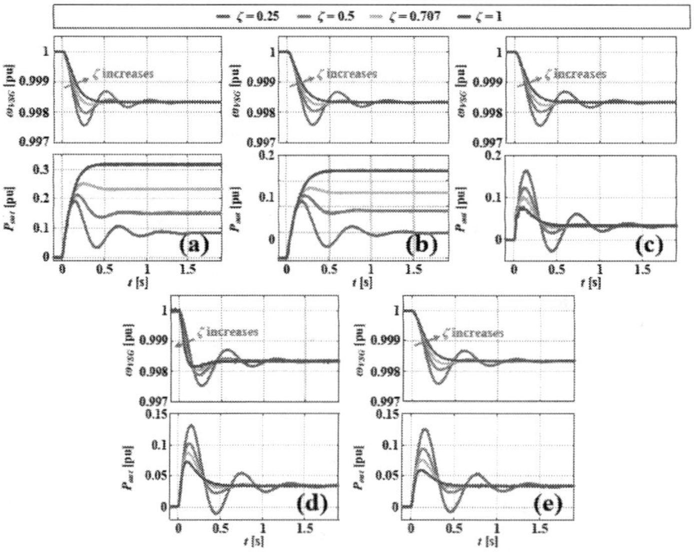

Fig. 2. Dynamic responses of the power converter under decreasing grid frequency and different ζ: (a) VC-VSG, (b) RFF-VSG, (c) PFD-VSG, (d) CC-VSG with feedback loop, (e) modified CC-VSG.

© The Authors, published by EDP Sciences. This is an open access article distributed under the terms of the Creative Commons Attribution License 4.0 (https://creativecommons.org/licenses/by/4.0/).

The reported study was funded by the Russian Science Foundation, project number 24-29-00004.

References

1. O.V. Shepovalova, Y.D. Arbuzov, V.M. Evdokimov, P.V. Ilyushin, K.V. Suslov, Energies **16**, 3 (2023)
2. R. Ufa, A. Gusev, A.A.Z. Diab, A. Suvorov, N. Ruban, M. Andreev, A. Askarov, V. Rudnik, O. Abdalla, Z.M. Ali, A. Ibrahim, R. Aboelsaud, Energy Rep. **6** (2020)
3. R. Ufa, M. Andreev, N. Ruban, A. Suvorov, A. Gusev, I. Razzhivin, A. Askarov, Y. Bay, A. Kievets, N. Lozinova, O. Suslova, Electr. Eng. **101**, 1 (2019)
4. S. Anttila, J.S. Döhler, J.G. Oliveira, C. Boström, Energies **15**, 15 (2022)
5. A. Suvorov, A. Askarov, A. Kievets, Electr. Eng. **105** (2023)
6. P. Ilyushin, V. Volnyi, K. Suslov, S. Filippov, Energies **16**, 7 (2023)
7. M. Shadoul, R. Ahshan, R.S. AlAbri, A. Al-Badi, M. Albadi, M. Jamil, Energies **15**, 22 (2022)
8. A. Suvorov, A. Askarov A, A. Kievets, V. Rudnik, Electr. Power Syst. Res. **209** (2022)
9. X. Hou, H. Han, C. Zhong, W. Yuan, M. Yi, Y. Chen, *Improvement of transient stability in inverter-based AC microgrid via adaptive virtual inertia*, in Proceedings of the IEEE Energy Conversion Congress and Exposition, ECCE, 18-22 September 2016, Milwaukee, WI, USA (2016)
10. A. Suvorov, A. Gusev, M. Andreev, A. Askarov, Int. Trans. Electr. Energy Syst. **30**, 4 (2020)
11. A. Suvorov, A. Gusev, M. Andreev, A. Askarov, Electr. Eng. **101**, 2 (2019)
12. M. Chen, D. Zhou, F. Blaabjerg, IEEE Trans. Power Electron. **36**, 8 (2021)
13. A. Suvorov, A. Gusev, N. Ruban, M. Andreev, A. Askarov, S. Stavitsky, Int. J. Emerg. Electr. Power Syst. **20**, 1 (2019)
14. N. Ruban, A. Suvorov, M. Andreev, R. Ufa, A. Askarov, A. Gusev, B. Bhalja, IEEE Trans. Power Syst. **36**, 5 (2021)
15. X. Meng, J. Liu, Z. Liu, IEEE Trans. Power Electron. **34**, 6 (2019)
16. T. Shintai, Y. Miura, T. Ise, IEEE Trans. Power Deliv. **29**, 2 (2014)
17. L. Huang, H. Xin, Z. Wang, IEEE Trans. Power Electron. **34**, 6 (2019)
18. J.A. Suul, S. D'Arco, G. Guidi, IEEE Trans. Ind. Appl. **52**, 4 (2016)
19. Y. Yu, S.K. Chaudhary, J. Matas, L. Xu, G.D. Agundis Tinajero, J.C. Vasquez, J.M. Guerrero, *Accuracy Assessment of Reduced- and Full-Order Virtual Synchronous Generator Models Under Different Grid Strength Cases*, in Proceedings of the 48th Annual Conference of the IEEE Industrial Electronics Society, IECON, 17-20 October 2022, Brussels, Belgium (2022)
20. A. Suvorov, A. Askarov, Y. Bay, R. Ufa, Int. J. Hydrog. Energy **47**, 82 (2022)
21. A. Suvorov, A. Askarov, N. Ruban, V. Rudnik, P. Radko, A. Achitaev, K. Suslov, Mathematics **11**, 18 (2023)
22. A. Suvorov, A. Askarov, Y. Bay, B. Maliuta, A. Achitaev, K. Suslov, Int. J. Electr. Power Energy Syst. **147** (2023)
23. V. Mallemaci, F. Mandrile, E. Carpaneto, R. Bojoi, IEEE Trans. Ind. Appl. **59**, 5 (2023)
24. Y. Yu, S.K. Chaudhary, G.D.A. Tinajero, L. Xu, N.N.B.A. Bakar, J.C. Vasquez, J.M. Guerrero, IEEE Trans. Power Electron. **37**, 7 (2022)
25. M. Li, P. Yu, W. Hu, Y. Wang, S. Shu, Z. Zhang, F. Blaabjerg, IEEE Trans. Power Electron. **37**, 8 (2022)
26. A. Askarov, N. Ruban, Y. Bay, R. Ufa, Y. Malkova, A. Suvorov, Electr. Power Syst. Res. **234** (2024)
27. N. Ruban, V. Rudnik, A. Askarov, B. Maliuta, Int. J. Hydrog. Energy **48**, 73 (2023)

© The Authors, published by EDP Sciences. This is an open access article distributed under the terms of the Creative Commons Attribution License 4.0 (https://creativecommons.org/licenses/by/4.0/).

Control Algorithm of Energy Storage System Based on Virtual Synchronous Generator

Aleksey Suvorov[1*], *Alisher* Askarov[1], and *Vladimir* Rudnik[1]

[1]National Research Tomsk Polytechnic University, School of Energy & Power Engineering, 634050 Tomsk, Russian Federation

Abstract. Frequency control is one of the most important tasks in electric power systems. However, as a result of the ongoing introduction into modern power systems of inertia-free generating facilities based on power converters, mainly renewable energy sources, this task is becoming more complex. This specificity is especially severe in microgrids, which are characterized by significant frequency changes. An effective way to solve such a problem is the use of energy storage systems (ESS) with grid-forming power converters. For this purpose, the paper develops a new structure of the control algorithm based on a current-controlled virtual synchronous generator (CC-VSG), in which the damping of oscillations is implemented via a feedforward controller. Also, a proportional-integral control loop is added to the structure of the CC-VSG to control the state of charge of the ESS. A methodology based on the separation of bandwidths of different control loops is proposed for tuning the CC-VSG. To confirm the obtained results, time-domain simulation of the microgrid with ESS using the PSCAD software was performed.

1 Introduction

One of the most important tasks in electric power systems is a frequency regulation to ensure that its value is within the permissible limits, taking into account the requirements of grid codes [1,2]. Moreover, these requirements are different, depending on the current operating state in a particular power system. At the same time, as a result of the introduction of renewable energy sources, the task of frequency regulation becomes more complex and requires taking into account additional factors that are related to the process of variable power generation from renewables and their technological features [3–5]. For example, the permissible frequency deviations leading to unit shutdown by their own protections for most industrial inverters are ±3 Hz [4,6], which often differs from the requirements of grid codes. The stochastic nature of solar radiation and wind speed can cause significant frequency and power fluctuations in the electrical grid, which must be controlled and damped [7–10].

As a result, the application of microgrids as a part of modern power systems is seen as the most promising solution, which allows flexible adaptation to new conditions of power systems operation, and can also become an effective solution for leveling the problems associated with the renewables introduction [11]. The latter is due to the unique properties of microgrids, their architecture and control concept [12]. In general, a microgrid combines sources of distributed generation (DG), energy storage systems (ESS) and local load, which are coordinated with each other to ensure the required parameters in terms of power supply reliability and power quality [13–15]. At the same time, microgrids can operate both in parallel with the external grid if there is a connection point, and in island mode, with the possibility of transition between them [16].

Within a microgrid, renewable energy sources, such as photovoltaic or wind power plants, and conventional fossil fuel-based sources, such as diesel generators, can act as DG facilities. Nowadays, for microgrids and isolated power districts, the application of hybrid energy systems, in which diesel generators and renewables are used together, is the most widespread [17]. In this case, the diesel generator acts as a guaranteed source of power supply, which eliminates the problem of sharply variable generation of renewable energy sources when weather conditions change both during the day and in different seasons of the year. At the same time, within the framework of microgrids, deviations of grid frequency can be significant [18], since DG facilities are either low-inertia in the case of diesel generators, or have no inertia at all, which is typical for renewables, which are mainly connected to the grid through power converters – inverters [19]. In this regard, hybrid energy systems are often supplemented with ESS, which allows to increase the reliability of power supply to consumers, as well as to reduce the installed capacity of the guaranteeing source [20]. In order to ensure a normal frequency change profile in various operating conditions of microgrids, especially in their island operation, batteries, supercapacitors, hydrogen storages, etc. can be used [21].

In order to achieve the desired indicators of reliability and stability of microgrid operation, as well as the required indicators of power quality, it becomes necessary to develop effective control strategies for the hybrid energy systems in various operating conditions,

* Corresponding author: suvorovaa@tpu.ru

© The Authors, published by EDP Sciences. This is an open access article distributed under the terms of the Creative Commons Attribution License 4.0 (https://creativecommons.org/licenses/by/4.0/).

especially in island mode [22]. Currently, the transition from the conventional control strategy with a grid-following inverter to control with a grid-forming inverter is actively proposed [23–25], through which it becomes possible to provide a number of system services, including the introduction of virtual inertia, the ability of fast frequency response and participation in frequency regulation [26–28]. In addition, with such control, it is feasible to have continuous operation of ESS with renewable sources without conventional synchronous generation, for example, in case of load supply only by renewables or in case of emergency shutdown of diesel generator. In this connection, many studies develop algorithms for the inverter control system based on the virtual synchronous generator (VSG) [29,30], which also find application in practice in the form of experimental designs because of their undeniable advantages [31].

This paper considers a modified structure of current-controlled VSG (CC-VSG) with a feedforward controller [32], which has a number of advantages over the conventional voltage-controlled VSG structure, as detailed in [33,34]. At the same time, the CC-VSG is supplemented with the algorithm for controlling the storage state of charge (SoC) based on the PI controller in the structure of the inner active power control loop to ensure continuous control of the inverter based on the current operating conditions.

2 Structure of CC-VSG for ESS

The implementation of system services, i.e., the formation of frequency and voltage in the electric network, via the ESS is possible only with a grid-forming inverter. In this paper, for this purpose, the control strategy developed earlier by the authors on the basis of the CC-VSG algorithm is used. This algorithm has been reviewed in detail [33], where its comparison with conventional VSG structures has been performed and its advantages as well as its performance under stiff and weak grid conditions have been proved. However, damping in the CC-VSG model is realized by a virtual damper winding in the form of simplified differential equations. This approach leads to the need to consider third order equations, which affects the difficulty of adequately tuning the control loops, e.g., using the Vyshnegradskii's stability diagram [35]. In order to reduce the order of the CC-VSG model, this paper considers the application in the active power control loop of a feedforward controller [36], which improves the control system characteristics due to direct compensation of the disturbance, instead of the virtual damper winding equations. This damping controller is based on the feedforward proportional loop with the gain K_{FF}, the input of which is fed by the VSG frequency deviation, and the output signal is added to the VSG phase angle. The structure of the modified CC-VSG is shown in Fig. 1.

In addition to providing the usual VSG functions of inertial response and frequency control, a loop for restoring the ESS charge level to the set value after various disturbances was added to the active power control loop of the VSG-based ESS control system. This loop is based on a classical PI controller for continuous charge level control with parameters $K_{p,SoC}$ and $K_{i,SoC}$ (highlighted in green in Fig. 1).

At the same time, in order to tune the presented CC-VSG algorithm, a method based on the separation of bandwidths for different control loops is proposed, with the help of which it is possible to achieve independent functioning of control loops at different stages of the transient process. For the considered microgrid, the character of frequency change is related to the primary and secondary frequency control, as well as the recovery of the ESS charge level. Moreover, given the obvious tasks at each of these stages, the sequence of their execution should be exactly as presented. In this regard, the corresponding bandwidths should be similarly arranged: from the fastest for the primary regulation loop to the slowest for the charge level recovery loop.

Fig. 2 shows the structural diagram of the linearized model of the considered microgrid with diesel generator and ESS, which reflects the processes of changes in active power and frequency at parallel operation of VSG and diesel generator in isolated mode.

With the help of this model a theoretical analysis can be performed proving that the proposed solutions

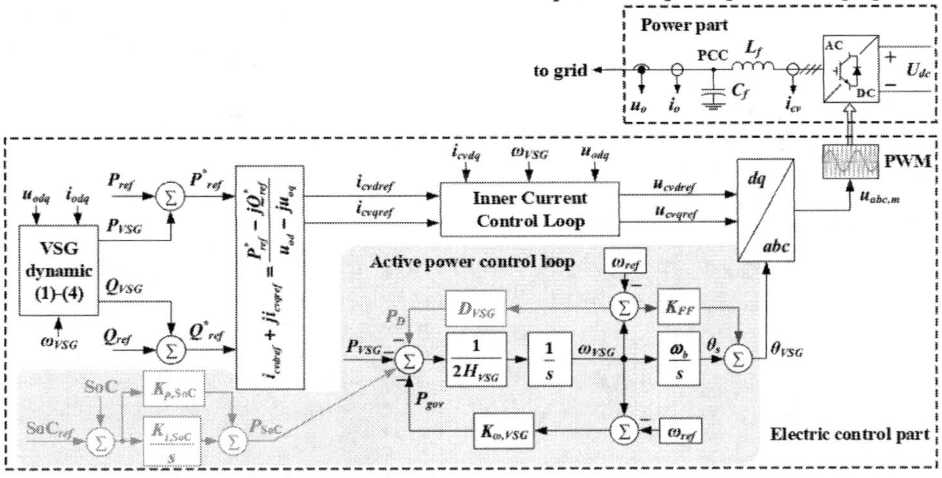

Fig. 1. Structural diagram of the modified CC-VSG algorithm.

© The Authors, published by EDP Sciences. This is an open access article distributed under the terms of the Creative Commons Attribution License 4.0 (https://creativecommons.org/licenses/by/4.0/).

 https://doi.org/10.1051/e3sconf/202458401017

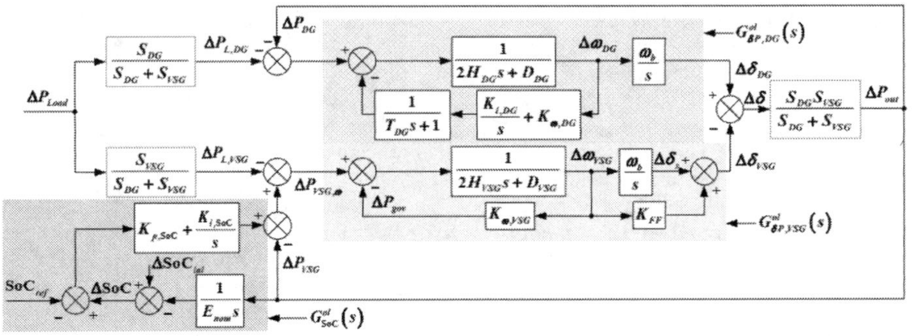

Fig. 2. Structural diagram reflecting the processes of active power and frequency changes in an isolated power system.

together allow to reduce the nominal energy intensity of the ESS, while maintaining the desired quality of the frequency change process. At the same time, when using the conventional approach to damping due to D_{VSG} (shown in red lines in Fig. 1), there is a need to find a trade-off between damping and ESS size. As a result, the overestimated damping properties affect a significant increase in the nominal energy intensity of the ESS, which is 11 times higher than in the case of using the approach proposed in this paper. A comparison of the transient characteristics when using the indicated two approaches to damping within the CC-VSG is shown in Fig. 3.

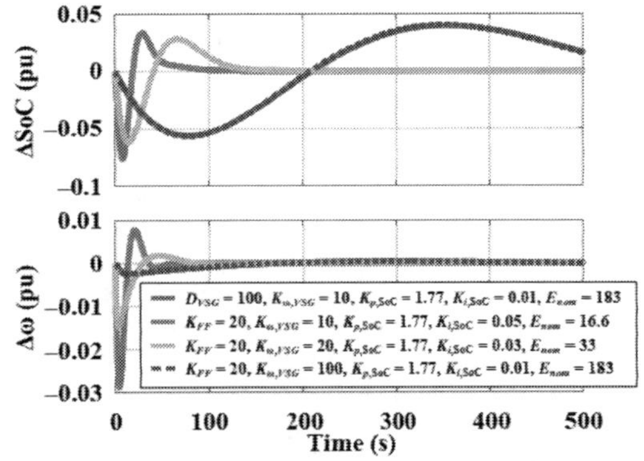

Fig. 3. Comparison of transient characteristics when using different damping approaches tuned according to the developed methodology for a specific permissible frequency deviation.

The developed new structure of CC-VSG and the approach to its tuning allow to achieve a wide control range and to choose the minimum energy intensity of ESS necessary to ensure the desired frequency nadir or zenith and rate of change of frequency. It is evident from the obtained curves that due to the CC-VSG with the feedforward controller and the control loop for the ESS charge level recovery any restrictions on the frequency deviation value are provided, up to the most severe ones ($\Delta\omega = 0.03$ pu – red curves in Fig. 3, $\Delta\omega = 0.015$ pu – yellow curves in Fig. 3, $\Delta\omega = 0.0025$ pu – purple curves in Fig. 3). For the conventional damping approach (blue curves in Fig. 3), such flexibility is not possible without degrading the response quality, resulting in excessive frequency control (desired frequency deviation value $\Delta\omega_{max} = 0.03$ pu and actual value $\Delta\omega = 0.0025$ pu).

3 Simulation results of microgrid with diesel generator and ESS

During the simulation studies on the detailed mathematical model of the system with diesel generator and ESS, the microgrid frequency change profile was evaluated in the result of load increase ($\Delta P_{load} = 0.3$ pu) in the case of the storage control using the proposed CC-VSG structure with the feedforward controller and the charge level recovery circuit. In this case, in order to carry out a comprehensive analysis of the effectiveness of the proposed solutions, several control scenarios were considered (Fig. 4).

1. Case 1 – operation of the diesel generator without ESS,
2. Case 2 – operation of the diesel generator and ESS with CC-VSG algorithm and without the SoC control;
3. Case 3 – operation of the diesel generator and ESS with CC-VSG algorithm and the SoC control, the parameters of which are matched with the bandwidth of the secondary frequency control in the microgrid;
4. Case 4 – operation of the diesel generator and ESS with CC-VSG algorithm and the SoC control, the bandwidth of which is equal to the bandwidth of the secondary control, while the damping factor for the charge level recovery loop remains the same ($\zeta \approx 1$).

Fig. 4a clearly shows that in the case of operation of only diesel generator there is a maximum frequency deviation ($\Delta f_{max} \approx 1.8$ Hz) and its rate of change (RoCoF ≈ 3.8 Hz/s), as well as prolonged frequency oscillations – the damping time is about 13 seconds. By installing ESS with CC-VSG in the microgrid, it becomes possible to significantly reduce the magnitude of frequency deviation by almost one third to $\Delta f_{max} = 1.22$ Hz and its rate of change by more than three times to 1.2 Hz/s (blue curves in Fig. 4b).

However, in case of absence of ESS SoC level control during frequency control and additional active power supply to the grid, as a result of the disturbance a natural battery discharge occurs ($\Delta SoC_{meas} = 11.41$ %) without restoration of the SoC level to the previous value. Based on the obtained expression of the expended energy of the ESS it is possible to analytically estimate the depth of discharge under the considered disturbance ($\Delta SoC_{calc} = 9.04$ %). As can be seen, the obtained values for ΔSoC are sufficiently close.

© The Authors, published by EDP Sciences. This is an open access article distributed under the terms of the Creative Commons Attribution License 4.0 (https://creativecommons.org/licenses/by/4.0/).

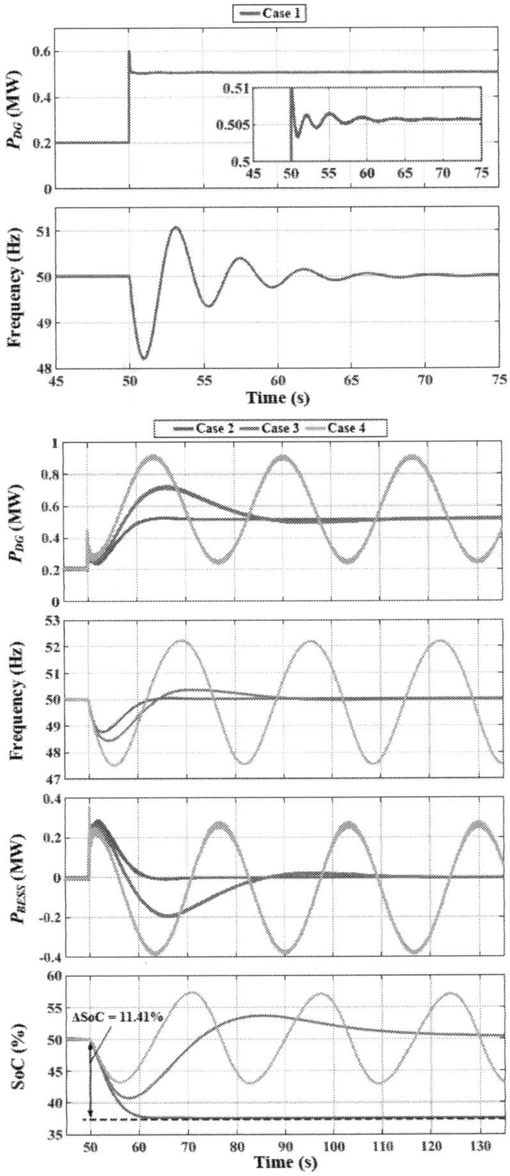

Fig. 4. Simulation results under step load increase in the microgrid.

By adding SoC control to the CC-VSG structure, it is achieved to maintain the storage SoC at a predetermined level (SoC_{ref} = 50 %) after the disturbance (red curves in Fig. 4b). As a result, the ESS is always able to participate in the microgrid state regulation. In this case, the value of Δf_{max} relative to the previous case increases slightly to 1.48 Hz. However, as can be seen, by applying the proposed recommendations for the tuning of the SoC control loop, it is possible to comply with the accepted requirements for the range of microgrid frequency deviation (±1.5 Hz relative to the nominal value). In addition, based on the analysis of the waveform for the current SoC, after the disturbance the maximum value of ΔSoC decreases to 6.77 %, allowing to reduce the nominal storage capacity. It should be noted that the assumed SoC_{ref} = 50 % is conditional and its choice depends on the specific power system under consideration and possible most severe disturbances in it, as well as on the charge and discharge limits for the used type of energy storage device.

Considering the fourth case (the bandwidths of the secondary control and the SoC control are equal), it can be seen that when the operation of different control loops of the diesel generator and ESS is superimposed, undamped frequency and power oscillations of significant amplitude occur, for example, for the frequency of the microgrid, the range is from 47.51 Hz to 52.22 Hz. In case of more severe disturbance, oscillations with increasing amplitude occur, which leads to instability in the microgrid. Thus, the simulation results prove the applicability of the developed methodology for matching the bandwidths of different control loops.

4 Conclusion

This paper presents the results of the development of a CC-VSG structure for controlling the grid inverter of ESS. In the developed CC-VSG structure, due to the use of a feedforward controller, it is possible to damp the oscillations without affecting the low-frequency part of the spectrum associated with the secondary frequency control. Therefore, the addition of a PI regulator to the CC-VSG structure to recover the ESS charge level does not lead to the occurrence of mutual impact with the damping loop. The proposed approach to CC-VSG tuning on the basis of matching the bandwidths of different control loops allows to provide effective recovery of the SoC with no negative impact on the process of frequency regulation of the microgrid as a whole, as well as to reduce the nominal required energy intensity of ESS. The presented simulation results have shown that in the absence of matching the settings of the control loops responsible for frequency regulation and SoC control, prolonged non-damped oscillations may occur, up to the instability of the ESS grid inverter operation.

The reported study was funded by the Russian Science Foundation, project number 24-29-00004.

References

1. B. Jin, Energy **273** (2023)
2. O.V. Shepovalova, Y.D. Arbuzov, V.M. Evdokimov, P.V. Ilyushin, K.V. Suslov, Energies **16**, 3 (2023)
3. J. Wen, S. Bu, F. Li, P. Du, Energy **228** (2021)
4. R. Ufa, A. Gusev, A.A.Z. Diab, A. Suvorov, N. Ruban, M. Andreev, A. Askarov, V. Rudnik, O. Abdalla, Z.M. Ali, A. Ibrahim, R. Aboelsaud, Energy Rep. **6** (2020)
5. R. Ufa, M. Andreev, N. Ruban, A. Suvorov, A. Gusev, I. Razzhivin, A. Askarov, Y. Bay, A. Kievets, N. Lozinova, O. Suslova, Electr. Eng. **101**, 1 (2019)
6. P. Ilyushin, V. Volnyi, K. Suslov, S. Filippov, Energies **15**, 23 (2022)

© The Authors, published by EDP Sciences. This is an open access article distributed under the terms of the Creative Commons Attribution License 4.0 (https://creativecommons.org/licenses/by/4.0/).

7. M. Farrokhabadi, D. Lagos, R.W. Wies, M. Paolone, M. Liserre, L. Meegahapola, M. Kabalan, A.H. Hajimiragha, D. Peralta, M.A. Elizondo, K.P. Schneider, C.A. Canizares, IEEE Trans. Power Syst. **35**, 1 (2020)
8. A. Suvorov, A. Gusev, M. Andreev, A. Askarov, Int. Trans. Electr. Energy Syst. **30**, 4 (2020)
9. A. Suvorov, A. Gusev, M. Andreev, A. Askarov, Electr. Eng. **101**, 2 (2019)
10. Y.D. Bay, A.A. Suvorov, A.S. Gusev, I.A. Razzhivin, A.B. Askarov, IEEE Access **9** (2021)
11. L. Meng, J. Zafar, S.K. Khadem, A. Collinson, K.C. Murchie, F. Coffele, G.M. Burt, IEEE Trans. Smart Grid **11**, 2 (2020)
12. P. Ilyushin, V. Volnyi, K. Suslov, S. Filippov, Energies **16**, 7 (2023)
13. A. Kulikov, P. Ilyushin, A. Sevostyanov, S. Filippov, K. Suslov, Energies **17**, 5 (2024)
14. A. Suvorov, A. Gusev, N. Ruban, M. Andreev, A. Askarov, S. Stavitsky, Int. J. Emerg. Electr. Power Syst. **20**, 1 (2019)
15. N. Ruban, A. Suvorov, M. Andreev, R. Ufa, A. Askarov, A. Gusev, B. Bhalja, IEEE Trans. Power Syst. **36**, 5 (2021)
16. V. Volnyi, P. Ilyushin, K. Suslov, S. Filippov, Energies **16**, 15 (2023)
17. B.E.K. Nsafon, A.B. Owolabi, H.M. Butu, J.W. Roh, D. Suh, J.-S. Huh, Energy Strategy Rev. **32** (2020)
18. N. Ruban, V. Rudnik, A. Askarov, B. Maliuta, Int. J. Hydrog. Energy **48**, 73 (2023)
19. K. Zeb, W. Uddin, M.A. Khan, Z. Ali, M.U. Ali, N. Christofides, Renew. Sustain. Energy Rev. **94** (2018)
20. E. Boyko, F. Byk, P. Ilyushin, L. Myshkina, K. Suslov, Energy Rep. **9** (2023)
21. I.A. Razzhivin, A.A. Suvorov, R.A. Ufa, M.V. Andreev, A.B. Askarov, Int. J. Hydrog. Energy **48**, 58 (2023)
22. N. Soni, S. Doolla, M.C. Chandorkar, IEEE Trans. Ind. Appl. **52**, 6 (2016)
23. R. Rosso, X. Wang, M. Liserre, X. Lu, S. Engelken, IEEE Open J. Ind. Appl. **2** (2021)
24. A. Suvorov, A. Askarov A, A. Kievets, V. Rudnik, Electr. Power Syst. Res. **209** (2022)
25. A. Suvorov, A. Askarov, A. Kievets, Electr. Eng. **105** (2023)
26. D. Curto, S. Favuzza, V. Franzitta, Energy **254** (2022)
27. U. Tamrakar, D. Shrestha, M. Maharjan, B.P. Bhattarai, T.M. Hansen, R. Tonkoski, Appl. Sci. **7** (2017)
28. Y. Cheng, R. Azizipanah-Abarghooee, S. Azizi, L. Ding, V. Terzija, Appl. Energy **279** (2020)
29. W. Sang, W. Guo, S. Dai, C. Tian, S. Yu, Y. Teng, Energies **15**, 17 (2022)
30. A. Suvorov, A. Askarov, Y. Bay, R. Ufa, Int. J. Hydrog. Energy **47**, 82 (2022)
31. H. Kikusato, D. Orihara, J. Hashimoto, T. Takamatsu, T. Oozeki, T. Matsuura, S. Miyazaki, H. Hamada, T. Miyazaki, Energy Rep. **9** (2023)
32. A. Askarov, N. Ruban, Y. Bay, R. Ufa, Y. Malkova, A. Suvorov, Electr. Power Syst. Res. **234** (2024)
33. A. Suvorov, A. Askarov, Y. Bay, B. Maliuta, A. Achitaev, K. Suslov, Int. J. Electr. Power Energy Syst. **147** (2023)
34. Mallemaci, F. Mandrile, E. Carpaneto, R. Bojoi, IEEE Trans. Ind. Appl. **59**, 5 (2023)
35. A. Suvorov, A. Askarov, N. Ruban, V. Rudnik, P. Radko, A. Achitaev, K. Suslov, Mathematics **11**, 18 (2023)
36. M. Li, P. Yu, W. Hu, Y. Wang, S. Shu, Z. Zhang, F. Blaabjerg, IEEE Trans. Power Electron. **37**, 8 (2022)

© The Authors, published by EDP Sciences. This is an open access article distributed under the terms of the Creative Commons Attribution License 4.0 (https://creativecommons.org/licenses/by/4.0/).

On Reliability of 110-500 kV GIS: Operational Experience

Evgeniy Korotchenko[1], *Irek* Galiaskarov[1]*, *Misrikhan* Misrikhanov[2], *Vladimir* Ryabchenko[3], and *Andrey* Shuntov[2]

[1] PJSC ROSSETI CIUS EES - CIUS Centre, Moscow
[2] Moscow Power Engineering Institute, Moscow
[3] JSC ROSSETI Scientific and Technical Centre, Moscow

Abstract. The accident rate of gas-insulated switchgears (GIS) is considered. The statistical sample covered 30 110-500 kV switchgear switchgears in a vast region of the European part of the country for the period 2008-2022. Their design features and technical characteristics are given. Significant and insignificant failures of GIS, including their structure, namely, distribution by elements of installations, technical and organizational causes of occurrence, dependence on the year of operation are subjected to analysis. The flow parameter (frequency) of GIS failures is determined, as well as the maximum possible time of GIS restoration after damage to the arc suppression chamber of the pole of the 500 kV circuit breaker. Comparative data are given with failure rate of GIS abroad and with gas-insulated circuit breaker cells in open switchgears.

Introduction

Gas-insulated switchgears (GIS) appeared on the world market in the late 1960s. Since then, they have held a strong position in the power systems of many countries. The advantages of GIS are well known: exceptional compactness, high reliability, earthquake resistance and safety of operating personnel, ease of installation and maintenance. For gas-insulated equipment there are no problems of inducing powerful electric fields and audible noise.

Since the 90s of the last century, CIGRE working groups have been publishing reviews on the reliability of GIS of 60 kV and above [1, 2]. The first review was for data up to 1990, the second for 1995, the third for the period 2004-2007, and finally the fourth for 2014-2017. However, the data in [1, 2] are averaged for electrical installations of different manufacturers in dozens of countries with differences in technical parameters of equipment and natural and climatic conditions of its location. In this regard, the reliability characteristics of GIS and the experience of their operation accumulated in domestic power systems are of interest.

Since the formation of the Unified National Power Grid (UNPG) in one of the zones of a large agglomeration of cities in the European part of the country in the period 2008-2022, the number of 110-500 kV switchgears has increased from eight to 30. They were constructed most rapidly in 2012-2013, when six electrical installations were commissioned at the same time. The use of GIS was connected with the need to replace physically and morally worn-out equipment at 220-500 kV UNES substations (the service life reached 50-60 years). It was necessary to significantly (two or more times) increase the installed capacity of substations, as well as to locate repair and production bases on them without increasing land allotments, and sometimes with the release of part of the occupied areas for transfer to local authorities.

The distribution of the number of switchgear units by nominal grid voltage is as follows: 110 and 500 kV seven units each, and 220 kV - 16. The vast majority of them were supplied by the world's leading manufacturers. The reason lies in the fact that the production of 500 kV switchgear in our country has not yet been mastered, and the voltage of 110 and 220 kV required cells for the rated tripping current of circuit breakers equal to 63 kA. The latter is also not supported by domestic manufacturers.

Reliability experience

The switchgears under consideration from different manufacturers were commissioned over a relatively long period of time (see above). Therefore, all other things being equal, they differed from each other in design. For example, of the seven 500 kV GISs, five units commissioned before 2013 had two breaks per breaker pole. In 2021, two more 500 kV GISs were commissioned, but in a more advanced single-break design. Of the 16 220 kV switchgears, only one, put into operation in 2008, is made in a double-break circuit breaker design. The rest of the 220 kV units, as well as all 110 kV switchgears are made in single-break design. It is worth noting the maximum number of circuit breaker cells in a single GIS, which is typical for areas with extremely high load density: 110 kV - 33, 220 kV - 48 and 500 kV - nine cells.

Of the 30 switchgears, only four 110 kV units commissioned in the last five years, including those with a rated tripping current of 63 kA, are of three- phase design, while the rest are of single-phase design. The

* Corresponding author: gim05032008@yandex.ru

© The Authors, published by EDP Sciences. This is an open access article distributed under the terms of the Creative Commons Attribution License 4.0 (https://creativecommons.org/licenses/by/4.0/).

circuit breakers of 22 switchgears have spring actuators and eight have hydraulic actuators (four each for 220 and 500 kV). The latter is due to the non-simultaneous improvement of GIS designs by different manufacturers at separate time intervals.

The types of applied GIS schemes covering all feeders of the voltage class under consideration are given in Table 1. In general, they correspond to the recommendations [3]. At 500 kV, so-called ring schemes are used, sectionalizing the network scheme through each feeder. The high proportion of ring schemes for 220 kV GIS is due to the large total capacity of (auto)transformers connected to the 220 kV switchgear and a significant number of outgoing lines. Thus, the main part of 500 kV substations in the region has two more stages of higher rated voltage: 110 and 220 kV. For example, two 500/220 kV autotransformers of 500 MV·A each, two 220/110 kV autotransformers of 250 MV·A each, and two 220/10 kV transformers of 100 MV·A each are connected to the 220 kV switchgear (a total of 1700 MV·A).

Table 1. GIS circuits

№№ n/a	Scheme	Number of circuits, pcs., at voltage, kV		
		110	220	500
1	triangle	-	-	1
2	Quadrilateral	-	6	3
3	One sectionalized busbar system	1	1	-
4	One sectionalized busbar system with (auto) transformers connected via a fork of circuit breakers	6	1	-
5	Bus transformer with one-and-a-half line connection	-	2	2
6	One and a half	-	6	1

Turning to the reliability assessment of GIS, we note that CIGRE reviews [1, 2] take into account the so-called major and minor failures. The former, as defined in IEC 62271-1, result in the loss of one or more functions of the equipment, including its control system; this is accompanied by an immediate change in the circuit state, e.g. emergency disconnection from protections, or an urgent (within 30 min) removal of the equipment from operation by personnel for unscheduled maintenance or repair. Accordingly, a minor failure is a failure of equipment other than a major failure and which does not cause serious disruption to the operation of the equipment.

Let us note that according to the domestic terminology in the field of reliability (see GOST R 27.102-2021), a major failure can be compared with a complete failure, and a minor failure with a damage, i.e. violation of the serviceable state of the object while maintaining its operable state.

Table 2 shows the number of failures in GIS for the period from 2008- 2022. It should be emphasized that for the entire period of observation no failures were detected in 110 kV GIS. These data were confirmed by an additional survey of the operating personnel servicing these electrical installations.

See Table 3 for the distribution of failures by GIS elements. At 220 kV the circuit breaker with drive (Table 3) accounted for half of the significant failures and the majority of the minor failures. At 500 kV, disconnectors and voltage transformers accounted for the bulk of significant failures, and busbars and busbars for minor failures.

Table 2. Number of failures in GIS

№№ n/a	Type of failure	Number of failures, pcs.,at voltage, kV		
		110	220	500
1	Significant	-	4	7
2	Minor	-	11	6
3	Total	-	15	13

Table 3. Distribution of failures by GIS elements

№ № n/a	Element	Number of failures, pcs., at voltage, kV:			
		220		500	
		signif icant	mi nor	signifi cant	minor
1	Switch with actuator	2	9	1	1
2	Disconnector	-	-	3	1
3	Earthing switch disconnector	-	1	-	-
4	Current transformer	-	-	-	1
5	Voltage transformer	1	1	2	-
6	Busbars and busbars	-	-	1	3
7	Surge arrester	1	-	-	-
8	Total	4	11	7	6

Technical causes of GIS failures are given in Table 4. For major failures the main cause was gas gap bridging, and for minor failures the main cause was actuator malfunction and gas leakage. In terms of organization, all significant failures were the result of the manufacturer's fault, except (Table 3) for the damage of busbars and busbars caused by poor quality installation.

The analysis of performed emergency recovery operations on these switchgears showed that the most time was required to replace the pole of a 500 kV circuit breaker when its arc suppression chamber was completely blocked. With any breakdown of the insulation of the GIS on site to restore the failed module is not possible - requires a complete spare part. The actual duration of 500 kV circuit breaker pole replacement

© The Authors, published by EDP Sciences. This is an open access article distributed under the terms of the Creative Commons Attribution License 4.0 (https://creativecommons.org/licenses/by/4.0/).

amounted to eight working shifts (four days, 96 h), while the reserve pole was at the substation. Such a long period of time is due to a large amount of gas-technological works to "buy" the faulty unit. For comparison, in [4] for 500 kV switchgears abroad, the total (with delivery to the site of the spare part) average restoration time of 216 h is given.

Table 4. Technical causes of failures

№№ n/a	Reason failures	Number of failures, pcs., at voltage, kV:			
		220		500	
		significant	minor	significant	minor
1	Insulation breakdown to earth (solid insulation)	-	-	2	-
2	Insulation breakdown to earth (gas insulation)	3	-	5	-
3	Gas leakage	-	1	-	4
4	Actuator malfunction	1	8	-	1
5	Others	-	2	-	1
6	Total	4	11	7	6

Table 3 shows that for 220 kV switchgear and switchgear switchgear of 220 kV, a large share of significant and minor failures came from the circuit breaker with actuator. This was brought by circuit breakers with hydraulic actuators. Failures manifested themselves in continuous difficult to eliminate oil leaks from the hydraulic systems of actuators. It should be noted that spring actuated circuit breakers had no failures during the observation period. The accident rate associated with the use of hydraulic actuators casts doubt on the expediency of their use. Especially nowadays, when works with the involvement of the chief engineer of the manufacturer's plant are not possible due to the imposed sanctions of unfriendly countries.

Another element of 500 kV GIS subject to high accident rate turned out to be voltage transformers: in the last two years of operation there were two cases at different substations. The cause is overlapping to the case in primary and secondary circuits.

One of the most important indicators of equipment reliability is the failure flow parameter ω (more precisely - the average failure flow parameter or failure rate). According to the terminology of IEC 62271-1, this is a significant failure, which is accompanied by an investigation, drawing up a corresponding act and carrying out emergency repair. A minor failure is referred to as an unscheduled repair or maintenance. The term "unscheduled repair" is due to the fact that the equipment is taken out for repair at an unscheduled dispatch request, being, in fact, a planned shutdown. The most typical examples of minor failures are a decrease in gas pressure in sealed compartments or oil level in the hydraulic actuator system of circuit breakers to the warning alarm settings.

As it is known, at restored elements the parameter ω is defined as a density of probability of occurrence of failures for the considered period, i.e. $\omega = m/(kT)$, where m - number of failures observed by k elements for time T. Table 5 summarizes the values of the failure flow parameter of the considered GIS.

As can be seen from Table 5, the failure rate of 500 kV GIS cells is an order of magnitude higher than that of 220 kV installations. In order to evaluate the reliability level, let us compare the data in Table 5 with the characteristics of switchgear and switchboards given in SIGRE reviews [1] - Table 6.

It follows from Tables 5 and 6 that the achieved failure rate of 110 kV GIS cells is extremely low (1/0,00026 ≈ 4000 years), so it is not surprising that no failures were recorded at the substations under consideration. There is an almost complete coincidence of the failure flow parameter of 220 kV GIS cells. As for the 500 kV installations, the domestic operating experience showed an order of magnitude worse results: ω=0,015 1/year against 0,00172 1/year in Table 6. Consequently, when using 500 kV switchgear should be particularly carefully approach to all stages of development of electrical installations of this voltage class, from design to commissioning, including means of monitoring partial discharges, which did not prevent a single failure of 500 kV switchgear.

Table 5. Frequency of significant failures of GIS cells

Voltage, kV	The number of switchgear, pcs.	Coli- honor cells, pcs.	Average number of priconnections, pcs.	Medium age cells, years	Number of failures, pcs.	Number of cells- years	Frequency failures, 1/year
110	7	144	20.6	9,0	0	1298	-
220	16	262	26,6	9,9	4	2582	0,0015
500	7	45	6,4	10,1	7	454	0,015

Table 6. Frequency of significant failures of GIS cells according to SIGRE data

№№ n/a	Nominal voltage, kV	Failure rate of GIS cells, 1/year, for periods:		
		Prior to 1995.	2004-2007.	2014-2017.
1	110	0,0068	0,00243	0,00026
2	220	0,0089	0,0037	0,0014
3	500	0,0131	0,005	0,00172

Nevertheless, GIS cells are noticeably superior to cells with gas- insulated circuit breakers in open

© The Authors, published by EDP Sciences. This is an open access article distributed under the terms of the Creative Commons Attribution License 4.0 (https://creativecommons.org/licenses/by/4.0/).

switchgear (GIS). Table 7 summarizes the failure flow parameters from previously published data. Comparison of the data in Tables 5 and 7 clearly demonstrates the noticeably higher (by about an order of magnitude) reliability of GIS cells.

Table 7. Frequency of significant failures of switchgear cells

№№ n/a	Nominal voltage, kV	Failure rate of switchgear cells, 1/year, according to the data:	
		[5]	[6]
1	110	0,06	0,0437
2	220	0,05	0,0355
3	500	0,14	0,1055

It is worth noting the distribution of GIS failures by years of operation. Out of 11 significant GIS failures (Table 2), 10 of them occurred in the first five years, and the "surge" of accidents - the second and fourth years of operation. Thus, within five years, the main part of defects in manufacturing and installation of equipment is manifested. This factor is important to take into account when drawing up the technical specifications for ordering GIS and specify the minimum warranty period from the start of operation of five years. As such, periodic scheduled overhauls are not foreseen for GIS. However, the manufacturers recommend extended repairs after 20 years of operation. Extended repair includes selective opening (disassembly) of some gas compartments (arc-fuse chamber, disconnector, measuring transformer, etc.). On the basis of which a conclusion is made on the necessity or sufficiency of the volumes of the modules to be opened. Then a decision is made on the scope of repair and restoration work for the entire installation, including the replacement (if necessary) of individual modules. The average age of 110-500 kV GIS installed in the region is 9-10 years. Therefore, in the foreseeable future, the power system will have to do a lot of work to assess the inter-repair life of elements of these electrical installations.

Conclusions

1. At present, the domestic power systems have accumulated certain experience in operation of 110-500 kV switchgears, which allows estimating their reliability level in terms of accident rate and comparing it with aggregated, representative, foreign data.

2. It is established that the failure rate of 110 and 220 kV GIS in the country and abroad is close. However, the reliability of 500 kV GIS over the given time interval demonstrates lower (by about an order of magnitude) indicators. This occurred despite the fact that 500 kV equipment was supplied by the world's leading manufacturers and the main cause of accidents - factory defect.

3. The majority of significant GIS failures are related to insulation breakdown: gas or solid. The failed module is not repairable and must be replaced. Therefore, the GIS spare parts list should cover all GIS components subject to failure.

4. All of the installed switchgears were equipped with advanced, expensive partial discharge monitoring systems. None of them fulfilled their functions and did not prevent insulation breakdowns. This fact requires closer examination.

5. The longest GIS restoration time amounted to four days and was associated with the replacement of a burned-out phase of a 500 kV circuit breaker. For other less severe failures (replacement of voltage transformer, disconnector, etc.), the restoration time was half as long, taking into account the necessary organizational measures (coordination of module replacement with the manufacturer, arrival of the chief engineer, delivery of the spare part, etc.).

6. In the design of new switchgear and switchboards it seems justified to exclude the use of hydraulic circuit breaker actuators, as they have shown low reliability in operation.

7. The majority of GIS failures occurred in the first three to four years of operation. Hence, five years is probably the most acceptable warranty period for the equipment.

8. Complete switchgear substations with gas-insulated switchgear do not require major (average) repairs during their entire service life. However, it will be possible to confirm this only after the individual GIS modules are opened and inspected for operability after 20 years of operation. This is to be done in the foreseeable future, which will make it possible to clarify the reliability parameters of GIS in terms of scheduled downtime.

References

1. CIGRE reliability survey on equipment / H. Ito, N. Uzelac, F. Richter etc. // CIGRE Science & Engineering. 2021, №23, 11 p. [Electronic resource]. - Access mode: https://cse.cigre.org/cse-n023/cigre-reliability- survey-on-equipment.html.
2. CIGRE fourth reliability survey on switching equipment // H. Ito, F. Richter, R. le Roux etc. // Colloquium 2023. Pap. 105. 8 p. [Electronic resource]. - Mode of access: https://cigre.org.uk/web-cont1001/uploads/CIGRE-fourth-reliability-survey-on-switching-equipment.pdf (date of address: 21.04.2024).
3. CTO 56947007–29.240.30.047–2010. Recommendations on the application of typical schematic diagrams of switchgear of substations 35-750 kV. [Electronic resource]. - Mode of access: https://docs.cntd.ru/document/1200088422.
4. Misrikhanov M.Sh., Mozgalev K.V., Shuntov A.V. About reliability of switchgear and switchgear with traditional insulation // Electric Power Plants. 2003. №11. p.
5. About reliability of 110-750 kV substation gas-insulated circuit breaker cells / Yu. A. Dementiev, M. Sh. Misrikhanov, E. I. Stolyarov et al. // Electric power stations. - 2011. - №1. p. 51-54.
6. Nepomnyashchy V. A., Daryan L. A. Reliability of equipment of electric networks 220-750 kV power systems. Moscow: NTF "Energoprogress", 2018, 124 p.

© The Authors, published by EDP Sciences. This is an open access article distributed under the terms of the Creative Commons Attribution License 4.0 (https://creativecommons.org/licenses/by/4.0/).

E3S Web of Conferences **584**, 01019 (2024) https://doi.org/10.1051/e3sconf/202458401019
RSES 2024

Placement of controlled shunt reactors in 330 kV electrical networks and study of their operating modes

Arif Hashimov[1], *Huseyngulu* Guliyev[2*], *Aytek* Babayeva[1]

[1] Azerbaijan Scientific–Research and Designed–Prospecting Institute of Energetics, 94 H.Zardabi str., Baku, Azerbaijan
[2] Azerbaijan Technical University, 25 H.Javid str., Baku, Azerbaijan

Abstract. The improvement of the method for selecting shunt reactors based on instantaneous power in order to eliminate a surplus of reactive power created by the charging capacities of power transmission lines in 330-500 kV elelctrical networks of the power system is considered. Based on the proposed method, the performance of controlled shunt reactors located in priority 330 kV nodes of a real power system is superficially described and their operating modes are studied. As a result of the study, in accordance with changes in the modes of shunt reactors in real time, the voltage in the 330 kV buses is maintained in a stable and acceptable range, as well as the increase in the voltage stability of the power system.

Introduction

In modern power systems, the creation of controlled or flexible AC power transmission lines that are part of intelligent electrical networks based on FACTS devices is prioritized and great importance is attached to this issue. [1-3]. For intelligent control of operating modes of power systems of this type, there is a need to use high-precision and efficient technical means of controlling flows of active and reactive power in high-voltage electrical networks.

In the recent period, various technical means are widely used to regulate voltage and reactive power in the power system, including generators, synchronous compensators, static thyristor compensators, switched reactors, static capacitor banks [4-7]. However, recently, controlled shunt reactors (CSR) related to first-generation FACTS devices, have been widely used. The CSR smoothly regulates the reactive power required for open-type devices at any voltage level, and the transformer design allows for installing it in any part of the power system. At that, the FACTS device ensures voltage stabilization at nodes of the electrical network, optimization of reactive power flows, increase in the transfer capability of power transmission, reduction of power losses, reduction of the number of switchings of load voltage regulation devices, used in transformers [8-12].

If shunt reactors are not installed in electrical networks with a voltage of 330 kV and higher, then in order to normalize the voltage mode in the power system load nodes in minimum load modes, it may be necessary to apply some ineffective operational measures that can lead to large losses.

As a result of the performed economic analysis it was established that due to the irrationality of the reactive power mode, additional power losses in high-voltage networks of the power system are at such a high level that, despite the rather expensive equipment, the installed CSR pays for itself in a shorter time [13].

Currently, in most power systems, the problem of eliminating the surplus of reactive power in minimum load modes is topical. It is obvious that the main reason for this situation is that the charging power of 330 kV lines exceeds the reactive power losses in them, which can lead to the increase in the voltage level to a dangerous level for the line insulation. From this point of view, preference is given to the CSR as a first-generation FACTS device that allows for eliminating the surplus of reactive power in high-voltage electrical networks and smoothly controlling reactive power flows, especially in 330-500 kV networks [14]. In other words, the relationship between reactive power losses in lines and their additional loading (charging) power is not constant and continuously changes depending on the mode. Therefore, in order to ensure a balance of reactive power in the power system at any given time, the power of the CSR must be regulated within wide range.

Based on studies conducted on real electrical network diagrams of the Azerbaijan power system, the issues of selection and placement of controlled shunt reactors in 330 kV networks have been considered [15-21]. At that, for the purpose of selecting the CSR and studying the network modes, the effective values of voltage and current are traditionally used. However, it should be noted that the use of instantaneous power values for smooth control of reactive power flows allows for more accurate regulation

* Corresponding author: huseyngulu@mail.ru

© The Authors, published by EDP Sciences. This is an open access article distributed under the terms of the Creative Commons Attribution License 4.0 (https://creativecommons.org/licenses/by/4.0/).

in real time. Therefore, in the work under consideration, it is important to develop a new approach to the selection and study of the CSR based on instantaneous power.

1 Traditional method for selecting the capacity of shunt reactors.

To determine the power of the CSR installed in the priority nodes of the power system, knowledge of the values of reactive power flows in the network is required. The total additional loading power of power system lines of different voltages can be determined from the following expression:

$$Q_{y,\Sigma} = \sum_{i=1}^{n} U_i^2 b_{0,i} \sum_{j=1}^{m} l_j = \sum_{i=1}^{n} q_{0,i} \sum_{j=1}^{m} l_j \tag{1}$$

or when considering only one node

$$Q_{y_i,\Sigma} = U^2 b_0 \sum_{j=1}^{m} l_j = q_0 \sum_{j=1}^{m} l_j, \quad i = \overline{1, n} \tag{2}$$

here b_0 – capacitive conductivity of 1 km line; q_0 – additional loading power of 1 km line; U_i – voltage of ith node; l_j – length of jth line.

The reactor power can be determined using the following formula [5,6]:

$$Q_r = P_{nat} \cdot \lambda \cdot \left(1 - \frac{P}{P_{nat}}\right) \cdot l \tag{3}$$

where l – line length; λ – wave length and is determined as follows: $\lambda = \omega/v$.

P_{nat} – line's natural power; P – power transmitted through the line; ω – angular velocity; v – speed of propagation of electromagnetic waves.

The reactor power is selected from the $P = P_{nat}$ condition , in this case

$$Q_r = P_{nat} \cdot \lambda \cdot l \tag{4}$$

For this purpose, the following formula [2] can also be used:

$$Q_r = P_{nat} \cdot tg\lambda \tag{5}$$

2 Selection of shunt reactors by instantaneous power

The traditional mentioned method and algorithm are based on the selection of CSR based on the effective values of voltage and current. However, it is known that this approach is associated with a certain precision and is not always justified [8]. To compensate for the surplus of reactive power in high-voltage electrical networks, the determination of the power of the CSR in real time based on instantaneous values of voltage and current gives more accurate results At that, the distorting effects of the CSR in the power system modes are also taken into account. Let's take into account the following instantaneous values of voltage and current in active power in expression (3):

$$\left.\begin{aligned} u &= U_m \sin(\omega t + \psi_u) \\ i &= I_m \sin(\omega t + \psi_i) \end{aligned}\right\} \tag{6}$$

In this case, from expression (3) we obtain:

$$q_r = P_{nat} \cdot \lambda \cdot \left(1 - \frac{p}{P_{nat}}\right) \cdot l =$$

$$= P_{nat} \cdot \lambda \cdot \left(1 - \frac{U_m I_m}{P_{nat}} \sin(\omega t + \psi_u) \cdot \sin(\omega t + \psi_i)\right) \cdot l \tag{7}$$

By summing the trigonometric product in brackets and performing the appropriate trigonometric transformations, expression (7) can be written as follows:

$$q_r = P_{nat} \cdot \lambda \cdot \left\{1 - \frac{1}{2P_{nat}} U_m I_m \left[\cos(\psi_u - \psi_i) + \cos\left(2\omega t + \psi_u + \psi_i\right)\right]\right\} \cdot l =$$

$$= P_{nat} \cdot \lambda \cdot \left\{1 - \frac{UI}{2P_{nat}} \left[\cos(\psi_u - \psi_i) + \cos(2\omega t + \psi_u + \psi_i)\right]\right\} \cdot l \tag{8}$$

Taking into account that $\psi_u - \psi_i = \varphi$ takes place in expression (8), as a result we obtain the following formula:

$$q_r = P_{nat} \cdot \lambda \cdot \left\{1 - \frac{UI}{2P_{nat}} \left[\cos\varphi + \cos\left(2\omega t + \psi_u + \psi_i\right)\right]\right\} \cdot l \tag{9}$$

or, knowing that $UI\cos\varphi = P$, the following expression can be written in a more compact form:

$$q_r = \lambda \cdot \left[P_{nat} - P + UI\cos(2\omega t + \psi_u + \psi_i)\right] \cdot l \tag{10}$$

As can be seen, the reactor power, expressed by instantaneous values of voltage and current (6), has got a spatial dependence. Expression (10) is much easier to implement, it more accurately describes the reactor power.

According to the proposed algorithm, RTU-180/330-U1 and RTU-120/330-U1 type controlled shunt reactors with a capacity of 180 MVAr and 120 MVAr were selected, respectively, for the 330 kV Goranboy distribution point and the 330/220/110/10 kV Yashma substation of the Azerbaijan power system.

© The Authors, published by EDP Sciences. This is an open access article distributed under the terms of the Creative Commons Attribution License 4.0 (https://creativecommons.org/licenses/by/4.0/).

3 Schemes of connection of CSR to the nodes of the Azerbaijan power system

3.1 Connection of CSR to Goranboy DP and operating modes

Based on the above method, the issue of determining the charging capacities and rated powers of shunt reactors for the 330 kV Yashma and 330 kV Goranboy nodes of the Azerbaijan power system was considered [3-7, 14-16]. Let's consider the node of the intrasystem 330 kV Goranboy distribution point. The scheme of connection of the RTU-180/330-U1 type 180 MVAr CSR to the DP is shown in Figure 1. As can be seen, the 330 kV Goranboy node includes the following lines: 330 kV Goranboy-1 (2xASO-300, 56.3 km), 330 kV Goranboy-2 (2xAKO-300, 84.2 km), 330 kV Goranboy-3 (2xAKO-300, 165.8 km), 330 kV 5th Mingechevir OHL (2xAKO-300, 33.8 km) and 330 kV 6th Mingechevir OHL (2xAKO-300, 33.3 km).To compensate for the additional loading capacity of all 5 lines, the USR is connected to the 3rd circuit of the one-and-a-half breaker scheme, where the Goranboy-2 PTL is connected.

The working superficial dependence $Q_r = f(P,t)$, constructed on the basis of expression (10) of the CSR, is presented in Figure 2.

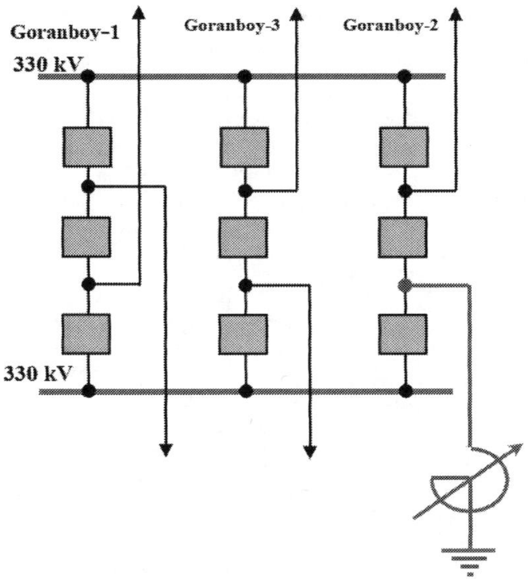

Fig. 1. Scheme of connection of CSR to 330 kV Goranboy DP

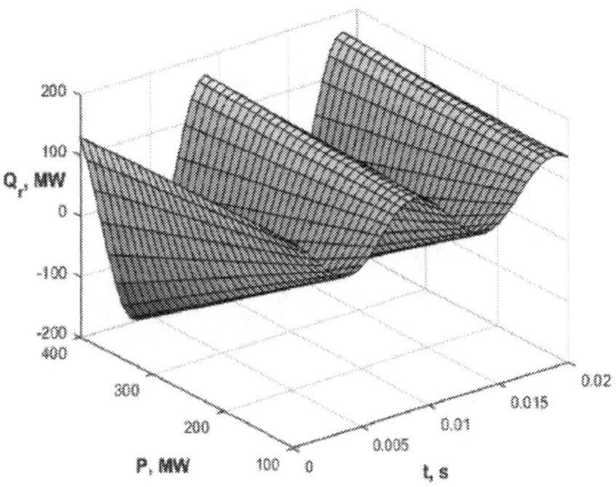

Fig. 2. Superficial description of the operation of the CSR, connected to the Goranboy DP

Table 1 shows the Additional loading powers generated by PTL and entering the distribution point are given in Table 1. As can be seen from the table, the total additional loading power generated by the 330 kV PTL at the distribution point is 152.69 MVAr. This power is large in magnitude and must be compensated in order to maintain constant voltage under various operating conditions.

Table 1. Additional loading powers of 330 kV PTLs entering the Goranboy distribution point

Seq. No.	PTL	Grade	Length km	Additional loading power, MVAr
1	Goranboy-1	2xASO-300/39	56,3	23,02
2	Goranboy -2	2xASO-300/39	84,2	34,43
3	Goranboy -2	2xASO-300/39	165,8	67,81
4	Mingechevir-5	2xASO-300/39	33,8	13,82
5	Mingechevir-6	2xASO-300/39	33,3	13,61

The actual operating mode of the CSR was monitored in different periods. Figure 3 shows The operation diagram of the CSR installed at the Geranboy DP and connected according to Figure 1 is presented in Figure 3.

© The Authors, published by EDP Sciences. This is an open access article distributed under the terms of the Creative Commons Attribution License 4.0 (https://creativecommons.org/licenses/by/4.0/).

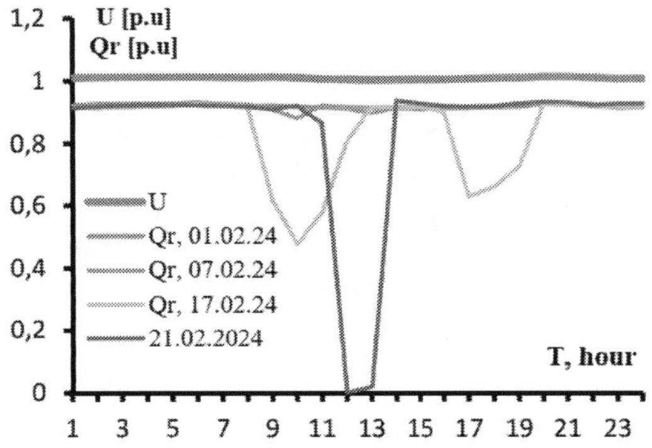

Fig.3. Diagram of the CSR connected to the 330 kV Goranboy DP

As can be seen from the Figure 3, despite the sharp changes in the mode during four days in February 2024, the USR operates in real time and stably maintains the voltage at the rated level. The trajectory of the CSR operation on the day in question is presented in Figure 4. As can be seen, regardless of the nature of the change in the mode, the CSR, performing its normal function along a complex trajectory, provides for the maintaining the voltage mode within acceptable limits.

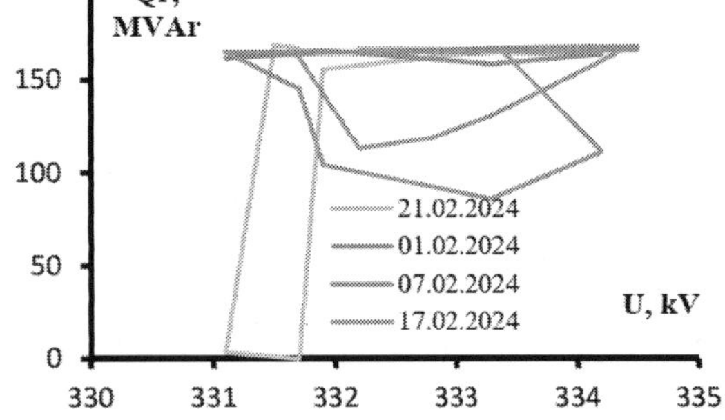

Fig. 4. The trajectory of the operation of the CSR connected to the Goranboy DP

3.2. Connection of CSR to the 330 kV Yashma Substation scheme and operating modes

Similarly, based on the mentioned method, the issue of determining the additional loading power and the rated power of the CSR for the connection to the 330 kV bus of the 330/220/110/10 kV Yashma Substation of the Azerbaijan power system was considered. The substation plays the role of an intersystem link and is of crucial importance in organizing parallel work with the Russian power system. The scheme of CSR connection to the 330 kV switchgear of the 330 kV Yashma Substation is shown in Figure 5.

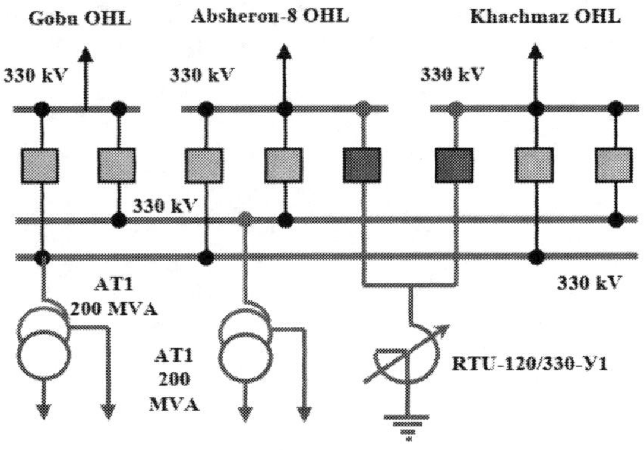

Fig. 5. Scheme of CSR connection to 330 kV buses of 330/220/110/10 kV Yashma SS

As can be seen from the figure, the 330 kV switchgear is made according to the "triangle" scheme and the RTU-120/330-U1 type CSR by means of two switches is connected to the busbar disconnector, to which the "Absheron-8" and "Khachmaz" overhead lines are connected. The 330 kV Yashma Substation is connected to the lines 330 kV Derbend-Khachmaz (2xACO-300, 110.2 km) and 330 kV Derbend (2xACO-300, 111.6 km) lines, and the 330 kV Absheron -330 kV Absheron-8 (2xACO-300, 46.7 km) node. In addition, the Gobu OHL is also connected to switchgear. The CSR plays an important role in ensuring the stability of the mode, especially when operating during power exchange in parallel mode with the Russian power grid. Table 2 shows The characteristics of the mentioned 330 kV OHLs connected to the 330 kV Yashma Substation and the charging capacities of the PTLs are given in Table 2. As can be seen from the table, the total additional load power generated by 330 kV power transmission lines is 126.78 MVAr and compensation must be carried out to maintain the voltage stable under different operating conditions.

Table 2. Additional loading powers of 330 kV PTLs entering the Yashma distribution point

Seq. No.	PTL	Grade	Length, km	Additional loading power, MVAr
1	Gobu	2xASO-300/39	41,5	16,97
2	Absheron-8	2xASO-300/39	46,7	19,1
3	Khachmaz	2xASO-300/39	110,2	45,07
4	Derbend	2xASO-300/39	111,6	45,64

The working surface, constructed on the basis of expression (10) of the CSR, is presented in Figure 6.

© The Authors, published by EDP Sciences. This is an open access article distributed under the terms of the Creative Commons Attribution License 4.0 (https://creativecommons.org/licenses/by/4.0/).

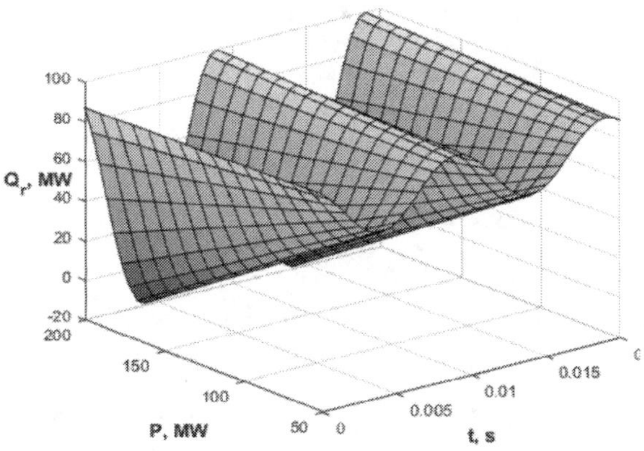

Fig. 6. Superficial dependence of the operation of the CSR connected to the 330 kV Yashma SS

The actual operating mode of the CSR connected to the 330 kV switchgear of the 330 kV Yashma Substation was monitored on different days. The diagram of operation of the CSR installed at the Yashma substation is presented in Figure 7.

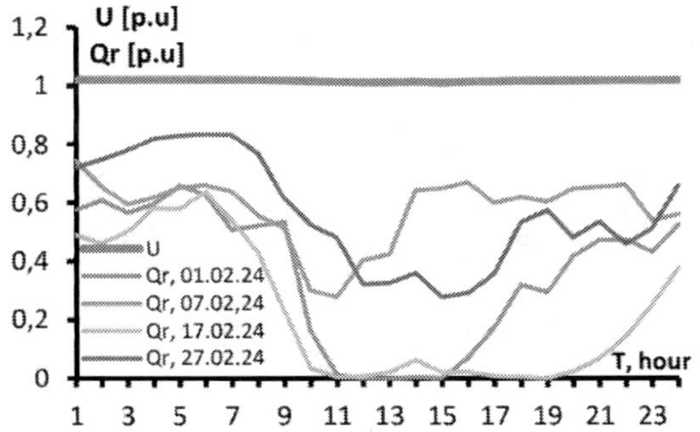

Fig. 7. Diagram of operation of the CSR connected to the 330 kV switchgear of the Yashma substation

As can be seen from the Figure 7, despite the sharp changes in the mode during four days in February 2024, the CSR operates in real time and stably maintains the voltage at the rated level. The trajectory of the CSR operation on the day in question is shown in Figure 8. As can be seen from Figure 8, regardless of the nature of the mode change, the CSR operates along a more complex trajectory, providing for the maintaining the voltage mode on the 330 kV buses within acceptable limits, and the stability of the parallel voltage operation mode is also maintained.

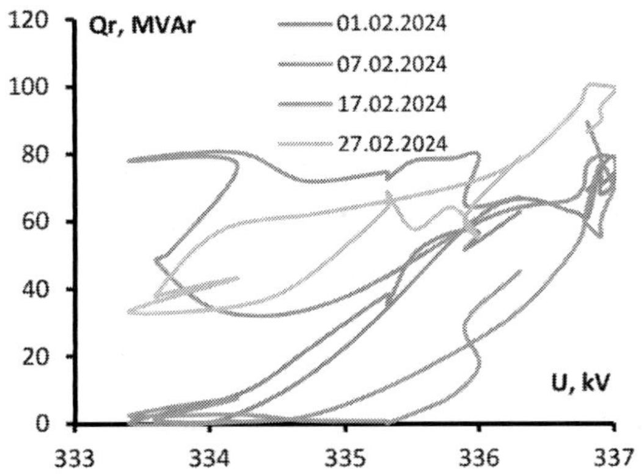

Fig. 8. Trajectory of operation of the CSR connected to the Yashma substation

Conclusion

1. A method for selecting shunt reactors under conditions of sharp mode changes within wide range, based on instantaneous power, is proposed. Based on this method, the rated powers of controlled shunt reactors in two 330 kV power grids of the Azerbaijan power system were determined and their placement in these nodes was ensured.

2. The results of the studies of the operating modes of the CSR installed at the Yashma substation and the Goranboy DP showed that operating within wide range, the CSR under conditions of sharp changes in the mode increases the voltage stability of the power system.

References

1. Dolgopolov A.G., Condratenko D.V., Ucolov S.V., Postolati V.M. Reactoare controlate shuntate pentru retele electrice. Problemele Energeticii Regionale, 3(17), 2011, c.2-7.

2. Magnetically controlled electric reactors / Edited by A.M. Bryantsev. Moscow, ZNAK, 2010, 288 p.

3. Hashimov A.M., Guliyev H.B., Babayeva A.R. Shunt reactors control algorithm using fuzzy sets theory, IJTPE Journal International Journal on Technical and physical problems of engineering (IJTPE), Issue 38, Vol. 11, No. 1, 2019, pp. 10-15.

4. Hashimov A.M., Rahmanov N.R., Guliyev H.B. Capacitor bank placement and control in electrical distribution systems using multiple objective fuzzy optimization method. IJTPE Journal International Journal on Technical and physical problems of engineering (IJTPE), Issue 34, Vol. 10, No. 1, 2018, pp.22-26

5. Hashimov A.M., Guliyev H.B., Babayeva A.R. Method and algorithm of control of shunt reactors of high-voltage power septenber network in maintenance modes based on fuzzy logic theory, 6 th International Conference on Modern Electric Power Systems (MEPS 2019), 9-12 2019,

© The Authors, published by EDP Sciences. This is an open access article distributed under the terms of the Creative Commons Attribution License 4.0 (https://creativecommons.org/licenses/by/4.0/).

Wroclav, Poland, Publisher:IEEE, Date Added to IEEE *Xplore*:06 April 2021, DOI: 10.1109/MEPS46793.2019.9395043

6. Hashimov A.M., Guliyev H.B., Babayeva A.R. Placement of shunt reactors in high-voltage network using fuzzy constraints, Rudenko International Conference "Methodological problems in reliability study of large energy systems" (RSES 2019). E3S Web of Conferences, Volume 139 (2019), Tashkent, Uzbekistan, September 23-27, 2019, pp.1-5

7. Hashimov A.M., Guliyev H.B., Babayeva A.R. Shunt reactors control algorithm based on fuzzy logic theory, Proceedings of the 7th International Conference on Control and Optimization with Industrial Applications (COIA-2020), Vol.1, 26-28 August, 2020 in Baku, Azerbaijan, p.209-212.

8. Mehmet T., Tuğçe D., Selva B., Rahmi İ.K., Burcu D., Mahmut A. A review of magnetically controlled shunt reactor for power quality improvement with renewable energy applications, Renewable and Sustainable Energy Reviews, Volume 77, September 2017, Pages 215-228, https://www.sciencedirect.com/science/article/abs/pii/S1364032117305105

9. Tao Zheng, Yanjie J. Zhao, Ying Jin, Peilu L. Chen, Fenfen F. Zhang. Design and Analysis on the Turn-to-Turn Fault Protection Scheme for the Control Winding of a agnetically Controlled Shunt Reactor, IEEE Transactions on Power Delivery, Volume 30, Issue 2, April 2015, pp.967-975,
https://ieeexplore.ieee.org/abstract/document/6939736

10. Kuchanskyy V. Application of controlled shunt reactors for suppression abnormal resonance overvoltages in assymetric modes. 2019 IEEE 6th International Conference on Energy Smart Systems (ESS), 17-19 April 2019, Kyiv, Ukraine,
https://ieeexplore.ieee.org/abstract/document/8764196

11. Aleksandrova M.I., Naumov V.A., Antonov V.I., Ivanov N.G. A Development of Shunt Reactor Controlled Energizing Theory, 2019 2nd International Youth Scientific and Technical Conference on Relay Protection and Automation (RPA), 24-25 October 2019, Moscow, Russia

12. Belyayev A.N., Serguei V.S. Steady-state and transient stability of 500 kV long-distance AC transmission lines with magnetically controlled shunt reactors, 2005 IEEE Russia Power Tech, 27-30 June 2005, St. Petersburg, Russia,
https://ieeexplore.ieee.org/abstract/document/4524464

13. Gusev A.S., Suvorov A.A., Sulaymanov A.O. Using Controlled Shunt Reactors for Voltage Stabilization on the Example of Real Electric Power System. IOP Conf. Series: Materials Science and Engineering 93 (2015) 012016, pp.1-6. https://iopscience.iop.org/article/10.1088/1757-899X/93/1/012016/pdf

14. Hashimov A.M., Guliyev H.B., Babayeva A.R. IJTPE Journal International Journal on Technical and physical problems of engineering, Issue 30, Vol. 9, No. 1, 2017.-pp. 18-22.

15. Pirverdiyev E.S., Hashimov A.M., Quliyev H.B., Babayeva A.R. Improved control algorithm for shunt reactors using fuzzy logic // Rudenko International Conference "Methodological problems in reliability study of large energy systems". Irkutsk, Russia, July 2-7, 2018, E3S Web of Conferences 58, 03015. https://doi.org/10.1051/e3sconf/20185803015, RSES 2018.

16. Ilyushin, P.V., Sukhanov O.A. The structure of emergency-management systems of distribution networks in large cities. Russian Electrical Engineering. 2014. Vol. 85. № 3. pp. 133-137. doi:10.3103/S1068371214030067.

17. Ilyushin, P.V. The analysis of dispersed generation influence on power system automatics settings and function algorithms // The Proceedings of Methodological Problems in Reliability Study of Large Energy Systems (RSES). pp. 1-5, July 02 – 07, 2018. Irkutsk, Russia. E3S Web of Conferences, Vol. 58. doi:10.1051/e3sconf/20185802001.

18. Ilyushin, P.V. Under-frequency load shedding strategies for power districts with distributed generation / P.V. Ilyushin, S.P. Filippov // The Proceedings of 2019 International Conference on Industrial Engineering, Applications and Manufacturing (ICIEAM). pp. 1-6. March 25-29, 2019. Sochi, Russia. doi: 10.1109/ICIEAM.2019.8743001.

19. Guliyev H.B. Management modes of reactive power compensation facilities in networks with renewable energy sources with distorting loads. IJTPE Journal International Journal on Technical and physical problems of engineering (IJTPE), Iss. 58, Vol. 16, No.1, March 2024, pp.14-20.

20. Yusifbayli N.A., Guliyev H.B. Intelligent Voltage Regulation on Base of Fuzzy Sets Theory. The 12th International Conference on Pobabilstic Methods Applied to Power Systems. PMAPS 2012, 10-14th June 2012, Istanbul, Turkey, p.447-453.

21. Yusifbayli N.A., Guliyev H.B., Aliyev A.G. Voltage Control System for Electrical Networks Based on Fuzzy Sets. Advances in Intelligent Systems and Computing, 2021, 1323 AISC, pp. 55–63.

© The Authors, published by EDP Sciences. This is an open access article distributed under the terms of the Creative Commons Attribution License 4.0 (https://creativecommons.org/licenses/by/4.0/).

Determination of optimal dimensions of hybrid AC/DC distributed generation system with renewable sources for autonomous power supply of remote locations

Nariman Rahmanov[1], *Huseyngulu* Guliyev[2*], *Famil* İbtahimov[2], *Zaur* Mammadov[1]

[1] Azerbaijan Scientific–Research and Designed–Prospecting Institute of Energetics, 94 H.Zardabi str., Baku, Azerbaijan

[2] Azerbaijan Technical University, 25 H.Javid str., Baku, Azerbaijan

Abstract. In this paper, an autonomous hybrid power system with direct voltage and alternating voltage busbars with connected renewable sources and traditional sources that support renewable sources in the demand coverage during periods of power shortage in the system is proposed for power supply of remote locations. The issues of synthesis of the optimal structure of an autonomous system consisting of PV and WT generators with an integrated system of electric energy batteries (EEB) and diesel- generator station, at which the high reliability of power supply in remote areas is provided, are being studied. To optimally determine the installed capacity of the elements of the PV–WT–EEB–DG hybrid system, a machine learning algorithm (TLBO) is proposed, by means of which, based on the proposed multi-objective function, the most advantageous values of unit powers for PV and WT stations are determined, as well as the number of batteries and diesel generators that ensure the demand coverage in the hybrid system under various options for PV and WT generation shortage.In order to improve the efficiency of solving the optimization problem taking into account the variability of meteorological conditions, manifested in random and uncertain variability of PV and WT generation, as well as in the stochastic nature of the electricity consumption process, the studies on clustering these processes and determining typical scenarios for an array of hourly measurements of PV, WT and load generation during the year were carried out in the paper. Using the k–means approach, the numbers of clusters for PV, WT and load for typical geographic regions are determined. The results of the analysis of the optimal compositions of the hybrid AC-DS system in the PV, WT, BEE, DG in these regions are presented.

Introduction

In the complex of problems of transformation of electric power systems, a large number of studies are devoted to the problem of development of distributed generation (DG). In the last decade, the DG has become even more attractive due to the widespread use of renewable sources. DG schemes in combination with renewable sources such as PV and WT generators can support full demand coverage in isolated operation mode.

At the same time, due to the intermittency and uncertainty of PV and WT power generation in the DG system, power supply stability disturbances may occur, and in the case of excess power generation, this may lead to its dumping [1-3]. Hybrid DG schemes with wind WT and solar PV generators and integrated grid of batteries and diesel power plant are considered as reliable autonomous power supply systems for rural and remote locations [4,5].

A very important problem for a hybrid distributed generation system with renewable sources is the determination of the optimal sizes of individual components – wind/solar generators/electric energy battery network, which correspond to the most advantageous technical (fulfilment of reliability indicators) and economic conditions (minimum total costs) of power supply for consumers in the studied area.

The unreliability of RES can be overcome by increasing the size of components (PV and WT generators), but this leads to the increase in the cost of the system. On the other hand, reduction of the sizes of components can lead to a loss of power supply (LPS), at which the power output from renewable sources is less than the load power. Therefore, optimal sizing of renewable sources and thus determining the exact quantity of their components is of great importance [6]. A hybrid PV-WT-DG-EEB microsystem is designed and successfully used to supply electricity to three remote villages [7]. In works [6, 7], as well as in many other works related to determining the optimal dimensions of a hybrid system with renewable sources, HOMER software is used to solve the problem. The program has some limitations in application, as it is mainly aimed at economic justification of the choice of a component of a hybrid system, the issues of system reliability depending on meteorological changes are not considered. In recent years, a number of algorithms

* Corresponding author: huseyngulu@mail.ru

© The Authors, published by EDP Sciences. This is an open access article distributed under the terms of the Creative Commons Attribution License 4.0 (https://creativecommons.org/licenses/by/4.0/).

for multi-objective optimization based on metaheuristic methods have been located, among these methods, the genetic algorithm has been widely developed [8, 9].

In this paper, the problem of determining the optimal component sizes of an autonomous PV–WT–batteries system is solved using an improved TLBO (JAYA) algorithm. Unlike existing algorithms for determining the optimal parameters of the hybrid system, in this study the TLBO algorithm is used taking into account meteorological variability. For the representation of the stochastic variability of wind speed, solar radiation and consumed electric power, their spatial-temporal analysis was carried out and typical scenarios for the optimal choice of the sizes of renewable energy resources in the proposed microsystem were established. The scenarios generated by the proposed model make it possible to approximate the initial scenarios and reduce the amount of data for an effective combination of hybrid system components.

The results of the research of the proposed optimization approach were obtained using the example of typical geographical locations in Azerbaijan.

1 Formulation of the problem of determining the optimal dimensions of a hybrid AC/DS (PV-WT-BES-DG) system for the purpose of technical provision of the system reliability

1.1 Hybrid AC/DC distributed generation system diagram PV – WT – battery – BG

A typical hybrid dual-bus AC/DC distributed generation system with renewable sources has been developed. To the AC subsystem are connected as follows: wind generators, load consisting of AC and DC consumers, and DG is also connected to the system, acting as an emergency source. Wind generator, electric energy battery, DC and AC load are connected to the DC subsystem.

The AC and DC subsystems are connected to each other by direct-acting and reverse-acting inverter. Depending on the power generation by sources in the AC and DC subsystems, various modes of load coverage in each subsystem and, in general, the total load in the entire hybrid system are possible: balance, excess and shortage. In balance mode, the total power generated by PV in the DS subsystem and the power generated by WT in the AC is equal to the total power of consumers in these subsystems. At excess of total power generation in the DC and AC subsystems relative to the total consumption and total power in the battery system, the excess output is rejected. In case of power generation shortage in the AC and DS subsystems, when the total load cannot be fully covered and the battery capacity is insufficient to cover this shortage, in this case the corresponding part of the load is disconnected from the power supply.

1.2 Reliability criteria of the autonomous AC/DS system

Maintaining the reliability of the isolated AC/DC system is very important when selecting technical conditions for continuous power supply depending on the power generation by renewable sources. These conditions are regulated by the following indicators [10]:

1. Probability of loss of power supply

$$\Delta \tilde{P}_{load} = \frac{\sum^{T} \Delta P_{load}(t)}{\sum_{t}^{T} (P_{loadDC}(t) + P_{loadAC}(t)} \quad (1)$$

where, $\Delta P_{load}(t)$ – the value of the disconnected power of the total load in the AC/DC system; t –period of shortage T = 8760 h.

2. Share of rejected generation $\mathrm{P}_{gen}(t)$, when the sum of generation of AC and DS subsystems exceeds the total demand for the load, and the EEB has an upper limit of its charging capacity, the excess generation must be tripped

$$\Delta \tilde{\mathrm{P}}_{gen} = \frac{\sum_{t=1}^{T} \Delta \mathrm{P}_{gen}(t)}{\sum_{t=1}^{T} (\mathrm{P}_{DC}(t) + \mathrm{P}_{AC}(t))} \quad (2)$$

A block diagram for calculating the most advantageous technical reliability indicators (1) and (2) for an isolated hybrid AC/DS system has been developed.

2 Application of k-means cluster method for generation of typical scenarios for random variations of solar radiation, wind speed and electricity consumption for typical geographical regions of Azerbaijan

The stochastic nature of the variability in power output from solar PV and wind turbine generators complicates their integration into the power grid. Currently, various approaches are proposed for decomposing a set of data of random changes of variables into a combination of a set of regular changes. In this paper, the k-means clustering apparatus is used for the generation of typical scenarios of solar radiation, wind speed and load. The obtained typical clusters will significantly reduce the consideration of many scenarios that would need to be formed from a large array of initial data. The k-means clustering technology is carried out as follows:

1. An array of time series of solar radiation (W/m^2), wind speed (m/sec), and load power (W) for 1 year is entered.

2. Clustering of stochastic processes of load, solar radiation, wind speed is carried out individually based on

© The Authors, published by EDP Sciences. This is an open access article distributed under the terms of the Creative Commons Attribution License 4.0 (https://creativecommons.org/licenses/by/4.0/).

hourly measurements for the entire year; the basic unit of clustering is one day (24 hours).

3. Based on the known dependence (spatial correlation) of the load and solar radiation, their joint clustering is carried out.

4. In accordance with the centers of the solar radiation–load clusters (is determined from the general set of load–solar radiation data), typical solar radiation–load scenarios are determined.

5. Accordingly, for each type of solar radiation-load and typical scenarios from wind data (defined in accordance with the centers of wind speed clusters), the relationships between solar radiation, load and wind speed are determined.

Based on the computer implementation of the above-described stages, the optimal number of clusters for stationary processes of solar radiation variability - consumption of electrical power and wind speed for the mountainous area of Azerbaijan is determined.

3 Modeling of clustering of solar radiation, load and wind speed variability for meteorological conditions of mountainous terrain of Azerbaijan

For modeling of clustering and identifying typical clusters, measurements of solar radiation, load and wind speed processes obtained for a remote area of the Kelbajar Region of the East Zangezur economic region of Azerbaijan were used in the form of hourly measurements throughout the year. Figures 1 (a, b), 2 (a, b), 3 (a, b) show Images of these processes in the interval of one year and one day, respectively, are shown in Figures 4 (a, b), 5 (a, b), 6 (a, b).

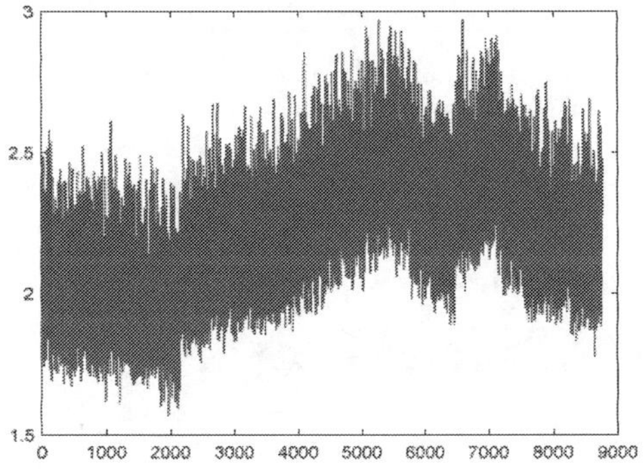

a)

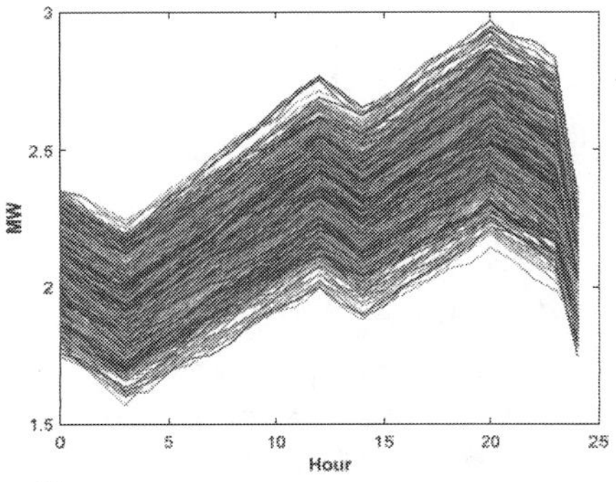

b)

Fig. 1. Profile of load changes by hours during the year (a) and day (b)

Comparative analysis of solar radiation and load processes both in the interval of one year or one day have a form closer to regular than the processes of speed change. Based on these considerations, clusters for the general load-solar radiation process were also considered to search for typical scenarios.

Using the algorithm for selecting the optimal number of clusters proposed in [10], the optimal numbers of clusters for load, solar radiation and wind speed for a period of 1 year for the Kelbajar region were determined. The trend of the percentage ratio of the dispersion of the results of clustering of the load process and the corresponding silhouettes of the clusters are shown in Figures 6,a and 6,b.

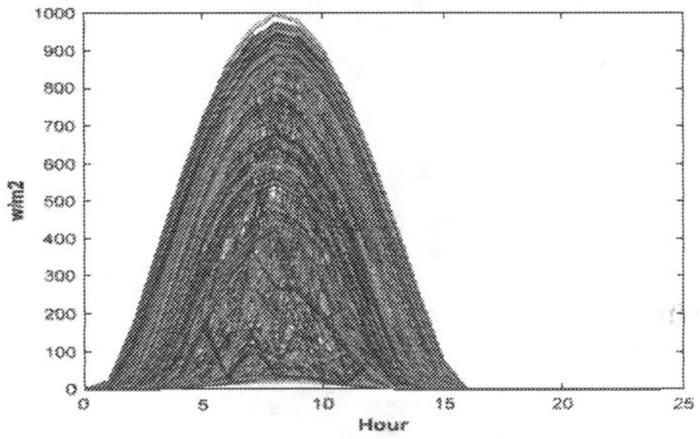

a)

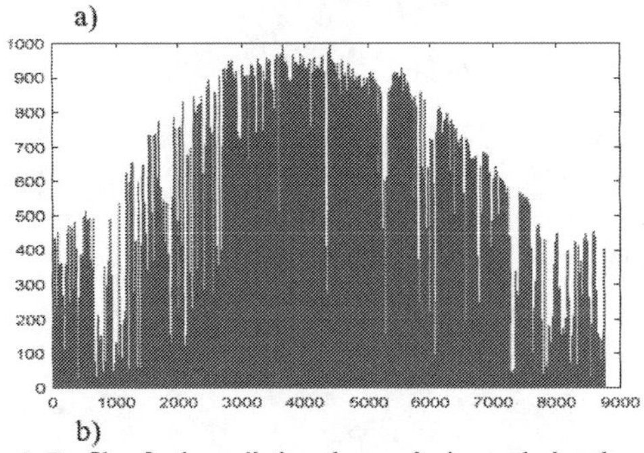

b)

Fig. 2. Profile of solar radiation changes by hours during the year (a) and day (b)

© The Authors, published by EDP Sciences. This is an open access article distributed under the terms of the Creative Commons Attribution License 4.0 (https://creativecommons.org/licenses/by/4.0/).

https://doi.org/10.1051/e3sconf/202458401020

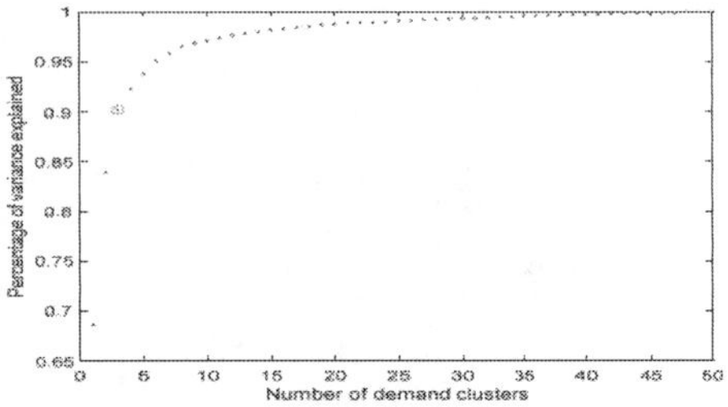

a)

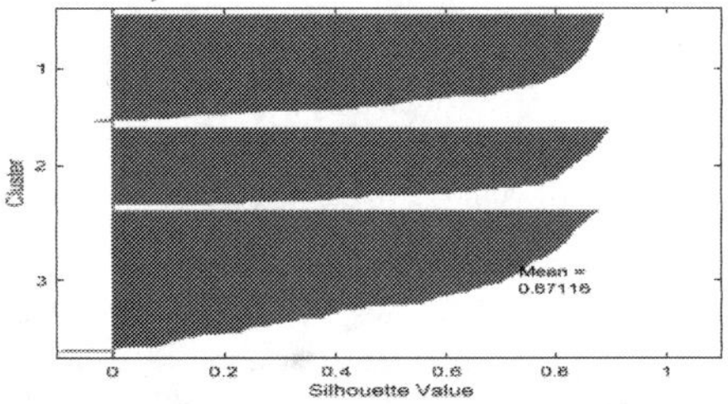

b)

Fig. 3. Optimal numbers of clusters for annual load variability (a) and corresponding silhouettes (outlines) (b)

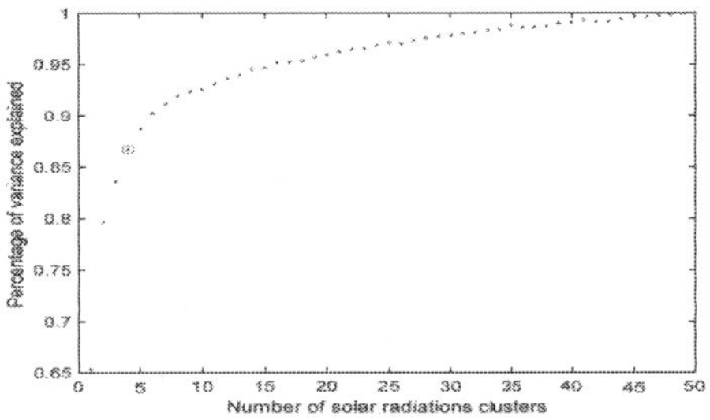

a)

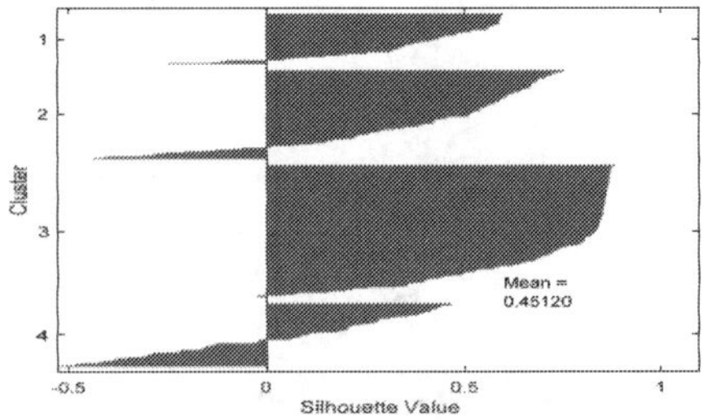

b)

Fig. 4. Optimal number of clusters for annual variability of solar radiation (a) and corresponding numbers of silhouettes (outlines) (b)

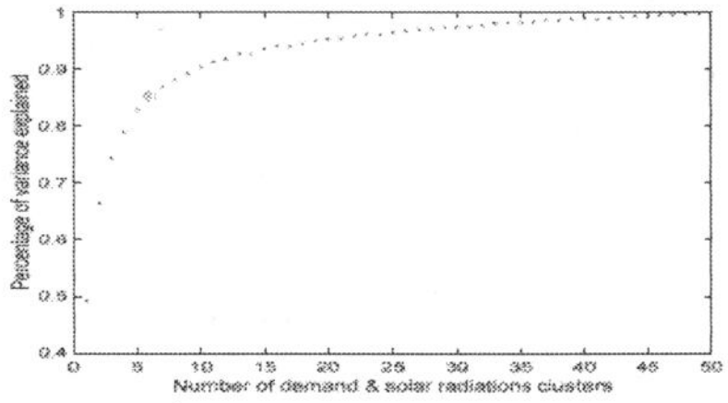

a)

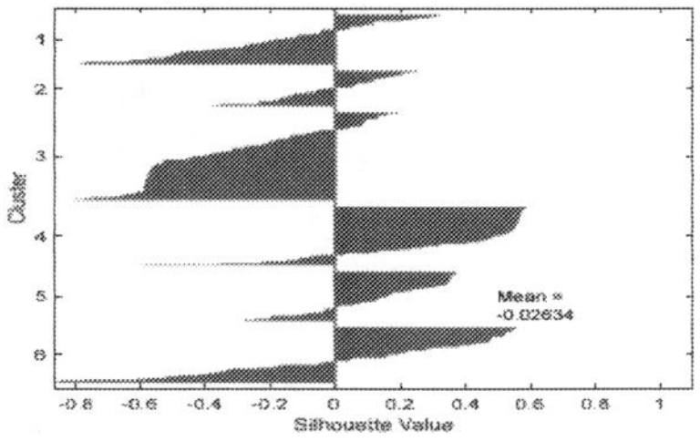

b)

Fig. 5. Optimal number of clusters for the solar radiation–load variability data set (a) and their silhouettes (outlines) (b) over the year

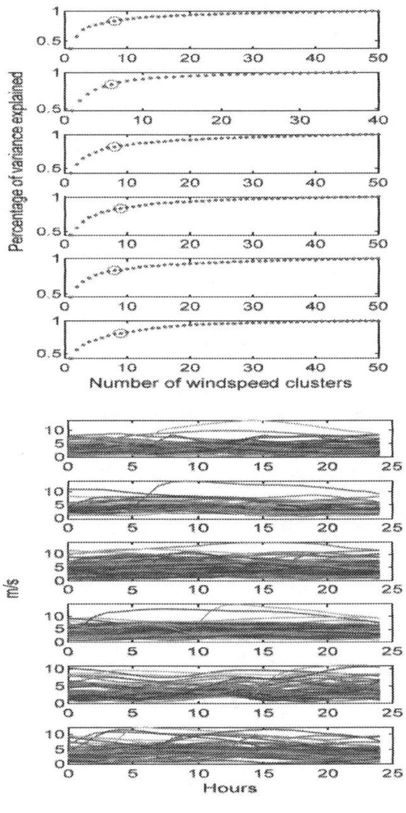

Fig. 6. Optimal number of clusters of load–solar radiation variability according to solar radiation variability

© The Authors, published by EDP Sciences. This is an open access article distributed under the terms of the Creative Commons Attribution License 4.0 (https://creativecommons.org/licenses/by/4.0/).

Thus, from the above Figure 1 - Figure 6 it is clear that the optimal number of scenarios for the data of load 3; solar radiation 4; solar radiation–load 6; load–solar radiation–wind data have six options of scenarios, which consist of the following number of scenarios: 6; 7; 8; 9; 8; 9. The identified numbers of scenarios serve as the initial base for the composition of renewable generating sources connected to the AC and DS subsystems of the hybrid system.

The optimal control of power generation in the AC and DC subsystems depends on the spatial-temporal state of the mode in the hybrid system, which is considered as a multi-objective problem.

4. Multi-criteria dimensionality optimization function of autonomous hybrid AC/DC system with integrated PV solar and WT wind generators

An important part of the problem of forming a scheme of an autonomous hybrid system with renewable sources is ensuring the reliability of its operation under conditions of uncertain changing meteorological data - solar radiation and wind speed, as well as a random process of power consumed by the load. The value of power generation in the AC subsystem is assumed to fully cover the load power P^e_{AC} connected to the subsystem, as well as the possibility of transmitting excess power up to 30% to the DC subsystem.

The amount of power generated by PV generators in the DC subsystem is assumed to cover the power of the load connected to the DC subsystem, as well as the possibility of generating additional power for transmission to the AC subsystem in case of its excess. In case of excess of the mode $P^W(t)$ for the period t, technically the reliability in the system can be maintained based on the fulfillment of the goal:

$$F_1\left(N^W, N^{PV}, N^{bat}, N^{DG}\right)=$$
$$\sum\left[P^W(t)-P^{PV}_{DC-AC}(t)-P^L_{AC}(t)\right]\Rightarrow \min$$

where, $N^W, N^{PV}, N^{bat}, N^{DG}$ – the number of wind turbines, solar cells, battery cells and diesel generators, respectively; $P^W(t)$ – wind turbine generation power, exchange between AC and DC buses, load connected to the AC bus at t time , respectively.

Accordingly, in the case of wind power shortage, the objective of optimal power control in the hybrid AC/DS system is described as a function

$$F_2\left(N^W, N^{PV}, N^{bat}, N^{DG}\right)=$$
$$\sum\left[P^W(t)+P^{PV}_{DC-AC}(t)+P^{DG}(t)+P^L_{AC}(t)\right]\Rightarrow \min$$

For conditions of excess or shortage of PV generation power (DC subsystem), power control optimization can be expressed in the form of the following objectives: excess power

$$F_3\left(N^W, N^{PV}, N^{bat}, N^{DG}\right)=$$
$$\sum\left[P^W(t)-P^{PV}_{DC-AC}(t)-P^{bat}(t)-P^L_{DC}(t)\right]\Rightarrow \min$$

Power shortage

$$F_4\left(N^W, N^{PV}, N^{bat}, N^{DG}\right)=$$
$$\sum\left[P^{PV}(t)+P^W_{DC-AC}(t)+P^{bat}(t)-P^L_{DC}(t)\right]\Rightarrow \min$$

Power from inverter outlet

$$P_{inv}(t)=\frac{P^L(t)}{\eta_{inv}}$$

where η_{inv} – inverter efficiency factor.

Cases of hybrid system operation (PV + Wind)

1) $\sum P_G$ exceeds the required load power (inverter input power). In this case, the excess energy is accumulated in the batteries, the remaining power after charging is not used, i.e. it is rejected.

2) $\sum P_G$ is less than power required for the loading. The energy shortage is covered by the storage device.

3) In case of equation $\sum P_G = P_{inv}$, the power (capacity) P_{bat} remains unchanged.

Table 1 shows the calculation results, when the objective function in the first case is minimum load tripping ($F=\min YA$) at $100<N^{bat}<20000$, $100<N^{PV}<60000$, $100<N^W<5000, C_{bat}=40000Wah, P^{PV}=125BW$ and $P^W=5000W$.

Table 1. Results of calculations at $F=\min YA$

N	N^{bat}, N^{PV}, N^W	*YA*	*IG*	$F(...)$
1	[15621,6893,3328]	0	0.2256633448	0
2	[7038,27790,2436]	0	0.3732130856	0
3	[9384,12836,3171]	0	0.3866004557	0
4	[12076,199,4069]	0	0.4159784484	0
5	[12433,33398,4241]	0	0.4875853297	0
6	[2068,51802,4291]	0	0.7508031625	0
7	[9220,8200,4361]	0	0.5338473061	0
8	[14504,55473,3022]	0	0.3311421121	0
9	[6096,35392,4867]	0	0.6783884689	0
10	[1939,28149,2969]	0	0.6308979325	0

© The Authors, published by EDP Sciences. This is an open access article distributed under the terms of the Creative Commons Attribution License 4.0 (https://creativecommons.org/licenses/by/4.0/).

As can be seen from the Table 1, the most rational option is obtained $F = \min YA$- the first option. Similarly, other options are considered.

Conclusion

1. For improvement of the reliability and increase the efficiency of an isolated power supply system with variable renewable sources, a method is proposed for determining the optimal sizes of this system, represented by a dual-bus AC/DC circuit. The solution to the problem of determining the optimal architecture of a hybrid power system is carried out on the basis of established scenarios of the system state for a year of observations of random variability of solar radiation, wind speed and stochastic variability of the load in the studied remote area.

2. An algorithm of the machine learning method for determining the optimal values of the placed power of wind generation and solar generator for all scenario options is proposed. The proposed method for determining scenarios and the algorithm for determining the optimal parameters of the hybrid system were tested in the planned power supply systems for typical geographic locations of Azerbaijan.

References

1. Rahmanov N.R., Kurbatskiy V.G., Guliyev H.B., Tomin N.B., Mammadov Z.A. Probabilistic assessment of power system mode with a varying degree of wind sources integration. E3S Web of Conferences, Volume 25 (2017), Methodological Problems in Reliability Study of Large Energy Systems (RSES 2017), Bishkek, Kyrgyzstan, September 11-15, 2017, pp.1-5.
2. Chen H.C. Optimum capacity determination of stand – alone hybrid generation system considering cost and realibility. Appl. Energy 2013, 103, 155 – 164.
3. Diaf S., Notton G., Belhamel M., Haddadi M., Louche A. Design and Techno – economical optimization for hybrid PV/Wind system under various metereological conditions. Appl. Energy 2008, 85, 968-987.
4. Yang H., Zhou W., Lu L., Fang Z. Optimal sizin method for stand – alone hybrid solar- wind system with LPSP technology by using genetic algoritm. Sol. Energy 2008, 82, 354- 367.
5. Khan A., Javaid N., Javaid S. Optimum unitsizing of stand – alone PV-WT battery hybrid system components using jaya. In : Proceedings of the 2018 iEEE 21 st. International multi Topic Conference, 2018, Nov. 1-2, Karachi, Pakistan, pp. 1 – 8.
6. Al Busaidi, A.S., Kazem H.A., Al- Bad H.A., Khan MF. A review of optimum sizing of hybrid PV – wind renewable energy system in Oman. Renewable Sustainable Energy Rev.
7. Mamaghani A.H., Escandon S.A.A., Najafi B., Shirazi A., Rinaldi F. Techno – Economic feasibility of fotovoltaic, wind , diesel and hybrid electrification systems for off-grid rural electrification in Colombia, Renew Energy, 2016, 97, 293 – 305.
8. Maleki A., Khajeh M.G., Rosen M.A. Two heuristic approches for the optimization of grid – connected hybrid solar – hydrogen systems to supply residential thermal and electrical loads. Sustain cities Soc. 2017 ;34: 278 – 92
9. Fathy A.A. Reliable methodology based on mine blast optimization algorithm for optimal sizing of hybrid PV – wind FL system for remote area in Egypt, Renew Energy, 2016, 95, 367- 380.
10. Rahmanov N.R., Kurbatsky V.G., Guliyev H.B., Tomin N.V., Yaqubov A.F. Investigation of the impact of integrating renewable sources with variable power generation, taking into account restrictions on the voltage stability limit. E3S Web Conf., Rudenko International Conference "Methodological Problems in Reliability Study of Large Energy Systems" (RSES 2023), **Volume 461**, 01023, 2023, pp.1-5
11. Ilyushin, P.V. Emergency and post-emergency control in the formation of micro-grids // The Proceedings of Methodological Problems in Reliability Study of Large Energy Systems (RSES). pp. 1-6. September 11 – 15, 2017. Bishkek, Kyrgyzstan. E3S Web of Conferences, Vol. 25. doi:10.1051/e3sconf/20172502002.
12. Ilyushin, P.V. Photovoltaic power plants participation in frequency and voltage regulation / P.V. Ilyushin, A.V. Pazderin, R.I. Seit // The Proceedings of 17th International Ural Conference on AC Electric Drives (ACED). pp. 1-7. March 26 – 30, 2018. Yekaterinburg, Russia. doi: 10.1109/ACED.2018.8341712.
13. Ilyushin, P.V., Eroshenko S.A. Features of implementing multi-parameter islanding protection in power districts with distributed generation units. The Proceedings of 2018 IEEE 59th International Scientific Conference on Power and Electrical Engineering of Riga Technical University (RTUCON). pp. 1-6. November 12 – 14, 2018. Riga, Latvia. doi: 10.1109/RTUCON.2018.8659857.

© The Authors, published by EDP Sciences. This is an open access article distributed under the terms of the Creative Commons Attribution License 4.0 (https://creativecommons.org/licenses/by/4.0/).

Issues of risk-oriented approach to electrical networks reliability management

Artem Vanin[1*], *Dmitry* Gabdushev[1], *Vladimir* Tulsky[1]

[1]National Research University "Moscow Power Engineering Institute", Moscow, Russia

Abstract. The features of implementing a risk-oriented approach in Russian regulatory documents are examined. A brief overview of foreign experience is provided. The need for and a method of incorporating fault statistics in the calculation of failure probability based on asset health indexes are demonstrated. Additionally, the necessity of considering the duration of power supply interruptions in assessing the consequences of equipment failure is shown, along with a proposed method for doing so.

1 Introduction

In accordance with the energy strategy of the Russian Federation until 2035 [1], a key measure to improve the efficiency of the power grid companies (PGC) involves the transition to risk-oriented management of production assets within the electricity network, using digital technologies. The asset management standard highlights that decisions regarding asset management should be grounded in risk assessments to ensure that assets function effectively [2]. The principles for implementing a risk-oriented approach in the electric power industry are defined in the Resolution No. 1401 by the Government of the Russian Federation [3], and further detailed in orders from the Ministry of Energy of the Russian Federation No. 676 [4] and No. 123 [5].

In the order No. 123 of the Ministry of Energy of the Russian Federation [5], technological risk of equipment failure is defined based on the widely accepted formula of the product of the probability of failure and its consequences. Therefore, calculating both the probability of a unit's failure and the resulting consequences in terms of cost are crucial tasks when making risk-based decisions.

In reliability theory, the probability of equipment failure over an arbitrary time interval $[t_1, t_2]$ is defined using the exponential function of the failure rate [6]:

$$P(t_1, t_2) = 1 - e^{-\int_{t_1}^{t_2} \lambda(t)dt}, \tag{1}$$

where the failure rate function is determined by statistical analysis of failure logs or laboratory tests results.

With a statistical approach, homogeneous samples of equipment are formed, where the number of failures is recorded within a specified operating time. A single reliability metric is calculated for the entire sample, which doesn't allow for differentiation between the elements of the sample. Since failures in power equipment occur relatively infrequently, forming representative samples, typically consisting of several thousand units, is necessary to obtain statistically significant estimates. This leads to a contradiction between the significance of the sample and its homogeneity. However, the advantage of such an approach is that it takes into account actual failure statistics of the equipment under real operating conditions.

To evaluate the probability of failure for an individual unit of equipment, the global practice involves using Asset Health Indexes (AHI).

The methodology for calculating the AHI in Russia is approved by the Order of the Ministry of Energy of the Russian Federation № 676 [4]. In the United Kingdom, the AHI calculation methodology is developed and approved by the national regulatory authority OFGEM (Office of Gas and Electricity Markets) [7]. The approaches used in Canada, Italy, and the Netherlands are outlined in references [8, 9, 10], respectively.

The AHI is an integrated indicator of technical condition, that combines the values of several other technical condition indicators into a single value, facilitating comparison and assessment [3]. The highest technical condition corresponds to a value of 100, while the lowest corresponds to 0. The assessment of technical condition indicators is carried out through visual inspections, measurements, or operational testing. According to the Russian methodology [4], each parameter of the technical condition is given a score from 0 to 4. The AHI of each functional part is determined by aggregating the scores of the technical condition parameters using weighting coefficients. The AHI of equipment unit is defined by combining the AHIs of its functional parts, taking their weighting coefficients into account. Similar principles of calculation are used in methodologies [4, 7, 8, 9, 10],

* Corresponding author: vaninas@mpei.ru

© The Authors, published by EDP Sciences. This is an open access article distributed under the terms of the Creative Commons Attribution License 4.0 (https://creativecommons.org/licenses/by/4.0/).

although the parameters considered and the weighting coefficients may differ.

Weighting factors can be determined through statistical methods or expert assessments. In Russia, since the system for collecting detailed failure information [11] was established shortly before the development of the AHI assessment methodology, the weighting factors were determined through expert assessments [12, 13].

Both Russian and foreign approaches typically distinguish between 4 to 5 ranges of AHI values, correlating to qualitative characteristics of equipment condition from worst to best. This helps rank equipment based on technical condition and approximately plan the scope and timing of maintenance activities, while not allowing for the application of numerical evaluations. To apply a well-developed mathematical framework from reliability theory, the AHI need to be transformed into failure probabilities.

2 Calculation of Equipment Failure Probability Based on health index

In the assessment methodology for the technical condition approved by the Ministry of Energy of the Russian Federation [4], the probability of failure is defined as a real number ranging from 0 to 1, reflecting the relative frequency observed in a series of observations or the degree of confidence regarding the failure of a particular functional unit or the main technological equipment as a whole. It is also stated that the probability of failure for equipment not covered by the methodological guidelines is determined based on existing failure statistics of similar equipment. Thus, it is emphasized that the probability of failure should represent the actual failure frequency of the equipment.

Formulas 1, 3, and 6 from the methodological guidelines on calculating the probability of failure for functional units and main equipment [5] establish the following correspondence between the probability of failure and the AHI:

$$P = 1 - \frac{AHI}{100} \qquad (2)$$

The current regulatory method for calculating the probability of failure is based on a scoring assessment of the technical condition and weighting coefficients determined by expert methods. This approach does not take into account the actual failure statistics, which vary depending on the type of equipment and operating conditions.

As an example, let us compare the probability of failure calculated according to the methodology [5] with reference data presented in V.A. Nepomyashchy's monograph [14].

According to [4], equipment with an AHI in the range of (85; 100] is classified as "very good," requiring only scheduled diagnostics. Consequently, according to formula (3), the probability of failure for such equipment will fall within the range of [0; 0.15).

In comparison, Table 1 presents the failure probabilities of various 110 kV equipment types over a year, calculated using formula (1) based on the average failure rates provided in V.A. Nepomnyashchy's monograph [14]. For the 110 kV overhead line, the calculation is based on the average length of this class of voltage lines, which is 30 km. From the presented values, it is evident that the probability of failure calculated according to the methodology [5] exceeds the statistical values by more than ten times.

TABLE I. AVERAGE FAILURE RATE AND PROBABILITY OF FAILURE FOR 110 KV EQUIPMENT

Equipment	λ, 1/year	Probability
Power Transformer	0,009	0,009
Oil Circuit Breaker	0,0032	0,0032
Gas-Insulated Circuit Breaker	0,0025	0,0025
30 km Overhead Line	0,04	0,039

The paper by the Canadian power grid company Kinetrics [8] highlights the need to compare actual failure rates and asset health indices by developing a function $\lambda = f(AHI)$ for each type of equipment. In the absence of necessary data, [8] suggests determining the function $\lambda = f(AHI)$ based on functions $\lambda = f(t)$ and $AHI = f(t)$. The relationship presented in [8] regarding the failure probability of a 110 kV transformer in relation to its technical condition index is illustrated in Figure 1a.

The formula used in [7] for calculating failure probability based on the asset health index is represented by the first three terms of the Taylor series for the exponential function adjusted by coefficients:

$$P = K \cdot \left[1 + C \cdot H + \frac{(C \cdot H)^2}{2!} + \frac{(C \cdot H)^3}{3!} \right] \qquad (3)$$

where H is the Asset Health Index, with a value of 0 for the best condition and 10 for the worst condition; K and C are coefficients. For values of $H \leq 4$, the failure probability is calculated for $H = 4$.

The coefficient C is the same for all types of equipment and is chosen so that the failure probability of equipment in the worst technical condition is ten times higher than that of new equipment. The coefficient K is determined for each type of equipment based on actual failure statistics and the distribution of AHI values among the equipment fleet.

Figure 1b shows the failure probability dependencies related to AHI for several types of equipment, constructed according to [7]. The obtained characteristics for 110 kV power transformers are consistent with the values in Table 1.

© The Authors, published by EDP Sciences. This is an open access article distributed under the terms of the Creative Commons Attribution License 4.0 (https://creativecommons.org/licenses/by/4.0/).

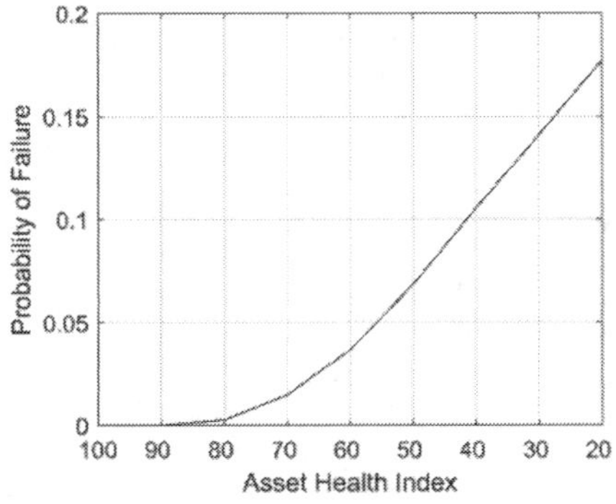

a)

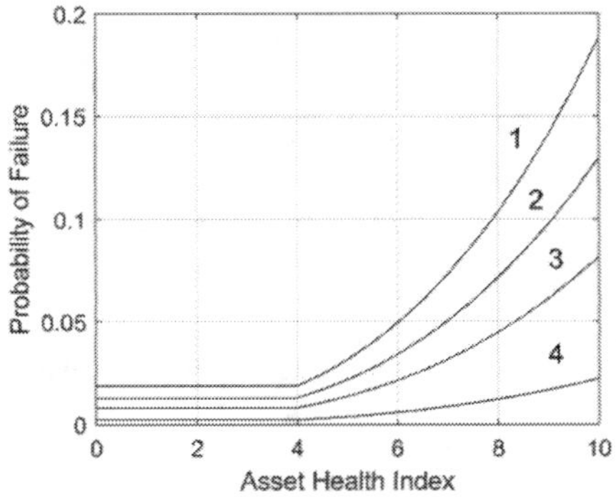

b)

Fig. 1. The dependence of failure probability on the Asset Health Index: a) For a 110 kV transformer according to [8], b) According to [7] for: 1 – 10 kV cable line (1 km), 2 – 110 kV transformer, 3 – 10 kV overhead line (1 km), 4 – 10/0.4 kV transformer.

To build statistical functions, it is necessary to take into account the AHI of each equipment unit throughout its entire lifecycle, as well as to record the current AHI of equipment that has failed in accident investigation reports. This allows to determine failure rates for each range of AHI values for a group of comparable objects:

$$\lambda_i = \frac{n_i}{\sum_{j=1}^{N} t_{ij}} \tag{4}$$

where n_i is the number of failures due to technical condition for equipment with an AHI falling within the specified range during the considered time period; N is the number of comparable objects in the examined group; t_{ij} is the duration that the j-th element's AHI remained within the defined range during the observed time period.

As an example, let's consider definition of the function $\lambda = f(AHI)$ in the form of exponential approximation. Let us denote the coefficients that define the shape and scale of the characteristic, as stated in [7], and choose the expression (2) as the parameter. The resulting function can be expressed as:

$$\lambda = K \cdot e^{C(1-AHI/100)}. \tag{5}$$

Given sufficient sample sizes of failure data across the entire range of AHI values, the parameters of function (5) can be easily fitted using software libraries.

In practice, however, for ensuring a high level of equipment operation, power grid companies may not generate statistics for low AHI values. In this case, the shape coefficient can be specified similarly to [7] through the ratio of failure intensities for specified AHI levels:

$$\frac{\lambda_1}{\lambda_2} = \frac{K \cdot e^{C(1-AHI_1/100)}}{K \cdot e^{C(1-AHI_2/100)}} = e^{C(AHI_2-AHI_1)/100} \tag{6}$$

$$C = \frac{100\, ln(\lambda_1/\lambda_2)}{AHI_2 - AHI_1} \tag{7}$$

As an example, Table 2 provides failure statistics for 110 kV transformers from the central region of Russia over a 3-year period. In this case, data is available only for equipment with AHI values ranging from 60 to 100. The distribution of equipment and failures across AHI values is uneven. The sample size for AHI = 60 is significantly smaller than for AHI values of 80 or 90.

TABLE II. FAILURE STATISTICS FOR 110 KV TRANSFORMERS

Health index	**60**	**80**	**90**	**100**
Number of Units	66	891	864	69
Number of Units × Number of Years	198	2673	2592	207
Number of Failures	4	23	13	1
λ (Failure Rate), 1/year	0,0202	0,0086	0,0050	0,0048

The calculated failure rates for each AHI level are shown in Figure 2 as points. The fitted function based on these values, in the form of (5), is represented by Curve 2 in the figure. With a small amount of data for low AHI values, the characteristics obtained are unstable. For instance, Figure 2a shows curves generated from the data in Table 2 with varying numbers of failures for transformers with AHI = 60. Curve 1 corresponds to 5 failures, while Curve 3 corresponds to 3 failures. To achieve stable characteristics, a large volume of statistical data for equipment with low AHI values is necessary.

As an alternative approach, the form of characteristic (5) can be defined using the ratio of failure intensities by expression (7). Curve 4 in Figure 2b was obtained under the condition that the failure intensity in a critical state is ten times higher than in a good condition:

$$C = \frac{100\, ln(10)}{85-25} = 3,84.$$

© The Authors, published by EDP Sciences. This is an open access article distributed under the terms of the Creative Commons Attribution License 4.0 (https://creativecommons.org/licenses/by/4.0/).

 https://doi.org/10.1051/e3sconf/202458401021

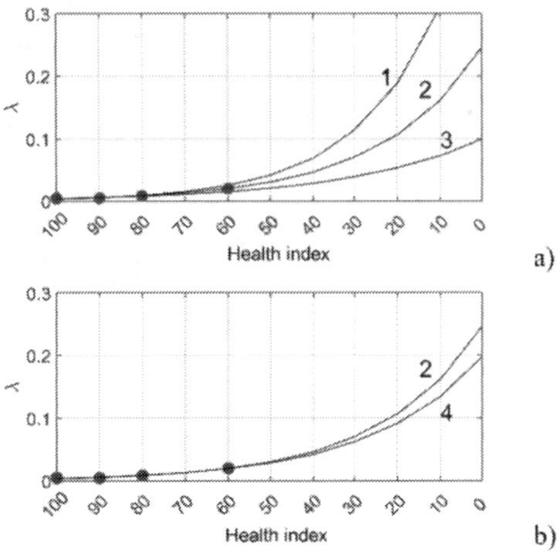

Fig. 2. Dependence of failure rate on asset health index.

Based on the obtained characteristics it is possible to determine the probability of failure during the year for equipment unit with a known health index by the expression:

$$P_i = 1 - exp(-K \cdot e^{C(1-AHI_i/100)}) \qquad (8)$$

The dependence of failure probability on the AHI obtained from the initial data is shown in Figure 3. This curve corresponds with the characteristics presented in Figure 1.

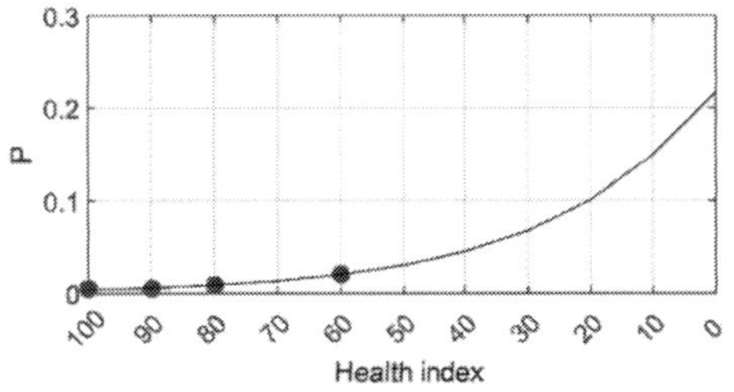

Fig. 3. Dependence of failure probability on asset health index.

3 Calculation of Equipment Failure Consequences

The second component of equipment failure risk is the consequences, measured in monetary terms of potential losses. The procedure for calculating these consequences (cost of failure - CoF) is outlined in Order № 123 by the Ministry of Energy [5]. The cost of equipment failure includes the cost of replacing or repairing the failed equipment, compensation for environmental damage, and compensation for harm to the health and safety of personnel. According to the methodology [5], the loss due to reduced tariffs and compensation payments to consumers is calculated only for consumers in Category 3, whose cumulative duration of power outages in a year exceeds the allowable value (72 hours). Under modern conditions, this condition is rarely met, resulting in the fact that the existence and duration of electricity supply interruptions are not accounted for in the consequences of equipment failure.

Figure 4 shows the dependence of CoF on the product of the duration of the outage and the number of supply points affected. The CoF are derived from accident investigation reports for the period 2020-2022 and are grouped by ranges of the product $T \cdot N$. The points on the graphs represent the average CoF, with the whiskers indicating the 10th and 90th percentiles.

From the graphs, it is evident that CoF, calculated according to the methodology [5], does not depend on the consequences in terms of the number of disconnected consumers and the duration of electricity supply interruptions. Under these conditions, the risk of equipment failure does not depend on whether it leads to an electricity supply interruption or not, nor does it depend on the speed of restoration of the electricity supply. As a result, the incentives to ensure reliability in electricity supply for individual consumers are formed through directive guidelines, often following mass outages and subsequent complaints, rather than within a risk-oriented approach.

Incentivizing power grid companies to improve the average reliability level (reducing the SAIFI and SAIDI) is implemented through setting planned values for these indicators [15]. A planned reliability indicator is considered achieved if the actual value does not exceed it by more than 30% [15, 16]. For each of SAIFI and SAIDI that do not meet this requirement, a reducing coefficient is applied to adjust the necessary gross revenue (NGR) by 0.6% of the previous year's NGR [15, 17]:

$$\Delta NGR_{saidi} = \Delta NGR_{saifi} =$$
$$= 0,3 \cdot 2\% \cdot NGR_{t-1} = 0,006 \cdot NGR_{t-1}.$$

Since the SAIFI is calculated as the ratio of the number of disconnected supply points during the year to their total number, and the SAIDI as the ratio of the total duration of power outages during the year to the total number of supply points, the limit values of these indicators correspond to the maximum allowable number and duration of outages.

$$SAIFI_{\text{limit}} = \frac{N_{\text{limit}}}{N_t} = 1,3 \cdot SAIFI_{\text{plan}} \rightarrow$$
$$\rightarrow N_{\text{limit}} = 1,3 \cdot SAIFI_{\text{plan}} \cdot N_t \qquad (9)$$

$$SAIDI_{\text{limit}} = \frac{TN_{\text{limit}}}{N_t} = 1,3 \cdot SAIDI_{\text{plan}} \rightarrow$$
$$\rightarrow TN_{\text{limit}} = 1,3 \cdot SAIDI_{\text{plan}} \cdot N_t \qquad (10)$$

where N_t is the maximum number of power supply points of PGC in year t (in units), $SAIFI_{\text{plan}}$ and $SAIDI_{\text{plan}}$ are the planned values of the reliability indicators.

© The Authors, published by EDP Sciences. This is an open access article distributed under the terms of the Creative Commons Attribution License 4.0 (https://creativecommons.org/licenses/by/4.0/).

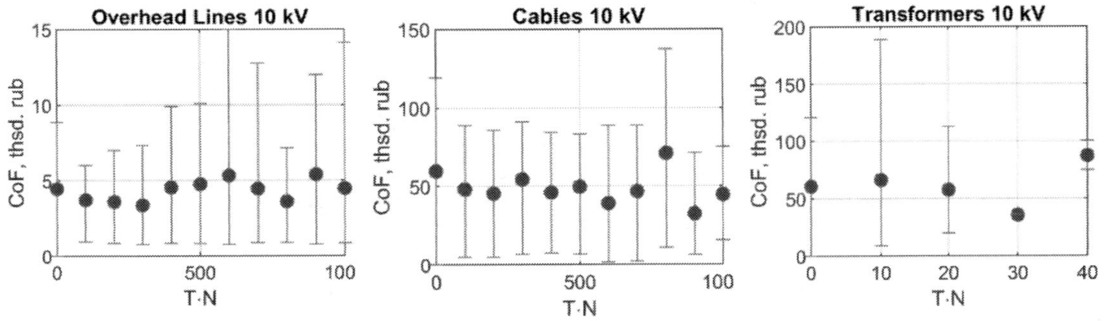

Fig. 4. Dependence of cost of failure (CoF) on the product of outage duration and number of affected supply points.

These limit values of for the quantity and duration of outages correspond to a decrease in the NGR:

$$TN_{\text{limit}} \cdot C_{saidi} = 0,006 \cdot NGR_{t-1}$$
$$N_{\text{limit}} \cdot C_{saifi} = 0,006 \cdot NGR_{t-1} \quad (11)$$

Thus, the cost of disconnecting one supply point and the cost of a one-hour power outage can be determined as a fraction of the decrease in NGR using the following expressions:

$$C_{saifi} = \frac{0,006 \cdot NGR_{t-1}}{1,3 \cdot SAIFI_{\text{plan}} \cdot N_t}$$
$$C_{saidi} = \frac{0,006 \cdot NGR_{t-1}}{1,3 \cdot SAIDI_{\text{plan}} \cdot N_t} \quad (12)$$

As an example, these values are determined for the power grid of the Central region of Russia based on 2023 data. For calculating the NGR decrease, the NGR value from the previous year is used. The data for 2022 are taken from the proposal on price (tariff) levels and long-term regulatory parameters for 2024 [18].

$$NGR_{2022}^{\text{actual}} = 57\ 093\ 080 \text{ thousand rubles}$$

The number of supply points in the examined power grid for 2023 amounted to 1 245 000. The planned values for SAIFI and SAIDI for 2023 are 0.2804 and 0.3683, respectively. Considering these values, the cost of disconnecting one supply point and the cost of hourly electricity interruption can be defined as a fraction of the decrease in NGR according to the following expressions:

$$C_{saifi} = \frac{0,006 \cdot NGR_{t-1}}{1,3 \cdot SAIFI_{\text{plan}} \cdot N_t} =$$
$$= \frac{0,006 \cdot 57\ 093\ 080 \text{ thousand rubles}}{1,3 \cdot 0,2804 \cdot 1245000} =$$
$$= 0,75 \text{ thousand rubles}$$

$$C_{saidi} = \frac{0,006 \cdot NGR_{t-1}}{1,3 \cdot SAIDI_{\text{plan}} \cdot N_t} =$$
$$= \frac{0,006 \cdot 57\ 093\ 080 \text{ thousand rubles}}{1,3 \cdot 0,3683 \cdot 1245000} =$$
$$= 0,57 \text{ thousand rubles}$$

These values can be used within a risk-oriented approach to account for the consequences of technological outages in the form of power supply interruptions. Thus, the cost of failure can be determined as follows:

$$CoF = CoF_{123} + C_{saifi}N + C_{saidi}NT \quad (13)$$

where CoF_{123} represents the consequences of equipment failure, calculated in accordance with Order № 123 by the Ministry of Energy, N is the number of disconnected supply points, and T is the duration of the power outage.

4 Conclusion

It is shown that the current Russian methodology for calculating the risk of equipment failure does not consider actual failure statistics and the consequences in terms of the number of affected consumers and the duration of power outages. Taking international experience into account, the need to align actual failure statistics with asset health indices is highlighted. The feasibility of addressing this issue under current conditions is illustrated with a calculation example.

A method is proposed for accounting for the duration of power supply disruptions and the number of disconnected supply points when calculating cost of failure. This approach considers the contribution to reliability indicators (SAIFI and SAIDI) and the share of potential NGR decrease due to not meeting planned targets.

References

1. Energy Strategy of the Russian Federation Until 2035. Approved by Government Decree № 1523-r of June 9, 2020.
2. GOST R 55.0.05-2016. Asset Management. Enhancing Safety and Reliability of Assets. Requirements. Moscow: Standardinform, 2016.
3. Government Decree № 1401 of December 19, 2016 on the Comprehensive Definition of Technical and Economic Indicators of Electric Power Facilities, including Indicators of Physical Wear and Energy Efficiency of Electrical Network Assets, and Monitoring of Such Indicators.
4. Order of the Ministry of Energy of the Russian Federation № 676 of July 26, 2017 "On Approval of the Methodology for Assessing the Technical

© The Authors, published by EDP Sciences. This is an open access article distributed under the terms of the Creative Commons Attribution License 4.0 (https://creativecommons.org/licenses/by/4.0/).

Condition of Major Technological Equipment and Power Transmission Lines of Electric Power Stations and Electrical Networks".

5. Order of the Ministry of Energy of the Russian Federation № 123 of February 19, 2019 "On Approval of Methodological Recommendations for Calculating the Probability of Failure of Functional Units and Major Technological Equipment and Assessing the Consequences of Such Failures".
6. GOST R 27.010 – 2019 Reliability in Engineering. Mathematical Expressions for Indicators of Failure Rate, Readiness, and Maintainability.
7. GB DNO Groups, Health & Criticality - Version 2.1, 2021. DNO Common Network Asset Indices Methodology.
8. Naderian, A., Piercy, R., Cress, S., Service, J., Wang, F., 2009. An Approach to Power Transformer Asset Management Using Health Index. IEEE Electrical Insulation Magazine, 25 (2), 2.
9. Scatiggio, F., Pompili, M., 2013. Health Index: The TERNA's Practical Approach for Transformers Fleet Management. IEEE Electrical Insulation Conference (EIC), 178–182.
10. Vermeer, M., Wetzer, J., van der Wielen, P., de Haan, E., de Meulemeester, E., 2015. Asset Management Decision Support Modeling, Using a Health and Risk Model. PowerTech, IEEE. Eindhoven, 1–6.
11. Order of the Ministry of Energy of the Russian Federation № 90 of March 2, 2010 (as amended on July 27, 2017) "On Approval of the Form of the Act on Investigation of Causes of Accidents in the Electric Power Industry and the Procedure for Its Completion".
12. Grabchak, E.P. Assessment of the technical condition of energy equipment in the context of the digital economy. Reliability and Safety of Energy. 2017;10(4):268-274.
13. Vodenikov, D. Technical Condition of Electrical Equipment: Assessment Methods. // Electric Power. Transmission and Distribution. Special Issue "Rosseti" № 4(15), 2019 – P. 8-11.
14. Nepomnyashchy, V.A. Reliability of Power System Equipment. Publishing House of the Journal "Electric Power. Transmission and Distribution", 2013, 196 pages.
15. Order of the Ministry of Energy of the Russian Federation № 1256 of November 29, 2016 "On Approval of Methodological Recommendations for Calculating the Reliability and Quality Indicators of Supplied Goods and Services for the Organization Managing the Unified National (All-Russian) Electric Grid and Territorial Network Organizations".
16. Order of the Ministry of Energy of the Russian Federation № 976 of October 18, 2017 "On Approval of Basic Values for Reliability Indicators, Allowable Deviation Coefficients for Actual Reliability Indicators from Planned Values, and Maximum Dynamics of Improvement of Planned Reliability Indicators for Groups of Territorial Network Organizations with Comparable Economic and Technical Characteristics and/or Operating Conditions, Using the Analog Comparison Method".
17. Order of the Federal Tariff Service of the Russian Federation № 254-e/1 of October 26, 2010 "On Approval of Methodological Recommendations for Calculating and Applying Reducing (Increasing) Coefficients to Ensure Compliance of Tariff Levels Established for Organizations Conducting Regulated Activities with the Level of Reliability and Quality of Supplied Goods and Services" (Registered with the Ministry of Justice of the Russian Federation on November 13, 2010, № 18951).
18. Proposal for Tariff Levels, Long-Term Regulation Parameters (type of price (tariff): service for electricity (capacity) transmission for 2024 by Public Joint-Stock Company "Rosseti Moscow Region". https://rossetimr.ru/upload/iblock/fd6/5bb2xbf748mr88rinozs6tbvof37f2io/predlogenie-M-2024.pdf.

© The Authors, published by EDP Sciences. This is an open access article distributed under the terms of the Creative Commons Attribution License 4.0 (https://creativecommons.org/licenses/by/4.0/).

G factor of the ESG agenda in the context of sustainable development of Russian electric power companies

Galina I. Sheveleva*

Melentiev Energy Systems Institute, Siberian Branch of the Russian Academy of Sciences, Irkutsk, Russia

Abstract. Reducing associated investment risks in the context of the modern ESG agenda requires a certain transformation of corporate governance in Russian power generation companies. In order to achieve this objective, a correlation analysis was carried out between the generalized G components of the ESG agenda and the principles outlined in the Russian Corporate Governance Code (Code). The priority G components of authoritative international rating agencies and the largest foreign companies, leading national rating agencies, and the Bank of Russia were examined. The study involved the analysis of Russian power generation companies that are voluntarily enhancing their corporate governance practices. The results of a general assessment of their corporate governance and additional analysis of the fulfilment of the criteria for each section of the Code are presented. The data obtained are summarized and proposals for further transformations of corporate governance in these companies are formulated. These proposals are highly relevant in the context of sustainable development of these companies. They concern the reformatting of the ownership structure and business ethics into separate G components with an appropriate set of evaluation criteria, and the introduction of modern compliance control.

Introduction

The modern trend in sustainable development is the ESG agenda (Environmental, Social, and Corporate Governance). The importance of the G factor (corporate governance) in this agenda compared to environmental (E) and social (S) factors is clearly and undeservedly underestimated, particularly in generation companies in the electric power industry (hereinafter referred to as generation companies). Corporate governance is generally considered fundamental for the implementing these factors. As a rule, ineffective corporate governance lies behind every environmental or social violation [1]. Risks and opportunities of corporate governance are critical in decision making, including "E" and "S" decisions. They can only increase with the further development of social, political, and cultural relations [2-7]. Both an increase in the importance of corporate governance in generation companies and a certain revision and (or) reformatting of its G components are required.

The purpose of the study is to find the avenues to transform corporate governance in these companies in the context of their sustainable development under unprecedented anti-Russian sanctions and the lack of alternatives to the domestic borrowing market.

The G factor of the Corporate Governance Code of the Organization for Economic Co-operation and Development (OECD) [8], the Russian Corporate Governance Code (hereinafter referred to as the Code) are analyzed [9]. A correlation analysis of the principles of Code G components of the draft model methodology is performed for ESG ratings of the Bank of Russia and its recommendations to boards of directors on how to consider ESG factors [10-11]. The main G components in the ESG ratings of leading international rating agencies (FTSE Russell, ISS, S&P Global Ratings, MSCI ESG Rating, RAEX Europe ESG, Vigeo Eiris, RepRisk, Ernst&Young, DJSI World, GRI) and national rating agencies (PWC (TeDo), RAEX, Expert RA, NRA, ACRA, AK&M) are identified [12-22]. A comparative analysis of the significance of these G components was performed.

The underestimation of "Anti-Corruption principle," highlighted as a priority in the UN Global Compact, is emphasized [23]. The most significant G components of ESG reporting of the largest foreign companies are identified [24]. The underestimation of the ownership structure in generation companies as the G component is emphasized. The reason is its opacity, instability, lack of restrictions for owners, and the presence of offshore zones. Associated investment risks are identified, including those arising from the lack of business ethics and modern compliance control in the G-components. An assessment based on the updated methodology of the Bank of Russia confirms the low practice of implementing the principles and recommendations of the Code in the majority of generation companies. Significant differences in their compliance with the Code criteria are identified [25-26].

All the data obtained are summarized and further steps for transforming corporate governance in generation companies are outlined to support their sustainable development. The transformations mainly concern the expansion and reformatting of the G-

*Corresponding author: sheveleva@isem.irk.ru

© The Authors, published by EDP Sciences. This is an open access article distributed under the terms of the Creative Commons Attribution License 4.0 (https://creativecommons.org/licenses/by/4.0/).

components and the criteria for their evaluation, primarily in relation to the ownership structure, corporate behavior, and compliance control.

1 G factor in the international and national ESG agenda

The study of the G factor of the international and national ESG agenda notes a certain similarity between the principles of the OECD Code and the Russian Code. This similarity is mainly due to the development of the principles of the Russian Code on the basis of OECD best corporate practice standards. Their fundamental differences mainly stem from different institutional environments. Most likely, for this reason, the Russian Code does not contain a principle of compliance of corporate governance with the requirements of the rule of law, effective supervision, and enforcement. There is no principle of proper redress for violations of shareholder rights.

The G components of the national ESG agenda in one form or another were previously present in the principles and (or) criteria of the Code. The differences are mainly in the emphasis on their significance. This is confirmed by the comparison of generalized G components of the Bank of Russia, in terms of the consideration by the board of directors of ESG factors, and the Bank's draft model methodology for ESG ratings with the principles of the Code.

Certain differences are identified when comparing the G components of the ESG agenda and the principles of the Code. In particular, the ESG agenda highlights the high significance of "the practice of implementing the principles and recommendations of the Code," "share capital structure," and "anti-corruption and anti-fraud measures." Moreover, they are presented as G -components, and not in the form of inconspicuous criteria for compliance with individual G components of the Code.

Good corporate governance is not limited to the G components of legislation and the Code. Many of these G components are outside the legislative sphere and are partly reflected in the ESG ratings of reputable international and national rating agencies. ESG ratings are highly valued by investors and help them identify opportunities and associated risks for the sustainability of companies in the long term. ESG ratings also play a key role for companies in building internal processes in accordance with the best corporate governance practices. A place in the overall ESG rating or in individual "E," "G," and "S" ratings is usually considered one of the main factors in the investment attractiveness of companies, their certain competitive advantage [4]. The significance of the identified G components in the ESG ratings of international and national rating agencies is determined. It is measured by the number of agencies using these G components in corporate governance assessments.

G components are considered in a generalized form, partly to facilitate their comparative analysis. In particular, "Corruption, bribery, extortion, money laundering, and fraud" are defined as "Corruption." "Structure of the Board of Directors" encompasses "Structure of the Board of Directors and its committees, diversity of the Board of Directors, etc." Corporate behavior is represented by separate components: business ethics, antitrust practices, and taxes.

Below are the differences identified in the emphasis on the significance of the same G components for international and national rating agencies? The most significant G components from the ESG reporting of the largest foreign companies are indicated.

1.1 The significance of G components in international ESG ratings

The highest priority G component in the ESG assessments of all international rating agencies is the structure of the board of directors. Business ethics and executive compensation are assessed only in six out of nine agencies. Only four or five agencies assess shareholder rights, tax payment, risk management, audit, and transparency. Antitrust practices and political activity, corruption and ownership structure are assessed by only two or three international rating agencies.

1.2 The significance of G components in national ESG ratings

National ESG ratings are distinguished by a relatively narrower coverage of G components and a different approach to their assessment. The structure of the board of directors and risk management were considered the most preferable for all national rating agencies and the Bank of Russia. Transparency and ownership structures were highly rated. Audit, tax payment, corruption, business ethics, and shareholder rights were less in demand. They were explicitly present in the assessments of two or three national rating agencies. Antitrust practice, political activity, and audit (except for the National Rating Agency) were completely absent in national ESG ratings.

1.3 The significance of the G components in ESG reporting of foreign companies

Additional analysis of the G components of ESG reporting of the largest foreign companies revealed that stakeholders showed the greatest interest in the board of directors, transparency, ethical behavior, and enhanced control [24]. These G components are actually the top-line factors in the ESG ratings provided by international rating agencies. The "Anti-corruption" policy proved to be clearly undervalued by the international and national ESG agenda, although in the UN Global Compact it is among the main ten principles of sustainable development [23].

© The Authors, published by EDP Sciences. This is an open access article distributed under the terms of the Creative Commons Attribution License 4.0 (https://creativecommons.org/licenses/by/4.0/).

2 The G factor of the ESG agenda in Russian generation companies

The main G components of the corporate governance factor of Russian companies (including electric power companies), it would seem, are exhaustively formulated by the regulator (Bank of Russia) in the Code and in its further numerous recommendations on the implementation of standards of best corporate practice. However, further expansion and reformatting of these components is required in generation companies to take into consideration the identified differences in the G components of modern ESG-agenda and principles of the Code, as well as certain sectoral specifics. Primarily, this pertains to the ownership structure, business ethics, and compliance principles.

2.1 Ownership structure

The ownership structure in the context of sustainable development of generation companies should be considered as an independent G component (with a certain set of criteria for its assessment). This is mainly due to opacity, instability, conflicts of interest, lack of restrictions for the owners of these companies, and the presence of offshore zones. This is partly confirmed by our generalized data on the rotation of ownership in these companies, including the data from the leading news agency Bigelectricpowernews and the business newspaper Kommersant.

Italian Enel sold shares of PJSC Enel Russia (now PJSC EL5-Energo) to PJSC Lukoil (26.9%) and the investment fund Gazprombank-Freesia (29.5%). Later, PJSC Lukoil increased its share to 63.8% by purchasing shares of UROC Limited (7.4%). The share capital of PJSC EL5-Energo also included LLC PPIT-7 (5.5%). The Russian generation assets of OJSC Fortum (renamed Forward Energo) (98.2%) and PJSC Unipro (83.7%) belonged, respectively, to the Finnish concern Fortum and the German government (through Uniper). They were transferred to the management of the Federal Property Management Agency.

The Federal Property Management Agency increased its share in the share capital of PJSC RusHydro to 62.2%. PJSC VTB Bank, on the contrary, reduced it from 13% to 12.4%. Rusal (9.5%) became a shareholder of PJSC RusHydro. The share of the Finnish Fortum Power and Heat Oy increased slightly (from 29.5% to 29.99%) in the share capital of PJSC TGC-1. JSC Rusatom Infrastructure Solutions (a division of the state corporation Rosatom) acquired shares of PJSC Quadra (82.47%), later increasing this share to 95.02%. PJSC Quadra - Power Generation has lost its public status and has become JSC Quadra - Power Generation.

The presence of En+ Group (O. Deripaska) in JSC EuroSibEnergo increased to its maximum level. More than 83.4% of the shares of PJSC TGC-2 in 2023 were withdrawn in favor of the state from Kostroma Kogeneratsiya Ltd., Janan Holdings Limited, Raltaka Enterprises Ltd., Litim Trading Limited, Kores Invest LLC, and Dolgovoye Agentstvo LLC (a subsidiary of PJSC TGC-2). This seizure is mainly due to the activities of its owner (L. Lebedev). He bought PJSC TGC-2 with illegally obtained income and registered it in the name of foreign companies in the British Virgin Islands in order to hide his participation in the ownership and management of this company.

The generation assets of PJSC TGC-14 (more than 92.2%) were purchased by JSC Far Eastern Management Company. Its main shareholder is K. Lyulchev (former top manager of Russian Railways). The beneficiary of PJSC T Plus, which united the generation assets of OJSC TGC-5, OJSC TGC-6, OJSC TGC-9, and OJSC TGC-7, was V. Vekselberg. At the beginning of 2019, he owned the Renova Group (39.6%), Brookwid Trading Limited (Cyprus) (20.5%), and Gotelia Management Limited (Cyprus) (12.7%), which were declared as the main shareholders of PJSC T Plus. Merol Trading Limited (Cyprus) (11.7%) was also one of these shareholders and was affiliated with Renova Group.

The ownership structure plays a crucial role as a G component because it determines the structure and overall functioning of the board of directors. Under a relatively weak institutional environment, the board of directors "de jure" has an overwhelming number of powers to safeguard the rights and interests of shareholders, including enhanced control over their protection. Unfortunately, "de jure" in generation companies often does not coincide with "de facto" [25-26].

2.2 Business ethics

Business ethics is absent both from the G components of the draft model methodology for ESG ratings of the Bank of Russia and from its updated methodology for assessing compliance with the Code. It is imperative to have business ethics as a G component of the sustainable development of generation companies (with a set of corresponding criteria) alongside "corruption control," which is currently a criterion in the Code. It is no coincidence that in the ESG ratings of reputable international and national rating agencies, business ethics is one of the priority G components, just like "corruption control" is in the UN Global Compact.

Lack of business ethics and proper compliance control in G scores has led to an increase in investment risks in generation companies, primarily in transactions with related parties and expropriation of minority shareholders, overpayment for resources or assets, underpricing when selling products or assets. In particular, the former general director of PJSC TGC-2 and the ex-head of Arkhoblenergo, in violation of procurement legislation, entered into contracts at inflated prices for the supply of fuel for the needs of JSC Arkhoblenergo. They deliberately destabilized the activities of PJSC TGC-2 with the aim of bankruptcy and appropriation of property belonging to the Arkhangelsk region. The assets of this company were transferred offshore by its beneficial owner L. Lebedev, creating a real threat of bankruptcy proceedings and the complete cessation of the activities of PJSC TGC-2.

© The Authors, published by EDP Sciences. This is an open access article distributed under the terms of the Creative Commons Attribution License 4.0 (https://creativecommons.org/licenses/by/4.0/).

Through fraudulent schemes, M. Abyzov and his accomplices stole large sums of money from JSC Sibeko and OJSC RES. They secretly shared insider information with a potential buyer, keeping it hidden from shareholders and regulatory authorities, to ensure a profitable sale of their stake in JSC Sibeko. Advisor to the General Director of JSC TGC-11 (part of PJSC Inter RAO) was arrested for receiving a bribe of RUR 3.5 million. A criminal case was opened against the director of capital investments, state construction supervision, and state examination of the state corporation Rosatom [27-30].

2.3 Compliance control

Compliance control is a critical component of the current national ESG agenda. Its capabilities are much broader in ensuring the fulfilment of the criteria for compliance with the G components, which are significant for the sustainable development of generation companies.

The results of the latest assessment of fulfilling the criteria for compliance with the principles of the Code with their status ("fulfilled," "partially fulfilled," "not fulfilled") are presented in Figure 1, based on the 2021 results. They were obtained primarily from the reports "On Compliance with the Principles and Recommendations of the Code" from the official websites of these companies.

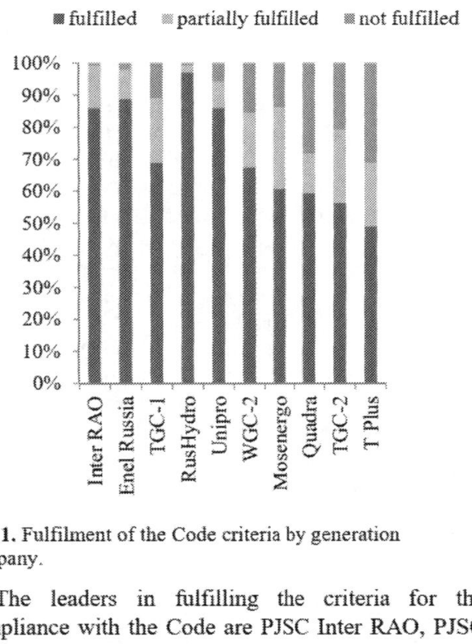

Fig. 1. Fulfilment of the Code criteria by generation company.

The leaders in fulfilling the criteria for the compliance with the Code are PJSC Inter RAO, PJSC RusHydro, PJSC Enel Russia, and PJSC Unipro. The outsiders are PJSC TGC-2, PJSC Quadra, and PJSC T Plus. One of the main reasons for these differences is the absence of modern Compliance control in outsider companies. This is a completely different type of monitoring of corporate governance's compliance with national legislation and the Code. Modern compliance control allows identifying risks at the early stages of their manifestation and preventing associated economic damage. It provides more real protection against corruption and conflicts of interest. Additionally, it includes a range of components to ensure the economic security of the company. Compliance control policy, unfortunately, was developed and documented internally only by companies leading in corporate governance.

Conclusion

The modern ESG agenda requires increased significance of the G factor and transformation of its components. To this end, additional G components and partially reformatted Code criteria are proposed. They mainly relate to ownership structure, business ethics, and compliance control. These G components are especially relevant in the context of sustainable development of generation companies. They are formulated based on a summarized wide array of data, including updated ESG recommendations of the Bank of Russia, priority G components of international and national ESG ratings, ESG reporting of the largest foreign and domestic companies, and characteristic sectoral features of corporate governance. These proposals will enable potential investors to re-evaluate corporate governance in generation companies to support their sustainable development.

The research was carried out under State Assignment Project (No. FWEU-2021-0001) of the Fundamental Research Program of the Russian Federation 2021-2030.

References

1. Defining the "G" in ESG Governance Factors at the Heart of Sustainable Business. Available at: https:// www3 weforum.org/docs/WEF _Defining_the_G_in_ESG_2022.pdf (Accessed: 12.11.2023)
2. What is "G" in ESG? Available at: https://www.spglobal.com/en/research-insights/articles/what-is-the-g-in-esg (Accessed: 12.11.2023)
3. ESG is underpinned by a strong "G". Available at: https://kpmg.com/be/en/home/insights /2023/02/rr-esg-is-underpinned-by-a-strong-g.html (Accessed: 12.12.2023)
4. ESG principles as one of the main trends in the development of the energy industry. Available at: http://www.itp.nsc.ru/news/esg - principle _ kak_odin_iz_glavnyh_trendov_razvitiya.html (Accessed: 12.08.2023)
5. "ESG" Best Practice - "G" Governance. Available at: https://www.gfoa.org/materials/esg-best-practice-g-governance (Accessed: 29.01.2024)
6. Abhishek Parikh, Divya Kuvari at all, Cleaner and Responsible Consumption, **8** (2023)

© The Authors, published by EDP Sciences. This is an open access article distributed under the terms of the Creative Commons Attribution License 4.0 (https://creativecommons.org/licenses/by/4.0/).

7. Leo E. Strine, Jr., Justin L. Brooke, Kyle M. Diamond, and Derrick L. Parker Jr., It's Time to Focus on the "G" in ESG. Available at: https://hbr.org/2022/11/its-time-to-focus-on-the-g-in-esg (Accessed: 26.01.2024)
8. Principles of corporate governance G20/OECD (OECD Publishing, Paris, 2016). Available at: http://dx.doi.org/10.1787/9789264252035-ru (Accessed: 15.01.2022)
9. Code of Corporate Governance. Bulletin of the Bank of Russia, **40** (2014)
10. Model methodology for ESG ratings. Available at: http://www.cbr.ru/Content/Document/File/144085/Consultation _ Paper _17012023.pdf (Accessed: 02.21.2023)
11. Information letter of the Bank of Russia of December 16, 2021. No. IN-06-28/96. Available at: https://cbr.ru/issuers_corporate/acts/ (Accessed: 02.17.2022)
12. Integrating ESG into investments and stewardship. Available at: https://research.ftserussell.com/products/downloads/FTSE-ESG-Methodology-and-Usage-Summary-Full.pdf (Accessed: 01.25.2024)
13. Environmental, social and governance scores from LSEG. Available at: https://thesource.refinitiv.com/thesource/getfile/index/f870ed69-807b-483d-8df1-ba782754a398 (Accessed: 25.01.2024)
14. ESG Evaluation Sustainable Practices. Sustainable Returns. Available at: https://www.spglobal.com/_assets/documents/ratings/ESG/ESG_evaluation_brochure_digital.pdf (Accessed: 01.25.2024)
15. Methodology for assigning environmental, social and governance Ratings to corporates. Available at: https://raexpert.eu/files/Methodology_ESG_Corporates _V3.pdf (Accessed: 01.25.2024)
16. Updated ESG ranking RAEX-Europe Available at: https://raex-rr.com/files/analytics/RAEX_presentation.pdf (Accessed: 02/05/2024)
17. RepRisk methodology overview. Available at: https://www.reprisk.com/news-research/resources/methodology (Accessed: 25.01.2024)
18. ESG Assessments. Available at: https://ESGrisk.ai/ESG-assessments/ (Accessed: 25.01.2024)
19. Environmental, social & governance index (GRI). Available at: https://www.coats.com/ru/sustainability/reporting (Accessed: 04.18.2024)
20. Methodology for assigning ESG ratings. Available at: https://www.ra-national.ru/wp-content/uploads/2023/12/metodologija-prisvoenija-esg-rejtingov-versija-5.0_vernaja-1.pdf (Accessed: 02.05.2024)
21. ESG ratings. Available at: https:/akmrating.ru/esgreyting/ (Accessed: 02.05.2024)
22. S.D. Grishankova, ESG ratings in (ESG transformation as a vector of sustainable development: In three volumes. Ed. by K.E. Turbina, I.Yu. Yurgens, Vol. 2, Moscow, Aspekt Press Publ., 2022)
23. UN Global Compact: finding solutions to global problems. Available at: https://www.un.org/ru/36167 (Accessed: 04.15.2024)
24. Environmental Social & Governance Report. JPMorgan Chase & Co. Available at: https://www.jpmorganchase.com/content/dam/jpmc/jpmorgan - chase - and - co/documents/jpmc - esg - report -2020.pdf (Accessed: 02.25.2022)
25. G.I. Sheveleva, ECO, **7** (2023). DOI: 10.30680/ ECO0131-7652-2023-7-51-70
26. G.I. Sheveleva, *Transformation of corporate governance in the face of modern challenges*, in Proceedings E3S Web of Conferences, Vol. 461 (2023). DOI: https://doi.org/10.1051/e3sconf/202346101021 (In Russian)
27. Former general director of TGC-2 Pinigina and ex-head of Arkhoblenergo Zaikin received 5 years and 4.5 years in prison. Available at: https://www.bigpowernews.ru/print/112344.phtml (Accessed:12.28.2023)
28. The beneficial owner of PJSC TGC-2, Lebedev, committed an illegal withdrawal of assets. – Rosfinmonitoring. Available at: https://www.bigpowernews.ru/print/111143.phtml (Accessed: 10.16.2023)
29. The defendants in the Abyzov case were sentenced to terms ranging from 5 to 12 years. Available at: https://www.bigpowernews.ru/print/112231.phtml (Accessed: 12.22.2023)
30. In Omsk, the adviser to the general director of JSC TGC-11 was arrested. Available at: https://www.bigpowernews.ru/print/111045.phtml (Accessed: 10.06.2023)

© The Authors, published by EDP Sciences. This is an open access article distributed under the terms of the Creative Commons Attribution License 4.0 (https://creativecommons.org/licenses/by/4.0/).

 https://doi.org/10.1051/e3sconf/202458401023

Methodology for Analysis of System Reliability When Planning States and Balances of Electric Power Systems

Dmitry Krupenev[1,2*]

[1] Melentiev Energy Systems Institute of Siberian Branch of the Russian Academy of Sciences, Department of Energy Security, Lermontov str., 130, Irkutsk, 664033, Russia
[2] Irkutsk National Research Technical University, Lermontov Str. 83, Irkutsk, 664074, Russia

Abstract. The article presents the main provisions of the methodology for analyzing the system reliability of electric power systems when planning their states and balances. The type of system reliability for solving problems when planning the states and balances of electric power systems is called planned reliability. A substantive statement of the problem of assessing planned reliability is formulated, where the main requirements for taking into account the factors influencing the determined indicators are noted. In the general structure of the methodology, based on the sequential Monte Carlo method, three main stages are distinguished: the stage of forming pseudo-random processes of functioning of electric power systems, the stage of analyzing the states of functioning of electric power systems, the stage of calculating the indicators of planned reliability.

1 Introduction

Management of electric power systems (EPS) includes three main stages: operational management, planning of states and balances (work) and long-term design. Planning of EPS states and balances is usually performed for up to one year. Within the framework of this planning, a number of tasks arise that require an assessment of the system reliability of the EPS. The type of system reliability assessed for the tasks of planning the operation of the EPS is called planned reliability [1]. As with the assessment of each type of system reliability, when assessing planned reliability, it is necessary to maximally reflect the specifics of the functioning of the EPS for the period of work planning, taking into account the uncertainty in the initial data for this period. The result of the assessment in the form of planned reliability indicators should give the most complete characteristic of the EPS, which is necessary for solving the problems of synthesis of planned reliability.

The methodology for assessing each type of system reliability has both similar features and differences. When assessing planned reliability, it is important to reflect those factors influencing the reliability of the EPS that can be adequately (with a high degree of reliability) taken into account during the planning period of the EPS operation.

The purpose of the article is to present a methodology for assessing the planned reliability of EPS with a description of specific features that must be taken into account to reflect the operating states of EPS and power equipment when planning states and balances. The article consists of two sections. The first section presents the substantive and mathematical formulations of the problem of assessing the planned reliability of EPS. The second section presents the methodology for assessing the planned reliability of EPS.

2 Substantive statement of the problem

The following factors (requirements) can be noted that should be taken into account in the methodology for assessing the planned reliability of the EPS: analysis of the complete EPS model without breaking it down into reliability zones; consideration of the balance of active and reactive powers when analyzing the states of the calculated cases of the EPS; modeling random processes of EPS operation with the analysis of deficit and deficit-free states; consideration of hourly annual schedules of EPS power consumption taking into account the correlation dependencies of power consumption in different EPS states and at different hours; per-unit representation of generating sources taking into account the features of complex structures of units; modeling of renewable energy sources taking into account the correlation dependence of the activity of primary energy resources of different sources; modeling of the operating states of the ESS with the correct reflection of the ESS capabilities for the accumulation and generation of energy, taking into account the dynamics of ESS degradation and a decrease in efficiency, as well as failures leading to the failure to perform a given function; consideration of the maneuverability of generating units for the analysis of the real capabilities of the power system to compensate for a sharp change

* Corresponding author: krupenev@isem.irk.ru

© The Authors, published by EDP Sciences. This is an open access article distributed under the terms of the Creative Commons Attribution License 4.0 (https://creativecommons.org/licenses/by/4.0/).

in the power balance; accounting of water reserves in hydroelectric power station reservoirs and requirements of regulatory bodies regarding reservoir drawdown and tailwater regimes.

Based on the listed factors, the substantive formulation of the task of analysis (assessment, calculation) of planned reliability can be presented as follows: for a given forecast of electricity consumption for a period of up to 1-2 years, a known structure and composition of the EPS equipment (including taking into account plans for commissioning/decommissioning of equipment), it is necessary to determine reliability indicators characterizing the provision of electricity and power to consumers, reflecting the "weak" points of the EPS and allowing to justify rational actions in solving subsequent problems of synthesis of planned reliability, taking into account random processes of the behavior of power equipment, the random nature of fuel supplies to power plants, random fluctuations in electricity consumption, as well as the most correct power flow distribution.

The following are known: the state diagram of the EPS; the hourly schedule of power consumption of each load state for the period of work planning, with a reflection of the characteristics for each day of the week; the law and values of irregular fluctuations in loads in the states of the EPS; the composition, placement and parameters of generating elements; the composition and parameters of network elements; reliability characteristics of the energy equipment taken into account.

The following main tasks can be identified that can be solved at a higher quality level when using planned reliability:

1. Planning of repairs of power equipment of the EPS. Planning of repairs of EPS equipment depends on two main components. Firstly, the need for scheduled repairs, which can be determined on the basis of operating time and calendar period or on the basis of assessment of the technical condition. Secondly, the EPS capabilities to put the power equipment on scheduled repairs with maximum efficiency and without increasing the risk of systemic accidents above the critical assessment. The methodological apparatus of analysis of scheduled reliability of power systems is suitable for solving the second problem. As part of the assessment of scheduled reliability, the operation of the power system is simulated with sufficient accuracy for the tasks of work planning. Based on the indicators of scheduled reliability, which reflect the dynamics of change in the reliability of the power system for the planned period, a schedule of repairs of power equipment is formed.

2. Analysis of scheduled reliability for the period of planning the operation of the EPS. Such an analysis of balance reliability is carried out in various power systems. Numerical assessment of scheduled reliability will make it possible to make this analysis more profound. When bottlenecks in the EPS are identified, it is necessary to adjust the control actions to ensure the reliability of power supply and eliminate bottlenecks using available means. In addition, in the process of assessing the planned reliability, it becomes possible to take into account the specifics and operating modes of power plants at a fairly detailed level, including the risks of extreme water content for hydroelectric power plants.

3. Assessing the capabilities of the EPS and efficiency when integrating renewable energy sources, including high-power ones. One of the main problems in the electric power industry in recent decades is assessing the efficiency of RES integration. Such integration, due to the uneven nature of electricity and capacity generation by RES, can lead to problems in ensuring the system reliability of the EPS and the reliability of power supply. Assessing the planned reliability of the EPS with adequate modeling of RES will reflect the most critical random processes for the EPS, characterizing the operation of RES. The main issue in the problem under consideration is the ability of the EPS to compensate for the imbalance of power and electricity caused by the lack of a renewable source (sun, wind). Ultimately, the accuracy of RES modeling will determine the size of the generating capacity reserve, as well as the structure and capacity of the electrical grid.

4. Analysis of the efficiency of integrating the SNE into electric power systems. There are different types of SNE, which are characterized by such parameters as capacity, charge/discharge rate, operating principle. These parameters must be correctly reflected when assessing the reliability of the EPS. The specificity of the SNE operation lies in the continuity of the charge/discharge process. This process depends on the capabilities of the SNE and on the operating modes of the EPS, so it is important to reflect all the influencing and limiting factors when modeling the SNE. In addition, the methodology for assessing the planned reliability of the EPS is based on the analysis of random processes, which is important when modeling the operation of the SNE charge/discharge process. 5. Analysis of the need to attract electricity consumers to ensure the required level of planned reliability. The main idea of this task, first of all, comes down to the selection of electricity and capacity consumers for demand management. Demand management is one of the mechanisms for managing the reliability of the power system. In the process of selecting consumers, it is important to know which consumers can make the maximum contribution to ensuring the reliability of the EPS.

3 General scheme of the methodology for assessing the planned reliability of the EPS

As can be understood from the description of the initial data, the calculation scheme for assessing the planned reliability is formed on the basis of the state-by-state calculation scheme of the EPS. The calculation scheme must meet the requirements for the analysis and optimization of the steady-state of the EPS. In some cases, it is possible to equip some sets of homogeneous elements with the definition of the necessary characteristics of the equivalent element for assessing the planned reliability.

© The Authors, published by EDP Sciences. This is an open access article distributed under the terms of the Creative Commons Attribution License 4.0 (https://creativecommons.org/licenses/by/4.0/).

The methodology for assessing the planned reliability of EPS is based on the provisions for assessing other types of system reliability [2-5]. These provisions boil down to the analysis of a set of possible random states of EPS and statistical processing of the results of this analysis. As noted above, the assessment of planned reliability will have a number of features that we will reflect in the methodology. The methodology for assessing the planned reliability of EPS is based on the modeling of random processes of EPS behavior over the assessed period. A similar approach was used in a number of methodological developments for assessing the reliability of EPS [6-8]. This paper considers, in essence, its adaptation to the requirements for representing EPS and their elements when assessing planned reliability. The need for modeling random processes is due to a number of factors characterizing modern energy systems: the traditional presence of hydroelectric power plants (HPPs) with the specifics of water drawdown in reservoirs; the emergence of energy storage systems, in which each subsequent operating mode depends, in addition to other influencing factors, on the previous state; restrictions on the operating state of power plants, which consist of a limited control range of power maneuvering due to the technological features of certain generating units. Taking into account the specifics of the operation of such equipment as energy storage devices forms the requirements for the analysis of all modes of the energy system, both deficit and non-deficit.

A detailed block diagram of the algorithm for the method of assessing the planned reliability of the EPS based on the sequential Monte Carlo method is presented in Fig. 1.

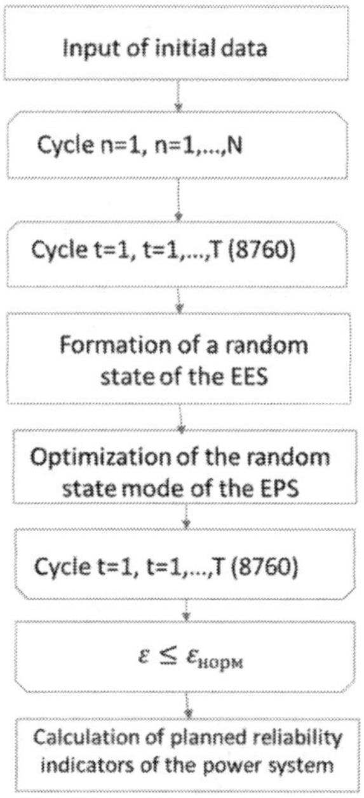

Fig. 1. An enlarged block diagram of the algorithm of the methodology for assessing the planned reliability of the EPS based on the sequential Monte Carlo method.

The algorithm of the methodology for assessing the planned reliability of the EPS includes two cycles, where one (internal) (*T*) cycle is an insertion of the second (external) (*N*). On the external cycle, a set of implementations of modeling random processes of the EPS functioning for the assessed period is formed. The size of the set of random processes depends on the required accuracy of the assessment ($\varepsilon_{\text{норм}}$).

In the internal cycle, the random process of EPS functioning is implemented taking into account the factors influencing the EPS reliability:

1. Modeling of a random process of EPS functioning for the analyzed period. At this stage, the formation of a flow of events in EPS is carried out, the occurrence of which at one time or another depends on internal and external factors of EPS functioning and their elements. The flow of events includes: failures and restoration of elements, charges and discharges of energy storage devices, as well as other random processes occurring in EPS.

To simulate the flow of failures and restorations, the distribution functions of the duration of intervals between events are specified, so to simulate the flow of failures, the distribution function of the duration of intervals between restorations and failures of power equipment $F(t_о)$ or the failure probability $P_о$ is specified, to simulate the flow of restorations - the distribution function of the duration of intervals between failures and restorations of power equipment $F(t_в)$ or the restoration probability $P_в$. These functions can generally be obtained on the basis of equipment reliability indicators: the failure flow parameter ω and the restoration intensity μ. As a rule, the distribution functions $F(t_о)$ and $F(t_в)$ have an exponential distribution law, but it is possible to specify a different distribution law. Depending on the assumptions made, the parameters ω and μ can remain constant or can change during the calculation period, depending on the internal and external operating conditions of the EPS and equipment. This means that the described flows of events are non-stationary. It is believed that after each emergency repair, the properties of the element are completely restored. The flow of events and recoveries of an EPS element is shown in Fig. 1.

For energy storage devices, in addition to the flow of failures and restorations, a flow of charges and discharges is formed. The flow of charges and discharges of energy storage devices is formed according to a different principle than the flow of failures and restorations. There are different types of energy storage devices [9]. Each type has unique characteristics related to power, capacity, response time, service life and cost. Thus, depending on the type of energy storage device, the following limitations are taken into account when modeling it: maximum and minimum storage device capacity, charge and discharge rate and some others specific to each type of storage device.

In addition to emergency repairs, which are modeled by forming a flow of element restorations, when assessing planned reliability, it is necessary to take into account planned repairs of power equipment. For a number of power facilities, planned repairs for the

© The Authors, published by EDP Sciences. This is an open access article distributed under the terms of the Creative Commons Attribution License 4.0 (https://creativecommons.org/licenses/by/4.0/).

planning period of balances and modes are strictly regulated and the start and end dates of specific objects being taken out for repairs are known. For the remaining part of the EPS power equipment, the repair of which must be carried out during the planning period (this is determined by standards or technical condition assessment), an optimization problem is solved to form a schedule of repairs of power equipment with maximization of the EPS reliability level [10].

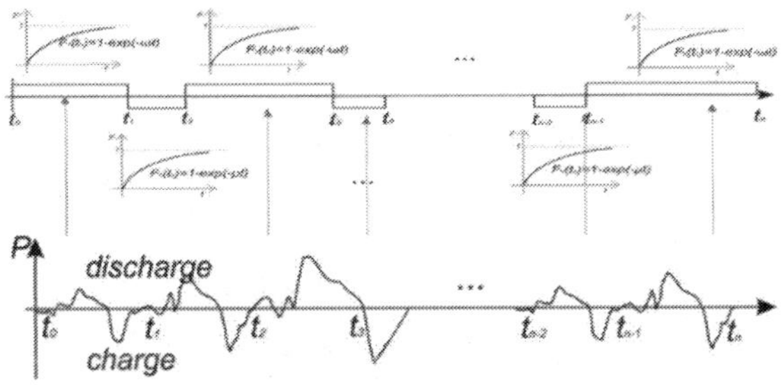

Fig. 2. The flow of events taken into account when modeling a random process of EPS operation.

2. For each section of the random process, the EPS operating state is calculated with the determination of the EPS state parameters and power deficits. Unlike the methods of assessing system reliability, in which only deficit modes are analyzed, the method of assessing planned reliability analyzes both deficit and non-deficit modes. This is necessary for adequate reflection of the parameters and states of energy storage devices, as well as other EPS parameters that can change in non-deficit states and affect deficit ones.

3. Determination of mathematical expectations (m.e.) of values of operating parameters and their deviations from the nominal value; m.e. of power deficit; m.e. of undersupply of electric power; probability of deficit-free operation; energy reliability characteristics of elements; m.e. of duration of electric power restrictions at consumers; functions of distribution of power deficit at consumers; frequency of charge/discharge cycles of energy storage systems.

4 Conclusion

Ensuring the reliability of electric power systems is the most important task in their management at different time stages.

To solve problems that arise when planning the operation (states and balances) of electric power systems, it is proposed to use a new type of system reliability, which is called planned reliability.

The methodology for assessing planned reliability is based on the sequential Monte Carlo method and takes into account a number of processes that critically affect the reliability of the electric power system.

Acknowledgements

The study was supported by the Russian Science Foundation grant No. 23-29-00435.

References

1. Krupenev D.S. Development of methodological foundations for the analysis of system reliability of modern electric power systems. Reliability and safety of energy. 2024, No. 2, pp. 114-121. (in Russian)
2. Krupenev, D.S., Domyshev, A.V. Assessment of the operational reliability of electric power systems based on the Monte Carlo method // Electricity, 2015, No. 2. – pp. 4–12. (in Russian)
3. Kovalev G.F., Lebedeva L.M. Reliability of electric power systems. Novosibirsk: Nauka, 2015. (in Russian)
4. Chukreev Yu.Ya., Chukreev M.Yu. Models for assessing balance reliability indicators in managing the development of electric power systems. - Syktyvkar: Komi Scientific Center of the Ural Branch of the Russian Academy of Sciences, 2014, 207 p. (in Russian)
5. Roy Billinton, Wenyuan Li Reliability Assessment of Electric Power Systems Using Monte Carlo Methods. Springer New York, NY, 1994, 352 p.
6. Mogirev V.V. Algorithm and program for calculating reliability indicators of electric power systems using the statistical modeling method (the "Potok" program). // Methodological issues in studying the reliability of large energy systems. Issue 4. Irkutsk: SEI SB RAS USSR, 1975. Pp. 24-35. (in Russian)
7. R.A. González-Fernández, A.M. Leite da Silva Reliability assessment of time-dependent systems via sequential cross-entropy Monte Carlo simulation. IEEE Trans. Power Appar. Syst., 26 (4) (2011), pp. 2381-2389
8. R. Billinton, P. Wang Distribution system reliability cost/worth analysis using analytical and sequential simulation techniques. IEEE Trans. Power Appar. Syst., 13 (4) (1998), pp. 1245-1250
9. S. Choudhury Review of energy storage system technologies integration to microgrid: types, control strategies, issues, and future prospects. J. Energy Storage, 48 (2022)
10. Krupenev D.S., Boyarkin D.A., Yakubovsky D.V. Formation of power equipment repair schedules based on planned reliability indicators in long-term planning of electric power systems. In the collection: Methodological issues of studying the reliability of large power systems. Proceedings of the 95th meeting of the International Scientific Seminar. Irkutsk, 2023. pp. 117-124. (in Russian)

© The Authors, published by EDP Sciences. This is an open access article distributed under the terms of the Creative Commons Attribution License 4.0 (https://creativecommons.org/licenses/by/4.0/).

Model for Minimizing Energy Consumption of Electric Power Systems Taking Into Account the Balance of Active and Reactive Power

Dmitrii Iakubovskii[1*], *Dmitry* Krupenev[1,2], and *Denis* Boyarkin[1]

[1] Melentiev Energy Systems Institute of Siberian Branch of the Russian Academy of Sciences, 664033 Irkutsk, Russian
[2] Irkutsk National Research Technical University, Lermontov Str. 83, Irkutsk, 664074, Russia

Abstract. The work covers the developed model of minimization of active power shortage taking into account the balance of active and reactive power. The motivation for the formation and testing of this model is the need to improve the adequacy of models of this class, as well as the need to implement it in the developed methodology for assessing planned reliability. The article specifies the requirements for the model and the process of its formation, which is based on the models of minimization of active power shortage and the optimal power flow format models. When forming the model, the required set of parameters was specified, namely, the modules of busbar voltages, phase angles of voltages, active and reactive resistance, and balance constraints of active and reactive power were formed. As a result of the analysis of experiments using circuits with 3 nodes, a difference in the result of the minimum power shortage (from 3.05 to 4.46%) was revealed compared to the problem based on the model of minimization of active power shortage without taking into account reactive power in balance constraints.

1 Introduction

In the modern conditions of operation of electric power systems, their constant enlargement and complication, effective planning of their development is a key factor for ensuring the reliability and stability of electricity supply to consumers. One of the basic operations necessary for conducting high-quality development planning is the adequacy assessment of the EPS [1 - 3]. In turn, the assessment methodology, which is based on the statistical testing method (Monte Carlo), includes several stages - the formation of random states of the system for a specific period, minimization of power shortage of these states and the calculation of reliability indicators as the main result.

An important part of the methodology is the capacity of power shortage minimization model, since the quality, correctness and complexity of the model directly affect the calculation results. Today, there are many different software packages [4 - 19] for adequacy assessment: ANTARES, Transmission Reliability Evaluation of Large-Scale Systems (TRELSS), Siemens PTI PSS/E TPLAN, CORAL, Grid Reliability and Adequacy Risk Evaluator (GRARE), PLEXOS, Multi-Area Reliability Simulation (MARS), GridView, NARP, MARELI; as well as ORION-M, POTOK-3, YANTAR and "Reliability". In most packages, calculations are carried out using flow, linear or linearized models. In some packages, cost minimization models are implemented, which differs from the original problem of minimizing capacity power shortage and, in turn, calls into question the solutions obtained in the framework of the adequacy assessment.

Existing models for minimizing the power shortage of EPS with linear losses have assumptions that do not fully account for losses during power flows. This problem was taken into account in [3] and a new model was developed, where power losses depend on the square of the transmitted power and is close in physical meaning to the real operation of EPS. However, such models do not fully take into account reactive resistance, different voltage levels and phase angles, which allows for calculations to be carried out for enlarged schemes, but with assumptions. Nevertheless, such an approach and other simplifications can lead to a distortion of the results of the system operation and their inconsistency with real conditions, including the accumulation of errors and mistakes in multiple calculations of various states.

Despite these assumptions, the existing methodology for adequacy assessment allows for basic calculations and analyzing EPS in an acceptable format. However, more comprehensive results can be obtained within the framework of assessing planned reliability taking into account random processes of EPS operation, where it is also necessary to carry out calculations using models for minimizing power shortage.

In view of the need to improve the adequacy of calculations for minimizing power shortage, as well as to increase the level of detail and compliance with the real physics of the process, there is a need to develop models taking into account active and reactive power.

* Corresponding author: dmitrii_iakubovskii@isem.irk.ru

© The Authors, published by EDP Sciences. This is an open access article distributed under the terms of the Creative Commons Attribution License 4.0 (https://creativecommons.org/licenses/by/4.0/).

As part of these studies, an analysis of existing models for calculating the steady state and models for minimizing costs taking into account reactive power was conducted [20-23]. As a result of the studies, a model for minimizing power shortage taking into account active and reactive power for assessing planned reliability was developed and presented in this paper. A comparative experiment and primary analysis of its operation relative to the existing model with a nonlinear formulation for adequacy assessment were also conducted.

2 Problem statement

Determination of the minimum power shortage is the basis of the methodology for assessing the planned reliability of the EPS using the Monte Carlo method, where each generated random process of EPS operation simulates its operation in accordance with the used model of minimization of power shortage (MPS). The development of such MPS models in adequacy assessment occurred in stages in the process of their use and based on the results of identified problems, such as physically incorrect distribution of power flows, the presence of ring flows, the need to take into account controlled cross-sections, active and reactive power. To date, several MPS models have been developed and improved [24-25], which differ in taking into account certain features of the functioning of the EPS and have varying degrees of adequacy and compliance with physical processes in the process of determining the power shortage. Using this experience, it becomes possible to effectively develop a new MPS model taking into account the specifics of assessing planned reliability.

The methodology for assessing planned reliability is based on the formation of a problem based on a model for minimizing power shortage using data (constraints) obtained from the block for generating specific random states of the system with subsequent optimization, i.e. searching for a minimum power shortage. In mathematical models, the parameters should include generator capacities and loads of all reliability zones/nodes, link capacities, and information on power flow losses. In general, the problem of distributing power across the entire system for a specific state is solved in such a way as to achieve a minimum power shortage, ensure a maximum covered load, taking into account power balances in reliability zones/nodes, and adequate, physically correct power flow distribution.

At present, a model with an improved formulation in a nonlinear form, without taking into account reactive power and controlled cross-sections, is used as a model for minimizing the power shortage for adequacy assessment. It is formulated as follows: "For known values of operational generator capacities, required levels of consumer loads, throughput capacities of EPS connections and power loss factors in EPS connections, it is necessary to determine the optimal flow distribution in EPS."

Mathematically, the problem is presented as follows:

$$\min_{y,x,z} \sum_{i=1}^{n} (\bar{y}_i - y_i), \tag{1}$$

subject to nonlinear balance constraints:

$$x_i - y_i + \sum_{j=1}^{n} (1 - a_{ji} z_{ji}) z_{ji} - \sum_{j=1}^{n} z_{ij} = 0, \quad i = 1, \dots, n. \tag{2}$$

as well as restrictions on the variables being optimized:

$$0 \le y_i \le \bar{y}_i, i = 1, \dots, n, \tag{3}$$

$$0 \le x_i \le \bar{x}_i, i = 1, \dots, n, \tag{4}$$

$$0 \le z_{ij} \le \bar{z}_{ij}, i = 1, \dots, n, j = 1, \dots, n, i \ne j \tag{5}$$

$$z_{ji} * z_{ij} = 0, i = 1, \dots, n, j = 1, \dots, n \tag{6}$$

where: x_i is the used capacity (MW) at node i, $\overline{x}_i$ is the available generating capacity (MW) at node i, y_i is the load covered at node i (MW), $\bar{y}_i$ is the load value at node i (MW), z_{ij} is the power flow from node i to node j (MW), $\bar{z}_{ij}$ is the transmission capacity of the transmission line between nodes i and j (MW), z_{ji} is the power flow from node j to node i (MW), $\bar{z}_{ji}$ is the transmission capacity of the transmission line between nodes j and i (MW) a_ji are the specified positive coefficients of specific power losses during its transmission from node j to node i, $j \ne i$, $i = 1, \dots, n$, $j = 1, \dots, n$.

The coefficients of specific power losses in the transmission lines during its transmission were determined in accordance with [3] and are calculated as follows:

$$a_{ji} = \frac{r_{ji}}{U_{\text{ном}}^2 cos\varphi_{ji}^2}, j = 1, \dots J, i = 1, \dots I, i \ne j \tag{7}$$

where: r_{ji} is the active resistance of the power transmission line between reliability zones j and i, Ohm/km; $U_{\text{ном}}$ is the nominal voltage of the power transmission line between reliability zones j and i, kV; $cos\varphi_{ji}$ is the average power factor of inter-zone connections between reliability zones j and i (usually taken to be equal to 0.9).

The presented MPS model (1) - (6) is based on power flow distribution models in the field of adequacy assessment and corresponds to the standard formulation of the transport problem. The main part of the model is formed due to the balance constraints of equalities that regulate the modeling of active power distribution during flows. Also, for the correct modeling of counter power flows, an additional constraint (6) is used, this allows for unambiguity in the direction of power flow in each operating mode of the system.

As shown in formula (7), the specific loss coefficients are determined only taking into account the active resistance and the average power factor of interzone connections, which undoubtedly simplifies the calculations. However, models of this type do not take into account reactive resistance, different voltage levels and phase angles, which allows for calculations for enlarged schemes with assumptions. Nevertheless, such an approach and other simplifications can lead to a distortion of the system's operating results and their inconsistency with real conditions, including the accumulation of errors and mistakes in multiple calculations of various states.

© The Authors, published by EDP Sciences. This is an open access article distributed under the terms of the Creative Commons Attribution License 4.0 (https://creativecommons.org/licenses/by/4.0/).

The constant process of complication and enlargement of electric power systems, the need to take into account their features and the transition to more detailed schemes for the correct solution of the problem of adequacy assessment of the EPS, as well as the problem of planning the development of the EPS, requires updating existing models and approaches taking into account additional system parameters and low-level calculations. First of all, the increase in the physical adequacy of the models should be carried out by adding the balance of active and reactive power, resistances and losses on power flows to the existing model (1) - (6). In this case, the target function should still be used to calculate active power shortages. The main corrections to the existing model should concern balance limitations, namely, calculations of power and losses on flows taking into account the possible difference in voltage, its phase angles and resistances.

Thus, an improved model should be formed within the framework of models similar to Optimal Power Flow used in the field of energy in various calculations. The developed model assumes that the minimum active power shortage should be designated as the target function:

$$\min \sum_{i=1}^{n} (P_{Di}^{max} - P_{Di}), \quad (8)$$

when implementing balance restrictions taking into account active and reactive power on the bus:

$$\sum_{g \epsilon G} P_i^g = \sum_{d \epsilon D} P_i^d + \sum_{j=1}^{n} P_{ij}^L, \quad (9)$$

$$\sum_{g \epsilon G} Q_i^g = \sum_{d \epsilon D} Q_i^d + \sum_{j=1}^{n} Q_{ij}^L, \quad (10)$$

auxiliary calculation of power during flows:

$$P_{ij}^L = \frac{1}{r_{ij}^2 + x_{ij}^2}\left[r_{ij}\left(v_j^2 - v_i v_j \cos(\theta_i - \theta_j)\right) + x_{ij}\left(v_i v_j \sin(\theta_i - \theta_j)\right)\right], \quad (11)$$

$$Q_{ij}^L = \frac{1}{r_{ij}^2 + x_{ij}^2}\left[x_{ij}\left(v_j^2 - v_i v_j \cos(\theta_i - \theta_j)\right) + r_{ij}\left(v_i v_j \sin(\theta_i - \theta_j)\right)\right], \quad (12)$$

and restrictions on variables:

$$P_i^{D,min} \leq P_i^D \leq P_i^{D,max} \quad (13)$$

$$P_i^{G,min} \leq P_i^G \leq P_i^{G,max} \quad (14)$$

$$Q_i^{D,min} \leq Q_i^D \leq Q_i^{D,max} \quad (15)$$

$$Q_i^{G,min} \leq Q_i^G \leq Q_i^{G,max} \quad (16)$$

$$\theta_i^{min} \leq \theta_i \leq \theta_i^{max} \quad (17)$$

$$v_i^{min} \leq v_i \leq v_i^{max} \quad (18)$$

where D is a set of consumers on the bus, load; G is a set of generators on the bus, generation; L is a set of connections; i, j are buses, nodes; P_i^G is the total active power of generators on the i-th bus; P_i^D is the total active power of the load on the i-th bus; Q_i^G is the total reactive power of generators on the i-th bus; Q_i^D is the total reactive power of the load on the i-th bus; θ is the phase angle of the voltage on the bus; v_i is the absolute value of the voltage on the i-th bus; $r_{i,j}$ is the active resistance of the line between nodes/buses i and j; $x_{i,j}$ is the reactive resistance of the line between nodes/buses i and j. Thus, this model will have to take into account the deep specifics of various EPS, and in turn will require correct data sets.

3 Experimental part

Experimental calculations were carried out for systems of various configurations, with different initial parameters. The performance test of the model, taking into account the balance of active and reactive power (8) - (18) and comparison of the obtained results with the model (1) - (6) was carried out on a system with 3 nodes. However, several test states of the system were used in the experiments, the main results were taken out in this work on the examples of two states, where the first state is shown in Figure 1.

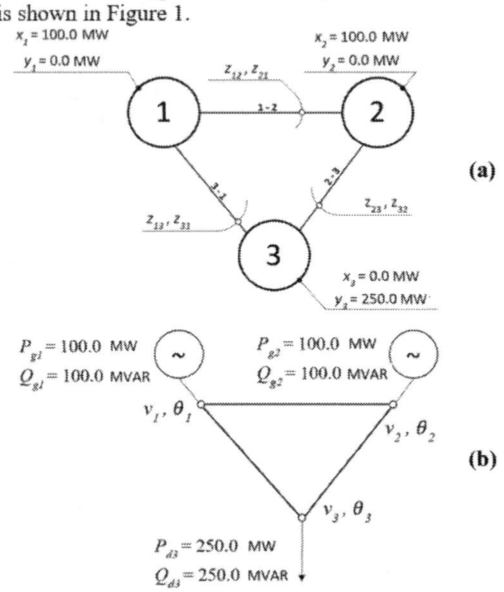

Figure 1: Test system #1 (a) – with 3 nodes for model (1)-(6), (b) – with 3 nodes for model (8)-(18)

Figure 1.a shows the system diagram used in conjunction with model (1)–(6), in this case it is assumed that each node combines the available generating capacity (x) and the load value at the node (y), and all losses within the node are zero, the nodes are also connected by links, each of which is represented by flows (z) in the forward and reverse directions. Figure 1.b shows the system diagram with two generators adjacent to nodes 1 and 2, respectively, and one load at the third node, all nodes are connected by three buses.

As a result of the calculations, the values of the objective function and the optimized variables were obtained for the problems based on two models. As a result of the optimization of the problem according to model (1)–(6), the power shortage was 51.842 MW, and two power flows z23 and z13 were involved. This power flow distribution is correct, since it was assumed that due to the surplus nodes 1 and 2 and connections with a capacity of 100 MW, the load in the 3rd deficit node will be provided. On the other hand, the problem based on the model (8) - (18) showed a minimum power shortage

© The Authors, published by EDP Sciences. This is an open access article distributed under the terms of the Creative Commons Attribution License 4.0 (https://creativecommons.org/licenses/by/4.0/).

at the level of 49.519 MW, which is 2.323 MW of difference, while all three connections are loaded.

These results demonstrate the operability of the developed model (8) - (18) and the difference in the minimum power shortage level of 4.48% for similar circuits shown in Figure 1. However, a significant addition to (8) - (18) is the ability to take into account reactive power in calculations, which affects the power flow distribution, and the analysis of the data obtained for the developed model showed the loading of additional lines. This feature of the result could be considered an incorrect power flow distribution for the problems of the model (1) - (6), since with an analytical solution and correct power flow distribution, power transmission through an additional node was not allowed. In the case of model (8) - (18), the presence of voltage level parameters, phase angles, active and reactive resistance can affect the power flow distribution in a similar way.

Another experiment was conducted using different data presented in Figure 2, where each node/bus was assigned both load and generation, which allowed testing the model for correct power flow distribution in the presence of its own generation on the node/bus and the possibility of reducing the deficit in/on neighboring nodes/buses.

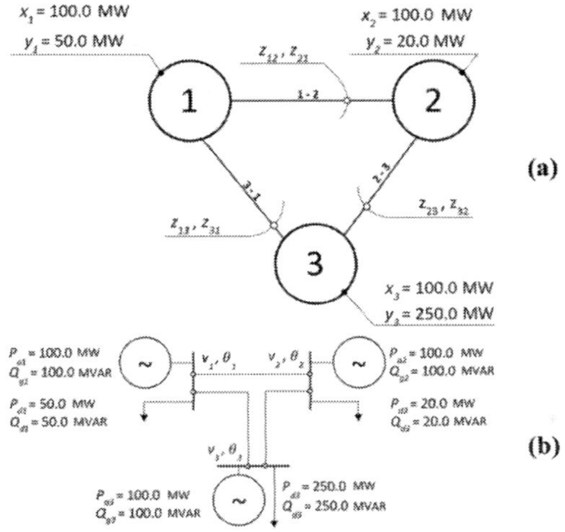

Figure 2: Test system No. 2 (a) – with 3 nodes for model (1)-(6), (b) – with 3 tires for model (8)-(18)

Based on the optimization results of test system No.2, the obtained values were 20.428 MW and 19.804 MW for problems (1)–(6) and (8)–(18) models, respectively, where the difference is 0.624 MW or 3.05%. During the optimization of problems for both models, the 3-1 and 2-3 flows were used to minimize the power shortage in node 3, but the 1-2 flow was also used, which is not directly necessary, since the generation was sufficient to provide the intra-nodal load. This problem of model (1)–(6) was previously solved by fixing the optimized level of the covered load and additional optimization by the sum of the power flow rate. It was also found that when solving the problem using model (8)–(18), only 30.196 MW out of 50 MW of the load is covered. In the future, it is planned to conduct an additional analysis of the operation of model (8) – (18), as well as to implement the necessary clarifications and improvements to increase its accuracy and adequacy.

4 Conclusions

The paper considers the problem of developing a model for minimizing active power deficit taking into account the balance of active and reactive power. The developed model was compared with a nonlinear (quadratic) model for minimizing active power deficit in electric power systems, but taking into account only the balance of active power. The developed model takes into account several additional parameters compared to the existing one, namely, bus voltage modules, voltage phase angles, active and reactive resistance. When developing the model, an approach to forming balance equations for minimizing active power shortage corresponding to Optimal Power Flow models was used. A circuit with 3 nodes was used to conduct experimental studies. Several tests were carried out for each system. Based on the results of the experiments, it was revealed that the results of solving problems based on the developed model differ from the results of solving problems based on the model used in adequacy assessment by 3.05-4.46%. Different flow distribution in the form of additional power flows was also revealed. This behavior is explained by the presence of reactive power accounting and dependence on additional parameters of voltage level, phase angles and resistances. In the future, tests will be carried out on larger-scale circuits, where the difference is expected to be more significant.

Acknowledgments

The study was supported by the Russian Science Foundation grant No. 23-29-00435.

References

1. Kovalev G.F., Lebedeva L.M. Model for assessing the reliability of electric power systems in long-term planning of their operation // Electricity. - 2000. - ¹ 11. - P. 17-24. (in Russian)
2. G. F. Kovalev. Reliability of electric power systems / G. F. Kovalev, L. M. Lebedeva; ed. N. I. Voropai. - Novosibirsk: Nauka, 2015. - 224 p. (in Russian)
3. Kovalev G.F., Lebedeva L.M. Complex of models for optimization of design state modes in assessing the reliability of electric power systems. Irkutsk: ISEM SB RAS, 2000, 74 p. (in Russian)
4. Review of the current status of tools and techniques for risk-based and probabilistic planning in power systems, Working Group 601 of Study Committee C4 //International Conference on Large High Voltage Electric Systems. March. 2010
5. Fernandez Blanco Carramolino, R., Careri, F., Kavvadias, K., Hidalgo Gonzalez, I., Zucker, A.

© The Authors, published by EDP Sciences. This is an open access article distributed under the terms of the Creative Commons Attribution License 4.0 (https://creativecommons.org/licenses/by/4.0/).

https://doi.org/10.1051/e3sconf/202458401024

and Peteves, E., Systematic mapping of power system models: Expert survey, EUR 28875 EN, Publications Office of the European Union, Luxembourg, 2017, ISBN 978-92-79-76462-2, doi:10.2760/422399, JRC109123.

6. Antonopoulos G; Chondrogiannis S; Kanellopoulos K; Papaioannou I; Spisto A; Efthimiadis T; Fulli G., Assessment of underlying capacity mechanism studies for Greece, EUR 28611 EN, Publications Office of the European Union, Luxembourg, 2017, ISBN 978-92-79-68878-2, doi: 10.2760/51331, JRC106307
7. RTE Antares, Antares Optimization problems formulation, URL: https://antares.rte-france.com
8. A. Gaikwad, S. Agarwal, K. Carden, N. Wintermantel, S. Meliopoulos, M. Kumbale, A Study on Probabilistic Risk Assessment for Transmission and Other Resource Planning, Electric Power Research Institute For EISPC and NARUC (NARUC-2013-RFP027-DE0316), 2015.
9. URREGO AGUDELO Lilliam, A novel method for the Approximation of risk of Blackout in operational conditions, Laboratoire Image-Signaux et Systèmes Intelligents / LISSI - EA 3956 (laboratoire) , 2016
10. Milorad Papic, Survey of Tools for Risk Assessment of Cascading Outages, IEEE GM, 2011
11. Ying-Yi Hong, Lun-Hui Lee Reliability assessment of generation and transmission systems using fault-tree analysis, Energy Conversion and Management 50 (2009) 2810–2817.
12. Siemens AG and Siemens Industry, Inc., Model Management Module for PSS®E, 2020
13. PSR – Energy Consulting and Analytics, OPTGEN User Manual, Version 7.4, 2019
14. ENTSO-E, Mid-term Adequacy Forecast 2018, Appendix 1: Methodology and Detailed Results, 2018
15. PLEXOS Market Simulation Software, URL: https://energyexemplar.com/solutions/plexos/
16. Kelvin Chu, MARS Multi-Area Reliability Simulation, EOP – On Demand Feature, General Electric Company, 2014
17. Panida Jirutitijaroen, Chanan Singh, Reliability and Cost trade-off in Multi-Area Power System Generation Expansion Using Dynamic Programming and Global Decomposition, IEEE Transactions on power systems, Vol 21,No 3, August 2006
18. J.McCalley, Module PE.PAS.U21.5 Multiarea reliability analysis // Electrical & Computer Engineering, Iowa State University, USA, – 81p.
19. Simulate security-constrained unit commitment and economic dispatch in large-scale transmission networks // ABB GridView, 2016.
20. Pandapower, URL: https://pandapower.readthedocs.io/en/v1.4.3/powerflow/opf.html
21. Kim, Seong-Cheol & Salkuti, Surender Reddy. (2019). Optimal power flow based congestion management using enhanced genetic algorithms. International Journal of Electrical and Computer Engineering (IJECE). 9. 875. 10.11591/ijece.v9i2.pp875-883.
22. Nusair, K.; Alasali, F. Optimal Power Flow Management System for a Power Network with Stochastic Renewable Energy Resources Using Golden Ratio Optimization Method. Energies 2020, 13, 3671. https://doi.org/10.3390/en13143671
23. Telyatnik, Andrey & Vaskovskaya, Tatiana. (2019). Acceleration of the method of sequential quadratic programming in the problem of optimization of steady-state modes of EPS. Bulletin of the Russian Academy of Sciences. Power Engineering.3-15. 10.1134/S0002331019040149. (in Russian)
24. Iakubovskii D.V. “Analysis of models for minimizing power deficit in the adequacy assessment of electric power systems” // System research in energy / Works of young scientists of the ISEM SB RAS, Issue 48. –Irkutsk: ISEM SB RAS, 2017. –131 p. (in Russian)
25. Iakubovskii D.V., Krupenev D.S., Boyarkin D.A. Model for minimizing the power deficit of electric power systems taking into account restrictions on controlled cross-sections // Data Analysis and Processing Systems. - 2021. - No. 2 (82). - P. 95–120. - DOI: 10.17212/2782-2001-2021-2-95-120. (in Russian)

© The Authors, published by EDP Sciences. This is an open access article distributed under the terms of the Creative Commons Attribution License 4.0 (https://creativecommons.org/licenses/by/4.0/).

Methodological Aspects of Reliability Assessment of Asynchronous Motors under Current Operation Conditions at Mineral Resources Enterprises

Victoria Romanova[1,2*], *Andrey* Batukhtin[1], *Sergey* Khromov[1], *Ahmad* Alzakkar[3]

[1]Trans-Baikal State University, Department of Energy, Chita, 30, Aleksandro-Zavodskaya Str., Russia
[2]Branch of the National Research University «Moscow Power Engineering Institute» in Volzhsky, Department of Energy, Volzhsky, 69, im.Lenin pr., Russia
[3]Albaath University, Faculty of Mechanical and Electrical Engineering, Homs, PP63+FW4, Damascus - Aleppo Str., Syria

Abstract. At present, the focus is on solving the urgent problem of reliable performance of power system elements and ensuring their efficient operation during mining. A methodological approach is considered that allows assessing the reliability level of low-voltage asynchronous electric motors in a coal mine with difficult geological and climatic mining conditions. The object of the study is asynchronous electric motors used at one of the enterprises of Siberian Coal Energy Company (SUEK) in Trans-Baikal territory. The main tool for implementing the research objectives was program MATLAB/Simulink. The purpose of the article is to study the influence of external operational factors on the service life of an asynchronous electric motor with a squirrel-cage rotor. A comprehensive solution to issues aimed at maintaining the reliability level of asynchronous electric motors is one of the priority areas, mainly aimed at ensuring the effective functioning of a coal mining enterprise.

1 Introduction

The efficient operation of a coal mining enterprise depends on the stable operation of its individual elements, in particular asynchronous electric motors (AM), which work as the electric drive for the vast majority of critical mechanisms. It should be noted that the role of low-voltage asynchronous motors in the technological chain of coal mining complexes is obvious and significant.

For example, the failure one of asynchronous electric motors leads to: disruption of complex technological processes, downtime of the technological chain, decrease in production volumes.

The specific nature of the work of coal industry enterprises is associated with complex mining, geological and climatic conditions of extraction. Therefore, the operation of electrical equipment in such conditions is due to the presence of an aggressive environment and increased dustiness [1]. In addtion, it is necessary to highlight the problem of low quality of electrical energy in electrical networks supplying enterprises in mineral resources enterprises.

Thus, as a result of the conducted studies of the quality of electrical energy in the electrical networks of the Trans-Baikal electric power system was established the quality indicators of electrical energy, such as [2, 3]: voltage asymmetry, nonsinusoidal voltage, voltage deviation, etc. Significant deviations from the regulated standards were noted for the following indicators: K_{2U}, $K_{U(3)}$, $K_{U(5)}$, $K_{U(15)}$, $K_{U(21)}$ [2, 4].

It is obvious that the specific features of the operation in the Trans-Baikal electric power system with a complex of unfavorable factors lead to increase in the accident rate of electrical equipment of consumers and decrease in the reliability and efficiency of electrical equipment operation.

The conducted studies of low-voltage asynchronous electric motors in the works of the authors [5-7] indicate that external influencing factors contribute to decrease in AM cooling, increase of individual parts temperature AM and overheat of the winding insulation.

In this regard, it becomes obvious that a violation of the thermal operating mode of electric motors, in the presence of difficult operating conditions, leads to a reduction in the service life of asynchronous electric motors. This study focuses on the dynamics of changes in the service life of low-voltage asynchronous electric motors depending on: electrical parameters (voltage asymmetry coefficient and voltage harmonic coefficient), environmental parameters (air temperature and humidity) and dust content.

2 Materials and Methods

The main objective of the study is to develop a methodological approach that allows us to assess the level

* Corresponding author: romanova181@mail.ru

© The Authors, published by EDP Sciences. This is an open access article distributed under the terms of the Creative Commons Attribution License 4.0 (https://creativecommons.org/licenses/by/4.0/).

of reliability of low-voltage asynchronous electric motors when operating in a coal mining complex with complex mining, geological and climatic conditions of production.

The main activity of the enterprises SUEK in Trans-Baikal territory is mining of brown coal. The technological process of the coal mining enterprise involves single-bucket walking excavators, which are necessary for the extraction of large volumes of rock during stripping operations and quarry dumps.

There are two types of electric motors that drive the main mechanisms of walking excavators:

- DC motors type MPE 450-900 for pulling, lifting and walking;
- DC motors type MPVE 400-900 for rotation.

DC motors are blown by asynchronous electric motors series AIR100S4 with a squirrel-cage rotor.

In accordance with the above, it is required to carry out a study and a detailed analysis of the operating modes of the asynchronous electric motors series AIR100S4 with P_n = 3 kW and nominal motor speed n = 1500 rpm [8].

As is known [9], the reliability indicators of asynchronous electric motors have an average service life more than 15 years and an operating time 40000 hours.

Besides, it is noted that the stable and trouble-free operation of the AM is possible under certain operating conditions that ensure an acceptable thermal mode of operation established by the enterprises.

The proposed methodological approach is presented in the following:

Step 1. Analytical review of operational factors that have the greatest impact on the thermal operating mode of low-voltage AM in a coal mine under difficult geological and climatic mining conditions.

Step 2. The equation of the thermal balance AM for the steady state with voltage distortions:

$$\sum \Delta P + \sum \Delta P_{add} = K_{ht} \cdot \sum S_s \cdot (\Delta T_s - \Delta T_{amb}), \quad (1)$$

$\sum \Delta P\ (W)$ - losses AM in condition there is no voltage distortion;

$\sum \Delta P_{add}\ (W)$ - additional losses AM due to voltage distortions;

$K_{ht}(W/m^2 \cdot {}^\circ C)$ - coefficient of heat transfer AM;

$S_s(m^2)$ - heat transfer surface area AM;

$\Delta T_s({}^\circ C)$ - temperature on the surface AM;

$\Delta T_{amb}({}^\circ C)$ - ambient temperature.

Step 3. Determination of power losses AM in condition there is no voltage distortion.

According to [10], power losses AM are calculated by:

$$\sum \Delta P = P_{stator} + P_{rotor} + \Delta P_{mec}, \quad (2)$$

$P_{stator}\ (W)$ - electrical losses in the stator winding AM;

$P_{rotor}\ (W)$ - electrical losses in the rotor winding AM;

$\Delta P_{mec}(W)$ - mechanical losses AM.

Electrical losses in stator windings are determined by:

$$P_{stator} = 3 \cdot I_1^2 \cdot r_1 \quad (3)$$

Electrical losses in rotor windings are determined by:

$$P_{rotor} = 3 \cdot I_2^2 \cdot r_2, \quad (4)$$

r_1, r_2 - active resistances of the stator and rotor phase windings at operating temperature (Θ_{op}):

$$r_1 = r_{1a}\left[1 + \alpha\left(\Theta_{op} - 20\right)\right]; \quad r_2 = r_{2a}\left[1 + \alpha\left(\Theta_{op} - 20\right)\right], \quad (5)$$

r_{1a}, r_{2a} - active resistances of windings at operating temperature($\Theta_{op} = 20°C$);

α - temperature coefficient, for copper ($\alpha = 0.004$).

The mechanical losses AM are determined by:

$$\Delta P_{mec} = 1.3(1 - D_a) \cdot \left(\frac{n_g}{10}\right) D_a^4, \quad (6)$$

$n_g(rpm)$ - nominal speed AM;

$D_a(m)$ - outer diameter AM.

Step 4. Determination of additional thermal power losses AM due to voltage distortion, which can calculate by Shidlovsky formula [11]:

$$\Delta P_{add} = \frac{r_1 \cdot I_s^2}{\eta \cdot cos\varphi}\left(2.41 K_{2U}^2 + 2\sum_{\nu=3}^{\infty} \frac{U_\nu^2}{\nu\sqrt{\nu}}\right), \quad (7)$$

$r_1(\Omega)$ - active resistances of stator phase windings;

$\eta(\%)$ - efficiency AM;

$K_{2U}(\%)$ - voltage asymmetry coefficient by the reverse sequence;

ν - ordinal number of the harmonic;

U_ν - ratio of ν-th harmonic voltage to the nominal voltage;

$I_s(A)$ - starting current AM.

Step 5. Determination of the change in the heat transfer process due to the power losses AM.

It is worth noting that the electric motor is cooled in two ways:

- direct heat transfer from the surface of the electric motor to the environment;
- heat transfer from the surface of the electric motor to the environment through a layer of dust.

Therefore, it is necessary to track the dynamics of changes in the heat transfer coefficient in the presence of dust on the electric motor housing by introducing a coefficient characterizing the amount of dust coverage of the asynchronous electric motor (K_{dc}).

Assuming that the dust layer as a single-layer flat wall of constant thickness [12], then:

$$Q = K_{dust}/h \cdot \sum S_s \cdot (\Delta T_s - \Delta T_{amb}), \quad (8)$$

$Q(J)$ - heat transferred through a layer of dust;

$K_{dust}(W/m^2 \cdot {}^\circ C)$ - coefficient of thermal conductivity of dust;

$h = 10\ mm$ - dust thickness [13].

Then the right side of equation (1):

$$Q = \frac{K_{dust}}{h} \cdot \sum S_s \cdot (\Delta T_s - \Delta T_{amb}) \cdot K_{dc} + K_{ht} \cdot \sum S_s \cdot (\Delta T_s - \Delta T_{amb}) \cdot (1 - K_{dc}) \quad (9)$$

In addition, taken into account the influence of air humidity on the operation of the motor, therefore, the heat transfer coefficient from the surface of the electric motor will be calculated by:

$$K_h = \frac{Q_h}{\sum S_s(\Delta\vartheta_{air})}, \quad (10)$$

Q_h - the amount of heat determined by:

$$Q_h = \Delta\vartheta_{air}(c_{air}V), \quad (11)$$

c_{air} - specific heat capacity of air and depends on air humidity (V_{air}), determined in [14];

$V\ (m^3/s)$ - required amount of cooling air;

$\Delta\vartheta_{air}$ - excess of cooling medium temperature.

$$V = m(n_k l_k + 100)\frac{n}{1000}\left(\frac{D_{ed}}{100}\right) \cdot 10^{-4}, \quad (12)$$

n - number of pole pairs;

m - coefficient of electric motor dimensionality.

If $n = 2$ then $m = 2.6$ & if $n = \{4,6,8,10\}$ then $m = 3.15$;

© The Authors, published by EDP Sciences. This is an open access article distributed under the terms of the Creative Commons Attribution License 4.0 (https://creativecommons.org/licenses/by/4.0/).

$n_{k,}l_k$ - number and length of radial ventilation ducts (in their absence($n_k l_k = 0$));

$D_{ed}(m)$ - external diameter AM.

Substitute all the equations obtained in (1):

$$3 \cdot I_2^2 \cdot r_2 + \left(\frac{3 \cdot I_2^2 \cdot r_2}{s} - \eta \cdot P_1\right) + 3 \cdot I_1^2 \cdot r_1 + \frac{r_1 \cdot I_s^2}{(\eta \cdot cos\varphi)(2.41K_{2U}^2 + 2\sum_{v=3}^{\infty}\frac{U_v^2}{v\sqrt{v}})} = \frac{K_{dust}}{h} \cdot \sum S_s \cdot (\Delta T_s - \Delta T_{amb}) \cdot K_{dc} + K_{ht} \cdot \sum S_s \cdot (\Delta T_s - \Delta T_{amb})(1 - K_{dc}), \quad (13)$$

s (mm) - electric motor slip.

Step 6. Determination of dynamics of changes in the service life AM by predicting its actual service life depending on the combination of influencing operational factors.

The change in the service life of an asynchronous electric motor:

$$z = e^{\frac{3 \cdot I_2^2 \cdot r_2 + \left(\frac{3 \cdot I_2^2 \cdot r_2}{s} - \eta \cdot P_1\right) + 3 \cdot I_1^2 \cdot r_1 + \frac{r_1 \cdot I_s^2}{(\eta \cdot cos\varphi)(2.41K_{2U}^2 + 2\sum_{v=3}^{\infty}\frac{U_v^2}{v\sqrt{v}})}}{\frac{K_{dust}}{h} S_S \cdot K_{dc} + S_S(c_{air}V)(1 - K_{dc})}} \quad (14)$$

The dynamics of change in the service life AM relative to the standard is calculated using the following formula:

$$\Delta Z = Z(1 - z) \quad (15)$$

Z = 15 years, standard service life of the electric motor.

After the procedure of preparing the necessary initial parameters, we proceed to the process of modeling the thermal mode AM.

Step 7. Modeling the thermal mode AM during its operation with the presence of external influencing factors (Fig. 1).

The developed model allows performing the procedure of simulating the operation AM under various specified operating conditions.

The modeling of the operating mode AM was performed by changing the following experimental parameters [15]:

- the value of the total harmonic voltage coefficient K_U = [4% ÷12%], range (2%);
- the value of voltage asymmetry coefficient by the reverse sequence K_{2U} = [0% ÷ 4%], range (1%);
- the value of ambient temperature ΔT_{amb} = [-30°C ÷ 30°C], range (10°C);
- the value of air humidity V_{air} = [20% & 60%];
- the value of the thermal conductivity coefficient of dust K_{dust} = [0.12 & 0.28].

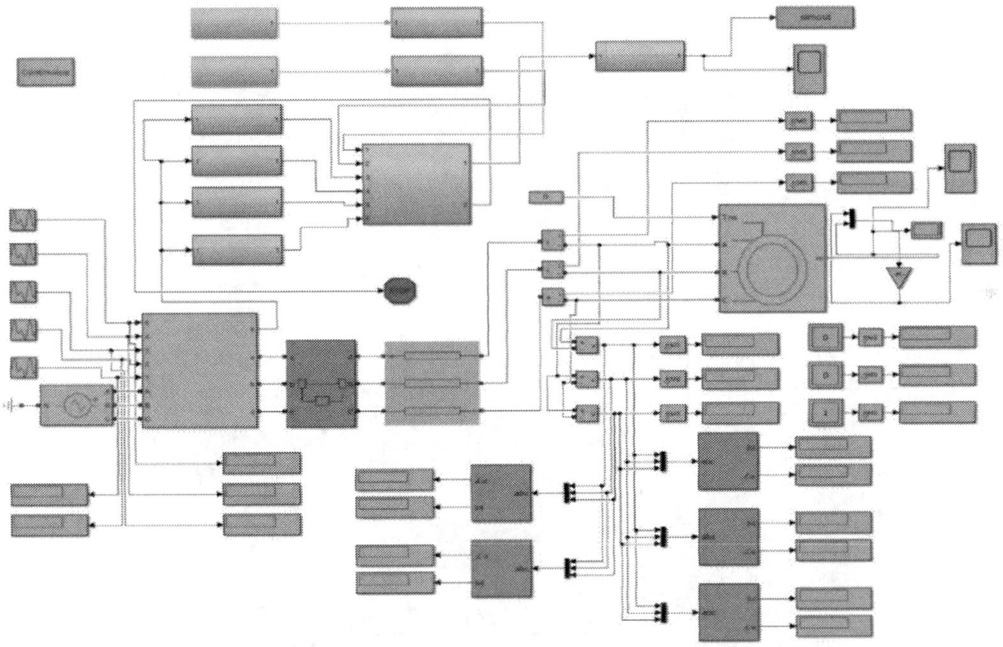

Fig. 1. Computer model of asynchronous electric motor implemented by MATLAB/Simulink.

In our study, the following conditions were adopted:

- the value of the total harmonic voltage coefficient consists of: $(K_{U(3)}, K_{U(5)}, K_{U(15)}, K_{U(21)})$, determined according to [16];
- the value of the thermal conductivity coefficient of dust depends primarily on the composition of the dust.

Therefore, two types of dust were considered in the study:

- coal dust formed as a result of overburden removal operations, with a thermal conductivity coefficient (K_{dust} = 0.12);
- coal dust with inclusions of quartz dust, with a thermal conductivity coefficient (K_{dust} = 0.28).

© The Authors, published by EDP Sciences. This is an open access article distributed under the terms of the Creative Commons Attribution License 4.0 (https://creativecommons.org/licenses/by/4.0/).

 https://doi.org/10.1051/e3sconf/202458401025

3 Results and Discussion

Based on the obtained modeling results, graphical displays (Fig.2, Fig.3) were constructed, which clearly demonstrate the dynamics of changes in the service life AM depending on the combination of operating factors.

The following results are found:

1. It should be emphasized that the impact of operational factors significantly reduces the level of reliability of asynchronous electric motors, reducing their service life. As a result, the quality indicator of electrical energy K_{2U} was identified, which has the greatest impact on the change of the service life AM.

The impact of dust on the service life AM depends on its physical and chemical composition. For example, the maximum value of service life variation is observed in the presence:

- coal dust with parameters ($K_{dust} = 0.12$, $V_{air} = 60\%$) was $\Delta Z = 4.65$ years;
- coal dust with inclusions of quartz dust with parameters ($K_{dust} = 0.28$, $V_{air} = 60\%$) was $\Delta Z = 4.56$ years.

It is noted that the maximum value of the service life variation is observed in the presence of coal dust with parameters ($K_{dust} = 0.12$, $K_{2U} = 4\%$, $V_{air} = 20\%$, $\Delta T_{amb} = 30°C$) was $\Delta Z = 4.67$ years.

Therefore, the degree of resistance of electric motors to the effects of dust depends on the composition of the dust, which determines the total value of its thermal conductivity.

2. The recommended operating mode AM, used in coal mines, is under the influence of factors with values: ($K_{2U} \leq 2\%$, $K_U \leq 8\%$; $20 \leq V_{air} \leq 50\%$; $-30°C \leq \Delta T_{amb} \leq 20°C$.

Therefore, critical values of the operating parameters of low-voltage asynchronous electric motors were identified.

3. It is necessary to adhere to the recommended operating parameters of electric motors during their operation. In the event of failure of electric motors, in order to maintain the production volumes of a coal mining enterprise, provision must be made for rapid restoration of the operability of electric drive mechanisms, which is ensured by the «hot reserve» of electric motors at the enterprise.

4. Stable and trouble-free operation of asynchronous electric motors undoubtedly depends on the conditions of their operation, therefore, the introduction of a system of organizational and technical influences aimed at maintaining the technical condition by correcting voltage distortions.

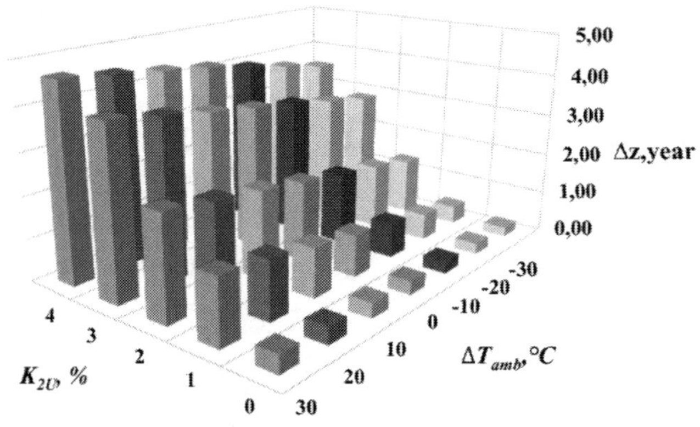

Fig. 2. Graph of the change in the service life AM, depends on K_{2U}, at different ambient temperatures with $K_{dust} = 0.28$, $V_{air} = 60\%$.

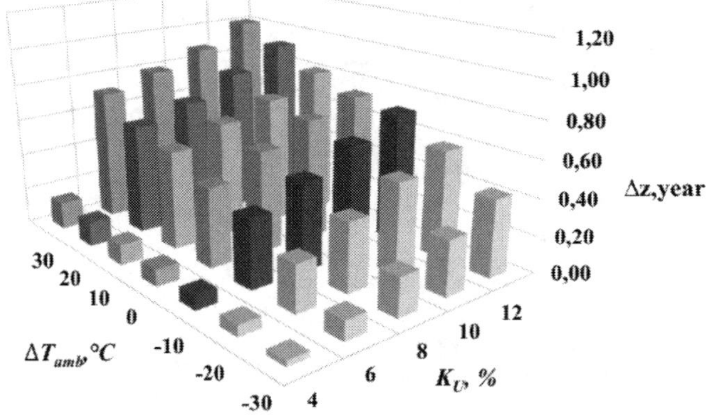

Fig. 3. Graph of the change in the service life AM, depends on K_U, at different ambient temperatures with $K_{dust} = 0.12$, $V_{air} = 20\%$.

© The Authors, published by EDP Sciences. This is an open access article distributed under the terms of the Creative Commons Attribution License 4.0 (https://creativecommons.org/licenses/by/4.0/).

Conclusion

The paper presents a conceptual vision of the implementation of a methodological approach that allows assessing the reliability level of low-voltage asynchronous electric motors when operating in a coal mining with complex mining, geological and climatic conditions.

The solution to the issues of reliable operation of a coal mining enterprise with asynchronous electric motors and ensuring their efficient operation is one of the priority areas, mainly focused on maintaining the stability of technological processes and maintaining the production volumes of a coal mining enterprise.

Acknowledgment

The authors team expresses its gratitude to S.V. Kovshov, Ph.D., Associate Professor, Deputy Director of the Institute of Targeted Educational Programs for Methodological Work of the St. Petersburg Mining University of Empress Catherine II for the assistance and valuable recommendations during the implementation of this study.

The research was carried out within the state assignment of Ministry of Science and Higher Education of the Russian Federation (theme № 123102000012-2 «Comprehensive study of aerodynamic characteristics of plasma systems of thermochemical fuel preparation», agreement № 075-03-2023-028/1 of 05.10.2023).

References

1. V. Afanasyev, Yu. Linnik, V. Linnik, Coal of Russia: status and prospects. M.: SIC INFRA-M, **271** (2017).
2. L. Kovernikova, Quality of electric energy, current state, problems, proposals for their solution. Novosibirsk: Nauka, **219** (2017).
3. GOST 32144 – 2013. Electrical power. Electromagnetic compatibility of technical equipment. Electrical power quality standards in the general purpose power supply systems. M.: Standartinform, 19 (2014).
4. O. Zapanov, L. Kovernikova, *On the power quality of electrical energy supplied to joint stock company «Aleksandrovsky mine»*, E3S Web of Conferences, Vol. 209 (07012), 6 (2020).
5. V. Romanova, K. Suslov, S. Khromov, *Multi-factor analysis of external effects influencing the operational reliability of asynchronous electric motors*, E3S Web of Conferences, Vol. 216 (01084), 4 (2020).
6. S. Deleanu, M. Iordache, M. Stanculescu, D. Niculae, *The Induction Machine Operating from a Voltage Supply, Unbalanced and Polluted with Harmonics: a Practical Approach*, 15th International Conference on Engineering of Modern Electric Systems (EMES). Oradea, Romania, 181 – 184 (2019).
7. Tabora J.M., de Matos E.O., Soares T.M., Tostes M.E., *Voltage unbalance effect on the behavior of IE2, IE3 and IE4 induction motor classes*, Proc. on VII Simpósio Brasileiro de Sistemas Elétricos (SBSE). Santo André, São Paulo, Brazil, 6 (2020).
8. http:// www.elektrikii.ru/publ/6-1-0-55.
9. I. Kopylov, Design of electric machines. M.: Yurait Publishing House, **828** (2024).
10. N. Shaitor, A. Velilyaev, *Calculation and analysis of temperature fields of asynchronous motors*, Power plants and technologies, Vol. 7, № 2, 112 – 118 (2021).
11. A. Shidlovsky, V. Kuznetsov, Improving the quality of energy in electric networks. Kiev: Naukova dumka, **268** (1985).
12. V. Kudinov, E. Kartashov, E. Stefanyuk, Technical thermodynamics and heat transfer. M.: Yurayt Publishing House, **454** (2024).
13. N. Sharov, R. Dudayev, D. Krishchuk, M. Liskova, *Dust suppression methods in coal mines of the Far North*, Perm Journal of Petroleum and Mining Engineering, Vol.19, №2, 184 – 200 (2019).
14. GOST IEC 60034-2A – 2012. Electric rotating machines. Part 2 - Methods for determining losses and efficiency of rotating electric machines during tests (excluding machines for traction vehicles). Measurement of losses by the calorimetric method. M.: Standartinform, 19 (2014).
15. E. Krasnoshchekov, A. Sukomel, Taskbook on heat transfer. M.: Energy, **288** (1980).
16. GOST 30804.4.7 – 2013 (IEC 61000-4-7:2009). Electromagnetic compatibility of technical means. General guidance on measuring instruments and measurements of harmonics and interharmonics for power supply systems and technical means connected to them. M.: Standartinform, 34 (2013).

© The Authors, published by EDP Sciences. This is an open access article distributed under the terms of the Creative Commons Attribution License 4.0 (https://creativecommons.org/licenses/by/4.0/).

Improving the reliability of the power system in postfault condition

Ivan Solovyev[1], *Alexey* Motovilov[1*], *Mikhail* Hvijuzov[1], and *Alexander* Emelyanov[1]

[1] Northern (Arctic) Federal University named after M.V. Lomonosov, Arkhangelsk, Russian Federation

Abstract. Priority tasks in electric power transmission are: the reliability of power supply to consumers and the efficiency of energy resources use. In postfault steady-state heavy modes of power system operation, these tasks are mainly solved by supervisory control. In order to improve the reliability of power supply, the authors propose an algorithm for assessing the weight of emergency actions, carried out by the power grid manager in postfault conditions. During them the power grid manager redistributes power flows in the electric grid. The application of the suggested algorithm allows to reduce the time for the manager to make a proper decision and choose the correct actions sequence. Methods, used to solve the problem: the theory of multifactorial experiment, methods of mathematical modelling, software "RastrWin3" and "SPSS Statistics". The regression function is used to simulate the process of changing the load current in the power grid elements. The suggested algorithm can be used in the practice of managing power system modes, by research organizations to improve the characteristics of modes and to develop power grids and power systems. The proposed algorithm complements existing methods for ensuring the reliability of an electrical power system and allows increasing their efficiency.

Keywords: power system, mathematical model, supervisory control.

Introduction

Currently, a number of requirements are imposed on the design and operation of electrical power systems (EPS), such as: reliability, balance of generation and consumption, provision the possibility to safety repair separate elements of the power grid, maintaining required active power flows via intersystem and main lines in normal mode without limitation of capacity and in postfault mode while some connections are disabled, taking into account allowable load of remaining equipment. Meeting these requirements provides an opportunity to ensure the reliability and survivability of the power grid, to achieve economical unified solutions.

The parameters of the power system in normal operation are limited by the regulated permissible values, the reserve volume of active power ensures that the mode can return to permissible parameters after acceptable disturbances.

While maintaining the operating mode of the power grid, there is always a possibility of technological disturbances that can subsequently lead to the cascade development of a fault, followed by a failure in power supply [1]. The main reason for the development of the fault is the insufficient transmission capacitance of the backbone power lines (power lines), determined by the parameters of the elements of the power grid circuit, in postfault modes.

Emergency control employs the current properties of the power system [2,3], such as the power regulation in power generation units, power grid topology change, as well as performing power supply restrictions for consumers while first two methods seem to be inefficient. In the event of disturbances in the power system, the stability is ensured by emergency control systems with a layered principle of construction and by the supervisory control operation.

The emergency electric power mode is characterized by parameters that go beyond the regulated requirements, and leads to the threat of equipment damage and disruption the power system stability. Emergency control automation, which has proven itself to be a reliable means of ensuring the reliability and survivability of energy systems, provides elimination of such emergency mode. This is proved by many years of operating experience. The control actions implemented by it are determined in accordance with the specified algorithms and settings on the basis of schemes and power grid mode information. It should be noted that usually the control actions performed by emergency control automation are determined preventively in case of failure of a critical power grid element, which ensures a successful transition to the postfault mode.

In the postfault mode (the mode after the end of power grid emergency disturbance), the actions of the supervisory and operational personnel are aimed at entering the mode into the permissible area and restoring the normal mode.

It should be noted that the supervisory control of the modern power system receives data from the

* Corresponding author: a.motovilov@narfu.ru

© The Authors, published by EDP Sciences. This is an open access article distributed under the terms of the Creative Commons Attribution License 4.0 (https://creativecommons.org/licenses/by/4.0/).

technological control system about the parameters of the power grid mode. The operative-information complex provides supervisory control personnel with information about changes in the power system mode, as well as about emergency signals. Provided information makes it possible to determine a plan for eliminating the violation of the normal mode, taking into account the available volume of possible emergency actions, their efficiency and time.

While predicting the output parameters of the electric power mode beyond the permissible values, the corresponding commands and instructions should be given by power manager in advance.

In order to restore normal mode and return power system in permissible conditions the power manager must compare the values of the actual parameters of the power mode with the regulatory requirements and determine a plan to eliminate the violation of the normal operating mode of the power grid. He has to analyze a large amount of incoming information about the operating mode of the power grid under time-limited conditions. The admissibility assessment of the power grid operation mode and the subsequent selection of emergency activities should be determined taking into account their efficiency and speed of implementation.

Consequently, relevant and important tasks are to analyze the influence of the power grid parameters on the distribution of power flows and to assess the effectiveness of postfault activities, which allow to prevent and eliminate violations of the power systems normal mode. Performing these actions increase the survivability and reliability of the EPS operation.

To solve this problem, an action algorithm is proposed, based on a numerical method for analyzing the power system scheme. It is assumed that the use of the developed mathematical model will make it possible to more quickly identify the risks of developing faults, determine a plan for eliminating violations of the normal mode, based on an assessment of the weight of postfault actions.

Methods

The technical feasibility issues of changing EPS parameters during electricity transmission control in postfault modes were studied using the methods of numerical analysis. Numerical methods make it possible to obtain a mathematical model of complex processes with mutual influence of mode parameters, based on the large amount of information, previously collected and (or) obtained online about the electrical system.

The structure of the suggested algorithm for assessing postfault actions:

1. Calculate the electrical mode, using known values of branch resistances, active and reactive powers of generation and load units. For the electric network, we select nodes and branches as a controlled parameters (variable factors of the regression function), which have a significant impact on the current load of the controlled power lines. The maximum and minimum value of the parameters are selected for the two levels of variation of factors. Then we calculate steady-state power transmission line operation modes for selected controlled parameters. We get an array of data, consisting of the current values in power lines, depending on variable factors.

2. Replace the examined object with a passive two-pole device that responds appropriately to a given input value.

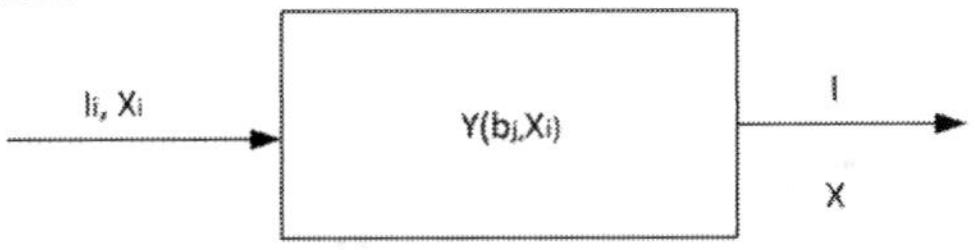

Fig. 1. The mathematical model of examined power grid element. I_i– calculated current values; X_i– variable factors; Y – regression equation; b_i– desired polynomial coefficients; I and X – required current values in power lines and variable factors.

The regression equation is formed - a mathematical model (Figure 1) of the steady state of the network element [10-13]. Calculated values of currents in controlled power transmission lines of the power system are taken as a response variable. Numerical processing of the data array, obtained at the previous step, is made to determine the coefficients of the objective function. For better accuracy of the regression equation and to take into account of the mutual influence of factors, this model is supplemented by terms, containing paired products of factors. The effect of factors interaction is considered in the following model:

$$y = B_0 + B_1X_1 + B_2X_2 + \cdots + B_kX_k + B_{12}X_1X_2 + B_{13}X_1X_3 + B_{(k-1)k}X_{k-1}X_k \quad (1)$$

The number of regression coefficients is less than the number of experiments (p < N), the plan is unsaturated. In general, a complete factor plan (CFP) allows you to take into account any order of interaction of factors.

The model reflects the dependence of the current load on the power lines on the values of the controlled parameters of the nodes and branches.

3. Determination of polynomial parameters from input and output signals. The coefficients bj of the objective function polynomial are determined based on factor analysis. In the case of PFP, the regression coefficients are determined by the formula:

$$b_j = \frac{1}{N}\sum_{i=1}^{N} Y_i \, X_{i,j} \quad (2)$$

Coefficients of pair interactions are calculated by formulas:

$$b_{\mathrm{iu}} = \frac{\left(\sum_{j=1}^{N} X_{\mathrm{ij}} X_{\mathrm{uj}} Y_j\right)}{\mathrm{N}} \quad (3)$$

The values of controlled parameters X_i, at which the current load from overloaded power grid elements is reduced, are determined using optimization methods.

Quantitative assessment of variable factors was carried out using the 14-node IEEE scheme (Figure 2) [4].

© The Authors, published by EDP Sciences. This is an open access article distributed under the terms of the Creative Commons Attribution License 4.0 (https://creativecommons.org/licenses/by/4.0/).

https://doi.org/10.1051/e3sconf/202458401026

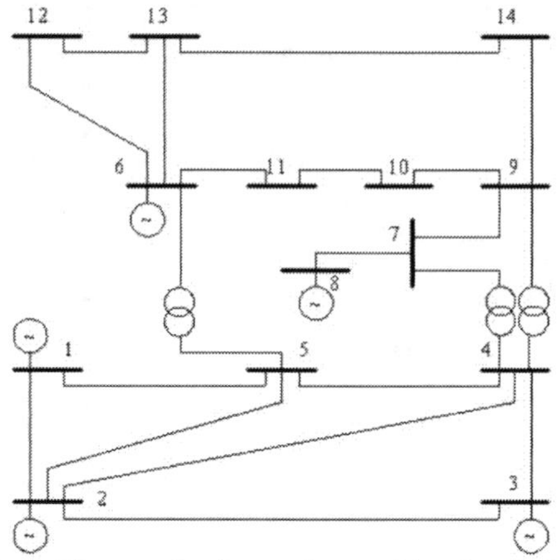

Fig. 2. The examined test scheme of the electrical power grid.

The electrical scheme consists of 5 power sources in the network nodes: 1, 2, 3, 6, 8. Power lines and autotransformers combine a network of different voltage classes into a single energy area. Load nodes: 4, 5, 7, 9, 10, 11, 12, 13, 14 are connected by ring and mixed circuits. It should be noted that load nodes 9, 10, 11, 12, 13, 14 are at a considerable distance from power sources, which imposes certain restrictions on maintaining the operating mode of the electrical network: there are some difficulties with voltage control, current loads both in the normal, repair, and postfault circuit.

In accordance with clause 1 of the algorithm, controllable and controlled parameters are selected: the active power of the generator in node 2, voltages in nodes 2, 3, 6, 9, as well as the operational state of the controlled power lines 1-5, 2-4, 2-5, 3-4 (on or off). For IEEE-14 circuit nodes where it is possible to change the voltage, the limitation on the highest operating voltage and the minimum permissible voltage is taken into account. For the generation value in node 2, the maximum possible value is taken into account. Thus, the number of independent factors $k = 9$, and, respectively, the number of experiments is $2^k = 512$ (steady-state modes).

The total number of paired interactions $C_k^2 = \frac{k(k-1)}{2} = 36$. Accordingly, the total number of regression coefficients is 46.

Results

The real power grid has been replaced by a bipole with equivalent parameters. The initial data for the analysis of the power grid throughput capacity are obtained in the form of load current dependencies in the monitored branches between nodes 1-5, 2-4, 2-5, 3-4 using PC RastrWin3 [5]. The values of variable factors are summarized in Table 1.

Table 1 – The levels of variable factors

Factors	Operational state of the line 1-5, 2-4, 2-5, 3-4	Active power generation, MWt	Voltage in node 2 & 3, kV	Voltage in nodes 6 & 9, kV
Minimal	On	0	200	100
Maximal	Off	40	252	126

The objective function was obtained with terms, containing paired factor interactions for each response value, as a result of completing clause 2 of the algorithm. Assessment of the main factors was performed in the SPSS Statistics software package [6]. The values of regression coefficients for normalized current values in the examined branches of the IEEE-14 circuit are summarized in Tables 2-5.

Table 2 – The resulting regression coefficients for the line 1-5.

The base coefficients		Coefficients of pared interactions								
			2	3	4	5	6	7	8	9
b_0	125,36									
b_1	-125,36	$b_{2..}$	-24,29	-22,88	0	0	0	0	0	0
b_2	24,294	$b_{3..}$	0	14,178	5,031	0	0	0	2,68	0
b_3	22,881	$b_{4..}$	0	0	4,251	0	0	0	0	0
b_4	0	$b_{5..}$	0	0	0	0	0	0	0	0
b_5	0	$b_{6..}$	0	0	0	0	0	0	0	0
b_6	0	$b_{7..}$	0	0	0	0	0	0	0	0
b_7	0	$b_{8..}$	0	0	0	0	0	0	0	0
b_8	-2,68	b_9	0	0	0	0	0	0	0	0
b_9	0									

Table 3 - The resulting regression coefficients for the line 2-4

The base coefficients		Coefficients of pared interactions								
			2	3	4	5	6	7	8	9
b_0	114,152									
b_1	38,680	$b_{2..}$	-38,68	18,036	5,581	0	0	0	0	0
b_2	-114,152	$b_{3..}$	0	0	0	0	2,212	0	0	0
b_3	30,023	$b_{4..}$	0	0	4,543	0	0	0	0	0
b_4	0	$b_{5..}$	0	0	0	0	0	0	0	0
b_5	0	$b_{6..}$	0	0	0	0	0	0	0	0
b_6	-2,21	$b_{7..}$	0	0	0	0	0	0	0	0
b_7	0	$b_{8..}$	0	0	0	0	0	0	0	0
b_8	0	b_9	0	0	0	0	0	0	0	0
b_9	0									

© The Authors, published by EDP Sciences. This is an open access article distributed under the terms of the Creative Commons Attribution License 4.0 (https://creativecommons.org/licenses/by/4.0/).

Table 4 - The resulting regression coefficients for the line 2-5

The base coefficients			Coefficients of pared interactions							
			2	3	4	5	6	7	8	9
b_0	106,92									
b_1	42,207	$b_{2..}$	16,665	-42,21	5,748	0	0	0	0	0
b_2	33,716	$b_{3..}$	0	-33,72	5,953	0	0	0	0	0
b_3	-106,92	$b_{4..}$	0	0	0	0	1,963	0	0	0
b_4	0	$b_{5..}$	0	0	0	0	0	0	0	0
b_5	0	$b_{6..}$	0	0	0	0	0	0	0	0
b_6	-1,963	$b_{7..}$	0	0	0	0	0	0	0	0
b_7	0	$b_{8..}$	0	0	0	0	0	0	0	0
b_8	0	b_9	0	0	0	0	0	0	0	0
b_9	0									

Table 5 - The resulting regression coefficients for the line 3-4

The base coefficients			Coefficients of pared interactions							
			2	3	4	5	6	7	8	9
b_0	22,459									
b_1	-3,396	$b_{2..}$	0	-3,655	3,396	0	0	0	0	0
b_2	-3,227	$b_{3..}$	0	-1,801	3,227	0	0	0	0	0
b_3	-2,955	$b_{4..}$	0	0	2,955	0	0	0	0	0
b_4	-22,459	$b_{5..}$	0	0	0	0	0	0	0	0
b_5	0	$b_{6..}$	0	0	0	0	0	0	0	0
b_6	0	$b_{7..}$	0	0	0	0	0	0	0	0
b_7	0	$b_{8..}$	0	0	0	0	0	0	-1,184	0
b_8	0	b_9	0	0	0	0	0	0	0	0
b_9	0									

All obtained models are presented in normalized values, for transition to calculations in natural values it is necessary to perform transformation, considering that:

$$x_i = \frac{\left(X_i - X_i^{(0)}\right)}{\Delta_i}, \qquad (4)$$

where Δ_i – factor variation interval. $\Delta_i = X_{i\max} - X_i^{(0)} = X_i^{(0)} - X_{i\min}$

$X_i^{(0)}$ – the main level of factor variation.

The statistical analysis of the obtained values was performed, based on the results of the numerical experiment. The results are presented in Table 6.

Table 6. The results of the statistical analysis.

	Branch 1-5	Branch 2-4	Branch 2-5	Branch 3-4
Dispersion s^2	18828,44	18105,53	17471,02	693,54
Sample average $\overline{y}$	126,75	113,14	108,42	24,75
Significance value q	0,01	0,01	0,01	0,01
Student's coefficient t (for degrees of freedom f=500 and q=0,01)	2,59	2,59	2,59	2,59
Maximum error value Δ	15,7	15,4	15,12	3,01

The resulting regression model was estimated by R-square factors (for branch 1-5 was 0.985; 2-4 — 0,976; 2-5 — 0,976; 3-4 - 0.810) and the Akayke information criterion (for branch 1-5 - 2913; 2-4 — 3147; 2-5 — 3131; 3-4 — 2534).

The impact of various factors on the controlled load of power lines was assessed in the format of a numerical experiment. A large number of calculations is performed before faults occur, while the system is in normal mode. During the emergency state under time constraints, while determining the plan of postfault actions, it is planned to use only the calculated before branch operational state coefficients in order to eliminate overloads and restore normal operation.

Determining the most significant factors that should be taken into account when planning the power systems' modes of operation allows to increase the efficiency and reliability of the operation of power systems, reduce the risks of possible faults. On the basis of the calculated coefficients of influence, it is possible to assess the significance of building new facilities and/or reconstruction of power grids in a wider range of parameters of the electric power mode to increase reliability in postfault modes.

Conclusion

The proposed algorithm improves the quality of analysis of the operating modes of the power grid, which allows to increase its reliability and more effectively restore the normal mode at minimum time. The analysis of the parameters of power grid elements and assessment of the impact of postfault actions are an additional means of improving the reliability of power systems.

© The Authors, published by EDP Sciences. This is an open access article distributed under the terms of the Creative Commons Attribution License 4.0 (https://creativecommons.org/licenses/by/4.0/).

References

1. N.I. Voropai, V.G. Kurbatskii, N.V. Tomin, D.A. Pnasetskii, D.N. Sidorov, A.V. ZHukov, D.N. Efimov, A.B. Osak, V.A. Spiriaev, A.V. Domyshev. The complex of intelligent means to prevent major faults in electric power systems. Novosibirsk: Izd-vo Nauka, 2016. 332 p. ISBN 978-5-02-038717-1
2. Motovilov, A. I. Online assessment of the electric power grid capacity of / A. I. Motovilov, I. I. Solov'ev // Izvestiia vysshikh uchebnykh zavedenii. Problemy energetiki. 2020. T. 22. no 3. pp. 51-59. DOI 10.30724/1998-9903-2020-22-3-51-59.
3. Motovilov, A. I. Increasing the throughput capacity of a heterogeneous electrical power grid by regulating the equalizing EMF / A. I. Motovilov, I. I. Solov'ev // Izvestiia vysshikh uchebnykh zavedenii. Problemy energetiki. 2022. T. 24. no 1. pp. 52-60. DOI:10.30724/1998-9903-2022-24-1-52-60.
4. IEEE 14 bus system. - URL: https://al-roomi.org/power-flow/14-bus-system (20.10.2023).
5. Software package "RastrWin3". User Manual.
6. Software package «SPSS Statistica». User Manual.

© The Authors, published by EDP Sciences. This is an open access article distributed under the terms of the Creative Commons Attribution License 4.0 (https://creativecommons.org/licenses/by/4.0/).

Methods for Calculating the Resistance of Grounding Devices with Backfilling Special Compositions

Mikhail Drako[1*], *Ivan* Pavlovich[2], and *Siarhei* Baraishuk[2]

[1]Scientific-Research Project Republican Unitary Enterprise Belenergosetprojekt (RUE Belenergosetprojekt), 5, 1st Tverdy lane, Minsk, 220038, Republic of Belarus
[2]Belarusian State Agrarian Technical University, 99, Nesavisimosti av., Minsk, 220023, Republic of Belarus

Abstract. Traditional analytical methods for calculating and designing of grounding devices (GDs) in heterogeneous soils can lead to results that do not correspond to those obtained in practice. The empirical coefficients used in calculations of grounding systems given in various literature sources do not always give an explicable and significant discrepancy with the same initial data. In this paper, some well-known methods for calculating the spreading resistance of GDs are considered and the obtained results are compared with experimental data. It is shown that the recommendations and algorithms for calculating the resistance of GDs, presented in the well-known reference literature and regulatory documentation, do not give full and correct description for grounding devices installed in heterogeneous soil. In particular, it is shown that such a factor as the proportional ratio of soils with different resistivity practically does not affect the final result. This fact may mislead specialists, since the results obtained may differ significantly from what is observed in practice after the installation of GDs. The study proposes a calculation method and shows a fairly good convergence of the results with experimental data, and defines a further direction in optimizing calculation methods GDs.

Introduction

Currently, in the Republic of Belarus, there is a quantitative increase in the work carried out on the construction of new and reconstruction of previously installed electrical installations. This is explained both by the increased requirements for protection against electric shock and by the introduction of stricter industry standards regarding the efficiency of the functioning of working grounding systems, especially for telecommunications devices and installations.

According to the requirements [1, 2], GDs of electric stations and substations (SS) must be periodically inspected. However, operating enterprises often ignore this requirement, which in the future can lead to serious consequences caused by a violation of the fundamentals of electrical safety in terms of grounding electrical equipment. The characteristic construction defects of GDs and lightning arresters identified by the authors during the examination of one of the 110/10 kV SS of the Belarusian energy system are given in [3].

There is an intensification of developments in the world in the direction of increasing the reliability and efficiency of grounding systems. Such developments are aimed at reducing the resistance and material consumption of GDs, as well as artificially lowering of seasonal coefficient (SC) [4–6]. The CS establishes the maximum change in the value of the resistivity of the upper layers of the earth in relation to its value when measured, which in a given climatic zone can take place during the year [1].

The influence of SC in calculating the resistance of GDs of small electrical installations (for example, distribution network facilities) is significant, except in cases of using deep grounding electrodes with a length of 10 m or more, for which the CS is approximately equal to 1 [1].

Based on the accumulated experience gained in measuring the parameters of more than a thousand electrical installations in the Belarusian energy system, it can be concluded that the use of traditional calculation and design methods for grounding devices [1, 2, 7] do not always lead to correspondence of the practical measured result to the expected design value.

Moreover, the empirical coefficients used in the calculations and obtained without computer and statistical processing can lead to discrepancies in the assessment results under the same initial design conditions. Further, using the example of experimental grounding electrodes, the convergence of the expected calculated values and measured values of real GDs is considered, and a calculation algorithm is proposed that takes into account the use of soil-substituting mixtures during the installation of GDs.

1 The importance of correct source data

The Technological Regulations of the Belarusian energy system [1, 2, 7] recommend take the electrical resistivity of soil (ERS) from the data of natural measurements when carrying out calculations and design of GDs, and

* Corresponding author: m.drako@besp.by

© The Authors, published by EDP Sciences. This is an open access article distributed under the terms of the Creative Commons Attribution License 4.0 (https://creativecommons.org/licenses/by/4.0/).

tabular data should be used only in the absence of such data.

In our opinion, the use of tabular data is highly undesirable due to the significant error introduced in the results of design calculations. Often the results of geological sounding which is usually conducted to a depth of up to 6 m are used as data on ERS, with further interpretation based on the known dependencies of ERS on the type of soil (as well as tabular data).

The use of reference data lead to errors in the calculations of the spreading resistance of GDs [8]. Thus, when commissioning the 330–750 kV OHL, the measured value of the spreading resistance of some supports exceeded the permissible limit specified by the standard [9] by 2–3 or more times. In this regard, a system for determining ERS at the locations of supports and SS has been implemented at RUE "Belenergosetprojekt", using the method of vertical electrical sounding (VES) [8].

It is also shown in [10, 11] that the experimentally determined value of ERS in the location of the projected SS and overhead lines, along with the data of engineering and geological surveys, allows us to take into account the factor of soil corrosion factor of GDs of electrical installations. Moreover, taking into account seasonal changes in the soil condition when considering the choice of grounding electrode according to the condition of corrosion activity leads to an increase in ERS by 1.53–2.27 times [8].

2 The experimental part

To verify convergence, a test cycle (measurements of spreading resistance) was carried out on previously mounted GDs consisting of vertical and horizontal grounding electrodes.

The control grounding electrode was constructed with backfilling of initial soil, while the experimental grounding electrode was constructed with backfilling of soil with a significantly lower electrical resistivity.

To ensure the correctness of further comparison of the considered methods, the ERS was determined experimentally using the VES method with interpretation to a single-layer soil and taking into account the CS. At the same time, individual measurements of spreading resistance of all elements of GDs construction were carried out. The calculations of the resistance of GDs and modeling were carried out using both mathematical packages such as Maple, Origin, and specialized software ZYM [12].

The composite vertical grounding electrode is made of two interconnected steel rods with a diameter of Ø 18 mm and a total length of 3 m. The horizontal grounding electrode is made of 4x40 mm steel strip; it is mounted at a depth of 0.5 m; with low ERS is 0.8 Ohms·m; CS of soil is 2 (measured in the laboratory); CS of replaced soil is 1.48 (measured in the laboratory).

The spreading resistance of the control GD was 380 Ohms, and of the mounted GD with replaced soil was 114 Ohms. The measurements were performed using the MRU-200 meter (Sonel, Poland) according to a validated measurement procedure.

The results obtained made it possible to make an objective comparison of the methods, estimating the error of each of them.

3 Comparison of calculation methods

In [13], edited by Yu.G. Barybin, it is proposed to calculate the spreading resistance separately for horizontal and vertical electrodes. The resulting resistance of the GDs is based on the assumption that this configuration of the electrode system can be considered as a parallel connection of conductors.

The resistance of one vertical electrode R_V is determined by the formula (1):

$$R_V = \frac{0.366\rho}{L_V} \cdot \left(lg\frac{2L_V}{d} + \frac{1}{2} lg\frac{4t_V + L_V}{4t_V - L_V} \right), \quad (1)$$

where ρ – ERS, Ohm·m;

L_V – the length of the vertical electrode, m;

d – the diameter of the vertical electrode, m;

t_V – the laying depth of the vertical electrode (the distance from the ground surface to the middle of the electrode).

The total resistance of the grounding electrode part, consisting of vertical electrodes that are electrically connected, without considering the resistance of the connecting strip:

$$R_{\text{GDs.V}} = \frac{R_V \cdot k_V}{\eta_V}, \quad (2)$$

where k_V=1.45 – the coefficient for vertical grounding electrodes, taking into account the variation of ERS depending on the climatic region;

η_V=0.51 – the utilization coefficient of vertical electrodes for the configuration "Pipes arranged along the contour" with the ratio of the distance between the electrodes to the length of the electrode L_G / L_V =3/3=1.

The resistance of horizontal electrodes is determined by the formula (3), and the resistance of the horizontal grounding electrode, taking into account shielding and climatic region, is determined by the formula (4):

$$R_G = \frac{0{,}366\rho}{L_{\text{SUM.G}}} \cdot lg\frac{L_{\text{SUM.G}}^2}{d \cdot T}, \quad (3)$$

$$R_{\text{GDs.G}} = \frac{R_G \cdot k_G}{\eta_G}, \quad (4)$$

where k_G=3.5 – the coefficient for horizontal grounding electrodes, taking into account the variation of ERS depending on the climatic region;

η_G=0.31 – the utilization coefficient of horizontal electrodes;

T – the laying depth of the grounding electrode, m.

The values of the utilization coefficients were calculated using the linear interpolation method for the average values of interval tabular data [13]. Total spreading resistance of the GDs:

$$R_{\text{GDs}} = \frac{R_{\text{GDs.V}} \cdot R_{\text{GDs.G}}}{R_{\text{GDs.V}} + R_{\text{GDs.G}}}. \quad (5)$$

The following results were obtained based on formulae (1–5): R_V= 188.03 Ohms, R_{GDsV}=534.62 Ohms, R_G= 207.9 Ohms, R_{GDsG}=2346.71 Ohms, R_{GDs}=405.42 Ohms.

© The Authors, published by EDP Sciences. This is an open access article distributed under the terms of the Creative Commons Attribution License 4.0 (https://creativecommons.org/licenses/by/4.0/).

We will perform a similar calculation using the methodology authored by A.A. Fedorov and G.V. Serbinovsky, presented in [14].

It should be noted that the calculation of the GDs is based on the same principles and assumptions as in [13]; however, there are some differences:

1. The purpose of the calculation is to determine the number of necessary vertical electrodes based on a specified value of R_{GDs};

2. The tabular data does not match for different climatic zones. In [14], the values of the coefficients are higher than those in [13].

3. The mathematical formulae for calculating the spreading resistance of single electrodes differ somewhat; in [14], those for vertical and horizontal electrodes are presented:

$$R_V = \frac{\rho}{2 \cdot \pi \cdot L_V} \cdot \left(ln\frac{2L_V}{d} + \frac{1}{2} \cdot ln\frac{4t_V + L_V}{4t_V - L_V} \right), \quad (6)$$

$$R_G = \frac{\rho}{2\pi \cdot L_{\text{SUM.G}}} \cdot ln\frac{L_{\text{SUM.G}}^2}{d \cdot \text{T}}, \quad (7)$$

It is obvious, from a mathematical point of view, formulae (1) and (6), as well as (2) and (7), are identical.

After calculating the formulae (2, 4, 6, 7) for the considered GDs according to the methodology described in [10], the following results were obtained: R_V= 177.16 Ohms, R_{GDsV}=503.7 Ohms, R_G= 196.7 Ohms, R_{GDsG}=2537.62 Ohms, R_{GDs}=396.8 Ohms.

In the work [15] by R.N. Karyakin, special attention is focused on the definition of ERS. The authors point out that the electrical conductivity of the soil, if it does not contain high concentrations of conductive substances, is determined by the amount of water in it, its mineralization, and the nature of the distribution of water in the rock, and also depends on temperature.

For water-saturated rocks, the effect of temperature on resistance is similar to the effect of temperature on the conductivity of a saturated water electrolyte in the soil, which is in good agreement with the data obtained in [16]. The work contains an indication of the need for improvement, taking into account changes in the ERS at the sites of the soil-substituting mixture, as well as permissible fluctuations in the resistance of the grounding electrodes depending on temperature, that were given in [18-20].

According to [15, 16], resistance changes caused by temperature in electrolytes are approximated by the formula:

$$\rho_T = \rho_{20} \cdot e^{-0.022 \cdot (T-20)}, \quad (8)$$

where ρ_T, ρ_{20} are the resistances at temperatures T and 20 °C, respectively.

According to the reference data given in [13], the second climatic region is characterized by an average long-term temperature (January) in the range from minus 14 °C to minus 10 °C. For calculations, we use an average value of minus 12 °C.

It was found that the ERS increased by 2.02 times. It is slightly higher than the correction factor for vertical rod electrodes K_V = 1.45, but also less than the correction factor for horizontal grounding electrodes
K_G = 3.5 [13]. According to [14], these coefficients are K_V = 1.5–1.8; K_G = 3.5–4.5.

It should be noted that as the depth of the soil increases, its temperature rises; therefore, using formula (8), it is possible to calculate the ERS at different depths depending on the soil temperature.

The problem lies in the fact that in the formula for calculating the resistance of the GDs in a single-layer soil, this difference in specific resistivity cannot be taken into account in any way. Therefore, this method can only solve the problem by switching to the method of calculating multi-layer soil, in which it will be possible to take into account the difference in the resistivity of layers of homogeneous soil at different temperatures. At the same time, when calculating the resistance of the GDs using the single-layer soil method, the resulting values will exceed those obtained using the multi-layer soil method. These values create a margin for an increase in resistance when designing GDs.

In general, the resistance of a complex grounding electrode, consisting of a horizontal strip with vertical electrodes placed at its nodes, is determined by the formula:

$$R_{\text{GDs}} = \frac{\rho_\text{T}}{\pi \cdot L_G} \cdot \frac{\lambda \cdot \text{C}_{11} \cdot \text{C}_{22} - \text{C}_{12}^2}{\text{C}_{11} + \lambda \cdot \text{C}_{22} - 2 \cdot \text{C}_{12}}, \quad (9)$$

$$\lambda = \frac{L_G}{n \cdot L_V} = \frac{3}{1 \cdot 3} = 1, \quad (10)$$

where C_{11}, C_{22}C, and C_{12} are coefficients for calculating grounding electrodes [15].

As a result of the calculation, it was found out that the spreading resistance of the GD in a single-layer soil was 93.61 Ohms. The obtained value was significantly lower than previously obtained using the methods outlined in [13, 14], and lower than the actual values obtained in practice (Table 1).

Table 1. Calculation results

Calculation method	Result, Ohm
Practical measurements	380
According to Yu. G. Barybin [13]	405.42
According to A. A. Fedorov and G. V. Serbinovsky [14]	396.8
According to R. N. Karyakin [15]	93.61

4 Calculation of GDs resistance with the condition of using soil-substituting mixtures

Currently, during the installation and repair of GDs, backfilling with so-called compensating mixtures or mixtures for optimizing the electrophysical parameters of GDs is actively used. This solution is quite well known [11, 17] and is used, among others, by the authors of this work. However, none of the previously mentioned methods allows us to assess GDs resistance when using such mixtures, as they have significantly lower resistance than the initial soil.

The issues of reducing GDs resistance by using soil treatment with moisture-stabilizing additives that are not aggressive to the grounding electrode material are considered in [18]. The most common and well-known method of assessing resistance when using soils with low

© The Authors, published by EDP Sciences. This is an open access article distributed under the terms of the Creative Commons Attribution License 4.0 (https://creativecommons.org/licenses/by/4.0/).

 https://doi.org/10.1051/e3sconf/202458401027

resistance is given in [18]. This method describes the calculation of the grounding electrode when it is filled with coal chips.

The spreading resistance of a single vertical electrode in coke fines is determined by the formula:

$$R_v = 0.37k_1 \cdot \frac{\rho}{l_V} \cdot \left(ln\frac{2l_V}{d_{eq}} + 0,5\, ln\frac{4t_V + 3l_V}{4t_V + l_V} + \frac{\rho_1}{\rho} \cdot ln\frac{d_1}{d_V} \right), \quad (11)$$

where k_1 – the freezing coefficient that takes into account seasonal soil temperature fluctuations for vertical grounding electrodes;

$d_{eq} - 0,95b$;

ρ_1 – electrical resistivity of the coke fines, Ohms·m;

d_1 – the outer diameter of the coke backfill, m.

The spreading resistance of a single horizontal electrode or connecting strip in coke fines at $l_g> > d_g$ and $t_g<<l_g/4$ is determined by the formula:

$$R_g = 0.37k_2 \frac{\rho}{l_g} \left(lg\frac{2l_g}{d_g} + lg\frac{l_g}{2l_g} + \frac{\rho_1}{\rho} lg\frac{d_1}{d_g} \right), \quad (12)$$

where k_2 – the freezing coefficient that takes into account seasonal soil temperature fluctuations for horizontal grounding electrodes.

$$R_{overall} = \frac{R_g R_v}{\mu_1 R_v + \mu_2 n R_g}, \quad (13)$$

where μ_1 – the utilization coefficient of extended grounding electrodes;

μ_2 – the utilization coefficient of vertical grounding electrodes;

n – the number of vertical grounding electrodes.

After calculating based on the initial data, the following results were obtained: R_V= 502.18 Ohms, R_G=243.26 Ohms, $R_{overall}$=163.88 Ohms, 43 % higher than the experimentally obtained data.

With partial replacement of the soil around the electrodes, the resistance of grounding electrodes will change slightly. It contradicts experimental studies and indicates the impossibility of applying this calculation method when using soils with low resistance or mixtures to optimize the electrophysical parameters of grounding, without introducing a correction coefficient. This coefficient should reflect the change in the resistivity of the soil directly in the near-electrode space.

Our previous studies [11, 17, 19–21] make it possible to empirically describe the reduction of GDs resistance as a value that depends on change in the ERS:

$$\rho_{eq} = \rho \cdot 0,98^{\frac{d_{mix}}{d_{GDs}}} \cdot \left(\frac{\rho_{mix}}{\rho}\right)^{\frac{1}{3\pi}}, \quad (14)$$

where ρ_{eq} – the required calculated resistivity, Ohms·m;

ρ – ERS, Ohms·m;

d_{mix} – equivalent diameter of the mixture around the grounding electrode, m;

d_{GDs} – diameter of grounding electrode, m;

ρ_{mix} – the resistivity of bulk soil or soil mixtures, Ohms·m.

This formula has been tested on samples with different ERS. Comparing it with the one presented in [18] for coke fines and making the necessary calculations, it can be seen that the resulting formula more accurately describes the effect of soil replacement on the overall GDs resistance.

After calculating the GDs with the proposed coefficient, it was learnt that the grounding electrode resistance is 123.07 Ohms.

Table 2. Comparison of results using calculation methods recommended for multilayer soils.

Calculation method	Result, Ohm	Deviation, %
Experimentally measured value	114	–
According to R. N. Karyakin [15]	93.61	17.9
According to the method described in [17]	163.88	43.8
Proposed method	123.07	7.9

When using low-resistance soils for backfilling, the volume of the initial soil in the near-electrode space of GD is partially replaced by soil or a mixture of soils with a much lower resistivity. The effectiveness of such a replacement depends on the resistivity of the initial soil, the resistivity of the bulk soil, and the volume of the replacement. However, using the method given in [11], we see that the calculated grounding electrode resistance depends extremely little on the volume of backfilling, which contradicts practical experience, since it is obvious that the parameter affecting the change in resistivity will be the area of contact between the mixture and the initial soil, and in the case of applying the mixture along the entire length of the electrode, it will be the contact perimeter. Thus, with an increase in the volume of backfilling, the grounding electrode resistance as a whole should also change. This is well aligned with our proposed formula for the equivalent resistivity of the soil. Applying it in conjunction with the methods outlined in [13, 14], calculated values match the experimental results within 10%.

Conclusion

The results of calculating the spreading resistance of the GD according to the methods described in [13, 14] align quite well. The main differences are observed in the fact that the algorithm in [13] is designed to calculate resistance based on a known configuration of the GD, while in [14] it is intended to determine the configuration of the GD based on a given value of the spreading resistance. These methods include various coefficients for calculating the ERS depending on the climate zone. Moreover, the mentioned methods do not allow for correct calculations of the GD's resistance for two-layer soil. The source [15] contains a method for calculating the GD located in a two-layer soil, but the resistance value of the GD in this case significantly exceeds the values obtained in [13, 14].

However, none of the above methods makes it possible to take into account changes in ERS based on the example obtained by the authors using empirical coefficients or those given in [20]. Using the formula obtained in the work together with the methods described in [13, 14], it allows us to describe

© The Authors, published by EDP Sciences. This is an open access article distributed under the terms of the Creative Commons Attribution License 4.0 (https://creativecommons.org/licenses/by/4.0/).

experimental results, while the convergence of the results with experimental data is within 10%.

The empirical formula for an effective ERS obtained on the basis of experimental data makes it possible to adapt known calculation methods for the use of both soils with low resistivity and soil-substituting mixtures, while also achieving sufficient convergence with experimental data.

References

1. Guidelines for the design of grounding devices for power plants and substations with a voltage of 35–750 kV: Corporate standard 09110.47.103-07, 2007, 76 p.
2. Guidelines for grounding at power plants and substations with a voltage of 35–750 kV: Corporate standard 09110.47.203-07, 2007, 48 p.
3. V.R. Kolik, M.A. Drako, O.A. Maiseyenka, Energy Strategy, **2**, (2014).
4. M.S. Reffin et all, Energies, **11** (2018). https://doi.org/10.3390/en11113187.
5. V.P. Androvitsaneas, et all, "Seasonal Variation and Timeless Evolution of Ground Resistance," 2018 IEEE International Conference on High Voltage Engineering and Application (ICHVE), Athens, Greece (2018). doi: 10.1109/ICHVE.2018.8641909.
6. D.E. Putra et all. International Journal of Advanced Technology and Engineering Exploration. **11** (2024). DOI:10.19101/IJATEE.2023.10102391.
7. Lightning protection of buildings, facilities and engineering communications: SN 4.04.03-2020, 2021, 95 p.
8. M.A. Drako, A.M. Korotkevich, E.P. Kovalev Energy Strategy, **1** (2015).
9. Electrical installations with a voltage of up to 750 kV: Technical code of common practice 339-2022. Introduced on December 20, 2022, 600 p.
10. M.A. Drako, A.M. Korotkevich, A.P. Ivanenko, Energy Strategy, **5** (2016).
11. M. Drako, S. Baraishuk, *Tendencies in the design of the grounding devices for the electrical installations of the Belarusian energy system,* E3S Web of Conf, **216** (2020). Rudenko International Conference "Methodological problems in reliability study of large energy systems" (RSES 2020). DOI: 10.1051/e3sconf/202021601067.
12. D.S. Shishigin, S.L. Shishigin, N.V. Korovkin, *AutoCAD application for LPS, grounding and EMC problems*. EMC 2015 Joint IEEE International Symposium on Electromagnetic Compatibility and EMC EUROPE, Dresden, Germany (2015).
13. Yu.G. Barybin, Handbook on the design of electrical networks and electrical equipment (Energoatomizdat, Moscow, 1991).
14. T.V. Ancharova, V.V. Kameneva, A.A. Katarskaya, Handbook on Power Supply for Industrial Enterprises: Electrical Equipment and Automation (Energoizdat, Moscow, 1981).
15. R.N. Karyakin, Grounding devices of electrical installations (Energoservice, Moscow, 2006).
16. IEEE Std 142-2007 IEEE Recommended Practice for Grounding of Industrial and Commercial Power Systems. Approved 7 June 2007, 225 p.
17. I.A. Pavlovich, S.M. Baraishuk, ENERGETIKA. Proceedings of CIS higher education institutions and power engineering associations, **66**(4) (2023). https://doi.org/10.21122/1029-7448-2023-66-4-322-332.
18. Guidelines for the Design, Construction, and Operation of Grounding in Wired Communication Installations and Radio Transmission Nodes (Svyaz Publishing House, Moscow, 1971).
19. I Pavlovich, S. Baraishuk, A. Skripko, *Ensuring the reliability of energy systems with the application of a new method of decreasing seasonal variations of ground resistance*, E3S Web of Conferences, **384** (2023) Rudenko International Conference "Methodological Problems in Reliability Study of Large Energy Systems". https://doi.org/10.1051/e3sconf/202338401014.
20. S. Baraishuk et all, *The use of hydrogels in mixtures to reduce the transient resistance of the soil – grounding device 5th International Conference on Energetics, Civil and Agricultural Engineering/* E3S Web of Conferences., **497** (2024) 03001/ ICECAE 2024. https://doi.org/10.1051/e3sconf/202449703001.
21. M.A. Drako, S.M. Baraishuk, I.A. Pavlovic, Power engineering: research, equipment, technology, **23**(1) (2021). https://doi.org/10.30724/1998-9903-2021-23-1-80-92.

© The Authors, published by EDP Sciences. This is an open access article distributed under the terms of the Creative Commons Attribution License 4.0 (https://creativecommons.org/licenses/by/4.0/).

 https://doi.org/10.1051/e3sconf/202458401028

Measurement and analysis of electric power quality indicators with targeted identification of the perpetrators of non-compliances on the example of the Belarusian energy system

Mikhail Drako[1*], *Aleh* Maiseyenka[1], and *Viachaslau* Kolik[1]

[1]Scientific-Research Project Republican Unitary Enterprise Belenergosetprojekt (RUE Belenergosetprojekt), 5, 1st Tverdy lane, Minsk, 220038, Republic of Belarus

Abstract. The existing problems of ensuring the established [1] standards of electric power quality indicators (EPQI) in electric networks of different nominal voltage are considered. The authors' practical experience in identifying the perpetrators of non-compliances in terms of non-sinusoidal voltage with the determination of their actual contributions (AC) based on the results of measurements made by their own accredited electrotechnical laboratory is summarized.

Introduction

The specificity of electric energy (EE) as a commodity lies in the fact that the quality of electricity (QE) in the electrical grid is generally influenced by all subjects of the electric power industry: power plants, electric networks, and consumers. At the same time, claims for improper QE in most cases are addressed to the energy supply organization (ESO) from consumers, who often themselves are the perpetrators of the deterioration of EPQI.

1 Regulatory and technical support

The fundamental regulatory and technical document in the field of QE is [1]. For long-term changes in the voltage characteristics [1] defines the EPQI and sets their norms; for random events, it provides reference values.

The scope of application [1] covers the transmission points of EE to users of electric networks.

In [2], the main provisions on the organization and conduct of EPQI control are established in order to determine their compliance with the standards established in [1].

The methods of measuring EPQI are defined in [3, 4]. In [3], the requirements for measuring EPQI are also given.

Methods for analyzing EPQI, determining the perpetrators of non-compliances and their AC are regulated in [5] - in the Republic of Belarus, [6] - in the Russian Federation.

[7] and [8], contain ESO recommendations for determining the requirements for connecting distorting installations.

2 Reasons for non-compliances with the EPQI. Targeted identification of perpetrators. Corrective (normalizing) measures

2.1 Frequency deviation

The non-compliance is practically not observed in the unified energy system and is not considered in this article.

2.2 Slow voltage changes

The perpetrators of non-compliances in terms of positive and negative voltage deviations are generally ESOs. For example, unacceptable negative voltage deviations occur when the electrical network has insufficient capacity. The most effective corrective measures are: strengthening the electrical network and improving dispatching control.

2.3 Non-sinusoidal voltage

Higher voltage harmonics in electrical networks are caused by the emission of higher current harmonics by distorting electrical receivers and converter devices with a nonlinear volt-ampere characteristic (VAC). The methodology of targeted identification of the perpetrators of non-compliances related to higher harmonics and calculating their AC, as mentioned above, is outlined in [5, 6]. The AC are compared to permissible contributions (PC). By default, the PC is equal to the limit norm [1]. The excess of the AC over the PC indicates that this source is the perpetrator.

* Corresponding author: m.drako@besp.by

© The Authors, published by EDP Sciences. This is an open access article distributed under the terms of the Creative Commons Attribution License 4.0 (https://creativecommons.org/licenses/by/4.0/).

Chokes, passive and active filters, and filter-compensating devices are used to combat higher harmonics. Separation transformers with one of the winding connections in a "triangle" configuration can be used for harmonics with orders that are multiples of the number three.

Passive filters are tuned to filter a certain higher harmonic, while active filters operate across the entire spectrum of higher harmonics. The filter-compensating device (FCD) combines the functions of filtering higher harmonics and compensating for reactive power.

2.4 Voltage asymmetry in three-phase systems

In low voltage electrical networks, voltage asymmetry most often occurs when single-phase consumers are incorrectly distributed in phases, when the operation mode is not full-phase during the single phase-to-ground fault (SGF), and due to improper network conditions, primarily contacts. In networks of 10(6) kV and above, different phase parameters can also cause unbalance (for example, different capacities due to non-transmission lines, PTL).

If non-compliances in asymmetry are detected, an audit of the nearest sections of the electrical network should be carried out and, for 0.4 kV PTL, a redistribution of single–phase consumers by phases should be carried out. If the asymmetry non-compliances persist after this, the perpetrators should be identified with the determination of their AC (using the same method as for higher harmonics, in accordance with [5, 6]).

Special balancing devices are used to eliminate asymmetry. For example, 6-20/0.4 kV transformers with Y/Z_H winding connection schemes, as well *as* Z/Z_H with a balancing device are effectively used in 0.4 kV networks.

2.5 Voltage fluctuations and flicker

The cause of flicker as a rule the sharply variable load of consumers with insufficient "network strength" to external influences and interference. The network strength is characterized by the short-circuit power (SC). The characteristic of a distorting source by the degree of its distorting effect at this particular point in the network is characterized by the SC ratio, calculated expressions are given, for example, in p. 3.10 [9].

Targeted identification of the sources and perpetrators of the flicker should generally be performed in the same way as for rapid voltage changes. It is recommended to rely on p. 4.6.3 [5]. At the same time, the more complex but effective methods proposed in [10] are of significant interest due to the wide spread of non-compliances related to flicker in electrical networks of all rated voltages. These methods are similar to the methods of identifying sources of higher harmonics and asymmetry in accordance with [5, 6]: the perpetrators and their AC should be determined by current interharmonics or current derivatives. In our opinion, the methods proposed in [10] require practical testing before widespread application.

2.6 Dips and overvoltages

The causes of dips and overvoltages can be, on the one hand, SGF and SC, disconnections of PTL and regulating devices in electric power grids of the ESO, on the other hand, abrupt load changes in consumer networks.

It is not possible to present a single formalized algorithm for identification of perpetrators. In our opinion, it is necessary to establish an empirical correlation between rapid voltage changes and events in the electric networks of ESO and consumers.

To prevent these non-compliances, it is necessary to strengthen the network scheme, increase its reliability of operation, and develop its active-adaptive characteristics.

If consumers are the perpetrators, work should be carried out with them to limit their electromagnetic effects that affect the occurrence of these non-compliances (with the possible application of standards [9, 11]).

2.7 Pulsed overvoltages

The causes of pulse overvoltages are lightning strikes and switching in electrical networks. In the specialized literature, electrostatic discharges and the consequences of a nuclear explosion are also called. The equipment is protected from pulsed overvoltages using special protective or limiting devices.

3 Measuring instruments

In [3] and [4], the measuring instrument (MI) refers to EPQI analyzers. At the same time, when connecting the EPQI analyzer to secondary circuits, not only the EPQI analyzer, but also the measuring current transformers (CT) and voltage transformers (VT) are MI and elements of the measuring channels. In this case, the requirements for CT and VT are not lower than for telemetry systems.

The characteristics of MI used for measuring higher harmonics and interharmonics are defined in [3].

When measuring short-term and long-term doses of flicker in accordance with [13], EPQI analyzers using a class F1 flickermeter should be used. Class F3 flickermeters can be used for monitoring EPQI. It is important to note that if the EPQI analyzer includes a flickermeter function, then the use of a separate flickermeter is not required.

The EPQI analyzer should provide measurement of the operating parameters of the electrical network: voltages and currents of higher harmonics (or corresponding non-sinusoidal coefficients), phase shift angles (primarily between currents and voltages) or (and) powers of higher harmonics.

For portable EPQI analyzers, a set of current clamps of various ratings must be provided to ensure proper measurement accuracy. Current clamps must be connected without breaking current circuits (according to

© The Authors, published by EDP Sciences. This is an open access article distributed under the terms of the Creative Commons Attribution License 4.0 (https://creativecommons.org/licenses/by/4.0/).

the principle of a Rogovsky coil or similar). It is also preferably that it is possible to download primary measurements (instantaneous values of phase currents and voltages) in a displayed format, for example, in EXCEL.

4 Measurement

As indicated above, the measurement of normalized EPQI is performed in accordance with the requirements of [1]. In addition, the measurements of non-normalized EPQI and operating parameters of the electric network required for the analysis of EPQI are regulated in [5] and [6].

When connecting EPQI analyzers to secondary current and voltage circuits, secondary circuits of CT and VT must be turned on for measuring or accounting for EE (without breaking the current circuits).

Currents, voltages and angles between them should be measured by the same MI. Based on this, it is advisable and convenient to connect the analyzer in the EE metering cabinet or telemetry cabinet, where there are current and voltage circuits.

During measurements, it is necessary to assess the reliability of measurement results. Unreliable measurement results may be obtained due to a number of reasons: incorrect operation of the MI, errors in the connection scheme, improper measurement conditions, errors in the analyzer configuration. In our practical activities, we use two main approaches to assessing the reliability of measurement results:

1) comparison of measured values with the same values obtained from the EM (electricity meter), telemetry equipment, and protection terminals;

2) assessment of the reliability of measurement results from the point of view of known physical laws and accumulated observation results.

According to the standards [5, 6], in order to identify sources of higher harmonics and voltage asymmetry at a certain point in the electrical network, simultaneous measurements should be carried out at all outgoing electrical connections with averaging for every 3 seconds. But, first, only a small part of the EPQI analyzers measures the current and power vectors of higher harmonics and averages them over a time interval of 3 s. Second, the number of electrical connections extending, for example, from a 10 kV power rail section can be about 12-20 or more. This means that the same number of EPQI analyzers will be required for measurements. Considering the high cost of these devices, as well as their simultaneous demand at different problem points of the electrical network, this requirement can be considered practically impossible.

In our practice, we are forced to apply the following approach. One EPQI analyzer is installed to monitor the EPQI for the entire measurement period, and its current circuits are installed for input into this power rail section or for electrical connection related to the alleged perpetrator. Cyclic monitoring of the EPQI is performed on the remaining electrical connections using 1–2 analyzers with the analyzer set for 5–10 minutes. Measurements made on all electrical connections within one cycle are taken as one-time. The calculations of AC for each connection case include the measurement results averaged over their own actual measurement intervals.

5 Analysis of measurement results. Emerging problems and possible solutions

In terms of the actual analysis of the EPQI and the presentation of its results, complex application issues also arise [5, 6]. For example, for a three-phase system, it is unclear whether to average the results of measurements and calculations by phases or to present them separately for each phase. In our opinion, we should do the latter, since the picture for different phases may generally be different.

For the analysis of EPQI, there is an urgent need for application software that can work both in delayed time mode and in the pace of the process.

6 Illustrative examples of completed work on EPQI measurement and analysis

6.1 The distorting source is a 330 kV shunting reactor controlled by magnetization (SRCM)

Such reactors have thyristor circuits with non-linear VAC in the control system and are therefore known to be distorting sources. Therefore, the technical requirements for the purchase included the following points:

1. Standards [1] should not be violated at all points of connection of the SRCM to the substation network (at the same time, the standards for 220 kV were applied to 330 kV);

2. The emission of higher harmonic currents in the 330 kV power winding with respect to the first harmonic should not exceed 3 % (taking into account the requirements of p. 2.1.17 [16]).

The measurements of EPQI revealed that the emission of higher harmonic currents in the power winding, as a rule, exceeded 3% and sometimes reached 12-14%. For a voltage of 330 kV, non-compliances were revealed in the non-sinusoidal coefficients of the 5th and 7th harmonics, as well as in the total non-sinusoidal coefficient. An analysis of the measurement results according to [5] showed that the distorting sources for the 5th and 7th harmonics are this SRCM and a similar SRCM installed at another substation. At the same time, for the 5th harmonic, the AC of the SRCM at this SS exceeded the PC, which means that, regardless of the other SRCM, this SRCM is the perpetrator. For the 7th harmonic, the AC of each of the two SRCM did not exceed the PC.

As a result of the claims made by the customer to the supplier based on the results of measurements and calculations, the SRCM supplier installed a higher harmonic filter in the compensation winding of the 330

© The Authors, published by EDP Sciences. This is an open access article distributed under the terms of the Creative Commons Attribution License 4.0 (https://creativecommons.org/licenses/by/4.0/).

kV SRCM as part of the fulfillment of warranty obligations. The control measurements made after the installation of the filter did not reveal any violations of the points of technical requirements.

6.2 Group of problematic 110 kV substations

The three studied substations (SS) are adjacent, two of them have intersystem overhead lines (OHL) with a voltage of 110 kV.

As a result of measurements on the 110 kV power rails of all three SS, non-compliances were established in the 5th and 9th harmonics. It was revealed that these non-compliances are caused by the combined effect of current emissions of these harmonics along the 110 kV intersystem lines and along individual 10 kV feeders (via transformers). Also, 110 kV SS connected to the branch line from an intersystem line was also revealed to be a source of the 5th harmonic.

Also, non-compliances in the 10th, 12th, 14th and 16th even harmonics were detected in the 1st 110 kV section of one of the SS. It is considered that the appearance of even harmonics is unlikely and is due to the use of rather specific electrical receivers. In this case, the presence of such electric receivers is not confirmed. There was a hypothesis that during non-working hours, along with a decrease in the load and an increase in voltage levels, the old transformer installed on the SS switches to a nonlinear magnetization mode (overexcitation), which, in turn, is accompanied by the emission of even harmonics. This hypothesis is supported by the monograph [14], which provides a mathematical description of the conditions for the emission of even harmonics in transformers.

6.3 There are no inconsistencies, but there is a harmful effect

In one of the areas of the electrical networks, the staff detected background acoustic noise in the transformers at the 10/0.4 kV TSS (transformer substation) feeding the pumps of the Water Utility. It was immediately assumed that this noise arises from higher harmonics emitted by the frequency-controlled drives of the pumps.

Measurements were conducted at two such TSS. As a result, non-compliances with the standards [1] were not identified. However, at the same time, very high levels of the components of the higher harmonic were established: (up to 100–110% relative to the 1st harmonic).

When using the current emission standards established in [9], the standards for 5, 7, 11, and 13 harmonics, as well as the permissible harmonic parameters, are exceeded several times.

In this case, a paradoxical situation has developed: there are no violations [1], but the harmful effect of higher harmonics is present. At the same time, there are violations of the requirements of the standard [9]. However, the action of [9] applies only to single devices, so the requirements of [9] are not legally applicable directly in this situation. In such cases, the requirements for limiting the emission of higher current harmonics can be stipulated in the Technical Specifications for power supply.

7 Own developments and research

For certain EPQI analyzers, we have identified errors in the measurement of phase shift angles. This is most likely related to the incorrect application of the method for determining the phase shift, given, for example, in [15]. This method involves determining the phase shift as the root of the equation $tg\varphi=A$. However, nowhere is the criterion for selecting one of the two roots specified. Therefore, we use the method of signal transition through zero. As a result, we have developed a specialized program for the calculation and analysis of EPQI. The source data consists of primary measurements, i.e., instantaneous values of phase currents and voltages ($\geq$1024 points per period of the fundamental frequency), obtained from the archives of EPQI analyzers. The program calculates normalized and non-normalized EPQI for any time interval, without averaging, which is important for the analysis of short-term phenomena and fast processes.

We plan to test the methods for address detection of flicker sources proposed in [10]. Since EPQI analyzers do not measure the required values, our program will be updated for this purpose.

In the future, experiments are planned to identify the sources of even harmonics in electrical networks. The experiments will be performed for the 110/10 kV transformer mentioned above and the 220/110/10 kV autotransformer.

Conclusion

To ensure proper QE in electrical networks at all levels, it is necessary to organize systematic approach to the control and monitoring of EPQI, analysis of measurement results, and conducting organizational and technical corrective (normalizing) measures.

It is required to revise the existing regulatory and technical documentation, taking into account the accumulated practical experience, and in certain cases, it is advisable to expand its scope of application. For example, the norms [1] can be applied not only to the points of transmission of EE from the ESO to the consumer.

References

1. Standards for electric energy quality in power supply systems of general purpose: GOST 32144-2013, 2016, 20 p.
2. Electric energy. Electromagnetic compatibility of technical instruments. Control and monitoring of the electric energy quality in power supply systems of general purpose: GOST 33073-2014, 2016, 52 p.
3. Electromagnetic compatibility of technical means. General guidelines on harmonic and interharmonic

© The Authors, published by EDP Sciences. This is an open access article distributed under the terms of the Creative Commons Attribution License 4.0 (https://creativecommons.org/licenses/by/4.0/).

measuring instruments for power supply systems and connected technical equipment: GOST 30804.4.7-2013, 2014, 40 p.
4. Electric energy. Electromagnetic compatibility of technical means. Methods of measurement of electric power quality indicators: GOST 30804.4.30-2013, 2014, 58 p.
5. Guidelines for controlling and analyzing the quality of electric energy in general purpose power supply systems. Part 2. Electric energy quality analysis: TCP 183.2-2009, 2009, 32 p.
6. Guidelines for controlling and analyzing the quality of electric energy in power supply systems of general purpose. Part 2. Electric energy quality analysis: GD 153-34.0-15.502-2002, 64 p.
7. Electromagnetic compatibility (EMC). Part 3-6. Standards. Assessment of electromagnetic emission standards for connecting installations that create interferences to power supply systems of medium, high and ultrahigh voltage: GOST IEC/TR 61000-3-6-2020, 2021, 54 p.
8. Electromagnetic compatibility (EMC). Part 3-6. Standards. Assessment of electromagnetic emission standards for connecting installations that create voltage fluctuations to power supply systems of medium, high and ultrahigh voltage: GOST IEC/TR 61000-3-7-2020, 2021, 58 p.
9. Electromagnetic compatibility. Part 3-12. Standards. Standards for harmonic components of current generated by equipment connected to low-voltage general purpose power supply systems with a current consumption of more than 16 A and not more than 75 A in one phase: GOST 30804.3.12, 2014, 28 p.
10. V.I. Biryulin, D.V. Kudelina, O.M. Larin, Power engineering: research, equipment, technology, **23,** (2021). https://doi.org/10.30724/1998-9903-2021-23-5-3-12.
11. Electromagnetic compatibility (EMC). Part 3-11. Standards. Limitation of voltage changes, voltage fluctuations and flicker in public low-voltage power supply systems for equipments with a rated current of not more than 75 A, subject to special connection conditions: GOST IEC 61000-3-11-2022, 2022, 24 p.
12. Electrical substations with a voltage of 35 kV and above. Standards of technological design: Corporate standard 33243.01.216-16, 2016, 198 p.
13. Electromagnetic compatibility of technical means. Part 4. Methods of testing and measurement. Section 15. Flickermeter. Functional and design requirements: GOST IEC 61000-4-15-2014, 2017, 40 p.
14. M.S. Libkind, Higher harmonics generated by transformers (Publishing House of the USSR Academy of Sciences, Moscow, 1962).
15. L.A. Bessonov, Theoretical foundations of electrical engineering. In 2 vol. Volume 1. Electrical Circuits (Urait Publishing House, Moscow, 2023).
16. Controlled shunt reactors for electric networks with a voltage of 110-500 kV. Typical technical requirements: STO 56947007-29.180.03.198-2015, 2015.

© The Authors, published by EDP Sciences. This is an open access article distributed under the terms of the Creative Commons Attribution License 4.0 (https://creativecommons.org/licenses/by/4.0/).

Research on Impact of Solar Power Plants on 110 kV Line Current Differential Protection Operation

Alexander Navolochny[1], *Olga* Onisova[1*], and *Alexander* Salmin[1,2]

[1]Chuvash State University named after I.N. Ulianov, Faculty of Energy and Electrical Engineering, Cheboksary, Russia
[2]Cheboksary Electric Apparatus Plant, Bureau of Relay Protection and Automated Process Control Systems

Abstract. The paper presents the research on the impact of renewable energy sources based power plants interfaced to power grids through inverters on differential protection of a line connecting such a power plant to an external grid. The paper considers the case of a weak power grid with a relatively high-capacity solar power plant. An approach to evaluate the IBR impact on differential protection in steady-state conditions under typical inverter control scenarios is suggested. The factors affecting the differential protection operation are revealed. The brief guidance concerning the application of differential protection in such power grids is given.

1 Introduction

Presently the share of renewable energy sources (RES) not including hydro power plants in some local regions of the Russian power system exceeds 10% of the total output generation capacity. In recent years a number of large RES power plants with a capacity of more than 100 MW have been commissioned [1]. The impact of RES on nearby power grids operation is becoming more significant.

In most cases, RES are connected to power grids through electronic power converters (invertors). Behaviour of such inverter based resources (IBR) during normal operation and power system faults is mostly dependant on technical specifications and control algorithms of invertors but not on generator electromechanical characteristics (as for conventional power plants). Typically, RES exhibits: low levels of short circuit currents that exceed rated currents no more than 10-30%, current-voltage phase relationships are different than for conventional power plants, lack of or low negative sequence current response during asymmetrical conditions [2].

The IBR characteristics listed above may have a negative impact on relay protection and automation operation reducing security and dependability of overcurrent and distance protections [3], leading to incorrect operation of negative sequence protection elements [4].

Considering field cases and research results it can be concluded that line current differential protection is the most preferable choice for protection of transmission lines connecting RES power plants to power grids. It is expected that RES have minimal impact on differential protection performance [3, 5]. However, differential protection operation in power systems with IBR may not be sufficiently studied; thus, some papers consider cases of decreased dependability of differential protection (low sensitivity, delayed operation, failure to operate) due to presence of IBR [6, 7]. It is essential to indicates major issues due to IBR characteristics that challenge differential protection; this will contribute to development of guidance to differential protection application in power grids with IBR and to differential elements settings.

2 Model of power system with IBR

The configuration of the study system is represented by figure 1. It is composed of a solar power plant tied to an equivalent grid source (S) via a 110 kV transmission line (L). The solar power plant includes a single equivalent IBR representing the aggregated capacity of multiple photovoltaic modules, the equivalent transformer 0,69/10,5 kV (T2) interfacing the IBR to the collector grid, the equivalent collector system (CS), the station transformer 110/10,5 kV (T1) with grounded neutral point at a high voltage side. The study system basic specifications are shown in table 1.

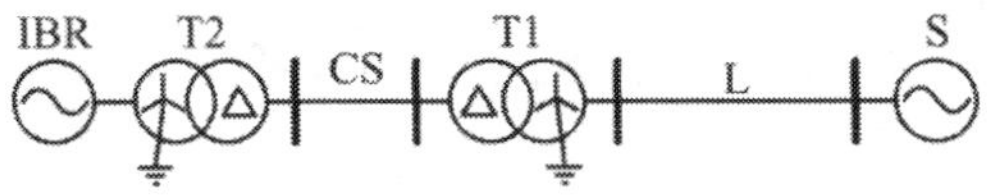

Fig. 1. Configuration of study system with IBR.

Typically, in practice a grid source is much stronger than an IBR source, due to this differential protection operation conditions are almost the same as on a single-infeed transmission line. Many papers analyze exactly such cases. Under the scope of this study, considering cases with approximately equal fault current contributions from an IBR and a grid source is of the most interest (weak grid source, long transmission line,

* Corresponding author: o.a.onisova@ya.ru

© The Authors, published by EDP Sciences. This is an open access article distributed under the terms of the Creative Commons Attribution License 4.0 (https://creativecommons.org/licenses/by/4.0/).

high-capacity IBR). Considering such a case, an IBR is expected to have a greater impact on differential protection performance.

Table 1. Basic specifications of the study system elements.

Parameter	Value
E.m.f. of grid source S, kV	110
Resistance of grid source S (reactive), Ohm	12,7
Line L lenght, km	80
Transformer T1 rated power, MVA	120
Equivalent resistance of collector system CS (active/ reactive), Ohm	0,011/0,009
Transformer T1 rated power, MVA	126
IBR rated power, MVA	100

For this study the IBR is represented by a nonlinear controlled current source (Figure 2). The source current is dependent on the algorithms and settings of the inverter control system. The current source is considered ideal; all the constraints on the magnitude and phase of currents come from the control system algorithms. The described model is mainly focused on steady-state conditions allowing for illustrative analysis of the IBR impact on short circuit conditions and for evaluating the quality characteristics of the differential protection performance under the various scenarios of the IBR control system operation. It should be noted that for detailed study of differential protection performance considering operation features of protection's software and hardware functions as well as for relays hardware-in-loop testing it is necessary to use a more detailed IRB model which is suitable for electromagnetic transients analysis in three-phase frame and considering fast transients within an electronic converter and its controls loops.

The mathematical equations for the IBR model are derived for a two-phase rotating direct-quadrature "d" and "q" axis; rotating frequency of the axis is the same as the grid frequency. The d, q – current components are dependent on the algorithm and settings of the inverter control. Different control scenarios are considered: active power level is the maximum of the plant power nominal rating; active power and reactive power are adjusted to a reference values; voltage support at the IBR connection point.

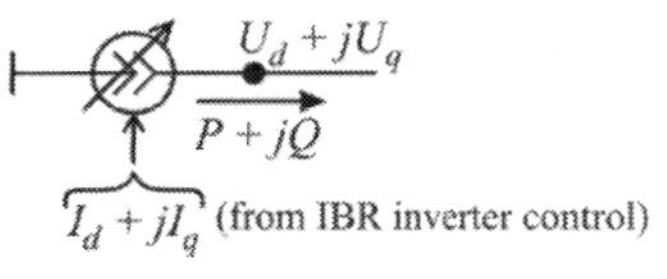

Fig. 2. IBR model.

In many practical cases, during power system normal operation and faults (including unbalanced) inverters contributes only three-phase symmetrical positive sequence currents; however, inverters may also be required to contribute a negative sequence current components during unbalanced events dependent on the negative sequence voltage measured at the inverter connection point.

As follows from the above, the source currents can be generally described by the following matrix equation:

$$\begin{bmatrix} I_{d1} \\ I_{q1} \\ I_{d2} \\ I_{q2} \end{bmatrix} = \begin{bmatrix} \frac{U_{d1}}{U_{d1}^2+U_{q1}^2} & \frac{U_{q1}}{U_{d1}^2+U_{q1}^2} & \frac{(1-U_1)\cdot U_{q1}}{U_1} & 0 & 0 \\ \frac{U_{q1}}{U_{d1}^2+U_{q1}^2} & \frac{-U_{d1}}{U_{d1}^2+U_{q1}^2} & \frac{-(1-U_1)\cdot U_{d1}}{U_1} & 0 & 0 \\ 0 & 0 & 0 & U_{d2} & -U_{q2} \\ 0 & 0 & 0 & U_{q2} & U_{d2} \end{bmatrix} \times$$

$$\times \begin{bmatrix} P_0 \\ Q_0 \\ K_U \\ K_2\cos\varphi_2 \\ K_2\sin\varphi_2 \end{bmatrix}, \quad (1)$$

where $U_1 = \sqrt{U_{d1}^2 + U_{q1}^2}$ – a module of the positive sequence voltage; I_{d1}, I_{q1} – the d,q-components of the positive sequence current; I_{d2}, I_{q2} – the d,q-components of the negative sequence current; U_{d1}, U_{q1} – the d,q-components of the positive sequence voltage; U_{d2}, U_{q2} – the d,q-components of the negative sequence voltage; P_0 – the reference value of the IBR active power; Q_0 – the reference value of the IBR reactive power; K_U – the droop gain for the reactive positive sequence current component control based on the positive sequence voltage deviation; K_2 – the droop gain for the negative sequence current component control based on the negative sequence voltage; φ_2 – the reference negative sequence current phase angle relative to the negative sequence voltage phase.

To perform further calculations, it is suitable to align the positive sequence voltage phasor at the IBR terminals with the "d" axis; in this case:

$$U_{d1} = U_1; \quad U_{q1} = 0; \quad U_{d2} = \mathrm{Re}\left(\frac{\underline{U}_2}{e^{j\arg(\underline{U}_1)}}\right)$$
$$U_{q2} = \mathrm{Im}\left(\frac{\underline{U}_2}{e^{j\arg(\underline{U}_1)}}\right), \quad (2)$$

where $\underline{U}_2$ – vector of the negative sequence voltage; $\arg(\underline{U}_1)$ – initial positive sequence voltage phase (before it aligning with d-axis).

Assuming (2) the initial equation (1) can be defined as the following simplified equation:

$$\begin{bmatrix} I_{d1} \\ I_{q1} \\ I_{d2} \\ I_{q2} \end{bmatrix} = \begin{bmatrix} \frac{1}{U_{d1}} & 0 & 0 & 0 & 0 \\ 0 & \frac{-1}{U_{d1}} & -(1-U_{d1}) & 0 & 0 \\ 0 & 0 & 0 & U_{d2} & -U_{q2} \\ 0 & 0 & 0 & U_{q2} & U_{d2} \end{bmatrix} \times$$

$$\times \begin{bmatrix} P_0 \\ Q_0 \\ K_U \\ K_2\cos\varphi_2 \\ K_2\sin\varphi_2 \end{bmatrix}, \quad (3)$$

© The Authors, published by EDP Sciences. This is an open access article distributed under the terms of the Creative Commons Attribution License 4.0 (https://creativecommons.org/licenses/by/4.0/).

The IBR phase currents can be described based on their d,q-components using the following equations:

$$\begin{aligned}
\underline{I}_A &= I_{d1} + jI_{q1} + I_{d2} + jI_{q2} \\
\underline{I}_B &= I_{d1}\frac{\sqrt{3}}{2} - I_{q1}\frac{1}{2} - I_{d2}\frac{1}{2} - I_{q2}\frac{\sqrt{3}}{2} + \\
&+ j(-I_{d1}\frac{\sqrt{3}}{2} - I_{q1}\frac{1}{2} + I_{d2}\frac{\sqrt{3}}{2} - I_{q2}\frac{1}{2}), \\
\underline{I}_C &= -I_{d1}\frac{1}{2} - I_{q1}\frac{\sqrt{3}}{2} - I_{d2}\frac{1}{2} + I_{q2}\frac{\sqrt{3}}{2} + \\
&+ j(I_{d1}\frac{\sqrt{3}}{2} - I_{q1}\frac{1}{2} - I_{d2}\frac{\sqrt{3}}{2} - I_{q2}\frac{1}{2}).
\end{aligned} \quad (4)$$

If d,q-current components from (4) are calculated according (2), (3), phase current angles will be obtained in a reference frame with positive sequence voltage aligned with d-axis.

There is no zero sequence current components in currents, because the IBR is tied to the grid via "star-delta" connected winding transformer.

3 Analysis of differential protection performance

A number of typical scenarios of the inverter control and anticipated differential protection performance will be discussed further.

The conventional percentage-restrained differential protection algorithm is taken; the criterion for the relay operation is following:

$$\begin{cases} I_{\text{DIFF}} \geq I_{\text{DIFF MIN}} \\ I_{\text{DIFF}} \geq K_{\text{RESTR}} \cdot I_{\text{RESTR}} \end{cases}, \quad (5)$$

where I_{DIFF} – the differential current which is a module of the phasor sum of the currents at the line terminals; $I_{\text{DIFF MIN}}$ – the minimum pick-up value of the differential current; K_{RESTR} – the slope coefficient of the protection restrained characteristics; I_{RESTR} – the restrained current.

The restrained current can be calculated in different ways; one of the ways – as a module of the difference of the current vectors at the line terminals. To ensure minimum protection sensitivity (sensitivity coefficient is more than 2 p.u.*) for single-infeed line operation taking into account the protection restrained characteristics, the slope coefficient should be accepted equal to 0,5 p.u.

4 The first control scenario. The inverter contributes maximum active power

In this case the following values describing the inverter control are used in the equation (3): $P_0 = 0,75$ p.u., $Q_0 = 0$ p.u., $K_U = 0$, $K_2 = 0$.

An internal two-phase fault on the transmission line L near the IBR side is simulated. The current phasors at the both line terminals are shown in figure 3. The currents contributed from the IBR for the unbalanced fault are three-phase symmetrical currents; the currents contribution from the grid is generally typical for an unbalanced event, however, the phase difference of the faulted phase grid currents is significantly less than 180° due to the IBR impact. The phase differences of the currents at the line terminals are 122° for "B" phase and 157,6° for "C" phase; this is not typical for faults in grids with conventional sources where the currents measured at the two terminals are nearly in phase. Because of the larger phase difference of the currents, the differential current becomes smaller, and the restrained current (with the accepted calculation approach) increases. This significantly reduces the protection sensitivity and can eventually results in protection failure to detect a fault.

The differential currents of the faulted phases for the considered case are illustrated by the dotted lines in figure 3; in this case the sensitivity coefficients are 1,16 p.u. for "B" phase, protection elements in "C" phase doesn't operate. It should be noted that if the IBR source is stronger (with respect to the grid source capacity), the protection elements in the both phases may not trip for the considered event and the inverter control scenario.

5 The second control scenario. The inverter control supports voltage at the IBR terminal

In this case the following values describing the inverter control are used in the equation (3): $P_0 = 0,75$ p.u., $Q_0 = 0$ p.u., $K_U = 6$, $K_2 = 0$.

The current phasors at the both line terminals for the BC-fault on the transmission line L near the IBR side are shown in figure 4. If the inverter contributes to voltage support (coefficient K_U increases), the symmetrical currents contributed from the IBR rotates clockwise. In this case the current phase's relationships at the line terminals are more suitable to the differential protection better performance. Thus, the phase differences of the currents at the line terminals are 6,9° for "B" phase and 88,5° for "C" phase; the sensitivity coefficients are 12,2 p.u. for "B" phase, 2 p.u. for "C" phase.

6 The third control scenario. The inverter contributes maximum active power with a negative sequence current component in a case of an unbalanced event

In this case the following values describing the inverter control are used in the equation (3): $P_0 = 0,75$ p.u., $Q_0 = 0$ p.u., $K_U = 0$, $K_2 = 3$, $\varphi_2 = -90^\circ$ (the negative sequence current leads the voltage by 90°).

The current phasors at the both line terminals are shown in figure 5. As previously, the BC-fault on the transmission line L near the IBR side is simulated. The currents contributed from the IBR are nonsymmetrical due to the negative sequence current component injected by the inverter; in this case the IBR currents look more

* calculated as the ration between actual differential current and protection pick-up differential current

© The Authors, published by EDP Sciences. This is an open access article distributed under the terms of the Creative Commons Attribution License 4.0 (https://creativecommons.org/licenses/by/4.0/).

typical for an unbalanced event. In comparison with the first scenario the line terminals currents phase relationship are more suitable for protection effective performance (the phase differences: 81,4° for "B" phase and 117,9° for "C" phase), the "C" phase differential current also increases due to current magnitudes difference. The sensitivity coefficients are 2,3 p.u. for "B" phase, 1,8 p.u. for "C" phase.

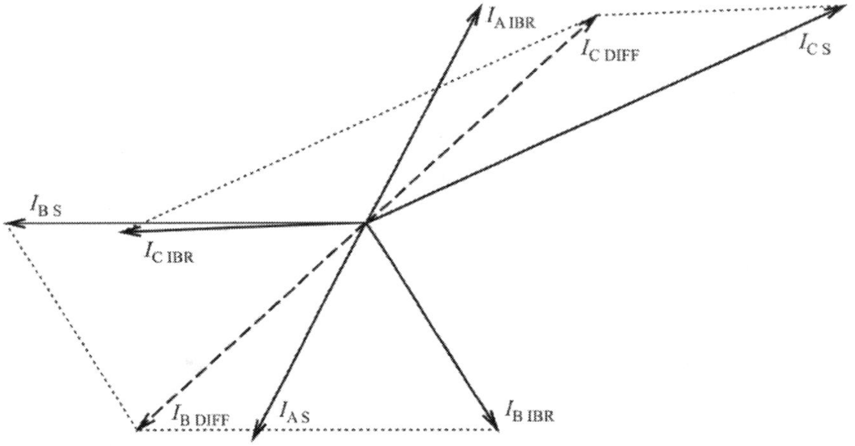

Fig. 3. The current phasors for two-phase fault (scenario 1).

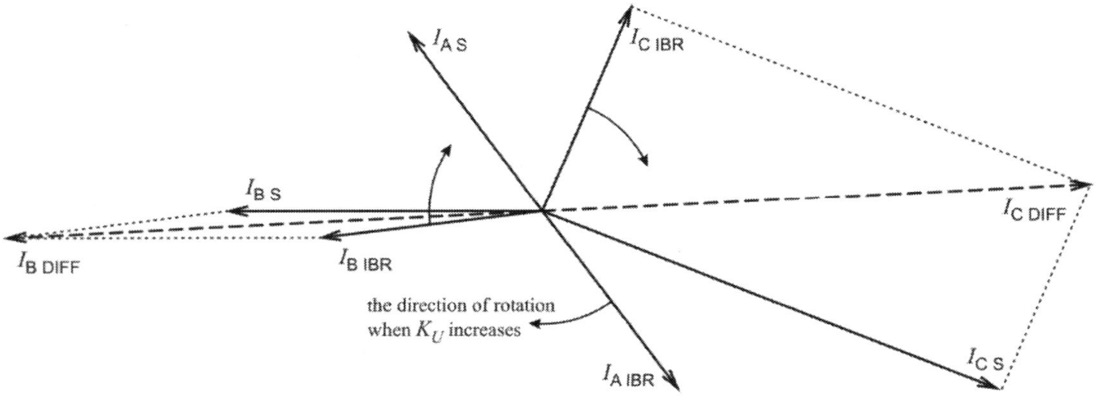

Fig. 4. The current phasors for two-phase fault (scenario 2).

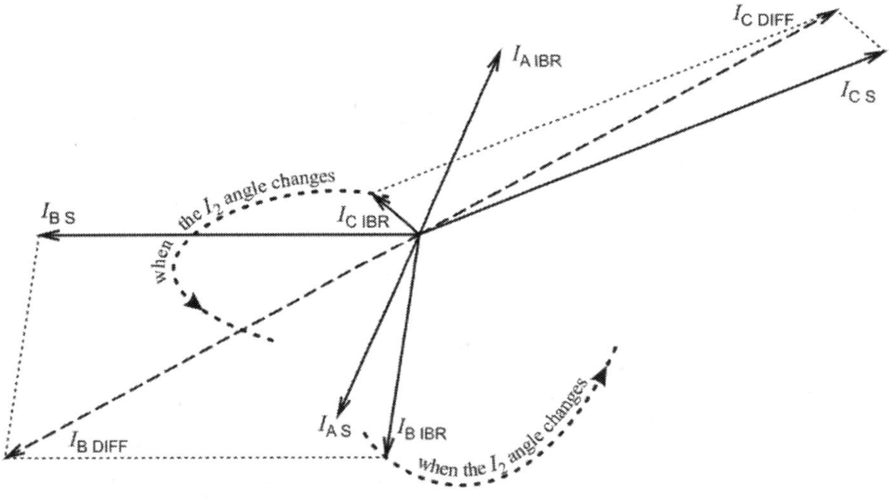

Fig. 5. The current phasors for two-phase fault (scenario 3).

© The Authors, published by EDP Sciences. This is an open access article distributed under the terms of the Creative Commons Attribution License 4.0 (https://creativecommons.org/licenses/by/4.0/).

It should be noted that the negative sequence current contributed by some inverters can have a different frequency than the negative sequence voltage [5]. In this case, the negative sequence current phase angle φ_2 in the equation (3) becomes variable. The faulted phase current phasors rotations due to φ_2 variations are illustrated in figure 5 by dotted bold lines. For a certain value of the φ_2 angle, the phase currents become close in magnitude and the phase difference between their phasors becomes large. Such a feature of the inverter injected negative sequence current can results in protection delayed operation (throughout φ_2 angle variations in the range of the unsuitable values) as well as in protection blocking elements activation, assertion of additional restriction algorithms activating secure protection settings and protection failure to operate.

7 Recommendation on the application of the transmission line differential protection

In the case, if grid source is much stronger than IBR source, the inverter control scenario doesn't significantly impact; differential protection operation conditions are almost the same as on a single-infeed transmission line.

In the case, if with the fault current contributions from IBR and grid source are approximately equal, it becomes necessary to perform faults studies considering inverter control algorithm and inverter control settings to evaluate differential protection performance and to pick up protection settings. For that, it is enough to perform steady-state short circuit analysis on one or a series of time intervals; the related calculations can be performed using engineered methods based on the math equations provided in the paper.

The research results presented in the paper revealed that the most unsuitable operation conditions of the differential protection occur when the inverter control is configured to contribute maximum real power and doesn't inject negative sequence current in a case of unbalanced events. The protection sensitivity may be significantly less than for single-infeed transmission line operation; in some cases, protection may even loss dependability. It should be noted, that the task of ensuring differential protection effective performance may be as well addressed in the other way – by assigning the requirements to inverter control.

The possible combination of the various factors (capacities of the IBR and grid source is close, inverter control scenario is unsuitable for ensuring protection sensitivity, there is no possibility to change inverter control and protection settings) can cause the need for solutions to improve the differential protection algorithm.

8 Conclusion

This paper presented the approach to evaluate IBR impact on differential protection in steady-state conditions under typical inverter control scenarios. By using the math equations provided in the paper, the differential protection operation is studied in the case of the power system composed of the IBR and the grid source of equal capacities; the factors affecting the differential protection operation are revealed and estimated; the case examples are given that illustrates the loss of protection sensitivity and protection failures to trip. The brief guidance is given concerning application of differential protection and considering IBR operation features to study protection operation and to pick up protection settings.

References

1. The Russian Renewable Energy Market: the Current Status and Prospects for the Development. Newsletter (The Renewable Energy Development Association, 2023)
2. *Fault Current Contributions from Wind Plants*, in Proceedings of the 68th Annual Conference for Protective Relay Engineers, 07 May 2015, TX, USA (2015). DOI: 10.1109
3. M. Nagpal, M. Jensen, M. Higginson, A. Bapary, *Protection Challenges and Practices for Interconnecting Inverter Based Resources to Utility Transmission Systems*, in Report of Working Group C32 of the System Protection Subcommittee (2020). DOI:10.13140
4. A. Haddadi, I. Kocar, E. Farantatos, J. Mahseredjian, *System Protection Guidelines for Systems with High Levels of Renewables: Impact of Wind & Solar Generation on Negative-Sequence and Power Swing Protection*, in EPRI Technical Report 3002010937 (EPRI, 2017)
5. R. Chowdhury, N. Fischer, *Transmission Line Protection for Systems with Inverter-Based Resources – Part I: Problems*, IEEE Transactions on Power Delivery **36(4)**:2416-2425 (2020). DOI:10.1109
6. Chowdhury R, McDaniel R, Fischer N. *Line Current Differential Protection in Systems with Inverter-Based Resources – Challenges and Solutions,* in Proceedings of the 76th Annual Georgia Tech Protective Relaying Conference, March 2023, Atlanta, Georgia (2023)
7. Li Y., Jia K., Bi T., Yan R., Li W., Liu B. *Analysis of Line Current Differential Protection Considering Inverter Interfaced Renewable Energy Power Plants*, in Proceedings of the IEEE PES Innovative Smart Grid Technologies Conference Europe, 26-29 September 2017, Turin, Italy (2017). DOI:10.1109

© The Authors, published by EDP Sciences. This is an open access article distributed under the terms of the Creative Commons Attribution License 4.0 (https://creativecommons.org/licenses/by/4.0/).

Influence of the temperature schedule of heating network and using differential pressure regulators on the reliability of heat supply systems

Julia Zvonareva[1*], *Yuri* Vankov[1], and *Irek* Sabirov[2]

[1]Kazan State Power Engineering University, 420066 Kazan, The Republic of Tatarstan, Russia
[2]NPO TN-Biotech LLC, 423450 Almetyevsk , The Republic of Tatarstan, Russia

Abstract. The methods of increasing reliability of heat supply systems are analyzed. A digital twin of the existing heat supply system has been created. Various modes operation of the system are modeled and control and decision-making system in the conditions operation of the heat network is considered on the example of one of heat supply systems of the Republic of Tatarstan. The indicators of reliability of heat supply to consumers at using these methods are determined. As a result of the application of control methods and increasing reliability of the thermal power systems, probability of failure-free operation and the coefficient of hydraulic stability of heat supply system were calculated, the influence of the temperature schedule and the use of differential pressure regulators (DPR) on the reliability of the system in general were analyzed. Special attention is paid to the hydraulic balancing of the heat supply system to increase her reliability.

1 Introduction

Reliability of heat supply is determined by the ability of projected and operating sources of thermal energy, heating networks and, in general, district heating systems to provide the required modes for a given time, parameters and quality of heat supply (heating, ventilation, hot water supply, as well as the technological needs of enterprises in steam and hot water) to provide regulatory indicators of the probability of failure-free operation [Pj], readiness factor [Kj], survivability [S] [1].

2 Measures to improve the reliability of district heating systems and methods of her calculation

Improving reliability of heat supply system during operation is a difficult arrangement. The basis of measures to improve reliability is laid at the design stage of heat supply systems. Many scientists from various universities and academic institutes are engaged in research on improving the reliability of district heating systems.

A method for assessing the reliability of a thermal network based on the definition of PFFO (Probability of failure-free operation), considering the duration of the service life of valves, was proposed at the Tomsk Polytechnic University [2].

The algorithm to implement the principle of increasing reliability of district heating networks through the development of a system model of a thermal network has been proposed at the Siberian Federal University [3].

Researchers of the Kazan State Power Engineering University have proposed a modernization of the method for calculating the failure rate of heat pipelines of decentralized heat supply systems [4].

3 Reliability indicators of the thermal network (main calculated dependencies)

The main method used to calculate the reliability of the heat network is described in the Guidelines for the development of heat supply schemes [5]. The aforementioned analysing tool of reliability of the system under study allows you to take into account the features and characteristics of the system itself, its complexity and structure, as well as the interdependence of components.

The failure rate of a heat pipeline λ, considering the time of its operation, is calculated as follows:

$$\lambda = \lambda^{init}(0.1 \cdot \tau^{exp\,l})^{(\alpha-1)}, 1/(km \cdot h). \quad (1)$$

where λ^{init} - the initial failure rate of the heat pipeline corresponding to the period of normal operation, $1/(km \cdot h)$;

$\tau^{exp\,l}$- network segment exploitation period, years;

α - coefficient considering the duration of operation.

* Corresponding author: skulinaun@mail.ru

© The Authors, published by EDP Sciences. This is an open access article distributed under the terms of the Creative Commons Attribution License 4.0 (https://creativecommons.org/licenses/by/4.0/).

https://doi.org/10.1051/e3sconf/202458401030

$$\alpha = \begin{pmatrix} 0{,}8\, for\ 0 < \tau^{expl} \le 3 \\ 1\, for\ 3 < \tau^{expl} \le 17 \\ 0{,}5 \cdot e^{\left(\frac{\tau^{expl}}{20}\right)}\, for\ \tau^{expl} > 17 \end{pmatrix} \quad (2)$$

The calculation of the failure rate of sections of thermal network with a service life of up to 25 years is performed according to the formula.

Sections of the network with a service life of more than 25 years are allocated to a separate group as potentially unreliable.

Based on an additional analysis of their condition, the sites that require priority shift are selected. For further calculations, it is assumed that the failure rate of these sections is equal to the failure rate of new sections, and not shifted sections – the maximum (that is, the same failure rate of sites with a service life of 25 years) [6].

The failure rate of the shut-off regulatory rebar (SRR) is assumed equal:

$$\lambda_{SRR} = 2{,}28 \cdot 10^{-7}, 1/h, \quad (3)$$

Parameter of the failure flow of sections of thermal network:

$$\omega = \lambda \cdot L,\ 1/h, \quad (4)$$

where L - length of heat network section, km.

4 Reliability of centralized heat supply system in Kazan

The total length of the heating networks of the Kazan district heating system is 929 km in two-pipe calculation, including hot water supply networks (HWS) – 42.8 km.

The heating networks of the zone of the unified heat supply organization are made according to a ring scheme. The hot water supply system is closed. The temperature schedule of the 115/70 ° C. heating networks is made with the straightening of the HWS system at 70 °C.

Figure 1 shows the percentage of all heating networks DHS (District heating system) of the city's, depending on the year of laying.

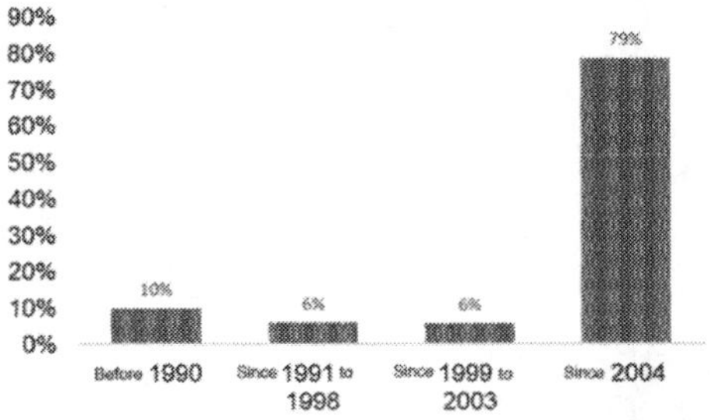

Fig. 1. Distribution by year of laying of heating networks of DHC of Kazan

As shown in the graph, 16% of the city's heating networks have a significant degree of wear and have been in operation for more than 25 years.

The dynamics of changes in failures in the main and distribution heat networks of the city over the past 5 years is shown on Figure 2.

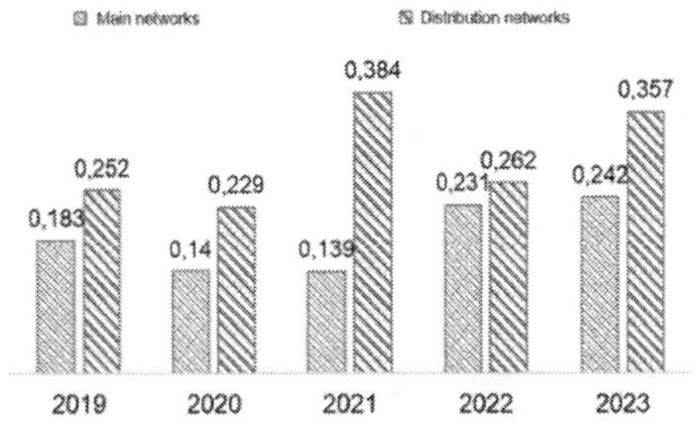

Fig. 2. Specific (related to the length of heating networks) number of failures in the city's heating networks, 1/km/year

Taking into account the current situation in the city's heat supply system and the given retrospective statistics for the purpose of possible methods to improve the reliability of the city's DHC on the example of one of boiler rooms, a study was conducted that includes an analysis of coefficient of hydraulic stability of the system under various changing conditions and an assessment of the reliability of DHC [7].

Fig. 3. The scheme of heating networks from the considered boiler house

The system selected for the study is unbalanced and has a PFFO index of 0.86. Existing condition in the studied system was adopted on the basis of calculations and retrospective analysis given in the heat supply scheme of the city of Kazan.

During the research used a mathematical modeling method which based on the created digital twin (Fig. 3).

In creating a digital twin of the existing city heat supply system, the ZuluThermo software and calculation complex was used, which is part of the Zulu geoinformation system (GIS Zulu) of «Polytherm» LLC, designed to perform thermal and hydraulic calculations of heat supply systems.

In addition, a study of influence of the temperature schedule on operation of the system was conducted, and proposals were made to improve reliability of the heat supply system using a differential pressure regulator (DPR).

The calculation of the reliability of the system is presented on the example of the calculated values of the PFFO of the heat pipelines from the boiler room that are part of the system without any measures (Fig. 4).

© The Authors, published by EDP Sciences. This is an open access article distributed under the terms of the Creative Commons Attribution License 4.0 (https://creativecommons.org/licenses/by/4.0/).

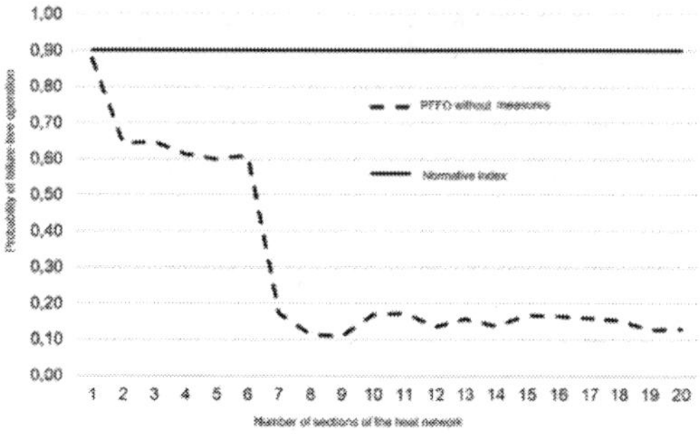

Fig. 4. The probability of failure-free operation of relative sections of the heat network without any measures

The nodal principle used allows to obtain detailed indicators distributed throughout the system. This makes it possible to identify «tight» places in the system and optimally determine the necessary conditions to improve its reliability.

To ensure the reliability of the heat supply system, it is necessary to correctly select the parameters of the heat network operation depending on climatic conditions and consumer's needs. This includes optimization of the thermal regime, the use of effective management and control technologies for the system, regular maintenance and diagnostics of equipment.

Proper management and maintenance of the temperature schedule at optimal values will help prevent accidents and ensure smooth operation of the system.

For the system under consideration, it was proposed to switch to an increased operating schedule of the 115/70 °C thermal network. This will reduce the volume of circulating coolant in the heat supply system and ensure the required temperature regime inside heated rooms, depending on their purpose.

In operating thermal networks, one of the problems is to identify the parameters of the disadjustment and timely decision-making on the adjustment of the hydraulic regime. That requires performing flow distribution calculations, and in case of deviation of real costs from the calculated values, carrying out adjustment measures. Essentially, the adjustment measures consist in linking the hydraulic resistances of heat-consuming installations in accordance with the operating mode of the network equipment and the required coolant consumption [8].

In the result of modeling various operating modes of the system and working out measures for hydraulic balancing, it was possible to achieve temperature stabilization inside heated rooms, reduce pressure at the source and reduce coolant consumption in the system, which in turn reduced operating costs for pumping coolant by an average of 18.5%.

An indicator of the stabilization of the hydraulic regime is the coefficient of hydraulic stability of the system, which characterizes the ability of the system to maintain a constant flow of coolant at subscriber inputs when working conditions change.

$$K = \frac{G_f}{G_c}, \quad (5)$$

where G_f - actual coolant consumption, t/h, G_c- calculated coolant consumption, t/h.

Violation of hydraulic stability negatively affects the reliability of the heat supply system, which is expressed by an increase in the failure rate [9].

The results of the PFFO calculation, after all the measures, are shown on Figure 5.

The results of the study showed that under the actual operating mode of the heat supply system, without carrying out measures for hydraulic balancing of the system and changing the temperature schedule, the coefficient of hydraulic stability is 0.8.

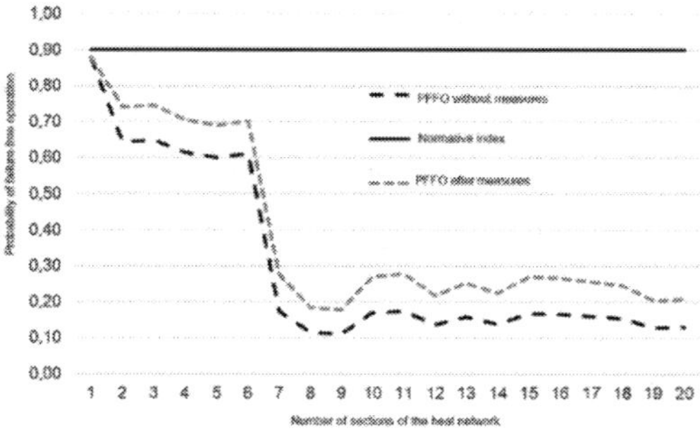

Fig. 5. The probability of failure-free operation of relative sections of the heat network after the measures in comparison with the actual operating mode

In modeling the operating mode for the increased temperature schedule and adjusting the parameters of the coolant at the consumer inputs, as well as changing the pressure at the heat supply source itself – the boiler room, the coefficient of hydraulic stability of the system was 0.96, which indicates stabilization of the operating mode and balanced flow distribution throughout the system.

The hydraulic balancing of the heat supply system and the change in the temperature schedule of the heat network allow to increase the efficiency PFFO of the system.

Besides to hydraulic balancing of the heat supply system by adjusting the parameters at the source of thermal energy and at consumers, an important role in terms of improving the reliability and efficiency of the system is played by the actual throughput of pipelines on the branches of the network and control of coolant flow.

For solving this issue, as an additional measure to increase the reliability of the heat supply system, it is proposed to establish dispatching control using a control and decision-making system and install differential pressure regulators with sensors in thermal chambers on the branches using the LoRaWAN wireless data transmission network.

Differential pressure regulators control and maintain the optimal pressure level and, accordingly, the flow rate of the coolant in the network, which prevents overheating and underheating of the system. Through to the use of pressure regulators, it is possible to reduce the probability of accidents and increase the efficiency of the heating network.

© The Authors, published by EDP Sciences. This is an open access article distributed under the terms of the Creative Commons Attribution License 4.0 (https://creativecommons.org/licenses/by/4.0/).

As the part of the implementation of the regional project "System for monitoring the condition SODC of the heating supply system based on through and through wireless communication technology" starting since 2021 TatAISEnergo LLC began building a LoRaWAN wireless data transmission network in three major cities of the Republic of Tatarstan: Kazan, Naberezhnye Chelny, Nizhnekamsk.

Over the past few years, TatAISEnergo LLC has carried out dispatching of a number of thermal chambers in the Kazan thermal networks and organized data transmission on temperature and pressure of the coolant in the pipeline based on LoRaWAN technology (Fig. 6).

Fig. 6. Operation of the base station LoRaWAN in Kazan

The control and decision-making system for the hydraulic balancing of the heat supply system, based on pressure control at the nodal points of the heat network, is a continuously functioning software package in conjunction with a dispatching control system [10].

The system allows you to monitor pressure surges online, determine the location of an emergency and eliminate problems on time, thereby increasing the reliability of heat supply.

The operation of the proposed pressure monitoring system in the system is carried out as follows. Pressure sensors, on command from the controller, measure the parameters, then transmit them back. LoRa transmitter transmits parameter values through LoRaWAN wireless channels received from sensors directly to the server via the LoRa receiver.

The management and decision-making system is a database for storing and processing the received data in real time. The operator of the control system (dispatcher) has instructions for making decisions depending on the situation, which already takes into account knowledge about all possible modes of operation of the thermal network at the moment and previous experience of personnel actions in similar situations.

In large cities of the Republic of Tatarstan, these systems are actively being implemented and have successfully undergone approbation. The actual experience of operating these systems in the Republic of Tatarstan shows that the response time to emerging emergencies is reduced from 30 to 60%, the "zone" of accident search and detection is narrowed, the period of disconnection of consumers and restoration of the operating mode of the heat supply systems is reduced by at least two times.

5 Conclusion

Based on the research results, it is possible to develop a strategy to improve the reliability of the heat supply system in conditions of changing temperature and hydraulic conditions. This may include upgrading equipment, improving the insulation of heating networks, developing backup action plans in case of extreme weather conditions, and other measures.

In general, the study of the influence of the temperature schedule on the heat supply system is an important step in ensuring the reliability and efficiency of engineering systems. It is only by analyzing and optimizing temperature control that it is possible to ensure a stable and uninterrupted supply of heat to consumers throughout the heating season.

Based on the existing experience of implementing a management and decision-making system, it is say for sure that the reliability indicator of the heat supply system, when implementing the DPR, tends to normative values.

The average time to eliminate an emergency situation and the time to turn off consumers is reduced by at least 1.5 times, thereby increasing the reliability of the heat supply system.

The main decrease in PFFO to a value below the standard is due to the significant service life of some of the longest sections of the thermal network.

With the introduction of a management and decision-making system, the probability of failure-free operation will increase significantly with increasing observation time.

Concerning hydraulic balancing and stabilization of the operating mode of the system as a whole, it should be noted that with an integrated approach of regime regulation of consumers themselves, together with the operation of differential pressure regulators with an installed control and monitoring system, the reliability of the heat supply system will be more than twice.

Subsequently, it is planned to study the effect of the operation of leak sensors and temperature control sensors on the efficiency of emergency detection and assess the relative reduction in the time to eliminate accidents.

References

1. SP 124.13330.2012. A set of rules. Heating networks. Updated version of SNiP 41-02-2003" (approved by the Order of the Ministry of Regional Development of Russia dated 30.06.2012 N 280)
2. 2. A.V. Savchenko, E. V. Kravchenko. *Assessment of the reliability of the thermal network*, in Proceedings of the IV All-Russian Scientific and Practical Conference «Thermophysical foundations of Energy Technologies», 2013, Tomsk, Russia, pp.179-182
3. M. V. Kolosov, Y. L. Lipovka, *Determination of the optimal pipeline replacement period*, Reliability and safety of energy, **14 №4** (2021), pp. 174-179
4. Yu.N. Zvonareva, Yu.V. Vankov, *Survivability indicators of interconnected engineering systems*,

© The Authors, published by EDP Sciences. This is an open access article distributed under the terms of the Creative Commons Attribution License 4.0 (https://creativecommons.org/licenses/by/4.0/).

in Proceedings of the IV International Scientific and Technical Conference "Energy-2019", **1** (2019), p.74

5. On approval of Methodological guidelines for the development of heat supply schemes, Order №212 of the Ministry of Energy of the Russian Federation dated March 5, 2019, URL https://docs.cntd.ru/document/553937025
6. D.V. Korpusov, A. A. Balashov, A. P. Korolev, *Improving the reliability of heat supply during the operation of a thermal network*, Bulletin of Tambov State Technical University, **29 №4** (2023) pp.594-604. doi:10.17277/vestnik.2023.04
7. Yu.N. Zvonareva, O.I. Zverev, *Improvement of heat supply systems through the introduction of AITP*, Bulletin of the Kazan State Energy University, **11 №1**(2019), pp. 10-18
8. V.P. Chernenkov, D.V. Popov, *Management of the hydraulic regime of thermal networks*, in Proceedings of the Far Eastern State Technical University, **134** (2003), pp.126-128
9. Yu.N. Zvonareva, K.S. Kuzborskaya, *Changing the parameters of heat supply systems during the phased implementation of AITP*, Bulletin of the Kazan State Energy University, **13 №2** (2021), pp.109-118
10. I.V. Prangishvili, System approach and management efficiency improvement, Nauka, 2005, p.422.

© The Authors, published by EDP Sciences. This is an open access article distributed under the terms of the Creative Commons Attribution License 4.0 (https://creativecommons.org/licenses/by/4.0/).

Improving protection of medium voltage networks

Sergey Piskunov[1, 2*], *Alexey* Mokeev[1, 2]

[1]Engineering Centre “Energoservice”, Kotlasskaya st. 26, Arkhangelsk, Russia
[2]Northern (Arctic) Federal University, Electroenergetics and Electrotechnics Department, Severnaya Dvina emb.17, Arkhangelsk, Russia

Abstract. This paper presents an efficiency analysis of medium-voltage distribution network protection and considers directions for improving protection characteristics based on synchronized phasor measurement technology (SPM). The authors propose protection algorithms based on two- and multi-sided measurements of voltage and current synchrophasors and cite the advantages of combining local protection with centralized protection and wide area protection.

1 Introduction

The reliability of medium voltage (MV) networks largely depends on the types of protection used. It is observed that 6-10 kV electrical networks, primarily urban and rural, often have protection relays based on overcurrent principles [1]. At the same time, due to the significant length and branching of MV networks, such protections generally do not provide the required sensitivity. In this regard, there is a need to develop new algorithms for protecting MV electrical networks.

One of the ways to improve the main characteristics (sensitivity, selectivity, tripping speed) of the MV network protection is the implementation of synchronized phasor measurement technology [2-3]. Synchrophasor technology provides advantages for improving simple types of protection on the overcurrent principle due to the implementation of two- and multi-sided measurements. In addition, SPM ensures the improvement of more complex protections, for example, distance and differential protections [4-5].

This paper considers 6-35 kV networks that operate with an isolated neutral or with a neutral grounded through an arc-suppression reactor. Such networks also require the implementation of selective protection against single-phase ground faults (SGF). Overcurrent SGF protection is widely used at present, but it has some disadvantages that are associated with its incorrect operation at low zero sequence (ZS) currents and in networks with arc-suppression reactors, where the SGF current decreases [6]. The authors propose several new SGF protection algorithms based on measuring the ZS current and voltage synchrophasors. These algorithms are intended for different types of protection: local feeder protection, centralized switchgear protection, wide area protection [7-8]. In some cases, a combination of these protections allows obtaining the most significant effect.

2 Analysis of MV network protection

Implementation of MV network protection requires consideration of some features, primarily due to the low level of urban and rural 6-10 kV network automation. Cable and overhead networks are traditionally radial and have a one-way power supply. Short-circuit protection is most often overcurrent protection [1]. To increase the reliability of the network, automatic transfer switching (ATS), and automatic reclosing (AR, for overhead lines) are used. Thus, non-selective action of protection is allowed in the presence of ATS and AR [1]. SGF protection, as a rule, is performed based on measurements of ZS current and voltage of fundamental frequency.

Networks with a rated voltage of 20-35 kV have other features. They have the same types of neutral grounding as 6-10 kV networks, but much less often have a radial design. These networks are mainly ring-shaped. Therefore, overcurrent directional protections do not always provide the required sensitivity here [1]. In some cases, this problem is solved by using distance protection.

The characteristics of overcurrent protection largely depend on the power system operating modes, which change significantly over time. Below are summarized data on short-circuit currents in urban and industrial distribution networks of several regions of Russia (Table 1).

The network objects whose data are presented in the statistics (Table 1) are 6(10)/0.4 kV transformer substations (TS), 6-35 kV distribution points (DP) and 35-110 kV step-down substations. In Table 1, the parameter k is the ratio of the three-phase short-circuit current in the maximum power system mode to the two-phase short-circuit current in the minimum mode.

* Corresponding author: s.piskunov@ens.ru

© The Authors, published by EDP Sciences. This is an open access article distributed under the terms of the Creative Commons Attribution License 4.0 (https://creativecommons.org/licenses/by/4.0/).

Table 1. Maximum and minimum short-circuit currents.

Type	Parameter	Rated network voltage, kV						
		urban network			industrial network			
1	$I^{(3)}_{k.max}$, кA	16.9	13.1	32.1	16.5	2.73	9.61	16.5
2	$I^{(3)}_{k.max}$, кA	1.97	0.98	1.2	15	1.06	4.95	-
1	$I^{(2)}_{k.min}$, кA	13.9	8.06	5.19	9.35	0.98	3.7	4.63
2	$I^{(2)}_{k.min}$, кA	1.7	0.82	0.24	6.06	0.63	2.13	-
	k_{max}	1	1.63	.	.	2.79	.	3.56
	k_{min}	1.16	1	.	1	1.68	.	-

*1 – location near power source, 2 – furthest network location

The maximum power system operating mode means that all possible power sources are simultaneously connected to the network and, accordingly, the short-circuit power is the highest. The minimum power system operating mode corresponds to those repair or emergency network circuits when the short-circuit current is minimal for the specified locations (Table 1).

Data analysis shows that short-circuit currents vary significantly depending on the location of the short-circuit (near the power source or in the furthest network location), network topology (urban or industrial), and rated voltage. In some cases, the coefficient k differs by several times, so the operating conditions for overcurrent protections change significantly. In addition, the coefficient k gradually increases with increasing rated voltage.

Based on the expressions for calculating the tripping current of overcurrent protection and assessing its sensitivity [1], it can be proven that, in general, for the correct operation of overcurrent protection, the following condition must be met:

$$\frac{|\underline{z}_{c.max}+\underline{z}_L|}{|\underline{z}_{c.min}+0.2\underline{z}_L|} \geq k_1,\ k_1=\frac{2}{\sqrt{3}}k_r, \qquad (1)$$

where $\underline{z}_{c.min}$, $\underline{z}_{c.max}$ - impedance of the equivalent power system in minimum and maximum mode, $\underline{z}_L$ – line impedance, k_r – protection reserve coefficient.

If we designate as λ the relative protection zone (from 0 to 1), then based on expression (1) we can obtain an equation describing the dependence of λ on the coefficient k, the maximum short-circuits current $I^{(3)}_{k.max}$ and the line length L:

$$\lambda^2+\lambda\frac{2kX_{cmax}X_\mu}{LZ_\mu^2}+$$

$$+\frac{(k_1^2k^2-1)X_{cmax}^2-2LX_\mu X_{cmax}-L^2Z_\mu^2}{k_1^2L^2Z_\mu^2}=0, \qquad (2)$$

where X_μ, Z_μ - line parameters per km.

Positive real roots of equation (2) characterize the protection zone depending on various conditions. Figures 1-2 show the corresponding values of λ for 10 kV cable and overhead lines for various combinations of k, $I^{(3)}_{k.max}$, L and the line cross-section (mm^2).

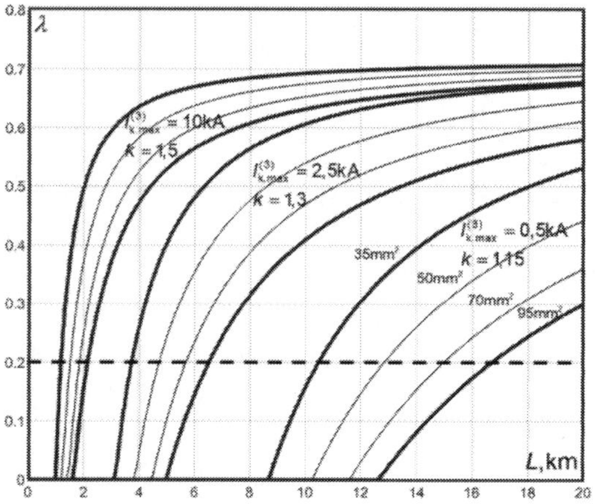

Fig. 1. Relative protection zone for 10 kV overhead lines

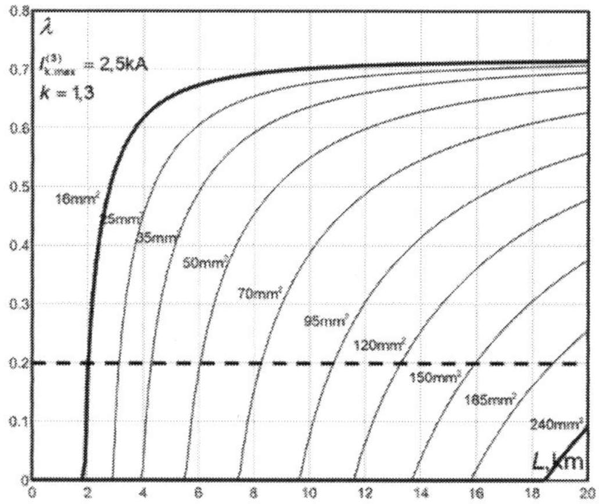

Fig. 2. Relative protection zone for 10 kV cable lines

Data analysis (Fig. 1-2) shows that for 10 kV networks there is a significant dependence of the parameter λ on the line cross-section; at low short-circuit currents it varies within wide limits. In addition, for many combinations of parameters k, $I^{(3)}_{k.max}$ and L the required condition $\lambda > 0.2$ is not met. Similar conclusions based on the analysis of equation (2) correspond to networks of other rated voltages. The lowest efficiency of overcurrent protection is observed for 35-110 kV overhead lines, since here the line impedance weakly depends on the line cross-section, and the coefficient k varies within the widest limits.

Expressions similar to (1) and (2) can be obtained for other types of current protection. In general, it can be concluded that the efficiency of overcurrent protection is quite difficult to ensure in MV networks with a branched structure, for example, when a long line has a short adjacent line, or when a line has taps. Increasing the number of current protection types can partially improve sensitivity [1] but does not solve the problem completely.

In modern conditions, the analysis of MV network protection requires considering such features as the development and significant increase in networks with distributed generation and renewable energy sources [9]. In such networks, it is practically impossible to

© The Authors, published by EDP Sciences. This is an open access article distributed under the terms of the Creative Commons Attribution License 4.0 (https://creativecommons.org/licenses/by/4.0/).

https://doi.org/10.1051/e3sconf/202458401031

ensure the effectiveness of overcurrent protection, as well as in 20-35 kV ring networks. In this regard, new approaches to the development of MV network protection are required.

3 Improving protection against phase-to-phase short circuits

Synchrophasor technology provides conditions for the development of new approaches and algorithms for protecting MV networks [2-3]. Based on current and voltage synchrophasors, network operating mode parameters can be calculated, and network element parameters can be assessed. In addition, it becomes possible to use sets of these parameters for faster and more accurate fault detection.

There are two main directions for improving protection based on synchrophasor technology: the first direction concerns the development of classical protection principles (overcurrent protection, differential protection); the second direction is based on the development of the synchrophasor technology theory, first of all, this is the analysis of differential equations of the protected object (line, transformer) in voltage and current synchrophasors during electromechanical and electromagnetic transient processes. This direction is especially promising for improving distance protection [4].

The advantage of the SPM technology for current protection is the possibility of using multi-sided measurements, when the protection uses measurements from different objects to form additional signs of recognizing the fault type and fault location. In many cases, a simple principle of logical selectivity is effective, when the presence of synchrophasor measurements of other substation feeders ensures the correct operation of overcurrent protection. In this case, for current protection with time delay, such principle can be the most effective, since in this case the protection does not have high speed, which allows using various communication channels to transfer data between protection devices.

Two- and multi-sided synchrophasor measurements allow implementing new differential protection algorithms. It is most in demand for providing protection for busbars and power transformers [5].

One of the most promising areas of synchrophasor technology is the improvement of distance protection. This can be classical local distance protection based on one-sided measurements or protection based on two-sided measurements of current and voltage synchrophasors.

Based on the analysis of the line differential equation in the transient synchrophasors, the authors proposed algorithms for improving the line distance protection [4]. The general expression for one-sided measurement:

$$\hat{\underline{z}}(t)=\frac{\dot{U}'(t)}{\dot{I}(t)+\underline{k}\dot{I}'(t)},\ \dot{I}'(t)=\frac{d\dot{I}(t)}{dt},\ \underline{k}=\frac{L_{\mu}}{\underline{z}_{\mu}}. \quad (3)$$

In many cases, the line short circuit is an arc fault. For distance protection, the nonlinear arc resistance introduces additional error into the line impedance estimate. The following expressions allow one to consider the effect of the arc resistance:

$$\hat{R}_{a}(t)=\mathrm{Re}\left[\hat{\underline{z}}(t)\right]-\mathrm{Im}\left[\hat{\underline{z}}(t)\right]\frac{R_{\mu}}{L_{\mu}},\ \hat{\underline{z}}_{1}(t)=\hat{\underline{z}}(t)-\hat{R}_{a}(t), \quad (4)$$

where $\hat{\underline{z}}_{1}(t)$ - estimation of line impedance without arc resistance for one-sided measurements.

If there are measurements of current and voltage synchrophasors at both ends of the line (side 1 and side 2), then the impedance of the line during a short circuit on side 1 will be estimated using the following expression:

$$\hat{\underline{z}}_{12}(t)=\frac{\dot{U}_{1}(t)-\dot{U}_{2}(t)+\underline{z}\dot{I}_{2}(t)+L\dot{I}'_{2}(t)}{\dot{I}_{1}(t)+\dot{I}_{2}(t)+\underline{k}\left(\dot{I}'_{1}(t)+\dot{I}'_{2}(t)\right)}. \quad (5)$$

Expression (5) allows us to proceed to estimating the distance to the fault location:

$$\hat{l}_{12}(t)=\frac{\dot{U}_{1}(t)-\dot{U}_{2}(t)+\underline{z}\dot{I}_{2}(t)+L\dot{I}'_{2}(t)}{\underline{z}_{\mu}\left(\dot{I}_{1}(t)+\dot{I}_{2}(t)\right)+L_{\mu}\left(\dot{I}'_{1}(t)+\dot{I}'_{2}(t)\right)}. \quad (6)$$

An important feature of the proposed algorithms (4) - (6) is the provision of an accurate and stable estimate of the line impedance when combining various electromagnetic and electromechanical transient processes (Fig. 3).

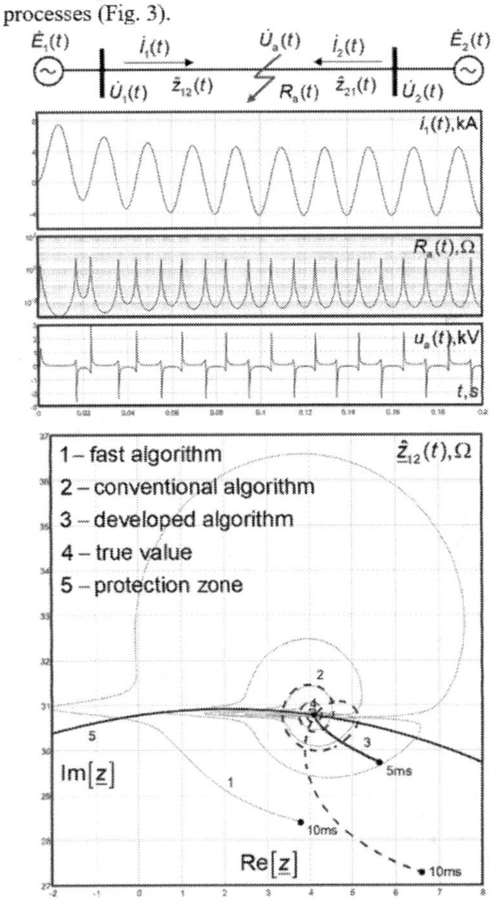

Fig. 3. Distance line protection based on synchrophasors

© The Authors, published by EDP Sciences. This is an open access article distributed under the terms of the Creative Commons Attribution License 4.0 (https://creativecommons.org/licenses/by/4.0/).

https://doi.org/10.1051/e3sconf/202458401031

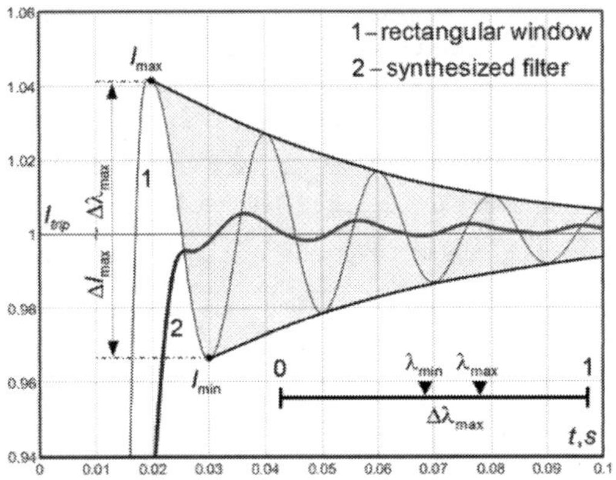

Fig. 4. Effect of signal processing on overcurrent protection

Fig. 3 shows the process of an arc short circuit in the line and the results of the line impedance estimation by different algorithms. Algorithm 1 (without averaging) assumes direct impedance calculation based on instantaneous values of voltage and current (the so-called "fast" algorithm, using two consecutive signal samples). This algorithm gives the largest error due to the presence of nonlinear arc resistance, which changes very quickly. Traditional algorithm 2 based on a filter with a rectangular window gives a more accurate estimate, but the time when the impedance estimate becomes stable is about 20-30 ms. In this case, the arc resistance can be taken as a constant value when selecting protection characteristics [10]. The developed algorithm 3 based on expressions (4) - (6) gives an accurate estimate of the line impedance already after 5-10 ms from the beginning of the short circuit.

For some power system time constants, a rectangular window filter can cause significant fluctuations in the estimated signal magnitude in the first 30-40 ms from the onset of a short circuit, which significantly affects overcurrent protection (Fig. 4). Therefore, the synthesis of digital filters for microprocessor protection devices and the use of synchrophasor technology [11] is a promising direction for improving the characteristics of both simple overcurrent protection and differential and distance protection.

4 Improving protection against single phase ground faults

Improvement of the SGF protection corresponds to various scientific directions [12-13]. One of the most promising ways of developing SGF protection algorithms is measuring ZS current and voltage synchrophasors [7-8].

Synchrophasor technology provides various principles of SGF protection: local, when synchrophasors are measured for one feeder; centralized, when synchrophasors of several feeders at a substation are measured and compared; wide area, when protection devices at different facilities can exchange synchrophasors with each other. All of the above directions for improving SGF protection are promising, since the conditions and level of network automation are always different.

The SGF protection on synchrophasors can have various algorithms that involve the assessment of certain parameters [7-8]. The simplest protection method is to measure the ZS current synchrophasors of the fundamental frequency. Their comparison by network segments ensures the operation of the centralized SGF protection. This protection method has been implemented at several 10 kV electrical network facilities and has proven its effectiveness [3].

However, the disadvantage of this method is less efficient operation of protection in networks with low capacity and installed arc-suppression reactors to compensate for the SGF current. Therefore, this method is more efficient in cable networks than in overhead ones.

A more advanced method of SGF protection involves the assessment of equivalent harmonic synchrophasors (EHSP) [7]:

$$\dot{I}_{0e}(t)=\sum_{m=1}^{M}\dot{I}_{0(2m+1)}(t)\,,\ \dot{U}_{0e}(t)=\sum_{m=1}^{M}\dot{U}_{0(2m+1)}(t)\,,$$

$$\dot{U}_{0me}(t)=\sum_{m=1}^{M}(2m+1)\dot{U}_{0(2m+1)}(t)\,, \qquad (7)$$

where $\dot{I}_{0e}(t)$, $\dot{U}_{0e}(t)$, $\dot{U}_{0me}(t)$ - equivalent harmonic synchrophasors of ZS current and voltage.

The advantage of this method (7) is its independent operation from the level and composition of harmonics and the possibility of implementing different principles of SGF protection (Fig. 5).

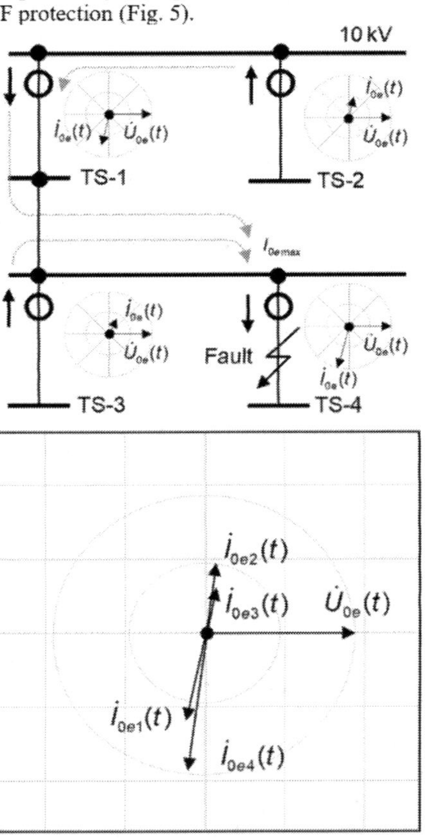

Fig. 5. SGF protection on equivalent synchrophasors

© The Authors, published by EDP Sciences. This is an open access article distributed under the terms of the Creative Commons Attribution License 4.0 (https://creativecommons.org/licenses/by/4.0/).

Fig. 5 shows that fault location is characterized by the equivalent current synchrophasor with the highest magnitude and a phase that differs from the phases of other network segments by 180°.

Another method of SGF protection on synchrophasors involves estimating the admittance or capacitance of the feeder [8]:

$$\widehat{C}_0(t) = \frac{\dot{I}_{0e}(t)}{j\omega_0 \dot{U}_{0me}(t) + \dot{U}'_{0e}(t)}. \quad (8)$$

This method is most suitable for local protection, since an assessment of the feeder capacity is necessary to select the protection setting relative to the theoretical line parameters. The admittance principle of the SGF protection is an analogue of the distance principle, only in this case it is not the impedance that is assessed, but the line admittance/capacity (Fig. 6).

Fig. 6 shows the SGF protection characteristic, which allows dividing the zone with the parameters of undamaged feeders and the zone with the parameters of the damaged feeder (fault zone). The admittance principle provides a flexibly adjustable protection characteristic depending on many influencing factors, for example, on the measurement error.

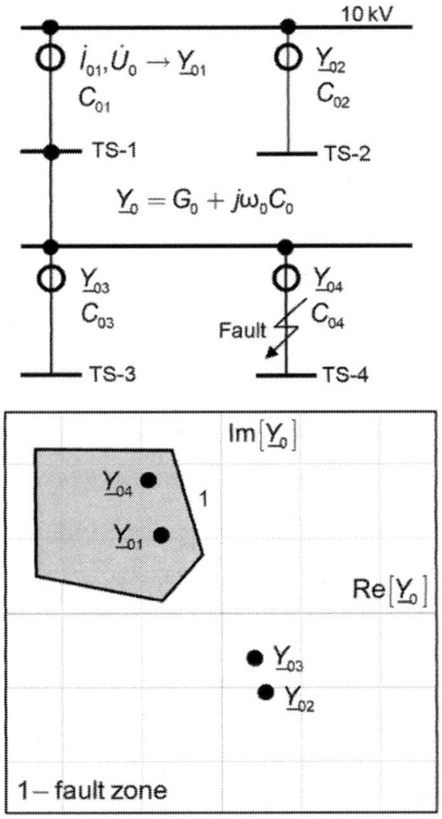

Fig. 6. SGF admittance protection

Synchrophasor technology also provides a centralized principle for the SGF localization system. This method has been successfully tested on mathematical models and on several electrical network facilities [2, 5].

5 Conclusion

Analysis of protection of medium-voltage electric networks shows that common overcurrent protections in many cases do not provide the required level of sensitivity. In addition, the development of networks with distributed generation and renewable energy sources requires the use of more advanced protections.

Synchrophasor technology is one of the promising areas for improving MV network protection. The research results confirm that measuring current and voltage synchrophasors in protection devices provides significantly better characteristics of overcurrent protection and allows the use of more advanced protection, such as differential and distance protection.

References

1. A.M. Fedoseev, M.A. Fedoseev, Protection of electric power systems, M., Energoatomizdat (1992).
2. S.A. Piskunov, A.V. Mokeev, D.N. Ulyanov, ES3 Web of Conferences (2023).
3. A.V. Mokeev, S.A. Piskunov, RPA, **1** (2023).
4. A.V. Mokeev, S.A. Piskunov, RPA, **1** (2024).
5. A.V. Mokeev, et al., Vestnik KGEU, **3** (2020).
6. V.A. Shuin, A.V. Gusenkov, Single phase ground fault protections in 6-10 kV electrical networks, M., Energoprogress (2001).
7. S.A., Piskunov, A.V. Mokeev, D.N. Ulyanov, 6th Int. Sc. and Tech. Conf. RPA, IEEE (2023).
8. S.A., Piskunov, A.V. Mokeev, D.N. Ulyanov, REEPE, IEEE (2024).
9. P.V. Ilyushin, NTS EES, Moscow (2022).
10. I.T. Titov, A.V. Bychkov, N.A. Doni, NGTU, **2** (2020).
11. Mokeev A.V., Protection and automation of power systems, Saint Petersburg (2017).
12. V.F. Lachugin, A.A. Ivanov, A.A. Belyanin, RPA, **3** (2012).
13. V.A. Shuin, Elektricheskie stancii, **7** (2018).

© The Authors, published by EDP Sciences. This is an open access article distributed under the terms of the Creative Commons Attribution License 4.0 (https://creativecommons.org/licenses/by/4.0/).

Transformer protection, automation and monitoring systems

Sergey Piskunov[1, 2*], *Alexey* Mokeev[1, 2], *Dmitry* Ulyanov[1]

[1]Engineering Centre “Energoservice”, Kotlasskaya st. 26, Arkhangelsk, Russia
[2]Northern (Arctic) Federal University, Electroenergetics and Electrotechnics Department, Severnaya Dvina emb.17, Arkhangelsk, Russia

Abstract. This paper presents the results of research on improving the protection, automation and monitoring systems of power transformers (PT) in 6-35 kV distribution networks. The authors have developed improved algorithms for differential and distance protection of a transformer based on measuring voltage and current synchrophasors, proposed additional criteria for blocking protection in the event of a magnetization inrush current (MIC), and considered the functions of the transformer monitoring and automation system based on synchrophasor technology.

1 Introduction

Statistics of primary equipment failures at substations show that power transformers with a high voltage of 6-35 kV have the greatest number of failures [1]. At the same time, existing transformer protection, automation and monitoring systems in distribution networks, in many cases, have limited functionality.

Analysis of power transformers and automation equipment for 6-35 kV networks shows that the most common nominal power of transformers is no more than 6.3 MVA. For transformers of lower power, overcurrent protection is mainly used, which does not provide the required tripping time in case of internal faults [2]. In addition, differential protection of transformers with a power of more than 6.3 MVA has low sensitivity in case of complex fault types (turn-to-turn short circuits, fire in steel) and must have a blocking during a magnetization inrush current [3]. In this regard, an important task is to improve the protection of a power transformer.

The paper examines the application of synchrophasor technology to improve transformer protection, automation and monitoring systems. First, this technology has advantages for the implementation of differential and distance protection principles [4-6]. In addition, the authors note that the transformer condition can be monitored by controlling its electromagnetic parameters based on voltage and current synchrophasor measurements [7]. This allows for timely detection and prevention of complex fault types.

Currently, the development of networks with distributed generation, including renewable energy sources (RES), causes the emergence of new tasks to improve the protection and automation of distribution networks [8]. Synchrophasor technology is a promising direction for solving these problems.

2 Power transformer analysis

The topology of distribution networks determines the presence of many taps supplying substations with power transformers of low rated power. Analysis of data from electrical networks of several Russian entities shows that the most common rated powers of transformers are 2.5 and 6.3 MVA, a significant portion of transformers have a power of 1 MVA or less (Fig. 1). In addition, a significant portion of transformers have a long service life, while substations have a low automation level, and transformer monitoring systems are absent in many cases.

In 6-35 kV networks, the share of electromechanical protection devices is more than 78 %, and the excess of their established service life is more than 59 % [9]. Electromagnetic current and voltage transformers are predominant, which is combined with a low integration level of protection systems and SCADA, and the incompatibility of the operating principles of existing protection with the requirements for protection in networks with distributed generation.

Overcurrent transformer protection, which is used in 6-35 kV networks, has limited capabilities and provides sensitivity only to faults in the high-voltage winding. Therefore, in order to provide transformer protection, special types of current protection with a significant time delay are added. The authors analyzed the expressions for calculating the tripping settings of overcurrent protection and the conditions for ensuring its sensitivity [2] to show that the protection characteristic significantly depends on the transformer impedance and the impedance of the equivalent power system in the minimum and maximum modes.

* Corresponding author: s.piskunov@ens.ru

© The Authors, published by EDP Sciences. This is an open access article distributed under the terms of the Creative Commons Attribution License 4.0 (https://creativecommons.org/licenses/by/4.0/).

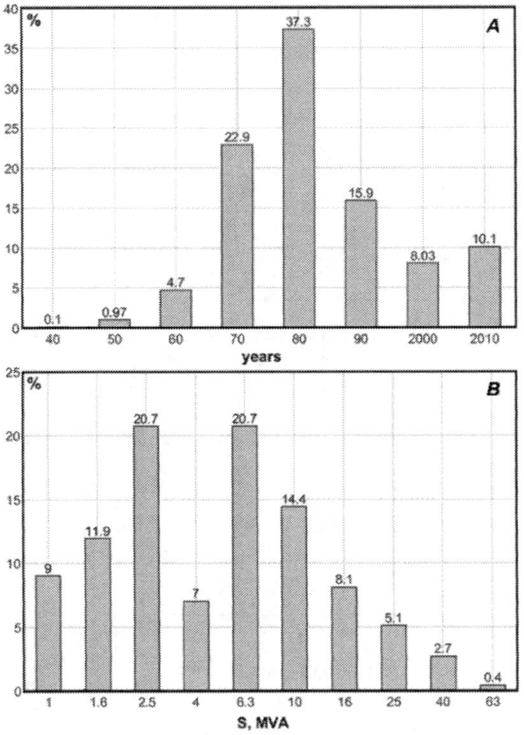

A – commissioning, B - rated power
Fig. 1. Statistics on power transformers 35-110 kV

The following expression represents this dependency:

$$\frac{|\underline{z}_{c.max} + \underline{z}_L + \underline{z}_T|}{|\underline{z}_{c.min} + \underline{z}_L|} \geq \frac{2}{\sqrt{3}} k_r k_{s.min}, \quad (1)$$

where $\underline{z}_{c.min}$, $\underline{z}_{c.max}$ – equivalent power system impedance in minimum and maximum mode, $\underline{z}_L$ – line impedance, $\underline{z}_T$ – transformer impedance, k_r – protection reserve coefficient, $k_{s.min}$ – minimum protection sensitivity coefficient.

There is an overcurrent protection type with a voltage setting start, but it also does not provide the required sensitivity characteristics. Distance transformer protection in networks up to 110 kV is rare, differential protection is used most often. However, differential protection is not sensitive to turn-to-turn short circuits and requires blocking at MIC.

In fact, at 35-110 kV substations, there are no transformer monitoring systems; in most cases, there are limited telemetry tools (Fig. 2, Table 1).

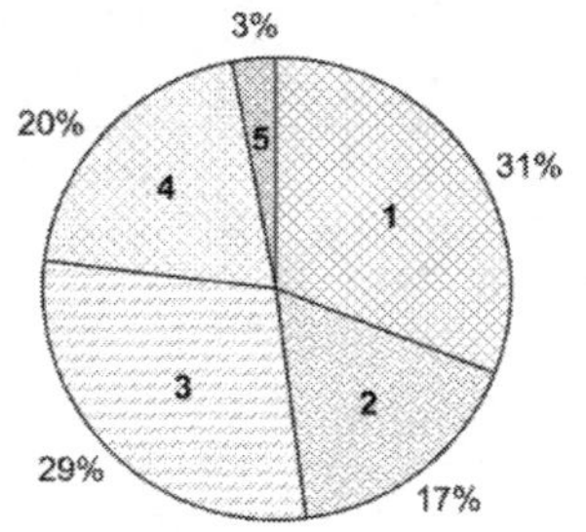

Fig. 2. Statistics on telemechanics of 35-110 kV substations

Table 1. Telemechanics levels.

Level (Fig.2)	Description
1	minimum telesignaling set ("General emergency", "Fault", "Ground fault", "Open")
2	minimum telesignaling set, telemetering for equipment ≥ 35 kV
3	full set of telesignaling and telemetering, remote control of high-voltage breakers 6-35 kV
4	full set of telesignaling and telemetering, remote control of high-voltage breakers of all voltage classes
5	full set of telesignaling and telemetering, remote control of all switching devices

The substation automation tools, which are connected with the power transformer control, also have a limited scope in distribution networks. The most frequently used is the automatic on-load tap changer (OLTC).

Thus, the power transformer analysis shows that distribution networks require a search for new principles and approaches to the development of protection, automation and monitoring systems, the use of new technologies and solutions, and the provision of measures aimed at increasing the observability and controllability of the network.

The development of modern communication technologies, mass production and reduction in the component's cost of intelligent electronic devices (IED) and microprocessor protection devices allows considering the use of synchrophasor technology for solving problems of distribution network automation. Based on these trends, the authors conducted research on the development of the synchrophasor technology theory and the expansion of its application in 6-35 kV networks [10]. The research results allow formulating the main directions for improving the power transformer protection, automation and monitoring systems.

3 Protection

Differential protection is one of the most reliable and widespread types of protection for power transformers. Modern microprocessor devices of differential protection provide adaptive differential current setting and can respond even to some complex fault types. However, the solution to the issue of differential protection blocking at MIC remains open, since the most widespread blocking principle by the relative value of the 2nd harmonic in the transformer current does not always ensure correct protection operation due to the presence of current transformer core saturation [11].

Electromagnetic current transformers (CT) with a steel core can saturate during a short circuit due to the presence of a significant aperiodic component of the current, even if their parameters were selected correctly. In this case, if the protection cannot act before the CT is saturated, it is significantly slowed down due to blocking by the 2nd harmonic (Fig. 3, B). Depending on the signal processing algorithms used, the protection can be blocked for the filter window length, since the 2nd harmonic level can also be large (Fig.3, A).

© The Authors, published by EDP Sciences. This is an open access article distributed under the terms of the Creative Commons Attribution License 4.0 (https://creativecommons.org/licenses/by/4.0/).

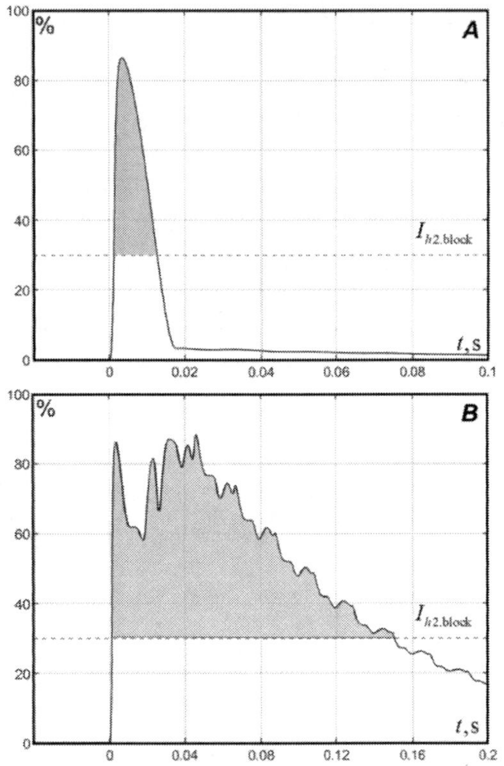

A - signal processing (short-circuit), B - CT saturation (MIC)
Fig. 3. 2nd harmonic relative value

Synchrophasor technology and electronic current transformers have significant advantages for improving digital transformer differential protection. Firstly, measuring transformer current synchrophasors based on electronic CTs without saturation provides a significant reduction in the protection unbalance current due to several factors: aperiodic component elimination, low CT error, and OLTC control. Electronic measuring current and voltage transformers are already used in 6-35 kV networks, and significant results have been obtained in this direction [12].

Secondly, measuring the transformer current synchrophasors $\dot{I}_1(t)$ and $\dot{I}_2(t)$ allows solving some issues related to blocking the differential protection at MIC and increasing its selectivity during external short circuits. The presence of angle measurements between $\dot{I}_1(t)$ and $\dot{I}_2(t)$ clearly determines the power direction in the transformer windings. Research shows that in case of transformer fault, an accurate angle estimate between the current synchrophasors can be obtained in a time shorter than the filter window length (for a rectangular window it is 20 ms).

Previously, the authors carried out research on the development of the synchrophasor technology theory to improve line distance protection principles [6]. Similarly, the analysis of power transformer differential equations shows that distance principles can also be effectively used to improve the transformer protection (Fig.4).

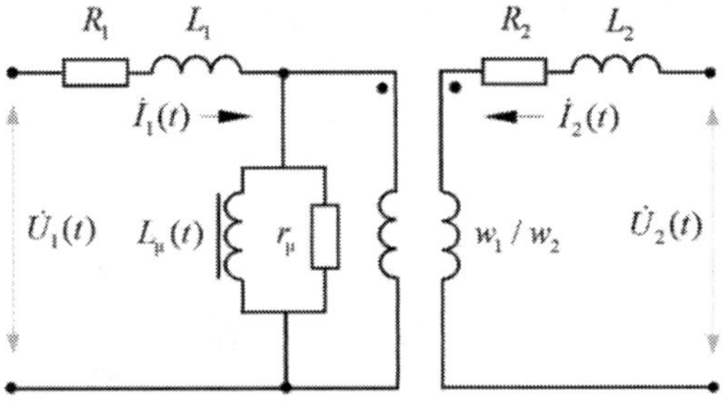

Fig. 4. Two-winding transformer equivalent circuit

Fig.4 shows an equivalent circuit of a two-winding power transformer, where its parameters are estimated by measuring the voltage and current synchrophasors. For transformer protection, the most important parameter characterizing its state is the impedance of the short-circuit branch $\underline{z}_{12}$. In addition, the authors consider the magnetizing inductance estimation $L_\mu(t)$ to provide an additional criterion for blocking the protection at MIC:

$$L_\mu(t) = \frac{\dot{U}_1(t) - \underline{z}_1\dot{I}_1(t) - L_1\dot{I}'_1(t)}{\dot{I}'_1(t) + j\omega_0\dot{I}_1(t) - \dot{I}'_2(t) - j\omega_0\dot{I}_2(t)}, \quad (2)$$

where $\dot{I}_1(t)$, $\dot{I}_2(t)$ - transformer current synchrophasors related to the high-voltage side, $\dot{U}_1(t)$ – voltage synchrophasor, $\underline{z}_1 = R_1 + j\omega_0 L_1$, $\dot{I}'_1(t)$, $\dot{I}'_2(t)$ - derivatives of current synchrophasors.

Based on expression (2), it can be shown that the minimum value $L_\mu(t)$ in case of transformer fault is less than the same value at MIC by more than 4-5 times for 35-110 kV power transformers. This allows using $L_\mu(t)$ as an additional criterion for blocking differential protection. In addition, the use of synchrophasor technology for analysing transformer differential equations, including for assessing inductance (2), is valid both in steady-state and in transient modes of the power system.

Fig. 5 shows the simulation of the MIC and short circuit in the high-voltage winding of a 35/6 kV transformer. The magnetizing inductance estimation according to expression (2) in different power system modes confirms the effectiveness of using this parameter as a protection blocking criterion.

Based on the research on improving the line distance protection [6], according to the equivalent circuit (Fig. 4), it can be shown that the expressions for estimating the line impedance during a short circuit for two-sided synchrophasor measurements are also valid for estimating the transformer impedance:

$$\hat{\underline{z}}_{12}(t) = \frac{\dot{U}_1(t) - \dot{U}_2(t) + \underline{z}_{12}\dot{I}_2(t) + L_{12}\dot{I}'_2(t)}{\dot{I}_1(t) + \dot{I}_2(t) + \underline{k}\left(\dot{I}'_1(t) + \dot{I}'_2(t)\right)}, \quad (3)$$

where $\underline{k} = L_{12} / \underline{z}_{12}$.

© The Authors, published by EDP Sciences. This is an open access article distributed under the terms of the Creative Commons Attribution License 4.0 (https://creativecommons.org/licenses/by/4.0/).

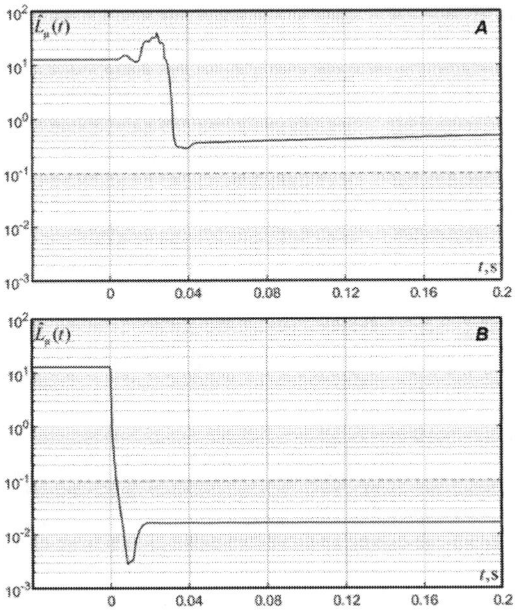

A – MIC, B – short circuit
Fig. 5. Magnetizing inductance estimation

Thus, the advantages provided by two-sided synchrophasor measurements in line distance protection [6] are valid for transformer protection. In addition, a promising approach is one that uses the estimation of the difference in impedances $\underline{\hat{z}}_{12}(t)$ and $\underline{\hat{z}}_{21}(t)$ by analogy with the unbalance current in transformer differential protection. The distance principle ensures high sensitivity of such protection to transformer fault.

4 Monitoring and automation

The differential protection of a transformer must ensure sensitivity to internal short circuits and be selective to external short circuits. Additionally, this protection can also be sensitive to complex fault types, such as turn-to-turn short circuits. However, the peculiarity of turn-to-turn short circuits is that even modern digital differential protection cannot be sensitive to such fault with a small number of shorted turns. In this case, the detection and prevention of turn-to-turn short circuits can be ensured by a transformer monitoring system [7].

Traditionally, a transformer monitoring system provides control of some parameters that are measured using special sensors and devices. However, this equipment is rarely installed in distribution networks, as it is expensive. The authors propose using measurements of the transformer voltage and current synchrophasors to operate its monitoring system. This approach provides control of many electrical transformer parameters. In addition, in this case, the monitoring system does not require additional measuring equipment.

The short-circuit impedance (Fig. 4) is estimated based on the transformer voltage and current synchrophasors in the normal mode of the power system using the expression:

$$\underline{\hat{z}}_{12}(t) = \frac{2\left(\dot{U}_1(t) - \dot{U}_2(t)\right)}{\dot{I}_1(t) + \dot{I}_2(t) + \underline{k}\left(\dot{I}'_1(t) + \dot{I}'_2(t)\right)}. \qquad (4)$$

Thus, the parameters (2) – (4) can be estimated both for the protection functions and for the transformer condition monitoring, which ensures the prevention of complex fault types, to which the transformer protection is not sensitive. In addition, electronic current and voltage transformers [12] ensure an accurate estimation of the transformer parameters both in the normal power system mode and during short circuits. This makes it possible to implement some monitoring functions in the transformer protection device.

The Engineering Center Energoservice LLC company has developed its own transformer monitoring system based on synchrophasor measurements. Currently, there is already a specialized device – a transformer condition monitoring module, which collects and processes transformer synchrophasor measurement data to evaluate the parameters and monitor its condition (Fig. 6). The device also generates a list of daily, weekly, monthly and annual reports on changes in the transformer condition. Any abnormal deviations in parameters are recorded on oscillograms.

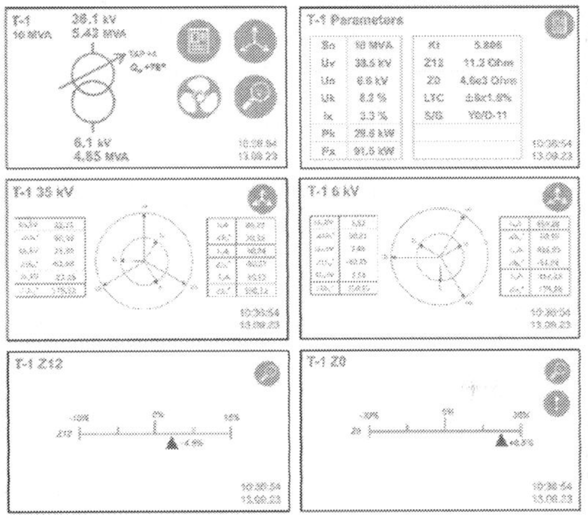

Fig. 6. Transformer monitoring module

Synchrophasor technology also has promising directions for transformer automation systems. One of the most significant functions of transformer automation is changing the transformation ratio using OLTC. Measuring voltage synchrophasors ensures control of switching duration, device drive resource, and allows recording cases of "self-running" and "jamming". Additionally, using synchrophasor technology, specialized OLTC algorithms can be implemented to ensure the required voltage level simultaneously for several consumers.

The principle of multi-sided synchrophasor measurements [6] provides additional conditions for automatic control of the 6(10)/0.4 kV transformer breaker, breaker failure backup device at large substations, automatic reclosing and automatic frequency transformer unloading. A special feature of synchrophasor technology is the possibility of creating

© The Authors, published by EDP Sciences. This is an open access article distributed under the terms of the Creative Commons Attribution License 4.0 (https://creativecommons.org/licenses/by/4.0/).

wide area systems, when the current and voltage synchrophasors of different feeders are compared either at the same substation or at different ones.

At present, the introduction of various generation sources at medium and low voltage necessitates the search for new algorithms for the operation of distribution network automation systems. It is certain that for distributed generation network automation systems, the use of synchrophasor technology has the greatest advantages [8], since there are new tasks for improving the devices for mode control automation and emergency automation. Therefore, measuring voltage and current synchrophasors in transformer automation systems provides for the expansion of their capabilities and stimulates the development of protection, automation and monitoring systems for distribution networks as a whole.

5 Conclusion

Improvement of protection devices largely depends on the efficiency of using new technologies, intelligent devices, and modern technical means. One of such areas for improving transformer protection, monitoring, and automation systems is the use of synchrophasor technology.

The paper reviewed the main results of the authors' research on improving transformer protection, monitoring and automation systems based on measurements of voltage and current synchrophasors. The developed algorithms and operating principles of these systems were tested on mathematical and physical models, as well as during pilot operation at distribution network facilities.

References

1. V.S. Bogomolov, Energiya edinoj seti, **2** (2013).
2. N.V. Chernobrovov, V.A. Semenov, Protection of electric power systems, Energoatomizdat (1998).
3. V.A. Andreev, Protection and automation of power supply systems (1991).
4. S.A., Piskunov, A.V. Mokeev, REEPE, IEEE (2021).
5. A.V. Mokeev, et al., Vestnik KGEU, **3** (2020).
6. A.V. Mokeev, S.A. Piskunov, RPA, **1** (2024).
7. S.A. Piskunov, A.V. Mokeev, D.N. Ulyanov, Elektroenergiya, **2** (2023).
8. P.V. Ilushin, A.L. Kulikov, RPA, **3** (2019).
9. Concept for the development of protection, automation and control systems for technological processes of electric grid complex, PJSC Rosseti, Order No. 286/186 (2022).
10. S.A. Piskunov, A.V. Mokeev, D.N. Ulyanov, RSES (2023).
11. D.G. Belkin, N.N. Markevich, Methods of adjusting differential protection of transformers from unbalance currents (2010).
12. D.N. Ulyanov, Relavekspo, Cheboksary (2023).

© The Authors, published by EDP Sciences. This is an open access article distributed under the terms of the Creative Commons Attribution License 4.0 (https://creativecommons.org/licenses/by/4.0/).

Methods for ensuring the efficiency and safety of energy system facility equipment based on digital technologies

Mahsud Sultanov[1], *Ilya* Boldyrev[1], *Maxim* Ivanitckii [1], and *Pavel* Menshikov[1]*

[1]National Research University «MPEI», Russia, 404110 Volzhsky, Lenin prospect, 69, Russia

Abstract. The results of developing a methodology for ensuring a high level of efficiency and safety of energy equipment are presented. Approaches for assessing and predicting parameters that determine the reliability of energy system equipment, and the structure of a cloud data collection and processing system, are proposed. The results obtained in this work can be used to ensure the efficiency and safety of energy system facility equipment.

1 Introduction

In addressing the issues of digital transformation at energy facilities, it is necessary to develop new approaches to data management (collection, storage, processing, forecasting) regarding the operating modes of energy systems. These approaches define the digital transformation of energy sector divisions and enterprises, ensuring high competitiveness in the electricity markets while considering the requirements of Russian legislation.

Thus, it is crucial to develop methods for ensuring the efficiency and safety of energy system facility equipment based on the improvement of the information flow distribution structure in the management system using advanced digital technologies.

1.1 Methodology for Ensuring High Efficiency and Safety of Energy Equipment

In solving the problems of digital transformation at energy facilities, it is necessary to develop new methods of managing data (collection, storage, processing, forecasting) on the operating modes of energy systems. These methods define the digital transformation of divisions and enterprises in the energy sector, ensuring high competitiveness in electricity markets in compliance with Russian legislation [1]. The main reasons hindering the degree of digitalization include the lack of workforce competencies, the high cost of digital technologies and tools, and existing complexities in regulatory acts governing the information security requirements of energy enterprises.

In this regard, the implementation of digital competencies should enable the ability and readiness to independently and responsibly make operational decisions and achieve necessary results when choosing new digital technologies aimed at improving the efficiency of operation and management of energy equipment using digital technologies.

The mechanisms of management schemes for generating and grid companies, which largely define the vertical hierarchy of interaction between the chief engineer of the enterprise and the main divisions, services, departments, and sections, must be considered. It should be noted that establishing a separate division overseeing digital transformation allows for more intensive development and implementation of digital tools. This significantly influences the conceptual issues of developing technological processes, organizational and technical tasks, and management, including the need for new digital competencies among the current employees of organizations.

In this case, the organization of the enterprise's activities does not undergo significant changes in its management structure; however, it promotes the emergence of new interaction mechanisms, the implementation of assigned tasks, and the optimization of timeframes for performing work [2].

Existing organizational management structures can be improved by utilizing a unified information platform that regulates the use and exchange of information for professional tasks. Moreover, the use of such digital tools belongs to new production technologies, forming communication processing in the hierarchical management structure of energy enterprises and systems.

Considering the technological priorities of generating companies, the main activities are aimed at the development of process automation technologies, remote control of objects, improving methodologies, tools, and business processes for operation, and developing a risk-oriented approach to managing power facilities, taking into account their actual technical condition [3, 4, 5, 6]

* Corresponding author:pavel-menshikov@vfmeil.ru

© The Authors, published by EDP Sciences. This is an open access article distributed under the terms of the Creative Commons Attribution License 4.0 (https://creativecommons.org/licenses/by/4.0/).

1.2 Methods of Data Analysis and Management for Control Systems, Monitoring, and Diagnostics to Ensure Reliability and Safety

Currently, automated control systems for technological processes (ACS TP) are widely implemented in energy sector enterprises. One of the most critical functions of ACS TP is collecting parameters and information about the current process state. Since abnormal modes are typically preceded by deviations in equipment operation over some time, i.e., a trend toward exceeding acceptable values, such phenomena can be identified at early stages.

Since ACS TP data are collected at intervals and time-bound, they represent time series that can be effectively controlled by statistical analysis methods [7]. Statistical methods enable processing the current technological process parameters to identify abnormal equipment operation modes at early stages. This approach prevents the occurrence of abnormal modes, increasing production stability and reducing equipment downtime. Therefore, using statistical methods in energy equipment diagnostics systems is advisable.

A detailed analysis of statistical methods is provided in [8]. Thus, considering the technological processes at thermal power plants (TPP), it can be concluded that statistical control and forecasting methods are appropriate for determining the current state of technological processes based on time series from continuously collected analog signals. The most suitable methods for TPPs are control charts and autocorrelation models, including ARMA models.

Statistical data processing is an important step before building a forecasting model, as it allows for more effective identification of regular and random components in time series to determine the model type - ARMA, ARIMA, or SARIMA.

Initial data analysis includes checking for outliers, missing values, anomalies, and other artifacts that may affect the quality of modeling. After this, the time series is evaluated for stationarity, i.e., whether it has constant mean and variance, which is a necessary condition for applying the ARIMA model.

Additionally, statistical data processing allows assessing the degree of autocorrelation and selecting the optimal values of ARIMA model parameters, such as the order of autoregression (p), the order of differencing (d), and the order of moving average (q), which significantly impact forecast accuracy.

Consider the steps of statistical research and initial processing of a time series using the example of the babbitt temperature parameter in the thrust bearing of a turbo generator (Babbitt T UPODSH-1). Fig. 1 shows a segment at the end of the time range where signs of non-stationarity appear.

The final segment is pre-emergency, so statistical processing of the time series was carried out during the normal operating mode of the steam turbine when the babbitt temperature did not exceed the permissible limits. Proof of the abnormal mode at the end of the time series is the autocorrelation function (ACF) and partial autocorrelation function (PACF) graphs.

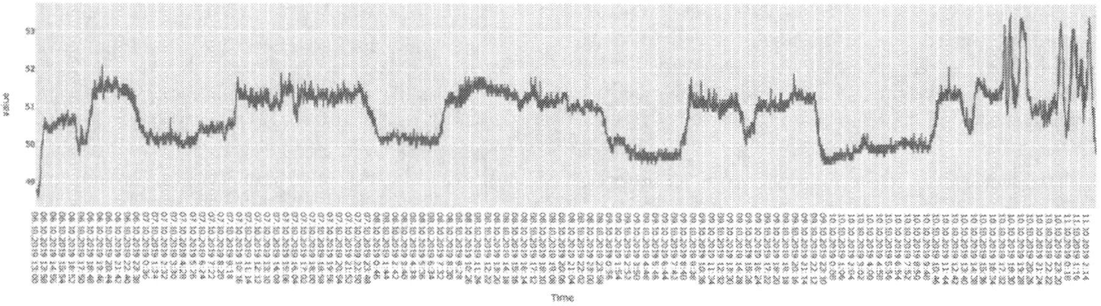

Fig. 1. Babbitt temperature in the thrust bearing of the turbo generator (Babbitt T UPODSH-1).

To assess the accuracy of the forecast for the «Babbitt T UPODSH-1» parameter, the RMSE, MAE, and SMAPE errors were calculated. The following results were obtained: RMSE = 1.312, MAE = 1.096, SMAPE = 0.88%. The SMAPE error indicates how much the prediction deviates from the actual data. The results of model testing for other significant parameters are presented in Table 1.

Based on the test results, it can be concluded that the forecast model works with high accuracy and reliability. The model predicts both stationary and non-stationary operating modes. The results of forecasting non-stationary modes can be used to prevent abnormal equipment operating conditions.

To implement the ARIMA model at TPPs, it is necessary to develop its architecture, including software and technical tools. The architecture of the ARIMA-based forecasting system in production may include the following components (Fig. 2):

- Data collection module from equipment;
- Data preprocessing module to standardize the data format and remove outliers;
- ARIMA forecasting module that uses data from the preprocessing module [9];
- Forecast results visualization module, allowing for real-time monitoring of technological parameter changes and decision-making based on forecasts.

© The Authors, published by EDP Sciences. This is an open access article distributed under the terms of the Creative Commons Attribution License 4.0 (https://creativecommons.org/licenses/by/4.0/).

https://doi.org/10.1051/e3sconf/202458401033

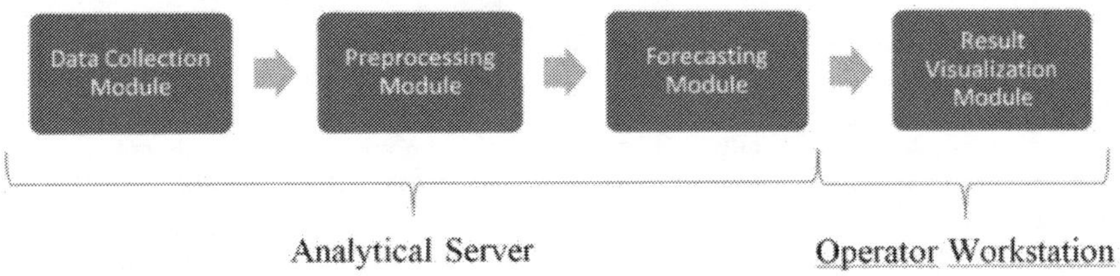

Fig. 2. ARIMA model-based forecasting system architecture.

2 Automated System for Monitoring and Accounting for Technical Condition of Equipment for Energy Enterprises Using Cloud Services

With the advancement of information technologies and cloud computing, methods for managing energy enterprises using cloud computing for economic efficiency and safety indicators of energy equipment are gaining interest [10]. In modern energy systems, monitoring equipment conditions is key to ensuring reliable and stable energy supply.

At present, most work on the implementation of cloud systems in the management of energy enterprises refers to foreign experience. To enhance security and trust among various participants in the data exchange and enterprise management systems, blockchain-based distributed ledger technology can be used. Blockchain technology offers significant advantages for the energy sector, including improved transparency of operations, data security and confidentiality, system efficiency and reliability, as well as seamless peer-to-peer transactions [11]. The proposed cloud-based management system for energy enterprises consists of the following elements with information exchange via an interface [12]:

Table 1. Results of ARIMA model testing on experimental data.

Parameter Designation	Forecast on Test Data		
T water before the OC TG	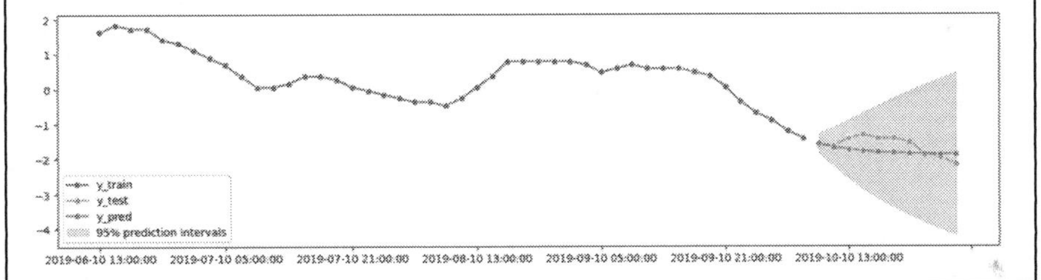		
	RMSE: 0.289	MAE: 0.239	SMAPE: 0.157.

1. A centralized data storage and processing system for the reliability and safety of energy equipment software implemented on a microservice architecture and deployed in the cloud [10, 14].
2. A decentralized ledger (blockchain) that interacts with the main system through a blockchain broker [15].
3. A private virtual cloud that connects with the local energy facility through an encrypted communication channel and interacts with the centralized system via a bastion access system.
4. A data testing and emulation system software for load and functional testing of the system [16].

For the decentralized ledger (private type), its consensus algorithm is implemented, meaning that only cryptographic hashes are published in the blockchain when triggering events occur, not the data itself.

During the testing and analysis of the developed system under the emulation of various types of market participants (generating companies, equipment manufacturers, software providers), the following shortcomings were identified:

1. The blockchain network uses foreign encryption algorithms, hash search, and digital signature (DS). This brings additional information security risks and does not comply with regulatory requirements [17], which mandate the use of Russian encryption standards for remote monitoring and diagnostics of energy equipment. To solve this problem, an analysis of the use of Russian encryption standards in the proposed blockchain network is recommended.

2. The proposed consensus algorithm [15] describes and implements general interaction functions between participants. However, when multiple system participants interact, undefined collisions may occur, making the system vulnerable. For example, there is no

© The Authors, published by EDP Sciences. This is an open access article distributed under the terms of the Creative Commons Attribution License 4.0 (https://creativecommons.org/licenses/by/4.0/).

check for a timestamp deviation from the current time, which allows sending a block with a date far beyond the present, effectively freezing the ledger. Ready-made solutions for such algorithms exist and can be integrated into the current consensus algorithm.

3. The improved consensus algorithm, considering the requirements of the centralized cloud-based system for managing the energy enterprise.

3 Conclusion

A detailed analysis of key directions for improving the organization of energy enterprise management systems in the context of modern digital technologies has been conducted.

The structure of information flows has been reviewed, and methodological principles for organizing the management system of energy enterprises have been established.

A predictive model for forecasting technological parameters in energy equipment diagnostic systems has been proposed. The predictive model is intended to forecast pre-failure equipment operating modes, enabling changes to the technological process to prevent emergencies. An analysis of the statistical characteristics of signals has been carried out, on the basis of which optimal parameters for the predictive model settings were obtained.

The architecture of a forecasting system based on the ARIMA model has been proposed for implementation at production facilities, including thermal power plants, using existing automated process control systems.

Technical solutions for managing energy enterprises using cloud services and technologies have been analyzed, taking into account domestic and foreign experiences in cloud technology implementation.

An improved methodology for managing energy enterprises using cloud servers has been proposed, featuring an enhanced consensus algorithm and the use of Russian encryption standards for the operation of the distributed ledger.

A comprehensive optimality criterion has been proposed for forming top-level management interventions, aimed at maximizing the efficiency of energy enterprise operations in terms of achieving systemic reliability and safety within the framework of digital transformation in a unified information-measurement management system.

This research was funded by Russian Federation public contract № FSWF-2020-0025 «Technique development and method analysis for ensuring power system object security and competitiveness based on the digital technologies».

References

1. F.V. Veselov, V.V. Dorofeev, Energeticheskaya Politika, Intelligent energy system of Russia as a new stage in the development of energy in the context of the digital economy, **5**, 43–52, (2018)
2. Rules for the technical operation of electric power stations and networks of the Russian Federation, August 13, 2018, No. 937 «On the approval of the rules for the technological functioning of electric power systems and on amendments to certain acts of the Government of the Russian Federation.»
3. M.S. Ivanitskii, M.M. Sultanov, V.M. Trukhanov, Proceedings of the 2nd 2020 International Youth Conference on Radio Electronics, Electrical and Power Engineering, REEPE 2020, Analysis of the influence of operating modes of heat generating plants on the energy and environmental safety of thermal power plants, p. 9059205, (2020)
4. M.M. Sultanov, V.M. Trukhanov, Y.A. Gorban, Proceedings of the 2nd 2020 International Youth Conference on Radio Electronics, Electrical and Power Engineering, REEPE 2020, High value heat-power equipment component test planning with time-series technique at single reliability level, p. 9059231, (2020)
5. E.K. Arakelyan, I.A. Boldyrev, Y.A. Gorban, Proceedings of the 2nd 2020 International Youth Conference on Radio Electronics, Electrical and Power Engineering, REEPE 2020, TPP generating unit technical and economic index accuracy increase, p. 9059234, (2020)
6. M.M. Sultanov, I.A. Boldyrev, A.A. Smirnov, Z.A. Lysakova, Software registration certificate «Unified educational portal system (digital console)», registration number: 2023664172, registered on: 03.07.2023. (2023)
7. M.M. Sultanov, I.A. Boldyrev, P.D. Menshikov, in Proceedings of the 2023 5th International Youth Conference on Radio Electronics, Electrical and Power Engineering, REEPE 2023, Moscow, Data processing model for equipment monitoring and diagnostics systems, pp. 1-4, doi:10.1109/REEPE57272.2023.10086875, (2023)
8. M.M. Sultanov, I.A. Boldyrev, M.E. Shevchenko, Reliability and Safety in Energy, Review of statistical methods for process control, **15(2)**, pp. 126–135. https://doi.org/10.24223/1999-5555-2022-15-2-126-135, (2022)
9. M.M. Sultanov, I.A. Boldyrev, M.E. Shevchenko, A.S. Kuznetsov, Software registration certificate «ARIMA-based forecasting software», registration number: 2023686473, registered on: 06.12.2023, (2023)
10. T. Kotsiopoulos et al., Computer Science Review, Machine learning and deep learning in smart manufacturing: The smart grid paradigm, **40**, p. 100341, (2021)

© The Authors, published by EDP Sciences. This is an open access article distributed under the terms of the Creative Commons Attribution License 4.0 (https://creativecommons.org/licenses/by/4.0/).

11. J. Bao et al., IEEE Systems Journal, A survey of blockchain applications in the energy sector, **15**, No. 3, pp. 3370–3381. (2020)
12. M.M. Sultanov, I.A. Boldyrev, A.A. Smirnov, P.D. Menshikov, Software registration certificate «Interface access software for energy system operation data», registration number: 2023684844, registered on: 21.11.2023, (2023)
13. M.M. Sultanov, I.A. Boldyrev, Y.A. Gorban, A.A. Smirnov, V.A. Yurov, Software registration certificate «Data collection, storage, and processing system for energy system operation data», registration number: 2022685054, registered on: 20.12.2022, (2022)
14. M.M. Sultanov et al., Proceedings of the 2021 3rd International Youth Conference on Radio Electronics, Electrical and Power Engineering, REEPE 2021, Development of a centralized system for data storage and processing on operation modes and reliability indicators of power equipment, pp. 1–5, (2021)
15. M.M. Sultanov et al., Proceedings of the International Symposium on Sustainable Energy and Power Engineering, Singapore: Springer, Joint use of distributed ledger technology and centralized software for storing power equipment reliability data, pp. 189–194, (2021)
16. Software registration certificate No. 2022685023, Russian Federation. «Software for emulating an energy object and data collection gateway», application No. 2022684337, registered: 20.12.2022, (2022)
17. Order of the Ministry of Energy of the Russian Federation No. 1015 dated November 6, 2018, «On the approval of requirements regarding the basic (mandatory) functions and information security of energy facilities in the creation and subsequent operation of remote monitoring and diagnostics systems for energy equipment in the Russian Federation», (2018)

© The Authors, published by EDP Sciences. This is an open access article distributed under the terms of the Creative Commons Attribution License 4.0 (https://creativecommons.org/licenses/by/4.0/).

Prospective power systems models development algorithms

Eugene Mashalov[1], *Vladimir* Neuymin[1], *Denis* Snegirev[1]*, *Andrej* Mihajlenko[2], *Daniil* Teplukhin[1], *Viktor* Klassen[1]

[1]JSC "STC UPS", Yekaterinburg, Sverdlovsk region, Russia
[2]JSC "SO UPS", Moscow, Russia

Abstract. This research aims to automate the creation of prospective models essential for long-term power system planning. Several algorithms and accompanying tools have been developed to transform an existing power system simulation model (SM) into a series of prospective simulation models (PSMs). This article outlines a general approach to SM conversion using a command system, enabling the binding of SM variants to specific planning periods and operating conditions (OC). Several tools to ensure PSM quality and adequacy, including forecasted power consumption balancing and security-constrained optimal power flow (SCOPF), are presented. Advanced techniques for assessing transient stability model quality and implementing hierarchical power system modeling are also integral parts of this development process.

1 Introduction

Accurate initial data and reliable simulation models (SMs) are essential for successful power system (PS) development planning. The accuracy of proposed technical solutions heavily relies on the correctness of this initial data and the SMs built upon it. Errors made during this stage can result in unreliable and inefficient PS operation. Traditionally, PS design authorities handled both the collection of initial data and the creation of SMs independently. The independent system operator (ISO) was tasked with verifying each project's prepared SM. However, a significant majority of issues arose from human error during this process of preparation and approval.

In 2022-2023, a number of regulation acts were adopted [1-5], aimed at creating a new effective planning system for long-term development in the electric power industry. A key innovation in this new system is the centralized initial data collection, verification, and PSM preparation for future planning periods. The responsibility for PSM development and ongoing updates has been granted to the Russian ISO, known as the "System Operator of the Unified Energy System".

2 The PSM generation task

The "Rules for the electric power industry long-term development documents creation and approval" [2] and "Methodological power systems development design guidelines" [4] establish key requirements for the PSM. These regulations cover long-term planning horizon, operational conditions, PSM components requirements and necessity of CIM notation for interoperability.

A software package (SP) has been developed to automate the creation of prospective simulation models (PSMs), addressing the requirements outlined above. This tool significantly reduces labor costs and minimizes the risk of errors during PSM development. The SP is capable of performing the following tasks:

1. Batch generation of 54 PSMs corresponding to the planning horizon of 6 years and 9 OCs on the basis of CIM.
2. Synchronized changes of PSMs set, selected by planning intervals and OCs.
3. Adoption of forecasted PS by regions consumption at bus level for the set of selected PSM.
4. Ensuring each PSM meets stability criteria, including convergent load flow (LF) with no constraints violations and transient stability simulation initial conditions consistency.

Tools have been developed to manage modifications to the existing SM, enabling the accomplishment of these tasks:

1. SM changes management using command system that provide binding to the planning intervals and OCs.
2. Forecasted consumption adoption at bus level, taking into account power losses.
3. Security constrained optimal power flow (SCOPF).
4. Transient-simulation model initial conditions validation.
5. Higher-level SM synthesis from lower-level SMs.

3 SM changes management using a command system

The command is a certain SM parameter change micro-algorithm. In the simplest case, the command is a replacement of the current value with the desired one. In more advanced cases, micro-algorithms may provide relative changes or parameter value calculation as a function of other SM parameters. The implemented command system is based on a standard data

* Corresponding author: snegirev@niipt-ems.ru

© The Authors, published by EDP Sciences. This is an open access article distributed under the terms of the Creative Commons Attribution License 4.0 (https://creativecommons.org/licenses/by/4.0/).

manipulation mechanism: CRUD: Create, Read, Update, and Delete.

The SM is stored in relational database form as a set of named tables linked by key fields. The fields of tables have fixed types. Delete and Update commands are trivial. The Create operation requires extra storage. For all SM tables supporting CRUD, there are extra companion command tables. These tables are identical in structure to the main SM tables but have additional fields, representing actions and their groups. Actions include the basic operations "Add", "Delete" and "Edit" and options for edit operations, such as: "Replace", "Increase" and "Multiply." Actions can be joined into the groups, allowing simultaneous application and binding to planning intervals and OCs.

The main SM table and the command table are linked by key fields. Commands are applied to records in the main SM table based on a matching of these key fields between the two tables. Key fields are complex in general. For instance, processing topological model branches requires the use of a triple key.

The relationship between the main SM table and the command table is shown on the left in Figure 1.

Main table fields can be selectively changed with actions from the command table, due to all fields of the command table having the attribute "Not set". Fields with such attributes are skipped by actions. In Figure 1 on the left, the operation of copying data fields is performed only for fields whose value differs from "Not specified" and thus does not have the "Not set" attribute.

The full set of commands describes SM changes necessary to convert it into PSM. But commands do not contain binding to planning intervals, OCs, order, and sequence of application to a particular PSM. The "Command Map" contains mapping and all extra rules between commands and PSM variants required. The "Object" field links commands to the map. All commands with matching "Object" will be applied in a single transaction. The relationship between command maps and commands is shown in Figure 1 on the right.

The "Command Map" table also links to planning intervals and up to 9 OCs. There are switches to selectively exclude transactions from desired variants.

4 Forecasted consumption distribution

The only data available on power consumption forecasts is related to regions of SM. The stability studies required to ensure PSMs quality operate on bus-branch network model, where each bus must have individual load quantities regarding its region forecast. Moreover, the forecast includes power losses, and thus, they must be adopted into the overall PSM power balance. An algorithm allowing distribution of the forecasted region consumption between region's buses and is developed. All bus loads are associated with load models that have power constraints according to CIM model. These constraints are used by the algorithm to construct bus load sensitivity to region consumption.

Let the buses be combined into some sets corresponding to regions. Let's denote the bus belonging to the set A as $P \in A$. At the same time, ΣP, ΣQ can be calculated for A. The bus load (without an index, one bus is considered) is represented by the sum of the loads:

$$P = \sum P_i. \quad (1)$$

For each load P_i, there are set of parameters: P_{iref} (base state load or expected load value), constraints P_{imin} and P_{imax}, load tangent $tg\varphi_i$, an attribute linking the load to future-planned equipment B_i and attribute which allows to set load to be constant – C_i.

For loads with $B_i > 0$, the calculation of P_i is performed:

$$P_i = \begin{cases} P_{i_{ref}} & , C_i > 0 \\ k_{Ac} P_{i_{ref}} & , C_i \le 0' \end{cases} \quad (2)$$

where k_{Ac} is the multiplier given for the buses set of A in OC of type c. Loads with $B_i > 0$ do not change in further calculations.

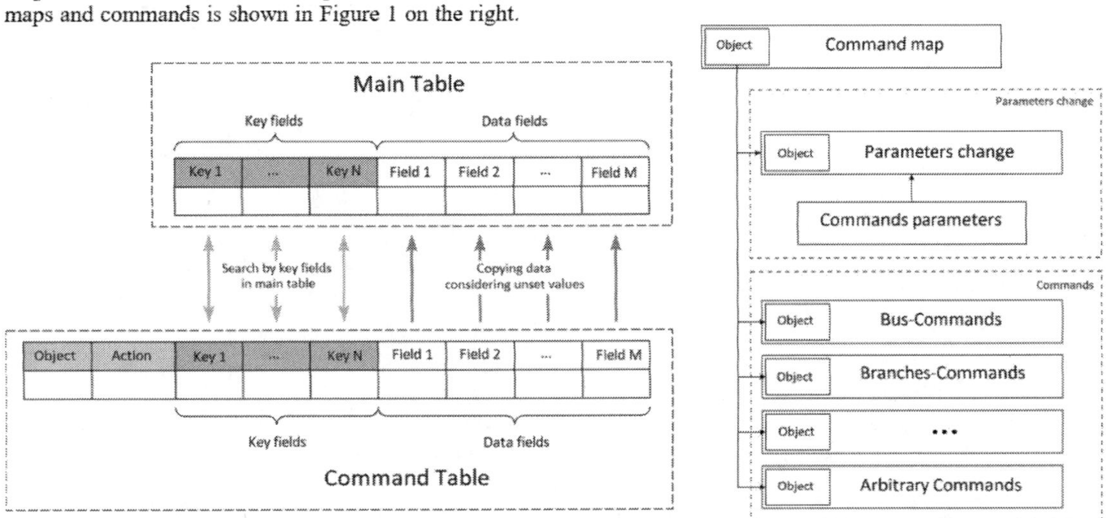

Fig. 1. Connection of main table and command table (left) and connection of "Command Map", "Parameter change" tables and tables commands (right)

© The Authors, published by EDP Sciences. This is an open access article distributed under the terms of the Creative Commons Attribution License 4.0 (https://creativecommons.org/licenses/by/4.0/).

For A, the constraints and the initial SM bus load point $\Sigma P_{i_{ref}}$ are calculated:

$$P_{A_{min}} = \sum_{P_i \in A} P_{i_{min}}, B_i \leq 0$$
$$P_{A_{max}} = \sum_{P_i \in A} P_{i_{max}}, B_i \leq 0 \quad (3)$$
$$\Sigma P_{i_{ref}} = \sum_{P_i \in A} P_{i_{ref}}, B_i \leq 0$$

As a result, for each P_i, $B_i \leq 0$, a dependence on P_A can be obtained, shown in Fig. 2.

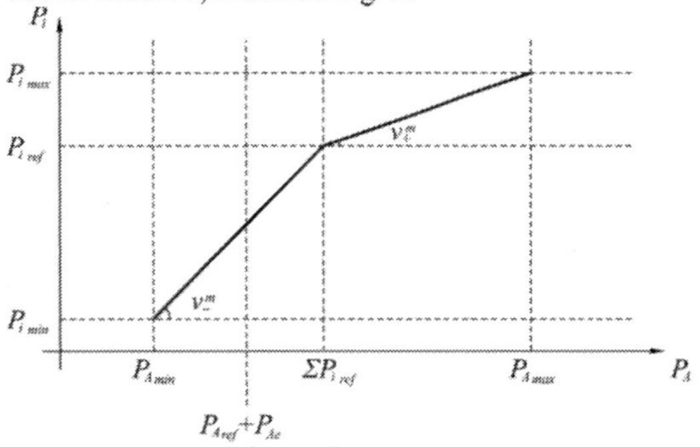

Fig. 2. Dependence of P_i on P_A

The goal is to achieve $|\Delta P_A^m| \leq \epsilon_A$, while:

$$\Delta P_A^m = P_{A_{ref}} - \Sigma P_{i_{ref}} + P_{A_e}^{m-1} \quad (4)$$

where $P_{A_{ref}}$ – given load of the set A; $P_{A_e}^{m-1}$ – deviation of the load A from $P_{A_{ref}}$, while $P_{A_e}^{m-1} = P_{A_{ref}} - P_A^{m-1}$; m is iteration number; ϵ_A – tolerance. While $P_i^0 = P_{i_{ref}}$.

Since $P_{i_{min}}$, $P_{i_{max}}$ and $P_{i_{ref}}$ are set for P_i, the change of P_i can be performed at different rates depending on the sign of (4).

As the load increases $\Delta P_A^m > 0$, P_i changes with the rate:

$$v_{i_+} = \frac{P_{i_{max}} - P_{i_{ref}}}{P_{A_{max}} - \Sigma P_{i_{ref}}} \quad (5)$$
$$P_i^{m+1} = P_{i_{ref}} + v_{i_+} \Delta P_A^m$$

When the load decreases $\Delta P_A^m < 0$, P_i changes with the rate:

$$v_{i\_} = \frac{P_{i_{ref}} - P_{i_{min}}}{\Sigma P_{i_{ref}} - P_{A_{min}}} \quad (6)$$
$$P_i^{m+1} = P_{i_{ref}} + v_{v_{i\_}} \Delta P_A^m$$

The rates of load change v_{i_+} and $v_{i\_}$ do not depend on the variables involved in the iterative process, and can be calculated for each load once.

There is an algorithm for iterative bus loads refinement:

1. $m = 0$
2. Take $P_i^m = P_{i_{ref}}$.
3. Calculate P_A^m.
4. Calculate ΔP_A^m, if $|\Delta P_A^m| \leq \epsilon_A$ – cycle is completed.
5. Increase m by 1. If $m > m_{max}$ – cycle is complete with an error or iteration count exceeding.
6. Calculate P_i^{m+1} and repeat from step 3.

As the cycle ends, bus loads are calculated as (1) for $B_i \leq 0$. Next, the LF calculation or SCOPF is performed to refine power losses and update consumption based on the C_A to consumption of A. In order to achieve this, $P_{A_{ref}}$ is calculated.

$$P_{A_{ref}} = C_A - L_A - LRC_A$$
$$LRC_A = \sum_{n \in A} P_n(V_n) - P_n \quad (7)$$

where L_A is the calculated losses of A; LRC_A is the consumption component, depending on the load response curves (LRC) associated with buses; $P_n(V_n)$ is bus load taking into account the LRC components.

Consumption balancing is performed in the outer loop of iterative process in m until a specified accuracy is achieved relative to C_A or a specified number of iterations. After consumption balancing, loads with $B_i > 0$ are added to the buses. The balance of A will be changed relative to targeted C_A, but this effect is considered appropriate for PSM.

5 Security constrained optimal power flow (SCOPF)

The following points must meet requirements for each of the PSMs, for all OPs in each planning interval:

- the voltage levels on the power facilities buses should be constrained;
- the current load of power lines and equipment should not be higher than the long-term permissible load current (LTPCL) in operational case and the active power flows through the system interfaces should not exceed the maximum load limit (MLL) in operational case.

To meet these requirements a security constrained optimal power flow (SCOPF) solution [6] is used. This algorithm provides is based on interior point method (IPM) optimization. The objective function is:

$$minF(x) = \sum_i f_i(x_i) =$$
$$= \sum_i a_i\left(x_i - x_i^{ref}\right)^2 + \sum_j b_j\left|x_j - x_j^{ref}\right|, \quad (8)$$

with constraints, representing operational limits of equipment in the form of inequalities:

$$x_i^{min} \leq x_i \leq x_i^{max}, \quad (9)$$

as well as constraints in the form of equalities, basically – power balances at buses:

$$\begin{cases} -\sum_j V_i V_j y_{ij} \sin(\delta_{ij} - \alpha_{ij}) - P_i = 0, \\ \sum_j V_i V_j y_{ij} \cos(\delta_{ij} - \alpha_{ij}) - Q_i = 0. \end{cases} \quad (10)$$

where a_i and b_j are the weighting coefficients determined empirically.

The objective function consists of two components: the first least square of deviations, the second – absolute values of deviations.

The least squares of deviation are used to smoothly change the value of:

© The Authors, published by EDP Sciences. This is an open access article distributed under the terms of the Creative Commons Attribution License 4.0 (https://creativecommons.org/licenses/by/4.0/).

- online generators power in order to maintain a power balance;
- the voltage references at generator buses within the constraints specified;
- the transformation ratios within the constraints specified (ratios are rounded to the closest tap).

Minimizing the deviation absolute values is used to account for the objective function:

- switching on the ready to run generators;
- switching on and off shunt compensators and other reactive power injectors.

6 Formation and initialization validation of the simulation model for the EMT

The current CIM representation of the SM is incomplete for transient simulation. To address this issue, it is necessary to conduct massive data collection, verification, and engineering to supply these data in CIM notation. The temporary solution is to achieve data for transient simulation from the native format of ISO software. To ensure data is transferred and converted appropriately, a short transient simulation with no disturbances for each PSM of 54 is performed. This way, the initial condition of the transient simulation can be verified on consistency. Due to rounding of data values, transient simulations can show low amplitude oscillations even for well-behaved cases. The goal is to ensure these oscillation decays, which indicates initial conditions acceptable.

The method of decay estimation is developed and applied in the Raiden SP. This method is now implemented into the RUSTab SP [7] to check transient decays.

The method is based on a sequential search for the state variables extremums. The search is performed at each integration step. After detecting the min to max transition, the max to min transition is expected. The resulting extremums are checked each component of state vector. When the next extremum is located, it is compared with the previous one of the same type. If there are conditions for decreasing the maximum or increasing the minimum, the counter of successive extremes of the component increases. If there are not such conditions, the counter is reset. Transient decay is reported when the counters of all components reach the set value D_{dc} (Decay detector cycles).

Fig. 3 shows the transient simulation result curve for the rotor angle δ of a synchronous machine with AVR and turbine governor. The arrows mark the extremums. The minimum increase is noted for all minima (each subsequent minimum exceeds the previous one). The decreasing maximum also noted for extremums with numbers 3, ..., 15. Therefore, the counter of the oscillation decrement of the δ component begins with an extremum of 3. For $D_{dc} = 3$, the decay of δ will be detected at the extremum with number 8 (three decreasing maxima and three increasing minima).

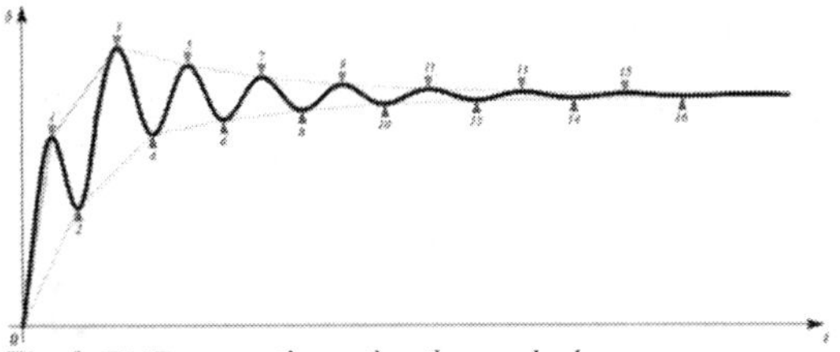

Fig. 3. EMT attenuation estimation method

This algorithm works also in the case of monotonous changes of state variables and is not prone to false-positive decays.

7 Hierarchical PSM synthesis

Due to the hierarchical structure of authorities the hierarchical approach in power system modeling is also adopted. An arbitrary number of lower-level PSM must be combined into one large, higher-level PSM. This process is called "synthesis". This process includes the removal of equivalent parts of lower-level PSMs. The parts to be removed identified as power system regions, neighboring with another lower-level PSMs.

Before synthesis, a list of lower-level PSMs is created. All buses from one of the matched neighboring regions are to be removed, and links between neighboring PSM is established by branches.

SM synthesis is carried out according to the following rules:

1. A list of branches linking PSMs is established. For each branch two buses in neighboring PSM are identified.
2. According to the branch attribute in the list, each bus is either included into higher-level PSM, or removed. The branch itself is also incorporated into the higher-level PSM with parameters from selected lower-level PSM.
3. All buses not in neighboring regions are included into the higher-level PSM as is.

8 Practical experience

Since January 2023, the practical application of the developed technology for generating and monthly updating PSMs has commenced.

In 2023, the System Operator of the United Power System received over 700 requests from engineering providers seeking PSMs for long-term development of the electric power industry. Each PSM provided relevant information for the month of the request concerning the existing and prospective state of the United Power System. To fulfil these requirements, over 72 000 PSMs were created in 2023 including lower-level PSM of local power systems.

The developed algorithms automate the processes of creating and updating PSMs, enabling the provision of engineering providers with PSMs that adhere to strict technical requirements concerning parameters of the power system, grid, and equipment within the timeframes stipulated by a Decree of the Government of the Russian Federation.

© The Authors, published by EDP Sciences. This is an open access article distributed under the terms of the Creative Commons Attribution License 4.0 (https://creativecommons.org/licenses/by/4.0/).

9 Conclusion

This article outlines approaches for constructing prospective power system simulation models and describes algorithms implemented within a software prototype capable of generating and updating these models. This prototype allows users to specify a desired planning horizon and the number of operational cases, facilitating PSM creation.

A developed command system enabling the generation of sets of changes for existing simulation models minimizes the risk of errors stemming from manually entered data. For the resulting simulation models, the application of a Security Constrained Optimal Power Flow ensures compliance with constrained voltage levels and current loads in existing and projected equipment. Additionally, it guarantees that operation conditions variables, including permissible active power flows in the interfaces remain within constraints.

The prospective simulation models generated using this developed software package and provided by the System Operator offer a versatile toolset for addressing diverse tasks: including long-term development planning of the electric power industry, investment planning, assessment of interconnection possibilities to electrical networks, and scientific research approaches.

References

1. Federal Law No. 174-FL dated June 11, 2022 "On Amendments to the Federal Law "On Electric Power Industry".
2. Decree of the Government of the Russian Federation No. 2556 dated December 30, 2022 "On approval of the Rules for the development and approval of documents for the progressive development of the electric power industry".
3. Decree of the Government of the Russian Federation dated December 30, 2022 No. 2557 "On Approval of the Rules for the Formation and Maintenance of information models of electric power systems and advanced simulation models of electric power systems for the purposes of long-term development of the electric grid".
4. Order of the Ministry of Energy of the Russian Federation dated December 6, 2022 No. 1286 "On approval of the methodological guidelines for the design of the development of power systems".
5. Order No. 82 of the Ministry of Energy of the Russian Federation dated February 17, 2023 "On Approval of the Procedure for Disclosure of Digital Information Models of Electric power Systems".
6. Mikhailenko A. F., Neuimin V. G., Satsuk E. I., Sacuk E. I. The use of technology for Security constrained optimal power flow in the formation of promising simulation models. Izvestiya STC of Unified Power System. No. 2(89). pp. 39-45. 2023
7. E. Mashalov, "Raiden. Power System Transient Stability Simulation Software Prototype," 2022 International Ural Conference on Electrical Power Engineering (UralCon), Magnitogorsk, Russian Federation, 2022, pp. 145-153, doi: 10.1109/UralCon54942.2022.9906601.

© The Authors, published by EDP Sciences. This is an open access article distributed under the terms of the Creative Commons Attribution License 4.0 (https://creativecommons.org/licenses/by/4.0/).

Development of a vector control system based on the DFSMC for the study of transients in the conditions of creating a digital twin of the power system

Igor Suprunov[1]*, *Nikolay* Novikov[2], *Alexander* Novikov[3]

[1] JSC "STC UPS", Moscow, Russia
[2] Rosseti R&D Center (R&D Center FGC UES, JSC), Moscow, Russia
[3] STC Energy Security, Moscow, Russia

Abstract. The article considers an event to increase throughput using smart grid technologies (FACTS) based on a combined vector control device for mode parameters (power flow) based on an asynchronous electromechanical frequency converter (DFSMC). A mathematical model of an electric power system with the proposed combined device has been developed, taking into account both electromechanical and electromagnetic transients. The results of the calculation of transients for the test circuit are presented, taking into account the implementation of the proposed mathematical model on the developed software and computing complex.

Introduction

Currently, there is an active development of renewable energy sources (RES) in the South UPS and Middle Volga UPS. In the future, until 2028, it is expected to introduce a large volume of renewable energy sources, which will lead to an increase in flows in controlled sections above the maximum power flow (MDP) and the need in the zone of influence of renewable energy to limit the capacity of power plants, turn off the load of consumers, and divide the electric network, which reduces the reliability of power supply.

In conditions of increasing load and immutability of local generation at a certain moment, the existing network will not allow the required power to be passed due to restrictions. In conditions of limited network capacity, its development is permissible through the use of short-term (efficiency is projected for a period of 5-10 years) measures. At the same time, taking into account the requirements for the power distribution scheme of renewable energy facilities, it is reasonable to resort to measures to increase capacity using smart grid technologies (FACTS).

The use of traditional devices based on power semiconductors is hampered by their high price, the complexity of manufacturing and servicing semiconductor devices for high currents and high voltages. To realize high-speed performance and create high-power devices, it is possible to combine low- and medium-power semiconductor devices and special alternating current machines with a phase rotor [1, 2]. Such devices make it possible to implement an asynchronous electromechanical frequency converter (DFSMC), which provides the possibility of vector control in power systems.

The construction of flexible inter-system communication based on the DFSMC requires significant costs. To increase the capacity of the control device, it is proposed to shunt the DFSMC with a reactive element [1 – 3], as which a transformer of a step-down substation can be used in electrical networks of various voltage classes (Fig. 1). This will reduce the cost of installing the device by using existing equipment on the network.

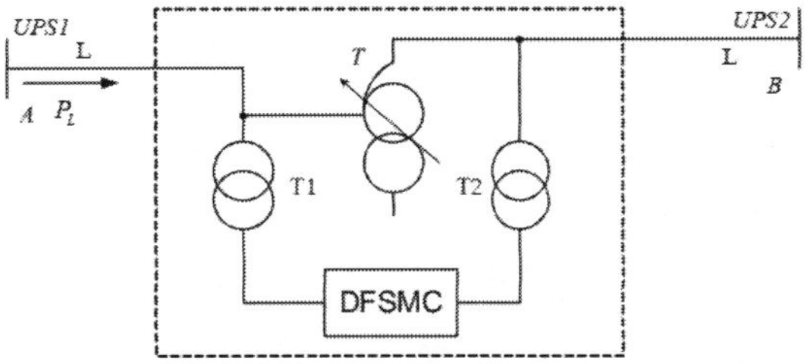

Fig. 1. Combined vector power flow control device with an EMF speaker and a transformer–power flow control device (UUPM)

A number of articles are devoted to the study of the behavior of the proposed combined vector control device for mode parameters in an electric network, as well as its effect on various transients in a complex multi-machine power system [3, 4]. However, in these articles, the accounting of synchronous (SM) and asynchronous synchronous machines (ASM) is carried out by simplified equations, and the elements of the electrical network are modeled by constant resistances.

*Corresponding author: suprunov-is@ntcees.ru

© The Authors, published by EDP Sciences. This is an open access article distributed under the terms of the Creative Commons Attribution License 4.0 (https://creativecommons.org/licenses/by/4.0/).

Currently, for the purposes of effective management, planning and development of the electric power system, it has become necessary to create its unified mathematical model with mutual synchronization with the real power system in real time – a digital twin. Such mutual synchronization implies the use of data, including vector measurements, to represent the parameters of quasi-steady-state and transient processes in the power system. A necessary condition for the use of digital twins is the development of mathematical models that correspond to the real behavior of a physical object in a wide range of observations, within acceptable fluctuations. For the possibility of creating a digital twin of an electric power system with the proposed combined vector control device for mode parameters, a mathematical model has been developed that takes into account both electromechanical and electromagnetic transients.

Development of a mathematical model of the EES for the calculation of transients

When creating a mathematical model, a complex EES is generally considered, which consists of n nodes and m branches (power lines and transformers (autotransformers), reactors), with synchronous machines and asynchronous synchronous machines connected to n_g nodes, and loads connected to n_l nodes. Synchronous machines are the equivalents of generators of power plants or the equivalents of power systems. Asynchronous synchronous machines can be the equivalent of station generators, or they can be part of an asynchronous electromechanical frequency converter (DFSMC). In the latter case, for a pair of ASMs, one general equation of rotor motion is written with the total inertia constant of the two rotors T_J.

A complete mathematical model of the EPS for calculating transients means such a mathematical model that meets the following requirements:

– both electromechanical and electromagnetic transients are taken into account;

– the frequency change in the transition process is taken into account;

– the effect of automatic control systems of the power elements of the EPS is taken into account for the entire duration of the transition process.

As a complete mathematical model of a synchronous machine, we will use the Park-Gorev equations without saturation, one excitation winding, one damper winding along the d axis, one damper winding along the q axis (3-contour Park-Gorev model) [5, 6], taking into account both electromechanical and electromagnetic transients.

A 4-contour Park-Gorev model is used to model AFM. Two circuits are defined along the longitudinal and transverse axes – the excitation winding and the damper circuit. A synchronous machine model is used with the addition of the excitation winding equation along the d axis [6 – 8]:

$$\frac{d\Psi_d}{dt} = -U_d - \omega \cdot \Psi_d - r_a \cdot I_d \tag{1}$$

$$\frac{d\Psi_q}{dt} = -U_q + \omega \cdot \Psi_q - r_a \cdot I_q \tag{2}$$

$$\frac{d\Psi_{fq}}{dt} = E_{qe} \cdot \frac{r_f}{x_{ad}} - r_f \cdot I_{fq} \tag{3}$$

$$\frac{d\Psi_{fd}}{dt} = -E_{de} \cdot \frac{r_f}{x_{aq}} - r_f \cdot I_{fd} \tag{4}$$

$$\frac{d\Psi_D}{dt} = -r_D \cdot I_D \tag{5}$$

$$\frac{d\Psi_Q}{dt} = -r_Q \cdot I_Q \tag{6}$$

$$\Psi_d = \frac{1}{\Omega_{nom}} \cdot \left(x_d \cdot I_d + x_{ad} \cdot \left(I_{fq} + I_D\right)\right) \tag{7}$$

$$\Psi_q = \frac{1}{\Omega_{nom}} \cdot \left(x_q \cdot I_q + x_{aq} \cdot \left(I_{fd} + I_Q\right)\right) \tag{8}$$

$$\Psi_{fq} = \frac{1}{\Omega_{nom}} \cdot \left(x_f \cdot I_{fq} + x_{ad} \cdot \left(I_d + I_D\right)\right) \tag{9}$$

$$\Psi_{fd} = \frac{1}{\Omega_{nom}} \cdot \left(x_f \cdot I_{fd} + x_{aq} \cdot \left(I_q + I_Q\right)\right) \tag{10}$$

$$\Psi_D = \frac{1}{\Omega_{nom}} \cdot \left(x_D \cdot I_D + x_{ad} \cdot \left(I_d + I_{fq}\right)\right) \tag{11}$$

$$\Psi_Q = \frac{1}{\Omega_{nom}} \cdot \left(x_Q \cdot I_Q + x_{aq} \cdot \left(I_q + I_{fd}\right)\right) \tag{12}$$

where Ψ_d, Ψ_q – are the projections of the flow coupling of the stator winding on the axis *(d, q)*;

U_d, U_q – projections of the voltage of the stator winding on the axis *(d, q)*;

ω – is the frequency of rotation of the rotor;

r_a – a is the active resistance of the stator windings;

I_d, I_q – projections of the stator current on the axis *(d, q)*;

Ψ_{fq}, Ψ_{fd} – projections of the flux coupling of the excitation winding on the axis *(d, q)*;

E_{qe}, E_{de} – forced EMF along the longitudinal and transverse axes. In the initial steady–state mode (before the disturbance) $E_{de} = 0$;

$r_f = r_{fd} = r_{fq}$ – active resistance of the excitation winding;

x_{ad}, x_{aq} – inductive resistances of the longitudinal and transverse reaction of the stator;

x_d, x_q – synchronous inductive resistances of a synchronous machine along the longitudinal and transverse axes;

I_{fd}, I_{fq} – the current of the excitation winding along the longitudinal and transverse axes;

Ψ_D, Ψ_Q – projections of the flow coupling of the damper contours on the axis *(d, q)*;

I_D, I_Q – currents in the damper circuits in the longitudinal and transverse axes;

© The Authors, published by EDP Sciences. This is an open access article distributed under the terms of the Creative Commons Attribution License 4.0 (https://creativecommons.org/licenses/by/4.0/).

r_D, r_Q, x_D, x_Q – active and inductive resistances of the damper circuits along the longitudinal and transverse axes;

Ω_{nom} – is the rated speed of the rotor.

According to the design, a classical asynchronous synchronous machine – ASM with two identical excitation windings located at an angle of 90° was considered [4, 7].

When considering ASM, several reference frames are usually introduced (Fig. 2) [7]. Two coordinate systems are connected to the rotor – the physical *(α, β)*, coinciding with the excitation windings A and B, and the logical *(d, q)*. During the calculation process, the angle between them remains constant. The equations of the machine are written in the system *(d, q)*, the system *(α, β)* is necessary to check the excitation limitations and form the necessary ratio of excitation currents. In the initial steady-state mode, it is assumed that the rotation frequency of the vector $\dot{E}_e$ relative to the rotor is also zero, that is, $\omega_f = 0$.

Two coordinate systems are also connected to the stator – (d_v, q_v) with respect to the support node (the axes of the rotor of the support machine) and the coordinate system *(x, y)* associated with the reference voltage, in the coordinates of which the control actions of the control system are formed. The voltage at the terminals of the ASM $-U_\text{г}$ itself is taken as the reference voltage in the calculations.

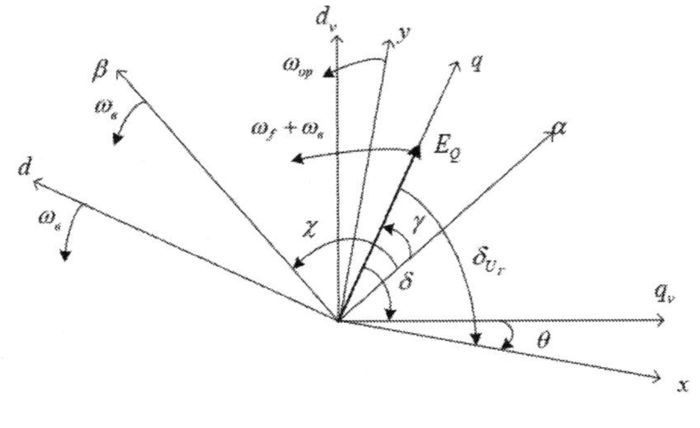

Fig. 2. – Vector $\dot{E}_Q = \dot{E}_e$ and the calculated coordinate systems in steady state (before the start of the disturbance)

The following designations and assumptions are accepted in the work:

– χ is the physical angle between the excitation windings A and B. An ASM with two identical excitation windings located at an angle of 90° is adopted, i.e. $\chi = 90°$;

– γ is the angle between the axes of the coordinate systems (α, β) and *(d, q)*. For ASM with two identical excitation windings located at an angle of 90°, $\gamma = 45°$;

– δ is the angle between the AFM rotor *(d, q)* and the axes of the support machine (d_v, q_v);

– θ is the angle between the reference voltage *(x, y)* and the axes of the reference machine (d_v, q_v);

– $\delta_{U_\text{г}} = \delta - \theta$ is the angle between the ASM rotor *(d, q)* and the reference voltage *(x, y)*;

– ϑ is the angle between the EMF excitation vector $\dot{E}_e$ and the axes of the support machine (d_v, q_v). In steady mode, it coincides with the angle δ.

The control signals of the regulators are formed in the axes of the reference voltage *(x, y)*. A voltage regulator signal is generated along the *x* axis (supports the set voltage on the ASM buses), and an electromagnetic torque regulator is formed along the *y* axis (sets the desired rotation speed of the vector $\dot{E}_e$ and the electromagnetic torque of the ASM).

The excitation is regulated by the deviation of the voltage at the terminals of the generator, the deviation of the sliding shaft of the DFSMC from the set (desired) value, as well as by the deviation of the control power P_y from the set value. The block diagram of the ARE of an asynchronous synchronous machine as part of a combined active power flow control device is shown in Figure 3 [2, 4].

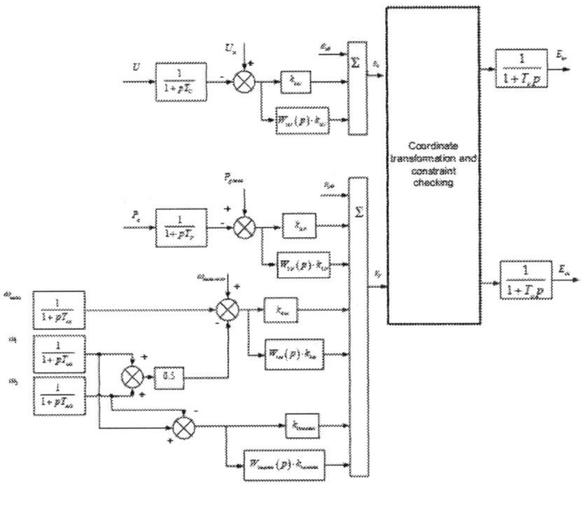

Fig. 3. – A block diagram of the ARE of an asynchronous synchronous machine as part of a combined control device for the flow of active power in a general form.

In the proposed control law implemented as part of the ARV (Fig. 3), during the control process, a machine operating in engine mode provides regulation of the active power of the device $P_\text{г} = P_{\text{уст}}$, frequency control at the terminals of the machines is assigned to the generator machine.

The output signals of the regulators E_x, E_y are converted into a rectangular coordinate system of the rotor *(d, q)*:

$$E_d = E_y \cdot \cos\delta_{U_\text{г}} - E_x \cdot \sin\delta_{U_\text{г}} \tag{13}$$

$$E_q = E_x \cdot \cos\delta_{U_\text{г}} + E_y \cdot \sin\delta_{U_\text{г}} \tag{14}$$

To check the limits on the excitation voltage, a transition is made to the coordinate system associated with the rotor windings (α, β):

$$E_\alpha = \left(E_d \cdot \cos(\chi - \gamma) + E_q \cdot \sin(\chi - \gamma)\right) / \sin\chi = \frac{\sqrt{2}}{2} \cdot \left(E_d + E_q\right) \tag{15}$$

© The Authors, published by EDP Sciences. This is an open access article distributed under the terms of the Creative Commons Attribution License 4.0 (https://creativecommons.org/licenses/by/4.0/).

$$E_\beta = \left(E_q \cdot \sin\gamma - E_d \cdot \cos\gamma\right)/\sin\chi = \frac{\sqrt{2}}{2} \cdot \left(E_q - E_d\right) \quad (16)$$

Substituting expressions (13), (14) in (15), (16), we obtain:

$$E_\alpha = \frac{\sqrt{2}}{2} \cdot \left(E_y \cdot \left(\cos\delta_{U_\Gamma} + \sin\delta_{U_\Gamma}\right) + E_x \cdot \left(\cos\delta_{U_\Gamma} - \sin\delta_{U_\Gamma}\right)\right) \quad (17)$$

$$E_\beta = \frac{\sqrt{2}}{2} \cdot \left(E_x \cdot \left(\cos\delta_{U_\Gamma} + \sin\delta_{U_\Gamma}\right) - E_y \cdot \left(\cos\delta_{U_\Gamma} - \sin\delta_{U_\Gamma}\right)\right) \quad (18)$$

The obtained values are limited to the values of the ceiling excitation, and the values E_α and E_β bordered by the ceiling of the excitation are further used in the differential equations of exciters:

$$\frac{\mathrm{d}E_{\alpha e}}{\mathrm{dt}} = \frac{1}{T_A} \cdot \left(-E_{\alpha e} + E_\alpha\right) \quad (19)$$

$$\frac{\mathrm{d}E_{\beta e}}{\mathrm{dt}} = \frac{1}{T_B} \cdot \left(-E_{\beta e} + E_\beta\right) \quad (20)$$

where T_A, T_B – are the time constants of the winding exciters A and B, respectively.

We switch back to the rectangular coordinate system of the rotor *(d, q)* to use the excitation voltage in the generator model:

$$E_{qe} = \left(E_{\alpha e} \cdot \cos(\chi - \gamma) + E_{\beta e} \cdot \cos\gamma\right) = \frac{\sqrt{2}}{2} \cdot \left(E_{\alpha e} + E_{\beta e}\right) \quad (21)$$

$$E_{de} = \left(E_{\alpha e} \cdot \sin\gamma - E_{\beta e} \cdot \sin(\chi - \gamma)\right) = \frac{\sqrt{2}}{2} \cdot \left(E_{\alpha e} - E_{\beta e}\right) \quad (22)$$

Regulators (ARV) are installed on synchronous machines, for which a detailed model is adopted, given in [5, 6]. Excitation is regulated by the deviation and the first derivative of the voltage at the terminals of the generator, as well as by the deviation and the first derivative of the frequency of this voltage.

Synchronous and asynchronous generators are driven by primary motors with an installed speed controller, for which a simplified model (23) is adopted.

$$\frac{dM_\text{т}}{dt} = \frac{\left(M_{\text{т}0} - M_\text{т} - M_\text{т} \cdot s_c\right)}{T_\text{т}}, \left(M_\text{т} = P_\text{т}\right) \quad (23)$$

The mathematical model of the EES takes into account electromagnetic transients in the elements of the electrical network. To account for them, power lines, transformers and reactors are represented using differential equations in which the voltages on inductive elements and currents in capacitive elements are represented as [6]:

$$U_L = L \cdot \frac{di_L}{dt}, i_C = C \cdot \frac{dU_C}{dt} \quad (24)$$

A mathematical model is compiled for a multi-machine EES scheme according to the specified methodology. To calculate transients, a computer program has been developed that implements the trapezoid method for the algebraization of differential equations at the integration step and the Newton method for solving a system of nonlinear algebraic equations and allows for the consideration of asynchronous machines, their excitation regulator and the device DFSMC a whole. When calculating all the resistances of the elements of the EES replacement circuit, which are at different stages of transformation, are reduced to the basic voltage stage.

Verification of the calculation results of electromechanical transients in the EES on the developed author's program and RastrWin

The circuit shown in Figure 4 was considered as a test.

A three-phase short circuit at a time of 0.1 seconds was considered. duration of 0.1 seconds. the beginning of the 220 kV line between Substation 8 and node 16 near node 8 and its subsequent localization without disconnecting the 220 kV line 8-16.

When modeling in RastrWin, the generators are set by the "3-contour Park-Gorev model" model, corresponding to the one considered in this paper.

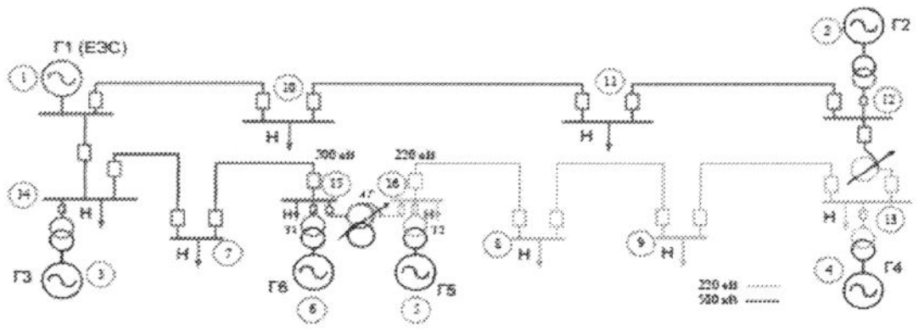

Fig. 4. – Diagram of the network under study

Figure 5 shows a comparison of graphs of changes in the angle of the rotor of the generator 5, which is most electrically close to the site of damage, in RastrWin and the author's program. It can be seen from the above graph that the maximum run-out of the rotor angle at a time of 0.25 seconds differs by 5 degrees, which is comparable to the error in setting the initial data during modeling.

Also, for the scheme under consideration, the limiting shutdown times of the three-phase short circuit at the beginning of the 220 kV line between Substation 8 and node 16 near node 8 for the author's program (0.771) and RastrWin (0.762 sec.) were determined. The difference in the time limits was 0.009 seconds. or 1.2%, which is also comparable to the error of specifying the initial data during modeling.

© The Authors, published by EDP Sciences. This is an open access article distributed under the terms of the Creative Commons Attribution License 4.0 (https://creativecommons.org/licenses/by/4.0/).

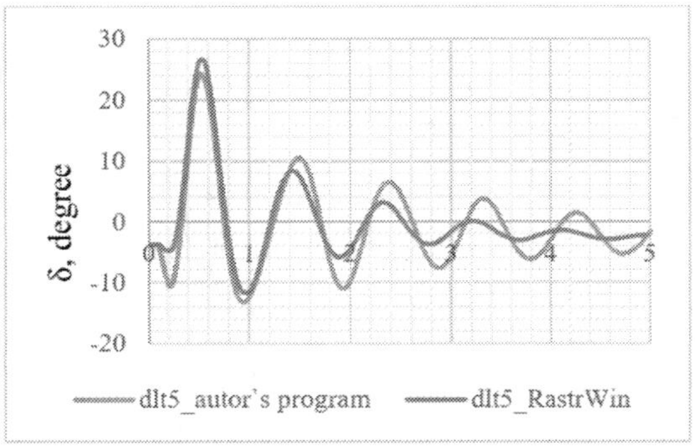

Fig. 5. – The angle of the rotor of the generator G5 when calculated in RastrWin and author's program, degrees

Analysis of the calculation results of electromechanical transients in the EES, taking into account the installation of a combined vector control device

The scheme shown in Figure 6 was considered as the initial one. The DFSMC is connected in nodes 16 and 15, an asynchronous synchronous generator and an asynchronous synchronous motor are installed, respectively.

If a short circuit occurs near points 15 or 16 on the lines leaving the power system, they are disconnected and the power supply of the local load of the DFSMC is carried out from the power system through an autotransformer and mechanically by means of a common converter shaft. Figures 7-8 show the results of calculating transients that occur during a three-phase short circuit at a time of 0.1 s with a duration of 0.1 s at the beginning of the 220 kV line between Substation 8 and node 16 near node 8 and its subsequent localization without disconnecting the 220 kV line 8-16.

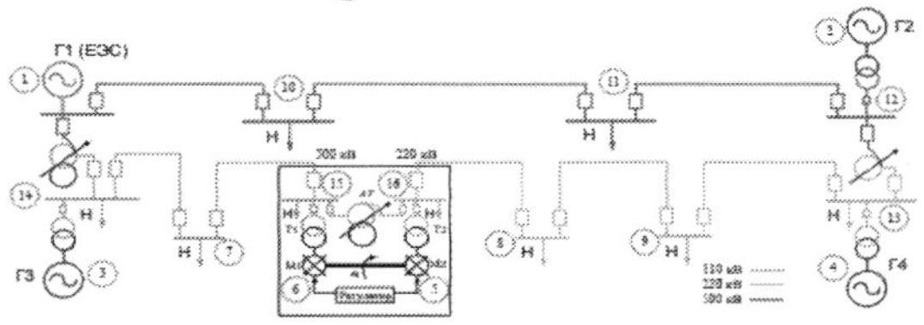

Fig. 6. – Diagram of the network under study

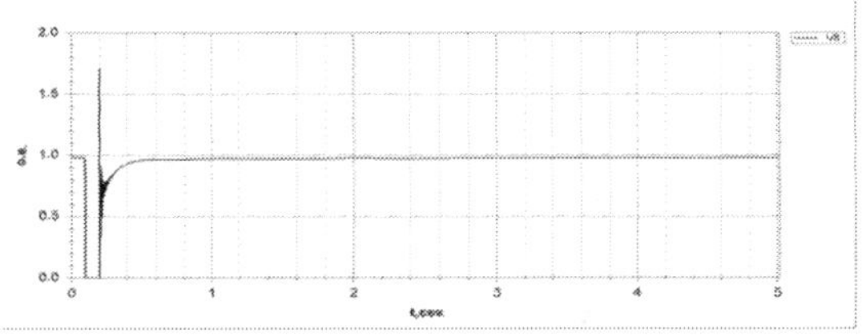

Fig. 7. – Voltage at node 8

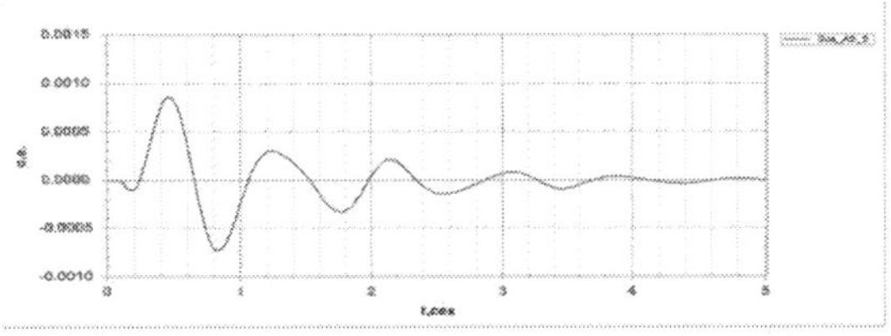

Fig. 8. – Sliding of the shaft of the DFSMC

The elimination of the short circuit does not lead to a change in the direction of the power of the DFSMC, which shows a positive sliding of the shaft (Fig. 8). There is no stability violation, the rotation of the EMF vector of the machines relative to the voltage vector at their terminals is not observed.

It should be noted that the tuning parameters of the proposed control laws are determined expertly for the network circuits considered in the work and require adjustments in the practical implementation of the device in the process of its specific design for the existing network based on the characteristic modes of its operation to ensure the required mode parameters. The choice of optimal tuning coefficients in the laws of control of DFSMC machines has not been considered in this work.

Conclusion

1. A device for vector control of mode parameters based on the DFSMC is proposed, and an algorithm for controlling the proposed device is developed to ensure its dynamic stability and control the flow of power and a mathematical model of the system for vector control of the parameters of the AC transmission mode based on the DFSMC as part of a combined device implementing the proposed control algorithm.

2. A mathematical model of an arbitrary power system structure has been developed for calculating transients, taking into account a combined device for vector control of the flow of active power based on an DFSMC speaker and an autotransformer.

3. A comparison of the results of calculations of transient modes in the developed and existing verified industrial software package (RastrWin) showed the reliability of the results obtained in the author's software package.

4. It is shown that the developed mathematical model, taking into account the implementation of the proposed control algorithm, makes it possible to ensure the stability of the device for vector control of mode parameters based on the DFSMC.

References

1. Zelenokhat N.I., Suprunov I.S., Dubonos V.R. Power flow control via heterogeneous intersystem communication with a combined control device // Energetik, 2015, pp. 40-43.

2. Zelenokhat N.I. Improving the efficiency of managing modes of flexible intersystem communication with EVPT // Bulletin of the MEI, 2005, **3**, pp. 66-73.

© The Authors, published by EDP Sciences. This is an open access article distributed under the terms of the Creative Commons Attribution License 4.0 (https://creativecommons.org/licenses/by/4.0/).

3. Suprunov I.S., Dvorkin D.V., Novikov N.L., Novikov A.N. Vector control of power transmission modes based on asynchronous synchronous machines // ELECTRIC POWER. TRANSMISSION AND DISTRIBUTION, 2021, **2(65)** pp. 84-93.

4. Suprunov I.S., Dvorkin D.V., Novikov N.L. Operation control of flexible intersystem communication based on asynchronous synchronous machines (part 2) // Methodological issues of reliability research of large energy systems. Issue 71. Reliability of energy supply to consumers in the context of their digitalization. Book 2, 2020, pp. 231-240.

5. Sharov Yu.V., Stroev V.A., Golov P.V. Various representations of ARV generators in calculations of transients // Bulletin of the MEI, 2007, **2**, pp. 77-81.

6. Stroev V.A., Golov P.V. Models for the study of transients in a complex regulated EES // Proceedings of the Russian Academy of Sciences. Energetika, 2010, **6**, pp. 66-74.

7. Vladimir Neumin, Vladimir Ivanov, et al. RUSTAB: User's Guide, TechSystem Group, 2020, 397 p.

8. Merkuryev G.V., Shargin Yu.M. Stability of power systems. Calculations Monograph. St. Petersburg: NOU "Energy Personnel Training Center", 2006, 300s.

© The Authors, published by EDP Sciences. This is an open access article distributed under the terms of the Creative Commons Attribution License 4.0 (https://creativecommons.org/licenses/by/4.0/).

Medium voltage digital instrument transformer

Dmitriy Ulyanov[1,2]*, *Roman* Plakidin[2]

[1]Northern (Arctic) Federal University, Russia
[2]Engineering center «Energoservice», Russia

Abstract. Instrument transformers with digital output provides significant advantages, including those related to increased reliability of power system operation. With the appearance of digital instrument transformers, the topology of the measurement channel and the functional purpose of its components change. This paper considers some function migration examples from protection devices to the primary converter (digital current and voltage transformer). Combined digital transformers expand the application area of distance protection principles, increase the fault location accuracy, as well as improve sensitivity and speed of single-phase earth fault (SPF) protection in 6-35 kV networks.

1 Introduction

Currently, primary instrument transformers with digital outputs are increasingly being used at electric power facilities. Their operation is regulated by international standards IEC 61869 and IEC 61850. Digital instrument transformers, in contrast to traditional electromagnetic transformers, have no secondary analog circuits and transmit instantaneous values of current and voltage signals in digital form in accordance with IEC 61850-9-2. The combined instrument transformer integrates two devices in one enclosure - a current transformer (CT) and a voltage transformer (VT). This reduces the phase error between the current and voltage vectors and, as a consequence, improves the performance of protection and automation.

2 Selection of instrument transformers for protection and automation

In electrical systems of all voltage classes, the measuring elements of protection and automation are connected to current and voltage transformers. The selection of their technical characteristics depends on the parameters of protected electrical equipment, network schemes and power system modes [1].

In addition, the secondary circuit total load of conventional current and voltage instrument transformers is also an important factor that influences the selection of their characteristics [2].

It is obvious that the reliability and correct functioning of protection and automation depends on the correct selection of current and voltage instrument transformers, and this affects the reliability of power systems.

Digital current and voltage transformers are predominantly designed on primary converters without iron cores, which inherently eliminates such negative effects as saturation and residual magnetization in current transformers and ferroresonance in voltage transformers [2].

Digital instrument transformers do not have transformation ratios since they form primary measurements of voltage and current at the digital output. In addition, their total measurement error does not depend on the secondary circuit load, because these circuits are absent. The instantaneous and RMS current and voltage values transmitted by digital combined instrument transformers are referenced to the primary values and, even in the absence of a local and/or global time synchronization system at the substation, provide high accuracy in measuring the relative angles of current and voltage for three-phase bay.

3 Digital instrument transformers with PMU function

Engineering center «Energoservice» company has developed a single-phase digital combined instrument transformer ECIT, which technical characteristics can significantly increase the reliability and speed of protection and automation systems. The device is designed for medium-voltage networks (up to 15 kV), the current measurement range is from 50 to 1000 A for accuracy class 0.5S, the maximum short-circuit surge current is 42 kA for accuracy class 5PR400.

A distinctive feature of this digital instrument transformer is the measurement in accuracy classes 0.5 and 3P in a wide range of rated voltages from 3000/√3 to 16000/√3 V without additional adjustment and calibration. It became possible due to the development of a special resistive-capacitive divider circuit.

The ECIT uses a Rogowski coil as the primary current converter, which eliminates the negative effect

* Corresponding author: d.ulyanov@ens.ru

© The Authors, published by EDP Sciences. This is an open access article distributed under the terms of the Creative Commons Attribution License 4.0 (https://creativecommons.org/licenses/by/4.0/).

of magnetic core saturation and ensures accurate current measurements over the entire range of transformer electrodynamic withstand. For ECIT, a special digital filter was synthesized to generate current synchrophasors, considering the features of Rogowski coil and with improved performance compared to the requirements for IEEE C37.118 class P [3]. This transformer can be connected to the process bus or directly to a protection device installed in this switchgear. The digital transformer ECIT has a relay contact for sending a signal to trip the high-voltage breaker. This ensures the implementation of built-in overcurrent protection. Thus, ICIT can provide both primary and backup bay protection.

ECIT can transmit up to three sampled values current and voltage streams with adjustable sampling rates from 4800 to 14400 Hz according to IEC 61850-9-2. In addition, an important ECIT feature is the function of calculating and transmitting voltage and current synchrophasor streams with a reporting rate of 50-400 Hz according to IEEE C37.118. This is a special development that modifies the Internet frame profile (IEC 61850-9-2) to transmit "fast" synchrophasors (complex amplitude and phase) instead of sampled values with a configurable frequency (8, 16, 32, 64 and 96 samples per period).

Digital combined instrument transformers with support for measuring voltage and current synchrophasors ensure the implementation of new algorithms and principles of protection and automation in distribution networks. Firstly, these are algorithms for distance line protection, busbar differential protection, power transformer protection, and centralized substation protection [4].

When developing ECIT, the company considered various requirements for the accuracy of synchrophasor estimation in steady-state and transient modes of power system operation, which is necessary for effective protection and automation algorithms. This confirms that ECIT, on the one hand, provides an accurate synchrophasor estimation with a frequency deviation of ± 5 Hz and different levels of the short-circuit current aperiodic component, and on the other hand, ECIT has a high speed of synchrophasor measurements. Such technical characteristics are possible due to the development of special digital signal processing algorithms that consider the frequency and transient characteristics of the integrated primary converters.

Let's consider some examples that demonstrate the ECIT features and confirm its technical characteristics.

Based on the provisions of standards [5-6] it is possible to distinguish the limit phase errors for protection accuracy classes CT and VT (Tables 1-2).

Table 1. Protection CT accuracy classes

CT accuracy class	Ratio error, ± %	Phase error, deg	Composite error at rated accuracy limit primary current
5P	1	60	5
10P	3	-	10

Table 2. Protection VT accuracy classes

VT accuracy class	Ratio error, ± %	Phase error, deg
3P	3	120
6P	6	240

Thus, the total theoretical measurement error in protection, which is caused by the CT and VT error, reaches significant values. The error value depends on the protection algorithm and the power system parameters. In some cases, this can lead to incorrect protection operation. Digital instrument voltage and current transformers provide significantly more accurate measurements, which is necessary for the effective protection operation.

Let's consider the influence of measurement error on the accuracy of line impedance estimation in distance protection.

In the simplest case, based on the traditional method of estimating the line impedance at short-circuit [7], this estimation error will be determined from the expression:

$$\hat{\underline{z}}(t) = \frac{\dot{U}(t)}{\dot{I}(t)} = \frac{\dot{U}_t(t)\left(1 \pm \varepsilon_U e^{j(\varphi_U \pm \Delta\varphi_U)}\right)}{\dot{I}_t(t)\left(1 \pm \varepsilon_I e^{j(\varphi_I \pm \Delta\varphi_I)}\right)} =$$

$$= \hat{\underline{z}}_t(t) + \Delta\hat{\underline{z}}(t), \quad (1)$$

where $\dot{U}_t(t)$, $\dot{I}_t(t)$ – theoretical values of current and voltage phasors;

ε_U, ε_I, $\Delta\varphi_U$, $\Delta\varphi_I$ – ratio and phase error CT and VT;

$\hat{\underline{z}}_t(t)$, $\Delta\hat{\underline{z}}(t)$ – theoretical value and error of the line impedance estimation.

Expression (1) shows that the error in estimating the line impedance can significantly exceed the errors of CTs and VTs taken separately, since the dependence is complex and may be nonlinear depending on the characteristics of CTs and VTs. In addition, the distance protection algorithm depends on the influence of various factors, so $\Delta\hat{\underline{z}}(t)$ can be significant.

Fig. 1 presents a comparison of the total measurement error for the electromagnetic current transformer model and ECIT measurement error, for which experimental data are presented.

Fig.2 shows a comparison of the distance estimation error to the short circuit location, which is caused by CT and VT measurement error.

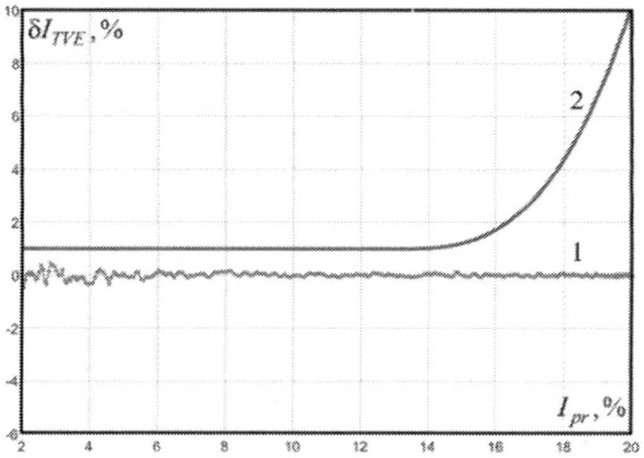

1 – ECIT, 2 – electromagnetic CT model
Fig.1. Total measurement error

© The Authors, published by EDP Sciences. This is an open access article distributed under the terms of the Creative Commons Attribution License 4.0 (https://creativecommons.org/licenses/by/4.0/).

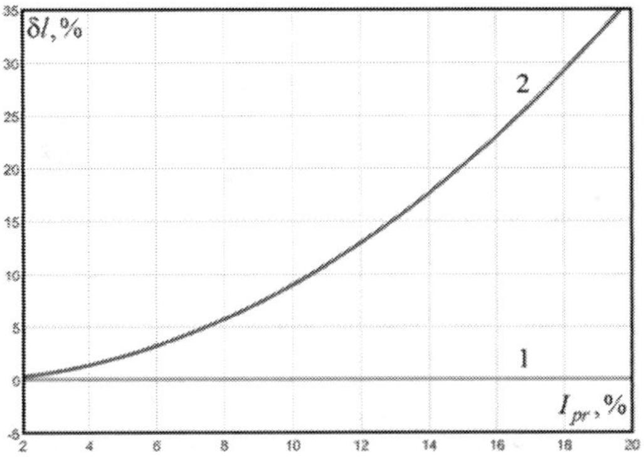

1 – ECIT, 2 – electromagnetic CT model
Fig.2. Distance estimation error

ECIT has high sensitivity for measuring zero sequence currents of fundamental frequency and harmonics. This is possible due to the measurements of phase currents from 0.1 to 1 A with a total vector error (TVE) of no more than 1.5%. Such measurements are necessary for effective SPF protection. ECIT provides accurate measurements of the zero sequence current magnitude and its phase relative to the zero sequence voltage. Accurate ECIT measurements provide an opportunity to implement special SPF protection algorithms and SPF location systems, including for networks with isolated and compensated neutral.

In such distribution networks, SPF protection is often based on the measurement of the fundamental frequency zero sequence current or on the zero sequence power direction estimation [8]. However, in networks with overhead lines, unlike cable lines, the SPF current can have small values (no more than 1-2 A), therefore the measured zero sequence currents of substation bays can be less than 1 A. This causes the difficulty in ensuring the SPF protection sensitivity.

For the majority of used electromagnetic zero sequence CTs, which are installed in the switchgear bay on the cable section, measurements less than 1 A are not specified and can have a significant error, which significantly complicates the implementation of SPF protection due to its low sensitivity. In addition, for overhead lines that have taps where reclosers are installed, conventional zero sequence CTs of toroidal design is difficult.

According to theory of electrical circuits, the zero sequence current can be expressed through the sum of the line phase currents:

$$\dot{I}_0(t) = \frac{1}{3}\left(\dot{I}_a(t) + \dot{I}_b(t) + \dot{I}_c(t)\right). \qquad (2)$$

Therefore, the availability of synchronized phase current measurements, which the ECIT can provide locally at the installation site, allows the implementation of zero sequence current measurements. And the availability of phase voltage measurements allows the realization of local SPF protection based on the calculation of the zero sequence power direction [9]. The high accuracy of ECIT compared to the characteristics of a conventional zero sequence CT provides higher sensitivity of protection at low currents (Fig.3).

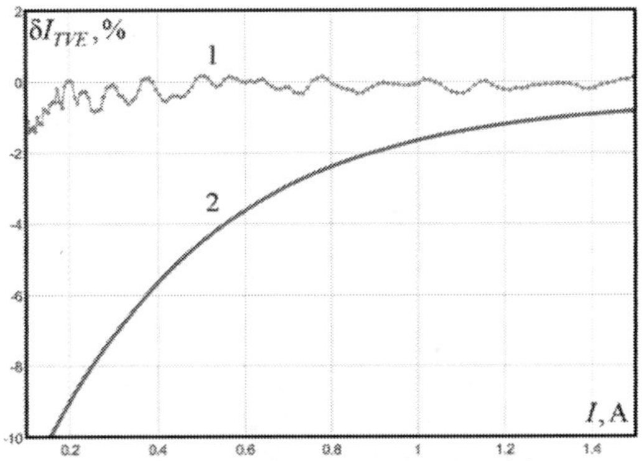

1 – ECIT, 2 – electromagnetic CT model
Fig.3. Zero sequence current measurement error

One of the most promising applications of digital current transformers is differential protection. Let's consider an example of estimating the unbalance current of power transformer differential protection, which shows how significantly the protection sensitivity depends on the measurement accuracy.

General expression for estimating the unbalance current of power transformer differential protection:

$$I_{ub} = \left(k_{aper}k_{uni}\varepsilon + \Delta U_{oltc} + \Delta f_{sch}\right)I_{ex.max} + I_{mag}, \qquad (3)$$

where k_{aper} – coefficient of aperiodic component of short circuit current,

k_{uni} – CT uniformity coefficient,

ε – composite error at rated accuracy limit primary current,

ΔU_{oltc} – coefficient that considers the change in the transformation ratio (on-load tap changer),

Δf_{sch} – component that considers the connection scheme of HV and LV current transformers for protection,

$I_{ex.max}$ – maximum external short circuit current,

I_{mag} – the magnetizing current.

The second term in expression (3) is the power transformer magnetization current I_{mag}. Its maximum value is determined by the magnitude of the magnetization inrush current when the transformer is switched on idling. In normal transformer operation, the magnetization current usually does not exceed 1-2% of the rated current. In the case of a short circuit, I_{mag} is small compared to the first term in expression (3), so in this mode it can be ignored.

The first term (3) determines the maximum transformer unbalance current during a short circuit at the terminals of the low-voltage winding. This term depends on several coefficients that determine the total protection measurement error. The maximum values of these coefficients are:

$k_{aper} = 2$, $k_{uni} = 1$, $\varepsilon = 0.1$,

$\Delta U_{oltc} = 0.16$, $\Delta f_{sch} = 0.05$.

Therefore, the maximum unbalance current due to the protection measurement error at external short circuit:

$$I_{ub} = (2 \cdot 1 \cdot 0.1 + 0.16 + 0.05)I_{ex.max} = 0.41I_{ex.max}.$$

© The Authors, published by EDP Sciences. This is an open access article distributed under the terms of the Creative Commons Attribution License 4.0 (https://creativecommons.org/licenses/by/4.0/).

Thus, the maximum unbalance current can be more than 40 % of the short-circuit current. This makes the differential protection insensitive to fault in the low-voltage winding, so in reality, various methods are used to reduce the unbalance current.

Modern microprocessor protection devices allow to partially reduce the coefficients from the first term in expression (3). If the protection uses a digital instrument transformer ECIT, the coefficients (3) will be minimal:

$k_{aper} = 1$ – aperiodic component filtering,

$k_{uni} = 0.5$ – uniformity of instrument CTs,

$\varepsilon = 0.005$ – no saturation effect,

$\Delta U_{oltc} = 0$ – OLTC position control,

$\Delta f_{sch} < 0.01$ – minimum digital signal processing error.

Minimum unbalance current considering the given coefficients:

$$I_{ub} = (1\cdot 0.5\cdot 0.005 + 0.01)I_{ex.max} = 0.0125 I_{ex.max}.$$

As a result, the unbalance current decreases to 1-2 % of the short-circuit current, which makes the differential protection sensitive to all internal short circuits in the transformer. In addition, theoretically, such protection can be sensitive to turn-to-turn short circuits.

ECIT measures current in a wide range, which corresponds to the ranges of measuring and protective windings of electromagnetic current transformer. This feature allows for simultaneous transformer protection and transformer condition monitoring. The transformer monitoring system is designed to prevent the progression of complex fault types to which the protection is not sensitive.

It should be noted that digital instrument transformers makes it possible to implement protection functions not only within various specialized devices, but also on the basis of highly redundant server platforms containing a large set of centralized protection and automation. Combining centralized and wide area protection and emergency control for specific tasks significantly increases the reliability of protection and automation devices at power facilities.

Replacing the instantaneous values in SV streams with synchrophasors in the process bus makes it possible to significantly increase the data transmission capacity and reduce the computational requirements for the protection and automation devices, since the input signal processing (calculation of the complex amplitude and phase) is already performed in the instrument transformer itself.

An important factor for increasing the reliability of protection and automation systems is the presence of self-diagnostics and warning alarms in the instrument transformer. ECIT provides constant self-diagnostics and can issue instantaneous warning alarms about the measurement quality to the installation control system. Continuous digital network communication provides automatic response to device failure and automatic algorithm reconfiguration, and any loss of ECIT communication is immediately signaled.

4 Conclusion

Instrument transformers with digital output, which emerged as a result of the development of information technologies, allows to increase the reliability of power systems during their operation and development due to the following possibilities:

- online self-diagnostics and provision of automatic integrity control of information exchange with protection, automation and control systems at power objects;
- absence of negative effects such as magnet core saturation and ferroresonance;
- simplification of the methodology for selecting types of measuring equipment, since there is no need for step-by-step selection of primary nominal values over a wide measurement range, considering the multiplicity factor and the load of secondary circuits;
- ensuring a high level of redundancy (including duplication) of protection and automation functions at all bay levels in an electrical system;
- increase of short circuit tripping speed in lines by distance principle and short circuit tripping in power transformers by differential principle;
- increases the precision of fault location.

References

1. A.M. Fedoseev, M.A. Fedoseev, Relay protection of electric power systems (1992)
2. Thomas R. et al. Non-conventional Instrument Transformers Enabling Digital Substations for Future Grid, IEEE/PES T&D (2016).
3. A.V. Mokeev. Filter synthesis for PMU, IEEE SIBCON (2017).
4. A.V. Mokeev, L.V. Perelygin, E.I. Khromtsov, *Combined primary measuring converters of current and voltage 6-35 kV*, New in Russian energy, **1** (2017).
5. IEC 61869-2:2012 Instrument transformers - Part 2: Additional requirements for current transformers
6. IEC 61869-3:2011 Instrument transformers - Part 3: Additional requirements for inductive voltage transformers
7. Shneerson E.M. Distance protection, Energoatomizdat, (1986).
8. V.A. Shuin, A.V. Gusenkov, *Protection against ground faults in 6-10 kV electrical networks*, Energoprogress, (2001).
9. S. A. Piskunov, A.V. Mokeev, D.N. Ulyanov, RPA, **1** (2023).

© The Authors, published by EDP Sciences. This is an open access article distributed under the terms of the Creative Commons Attribution License 4.0 (https://creativecommons.org/licenses/by/4.0/).

E3S Web of Conferences 584, 01037 (2024) https://doi.org/10.1051/e3sconf/202458401037
RSES 2024

Low-frequency oscillations analysis and big data

Aleksandr Popov[1*], *Kirill* Butin[2], and *Andrey* Rodionov[1]

[1]Engineering center «Energoservice», Russia
[2]Northern (Arctic) Federal University, Engineering center «Energoservice», Russia

Abstract. The increased observability of processes in electric power systems, provided by the widespread introduction of synchrophasor measurement technology, as well as the use of computational processing of measurement data arrays, has led to changes in the methodology for analyzing the power system operating modes. Along with the development of new approaches to modeling processes and received measurement signals, a class of data-based methods has emerged. In a number of recent studies, the analysis of synchrophasor measurement data at the large power system scale is considered as a big data problem. Despite the controversial nature of this point of view, the application of some big data methods in synchrophasor data analysis yields positive results. In the proposed report, an assessment of the volumes of initial and intermediate data in the analysis of forced low-frequency oscillations is performed. The effect of data explosion at intermediate stages of calculations is shown. The storage and transmission of these data can be replaced by their calculation in the presence of a storable and transmittable configuration of a generalized computational scheme. A method for compact visualization of an array of oscillation dynamic parameters is shown by the example of dissipating energy. In addition, an approach is presented for configuring streams of synthetic test data at the power system scale, which provides a desired result in the analysis of dissipative energy.

1 Introduction

Synchrophasor measurements technology allows the values of the parameters of the electric power system to be obtained in a single discrete time with a high degree of detail. Phasor Measurement Units (PMU) and Phasor Data Concentrators (PDC) of different levels form the basis of the Wide-Area Monitoring System (WAMS). WAMS has been developed in Russia since 2005. The number of installed PMUs continues to grow and currently exceeds 1000 [1]. The on-line analysis of WAMS data is aimed at increasing the situational awareness of the operational dispatchers about the current and predicted state of the power system. The processing of archived records is used to study the characteristics of the power system, to investigate the accidents, to develop, test and improve data analysis methods, etc.

One of the main applications of the WAMS data is the analysis of the oscillatory stability of the power system. High amplitude, poorly damped forced low-frequency oscillations (LFO) of operating parameters are pressing problem in the control of large power systems [2]. Many efforts have been made to find an effective solution to the problem of timely detection of the forced LFO source, i.e. to indicate the part of the power system that makes the greatest contribution to the development of the oscillatory process.

The complexity of power systems, the increase in their observability due to WAMS and synchrophasor measurements, and the intensive use of computational processing of measurement data arrays have become important factors in the current stage of analysis methodology development. Now, with the development of new approaches to modeling processes and received measurement signals, a class of data-based methods has emerged. A number of studies consider the analysis of PMU data at the scale of large power systems as a big data problem [3, 4, 5].

According to [6], the term “Big Data” refers to a wide range of technologies that include systems for storing and processing huge amounts of structured and unstructured data of different content, frequently updated and coming from different sources. The components of big data are volume, velocity, and variety. The uncertainty of the criteria for big data makes this topic controversial anyway. In the case of PMU data, the variety criterion is the main difficulty. This discrepancy to the definition is partially compensated by the problem of abnormal data [7] or the use of synchrophasor measurement technology in conjunction with the others [8, 9].

In any case, the involvement of some big data methods for the analysis of WAMS measurement information brings positive results. For this reason, the document [5] is of interest, which contains the 2022 report on the results of 8 projects aimed at exploring the possibilities of using big data methods for searching and classifying various events (including oscillations) in the PMU data archives. The U.S. Department of Energy was responsible for funding these projects. The participants were provided with training and test PMU data sets with

* Corresponding author: a.popov@ens.ru

© The Authors, published by EDP Sciences. This is an open access article distributed under the terms of the Creative Commons Attribution License 4.0 (https://creativecommons.org/licenses/by/4.0/).

a total volume of 25 Tbytes. The results of the projects are presented in about 60 publications. The work was performed in the following main areas:

1. Data storage and access.
2. Identification of data quality violations related to measuring instrumentation errors, quantization noise, inconsistency of measurement units, timestamps jitter, the use of obsolete values, and the status bits irrelevance.
3. Recovery of missing or bad data, in particular, by interpolating short-time omissions and using data from neighboring PMUs.
4. Event search and classification, data markup, event signature generation, simulation modeling for rare events. By the way, one of the conclusions of report states that when detecting the oscillatory modes, machine learning methods are inferior to digital signal processing ones.
5. Data visualization, considered mainly in terms of the possibilities of representing the extremely long time series. It is worth noting that in the conclusion of the report, visualization is in the first place among the main directions of the further work.

Here, the discussion of big data is complemented by the consideration of such information signals that are obtained by computation and take the form of data at certain stages of the LFO analysis process. These include the dynamic parameters of the oscillations extracted from the source signals and containing signs of participation in the oscillation process, as well as synthetic data used for software testing.

2 Dynamic parameters of oscillations

The processing of the data stream in the LFO analysis can be described by the following general scheme: extraction of the oscillatory components from the input measurement signals; calculation of the oscillatory dynamic parameters; decision on the participation and role of the considered objects of the power system in the oscillatory process development.

To compare the amounts of primary data and the signals extracted from them by calculations, consider Table 1. The first row corresponds to the minimum set of input data: the current and voltage synchrophasors from the PMU. Frequency, active and reactive power, which are commonly used in LFO analysis, can be obtained from the PMU or calculated from the voltage and current synchrophasors. The dynamic parameters of the oscillations include amplitude, full phase, frequency, rate of change of oscillation magnitude, some indicators of oscillation damping, and oscillation energy. These signals are inputs for subsequent processing steps and for decision support visualization. Oscillatory components are extracted from signals of frequency, current, voltage, active and reactive power. Monitoring is performed in 20 bands of the low frequency spectrum. In addition, let the signal value be represented by the data type float32 (4 bytes), according to the standard C37.118.2 [10], which regulates the structure of the frames of the online PMU stream. The amplitude and phase of synchrophasor are represented by two float32 values and occupies 8 bytes. The columns of the table correspond to one data frame, one second and one day at 50 frames/sec data rate.

Table 1. Data volume estimation in LFO analysis.

Quantities	Data volume		
	Data frame, 1 PMU	1 second, 1000 PMU	1 day, 1000 PMU
Current and voltage synchrophasors	16 b	~780 Kb	~64 Gb
Frequency, active, and reactive powers	12 b	~585 Kb	~48 Gb
Dynamic parameters of 20 LFOs	**3360 b**	**~160 Mb**	**~13.2 Tb**

In short, the amount of data increases a hundredfold at the stage of extracting LFO dynamic parameters.

In addition, the following takes place. Some general schemes describe the process of calculating LFO parameters, but specific implementations of the stages of such schemes form an ever-expanding set. The paper [11] investigates the dependence of the results of the search for LFO sources on the choice of a combination of algorithms. The result is also influenced by the values of various parameters of the algorithms used. Taking into account the combinatorial nature of the problem, the authors proposed the concept of generalized computational schemes, or plans [12]. The plan is a directed graph whose vertices correspond to works (computational stages) and arcs represent data flows. The parameter of each work is the worker - a reference to a specific implementation of a particular data processing algorithm. The configuration of the plan is recorded, for example, in JSON format, which ensures the unambiguousness of the result during repeated calculations. The plan is a functional program and has natural parallelism. This is important when processing large amounts of data.

The research and development of the concept of generalized computational schemes in relation to LFO analysis tasks is carried out by the authors in the development of the ES Phasor software [11, 13] and the Goplan parallel computing platform [12]. An important component of ES Phasor is the tool for optimizing computational schemes in the available solution space and on recordings of real LFO cases [11].

3 Dissipating energy flow

Many approaches and methods have been developed for LFO source detection (see review [2]). Currently, the Dissipative Energy Flow (DEF) method [14], which belongs to the group of energy-based methods, is the most effective in practice [15].

The value of W_{ij}^D in the running window is calculated for each PMU:

$$W_{ij}^D \approx \int 2\pi \Delta P_{ij} \Delta f_i dt + \Delta Q_{ij} \frac{d(\Delta V_i)}{\bar{V}_i + \Delta V_i}$$

Here, the symbol Δ indicates the operation of extracting the oscillatory component from the

© The Authors, published by EDP Sciences. This is an open access article distributed under the terms of the Creative Commons Attribution License 4.0 (https://creativecommons.org/licenses/by/4.0/).

measuring signal, P_{ij} denotes active power on the line $i - j$, Q_{ij} – reactive power, f_i – frequency, V_i – voltage, $\tilde{V}_i$ – average voltage on the point i.

The direction of the dissipating energy flow is determined by the angular coefficient of linear regression:

$$W_{ij}^D(t) = DE_{ij} * t + b_{ij}$$

The positive values of $W_{ij}^{D\prime}(t) = DE_{ij}$ indicate the flow of dissipating oscillation energy from the object (station, substation) to the power system, and negative values indicate the reverse direction of the dissipating energy flow. The maximum value of DE_{ij} thus corresponds to the LFO source (or the point closest to the actual source of oscillations if there is no data from the source is available).

4 Visualization of dissipating energy

Visualization is then one of most important components of PMU big data [5]. An informative representation of the measurement signal arrays, intermediate and final analysis results helps to increase the situational awareness of operational dispatch personnel [1] and can be included in the decision support tools. In addition, effective visualizations of oscillatory processes at the power system level contribute to further research in the field of WAMS data analysis.

With the data configuration in the above example, the initial measurement signals and the LFO dynamic parameters calculated in 20 spectral bands for 1000 data sources include more than 600 000 functions of time. It is obvious that a simple set of such a number of plots is not suitable for effective perception of the results of LFO analysis. In [13] the authors proposed time-domain, frequency-domain, amplitude-phase, value-velocity, and map visualizations of arrays of input PMU data, oscillation dynamic parameters, and LFO source detection results.

Here we consider the representation of the parameters of the dissipating energy of oscillations in the phase plane (Figure 1, Figure 2).

As mentioned above, to interpret the results of the dissipating energy calculation and to identify sources and receivers of oscillation energy, $W(t)$ is approximated by linear functions and the derivatives of these functions are compared. Accordingly, the rate of change of the dissipating energy is plotted on the abscissa axis and its value is plotted on the ordinate axis (it is convenient to compare such a plot with time plots).

The interpretation is as follows. The sources of oscillatory energy are located in the first quarter, and the receivers fall into the third quarter. The second and fourth quarters correspond to the trend of changing the role of the object from receiver to source and vice versa. Over time, the dot moves, leaving a fading trace that allows you to see the closets history of the values.

The case occurred in the southern part of the Russian power system in 2020. The oscillations lasted about three minutes and affected about 50 objects. Figure 1 shows the beginning of the LFO process. Figure 2 represents its evolution.

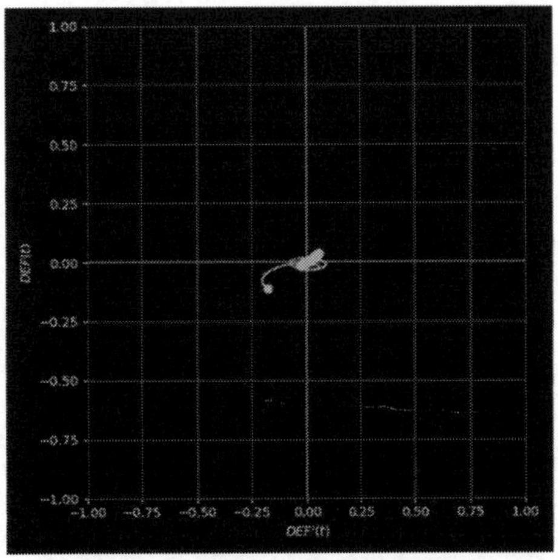

Fig. 1. Dissipating energy in phase plane: beginning of the LFO process.

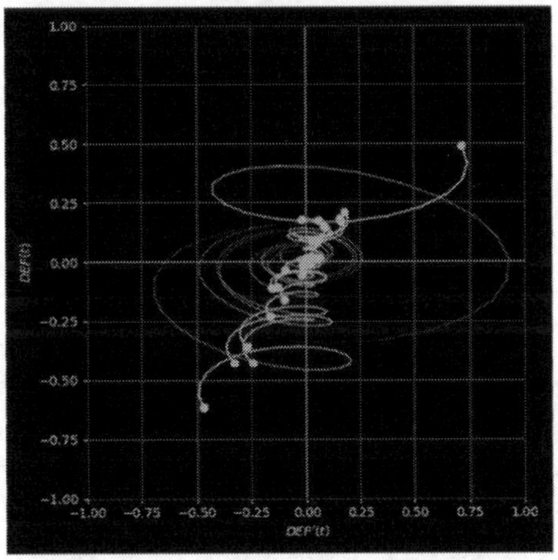

Fig. 2. Dissipating energy in phase plane: LFO process development.

The System Operator found out that the LFO source in this case is located in the neighboring state power system. The above figures also support the assumption that there are no data from the desired source of the LFO in the given data set: there is a receiver of dissipatiing energy, but there is no source (Figure 1); there is more incoming energy than outgoing one (Figure 2).

5 Test data generation

Engineering Center "Energoservice" company develops VirtPMU software. It is a library and a set of utilities for generating online synchrophasor data streams. The output of VirtPMU complies with the C37.118.2 standard [10]. Several ways of obtaining configuration and data frames are supported: mathematical modeling, archive record playback, retransmission of input stream.

The VirtPMU/vpmu-wams utility implements a simulation of the WAMS output.

It is important to note that simulation modeling of the power system is not meant here. When developing a method for configuring a data model for vpmu-wams,

© The Authors, published by EDP Sciences. This is an open access article distributed under the terms of the Creative Commons Attribution License 4.0 (https://creativecommons.org/licenses/by/4.0/).

the question was: how to compactly describe the data of a large-scale oscillatory process, ensuring on the one hand the uniqueness of the synthesized signals, and on the other hand obtaining the intended result of the analysis? The following approach was developed and implemented. First, the so-called setting profiles corresponding to the different roles in the oscillatory process are determined. The expected means of the measured parameters, the LFO parameters, the model of the response of the parameters to the oscillations, the parameters of the harmonic and stochastic noise are specified. Then the PMUs are listed and a profile is selected for each. This approach allows you to quickly configure the test data stream using, among other things, identification information for the PMUs in the real WAMS.

For the DEF method, the profiles “LFO source”, “DEF source”, “DEF receiver”, “not participating in oscillations” can be obtained by setting the phase shift of the active power oscillations relative to the frequency oscillations (see the DEF integral equation). Figure 3 shows plots of the generated signals.

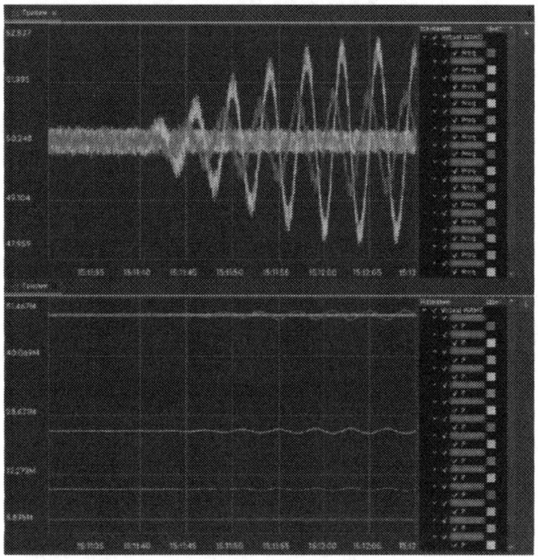

Fig. 3. Test LFO data example.

The SynphasorTest software [16] (Engineering Center “Energoservice”) is used to display the signals received via the C37.118.2 protocol. Plots of frequency and active power signals received from multiple connections are displayed.

The plots of the signals from the “LFO source” are shown in blue. The synphase oscillations of active power and frequency result in a positive value of the active component of the dissipating energy according to the expression for W_{ij}^{D}. A negative value of W_{ij}^{D} is obtained when the oscillations of active power and frequency are in antiphase. The amount of energy can also be adjusted by the value of the amplitude of the oscillations. The reactive component of the dissipated energy is not included in this example.

Figure 4 shows the results of the analysis of the test data obtained.

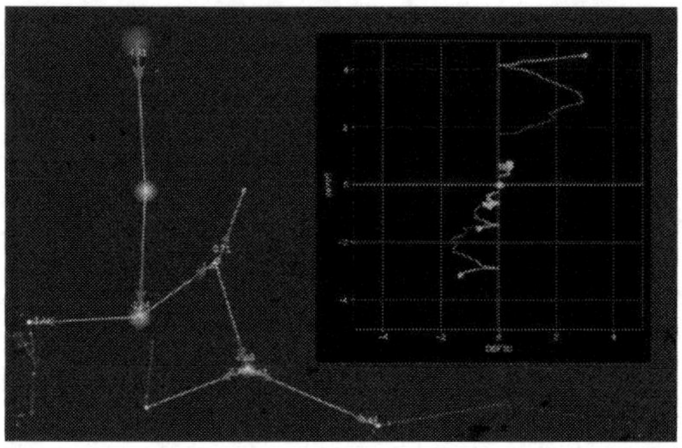

Fig. 4. Visualization of dissipating energy for the test data.

The power system fragment and the phase plane are placed together. The visualization is performed with the ES Phasor software [13] (Engineering Center “Energoservice”). The frame corresponding to the beginning of the LFO process is shown. Green color indicates transmission lines for which measurements are available, yellow color indicates lines with high LFO energy level. The red arrows show the outgoing oscillation energy; the blue arrows show the incoming energy. The glow shows the total outgoing oscillation energy; the red glow corresponds to the LFO source.

An additional feature of VirtPMU, useful when testing systems for processing a number of incoming PMU data streams, is the simulation of various problematic situations: frame delays, various types of data frame distortions, lack of time synchronization for individual devices, etc. In addition, it is possible to reproduce the real measurement data from the PMU archives or simulations in the single or multiple online streams. The PMU data record must be presented in COMTRADE format, Annex H [17]. Such a record can be extracted from the PDC.

6 Conclusion

Considering PMU signal analysis as a big data problem raises many issues. These include the organization of data storage and access, data quality investigation, recovery of missing data, use of high-performance computing, event search and classification, visual representation, etc.

This report introduces the topic of "secondary" data, i.e. information signals that are obtained solely by calculation, but also take the form of data for some stages of analysis. In the case of LFO source detection, the volume of such data is two orders of magnitude greater than the volume of the original measurements. The storage and transmission of these data can be replaced by their computation in the presence of a storable and transmittable configuration of a generalized computational scheme. The authors propose a concept and software implementation of such schemes. Using the example of the dissipating energy flow, a method of visual representation of a set of oscillation dynamic parameters is presented. In addition, an approach is presented for configuring streams of synthetic test data from large-scale LFOs that provide a desired result in dissipating energy analysis.

© The Authors, published by EDP Sciences. This is an open access article distributed under the terms of the Creative Commons Attribution License 4.0 (https://creativecommons.org/licenses/by/4.0/).

References

1. A.V. Zhukov, E.I. Satsuk, D.M. Dubinin, O.V. Zhuravleva, *Development of Dispatch Control Technologies in Russia on the Synchronized Phasor Platform*, in Proceedings of the International Scientific Seminar "Methodological Issues of Reliability Research in Large Energy Systems", **73**, pp. 465 – 474, (2022).
2. L.G. Meegahapola et al. *Review on Oscillatory Stability in Power Grids with Renewable Energy Sources: Monitoring, Analysis, and Control using Synchrophasor Technology*, in IEEE Transactions on Industrial Electronics, **68(1)**, pp. 519–531 (2020).
3. V. Roy et al. *Event detection from PMU generated big data using R programming*, in 2018 IEEE Conference on Technologies for Sustainability (SusTech), pp. 1 – 6 (2018).
4. I. Kosen et al. *UPS: Unified PMU-data storage system to enhance T+ D PMU data usability*, in IEEE Transactions on Smart Grid, **11(1)**, pp. 739 – 748 (2019).
5. S. Biswas et al. Big data analysis of synchrophasor data: Outcomes of research activities supported by DOE FOA 1861 in Pacific Northwest National Lab.(PNNL), Richland, WA (United States), PNNL-33548 (2022).
6. Big data in informatics, URL: https://bigenc.ru/bol-shie-dannye-v-informatike-e4a22a
7. P. Hart et al. *Application of big data analytics and machine learning to large-scale synchrophasor datasets: evaluation of dataset 'machine learning-readiness*, in IEEE Open Access Journal of Power and Energy, 9, pp. 386 – 397 (2022).
8. B. Mondal et al. *Application of PMU and SCADA data for estimation of source of forced oscillation*, in 2019 International Conference on Smart Grid Synchronized Measurements and Analytics, pp. 1 – 7 (2019).
9. J. Zuo et al. *Low-Frequency Oscillation Mode Source Identification with Wide-Area Measurement System*, in 2019 IEEE 3rd Conference on Energy Internet and Energy System Integration (EI2), pp. 1525 – 1539 (2019).
10. C37.118.2-2011 – IEEE Standard for Synchrophasor Data Transfer for Power Systems.
11. K. Butin, A. Popov, A. Rodionov. *Investigation the influence of the parameters of the computational scheme on detecting the source of low-frequency oscillations*, in E3S Web of Conferences, **384**, p. 01019 (2023).
12. A. Popov et al. *A. Framework for synchrophasor measurements data processing and the case studies of the low-frequency oscillations*, URL: https://www.naspi.org/sites/default/files/2021-04/D3S7_01_butin_energoservice_naspi_20210415.pdf
13. K.P. Butin, A.I. Popov, A.V. Rodionov. *Development of visual representations of low-frequency oscillations in power systems*, in Proceedings of the International Scientific Seminar "Methodological Issues of Reliability Research in Large Energy Systems", **73**, pp. 526 – 536 (2023).
14. S. Maslennikov et al. *Dissipating energy flow method for locating the source of sustained oscillations*, in International Journal of Electrical Power & Energy Systems, **88**, pp. 55 – 62 (2017).
15. B. Wang, S. Maslennikov. IEEE-NASPI Oscillation Source Location Contest-Case Development and Results (Tech. rep. National Renewable Energy Lab.(NREL), Golden, CO (United States), 2021).
16. I.A. Ugryumov. *Development of Software for Viewing and Reproducing Archived Data from the WAMS*, in Energy of the Arctic: Proceedings of the All-Russian Scientific and Technical Conference, pp. 69–73 (2023).
17. C37.111-2013 – IEEE/IEC Measuring relays and protection equipment – Part 24: Common format for transient data exchange (COMTRADE) for power systems.

© The Authors, published by EDP Sciences. This is an open access article distributed under the terms of the Creative Commons Attribution License 4.0 (https://creativecommons.org/licenses/by/4.0/).

Practical experience in monitoring low-frequency oscillations at electric power station of United Power System of Russia

Mikhail Govorun [1*], *Andrey* Zhukov [1], *Evgeniy* Satsuk [1], *Dmitriy* Dubinin [1], and *Nikita* Karimov [1]

[1]«System Operator of the United Power System», Joint-stock Company, Moscow, Russia

Abstract. The development of the Wide Area Measurement System (WAMS) in the United Power System of Russia has provided the opportunity to use phasor measurement unit (PMU) data to solve practical problems in operational dispatch management [1]. One of the most significant tasks is monitoring low-frequency oscillations (LFOs) to prevent and eliminate violations of the normal operating mode in the energy system of Russia [2]. Long-term, weakly damped, high-amplitude oscillations can lead to oscillatory instability in the power system, causing massive generation and load shutdowns, separation of the system into isolated parts, and cascading technological disruptions. Based on PMU technology, a powerful tool for monitoring low-frequency oscillations and identifying their parameters and sources in real-time has been developed and integrated into the control rooms of the System Operator. This article presents the fundamental principles of this monitoring system and describes the algorithms used to identify the sources of low-frequency oscillations. It also provides practical examples of low-frequency oscillation monitoring at power plants carried out by specialists from the System Operator.

1 Introduction

Synchronous oscillations can be classified into two types: systemic and forced. Systemic oscillations are caused by the dynamic properties of the synchronous machines that are connected by the common energy mode of the system [3]. Forced oscillations caused by external periodic forces that have a specific source or cause. One of the main causes of forced oscillations in power systems is the incorrect adjustment of system controls (automatic turbine controllers, automatic generator exciter controls). Another common cause of forced oscillations is the operation of generating equipment under certain conditions - within a certain range of active power output, with a suboptimal power factor value, in an area that is not recommended for operation (for hydroelectric power plant), etc. Additionally, the occurrence of oscillations may be caused by incorrect operation of the power electronics of stations powered by renewable energy sources, as well as inserts and direct current lines, or by the specific operating conditions of industrial facilities and load characteristics. In the case of hydroelectric power plants, increased vibration of individual components can also lead to forced oscillations [4]. The organization of continuous monitoring of LFOs parameters in the electrical system is essential to improve the reliability of power supply and the performance of generating equipment. By timely identifying unacceptable operating conditions, owners can extend the lifespan of their generating equipment, as well as prevent unexpected shutdowns caused by technological failures, which could lead to system accidents.

1.1 Methods of identifying the sources of forced oscillations

There are several main approaches to identifying the sources of forced oscillations:

• Methods based on the distribution of an electromagnetic wave: Some methods are based on analyzing the shape of the oscillating mode. With sufficient observability, the source of LFOs can be identified as an object with a previous peak of oscillation [5], and its phase lags behind that of damping objects [6]. Another group of methods relies on the velocity of wave distribution in the power system to determine the source of forced oscillation [7].

• Methods based on the estimation of the damping moment: In this group of methods, a synchronous machine with a negative damping moment is identified by the source of LFOs [8]. Location determination methods based on this principle have a clear physical meaning. However, in order to estimate the damping moment for each generator, it is necessary to simultaneously determine the values of the electromagnetic moment of the generator, the angle of rotor rotation, and the rotation frequency.

• Energy-based methods: A number of studies have shown that energy dissipation is consistent with the damping moment. In this case, the source of forced LFOs can be identified as an object that produces positive energy, i.e., makes a negative contribution to damping [9].

* Corresponding author: govorun-mn@so-ups.ru

© The Authors, published by EDP Sciences. This is an open access article distributed under the terms of the Creative Commons Attribution License 4.0 (https://creativecommons.org/licenses/by/4.0/).

1.2 Method based on Dissipation Energy Flow

The practical experience of monitoring LFOs in the energy system of Russia has shown that the most promising and efficient method for identifying the source of these oscillations based on real-time PMU data is the energy-based approach, which involves calculating the flow of dissipation energy. This method, known as the energy dissipation energy flow (DEF), allows for accurate calculation of energy flows at different points in the energy system. The dissipation flow of energy in a specific branch (ij) can be expressed as:

$$DEF_{ij} = DEF_P + DEF_Q, \quad (1)$$

$$DEF_P = \int(2\pi \cdot \Delta P_{ij} \cdot \Delta f_i \cdot dt), \quad (2)$$

$$DEF_Q = \int(\Delta Q_{ij} \cdot d(\Delta U_i) / (U_i^* + \Delta U_i)), \quad (3)$$

where $DEF_P = F(P, f)$ and $DEF_Q = F(Q, U)$ are the components of the dissipation energy flow, ΔP_{ij} and ΔQ_{ij} are the modes of oscillations of active and reactive power in the ij branch, δU_i and Δf_i are the modes of oscillation of phase voltage and frequency on bus i, and U_i^* is the average voltage value for the period under study.

The sign of the dissipation energy flow at the connection point of the generator indicates the following:
- Negative value - the generator damps oscillations.
- Positive value - the generator is a source of oscillations.
The sign of the energy propagating along the power transmission line indicates the direction of the *DEF* distribution through the electrical network:
- Negative value - *DEF* is directed towards the object's tires.
- Positive value - *DEF* is directed away from the object (the source is located on or behind the object relative to the power line).

The practical experience of using this method has confirmed that in order to identify the causes of oscillations, it is necessary to analyze the contributions of the active and reactive components of the *DEF* value, which is determined based on PMU data collected from generator connections. The active component reflects the response of the automatic turbine regulator (ATR) to LFOs, while the reactive component reflects the automatic excitation regulator's (AER) response. These findings are supported by international research.

To monitor LFOs, specialized software has been introduced into control rooms of System Operator. The following approach has been adopted in practice:
- The first step is to continuously monitor the electrical parameters of power plant connections using PMU data, including voltage frequency, active and reactive power, and phase voltage.
- The second step is to determine the parameters of LFOs at 1-second intervals. There is measuring the amplitudes and angles of oscillation for all parameters, and the frequency of oscillation in active power.
- At the third step, if the LFOs parameters are non-zero, the components and total *DEF* value for the analyzed connections can be calculated using an integration step of 200 milliseconds, according to the given formulas (1)-(3).

2 Monitoring of LFOs in the Russian energy system

There are some following examples of monitoring LFOs that were analyzed using the method of calculating dissipation energy flow.

2.1 LFOs at Smolensk Nuclear Power Plant

In December 2021, there were oscillations of active power at the Smolensk Nuclear Power Plant with a frequency of 1.09 Hz (Figure 1). The largest amplitude of these oscillations was 26.3 megawatts (MW), and the total duration was about two minutes.

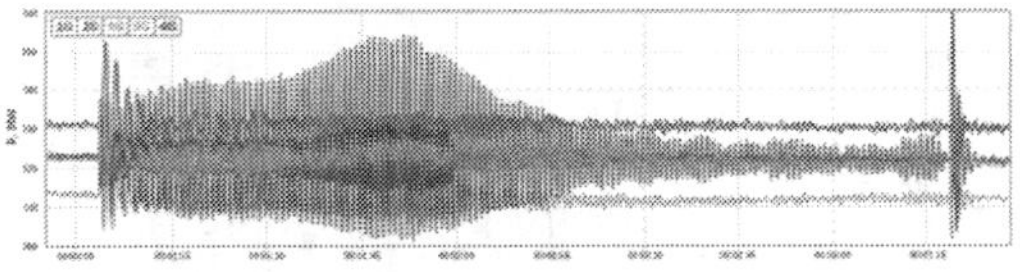

Fig. 1. The active power of the generators.

Based on calculations using the *DEF*, the source of the LFOs was found to be generator No.1, which makes a negative contribution to damping the oscillations, corresponding to a positive *DEF* value (Figure 2). Decomposing the total *DEF* into its components, DEF_P and DEF_Q, showed that both had positive values (Figure 3).

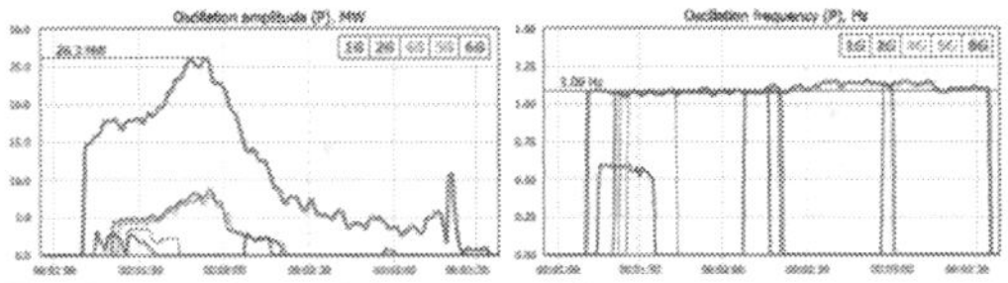

Fig. 2. Parameters of the LFOs of the generators.

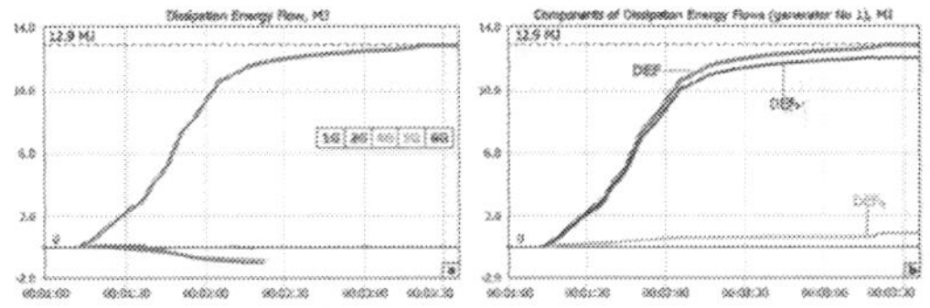

Fig. 3. a – *DEF* of the generators, b – components of the *DEF* of the generator No. 1.

The contribution of DEF_P to the total value was approximately 93% due to the large amplitude of the active power oscillations. Based on practical experience in analyzing, it appears that the ratio of components does not necessarily indicate the root cause of a problem. Considering the presence of DEF_Q and the specific operation of ATR, it can be inferred that the root cause of LFOs is likely to be an incorrect operation of AER.

2.2 LFOs at Belovskaya Power Plant

In January 2024, a protection system was triggered at Belovskaya GRES to prevent a reduction in insulation resistance in the excitation circuit of generator №1. As a result, the excitation system was automatically shut down, and the generator switched to a stable asynchronous operating mode, where it consumed reactive power from the power grid (Fig. 4).

© The Authors, published by EDP Sciences. This is an open access article distributed under the terms of the Creative Commons Attribution License 4.0 (https://creativecommons.org/licenses/by/4.0/).

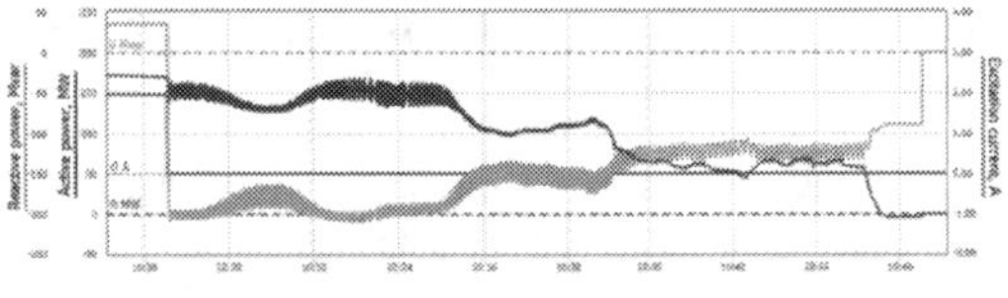

Fig. 4. Parameters of the generator of the Belovskaya station.

This mode is permissible for a short period of time, so the station was first unloaded from 170 MW to 100 MW, and only after 18 minutes, it was shut down by the protection system against asynchronous operation. According to the monitoring results of the electrical parameters of the generators at the Belovskaya power plant, an oscillation mode in active power was recorded.

The asynchronous operation mode of the generator without excitation is not steady state due to the resulting power oscillations. This can be explained by the dependence of the asynchronous torque on the slip, which leads to periodic changes in the $M_{as}(s)$.

In this mode of operation, the amplitude of the forced oscillations varied between 1.2 and 25 MW, and the frequency of oscillation varied between 0.29 and 1.01 Hz (Figure 5).

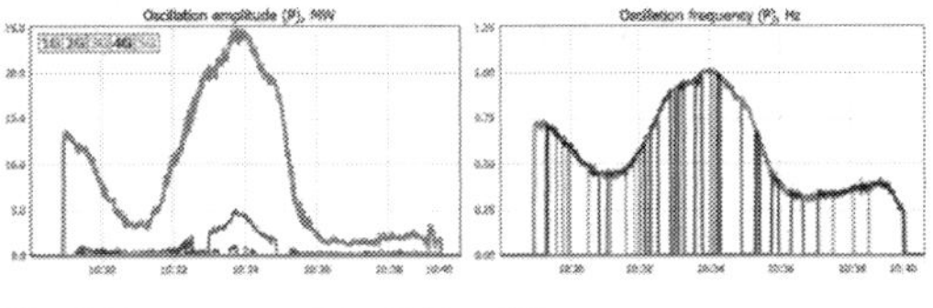

Fig. 5. Parameters of the LFOs of the generators.

The positive value of DEF_{1G} indicates that it is a source of fluctuations (Figure 6). The largest contribution to the positive value of *DEF* is made by the DEF_Q component, which suggests incorrect operation of the automatic voltage regulator (complete loss of excitation of the synchronous machine).

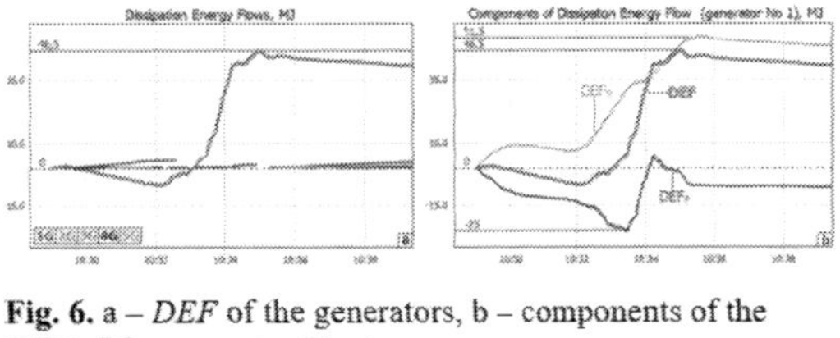

Fig. 6. a – *DEF* of the generators, b – components of the *DEF* of the generator No. 1.

2.3 LFOs at Surgutskaya-2 Power Plant

Analysis of the distribution of energy flows through the power system allows to search for the source of forced oscillations, even in cases where there is not PMU data from generator-source LFOs. For example, between December 9 and December 10, 2023, there were 7 cases of LFOs lasting 16-18 seconds with an amplitude of 27-58 MW and 2 cases lasting 70 and 85 seconds with an amplitude of 100 and 34 MW, respectively, at the generators of Surgutskaya-2 power plant (Figure 7). All generators of Surgutskaya-2 power plant dampened oscillations, which corresponds to a negative *DEF* value, indicating that the source is at another station (Figure 8a).

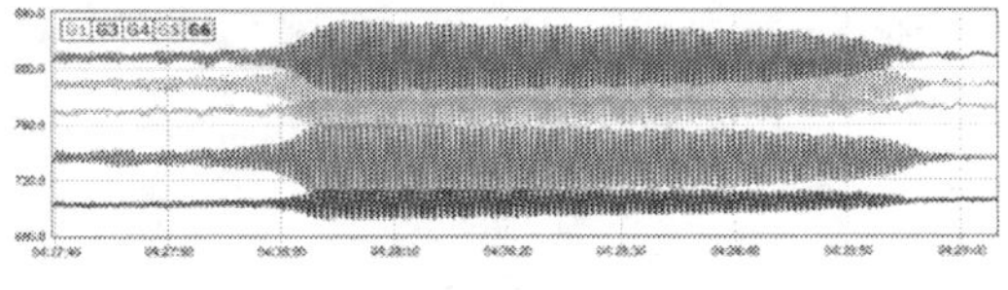

Fig. 7. Active power of generators of Surgutskaya-2 power plant.

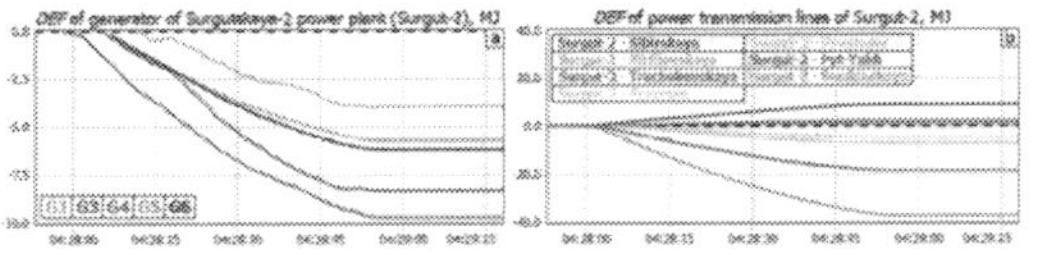

Fig. 8. a – *DEF* of generators of Surgut-2, b – *DEF* of lines of Surgut-2

To continue the search for the source of LFOs, we analyzed the *DEF* of lines of the station in question (Figure 8b). In this case, we can see that the flow of dissipative energy goes to Surgutskaya-2 power plant from substations Somkinskaya, Pyt-Yakh, Svyatogor, and Peresvet. Most of the incoming flow comes from Somkinskaya and Pyt-Yakh. The nearest electric power station with a galvanic connection to the Somkinskaya and Pyt-Yakh substations is the Surgutskaya-1 power plant. However, this power station only has PMU devices installed on one generator and three lines. The magnitude and sign of the *DEF* of the generator No.9 indicate that this machine dampens oscillations more effectively than the generators at the Surgutskaya-2 power plant (Figure 9a).

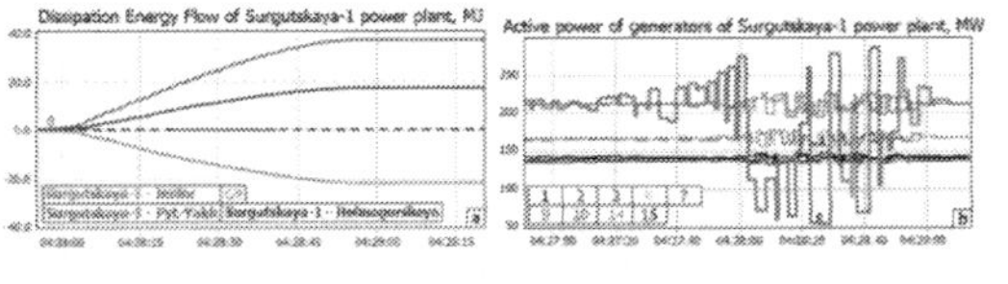

Fig. 9. a – DEF of Surgutskaya-1 power plant, b – active power of generators of Surgutskaya-1 power plant.

At the same time, the *DEF* along outgoing lines from the Surgutskaya-1 are directed away from the power station. Based on these facts, we can conclude that the source of the LFOs is one of the generators at Surgutskaya-1. After analyzing telemetry data, we found that the source of LFOs was generator № 7 at the Surgutskaya-1 power plant. The SCADA data on the values of active power of generators of Surgutskaya-1 power plant is shown in Figure 9b.

2.4 LFOs at Gusinoozerskaya Power Plant

In the period from November 2023 to January 2024, approximately 15 instances of forced low-frequency oscillations of the active power output of generator No.3 at the Gusinoozerskaya power plant were recorded, with an average duration of 60 minutes. The amplitude of these oscillations ranged from 3 to 4 MW, and their frequency was 2.44 Hz. It was initially assumed that the LFOs mode occurred exclusively during the generator's unloading from 170 MW to 125 MW. However, subsequent analysis revealed that fluctuations also occurred during quasi-steady-state modes and when the generator was being loaded.

© The Authors, published by EDP Sciences. This is an open access article distributed under the terms of the Creative Commons Attribution License 4.0 (https://creativecommons.org/licenses/by/4.0/).

Based on the value of dissipative energy flow, generator No. 3 was identified as the source of these forced oscillations ($DEF > 0$), with no reactive component present in the overall value ($DEF \approx DEF_P$). This suggests that the cause of the oscillations is due to the incorrect operation of the ATR during generator unloading, as well as during operation in the power range between 120 and 130 MW.

2.5 LFOs at Volzhskaya Power Plant

Continuous monitoring of the electrical parameters of the generators at the Volzhskaya Power Plant has allowed to establish a correlation between the output active power level and the occurrence of oscillatory modes. During the five-month monitoring period, several dozen occurrences of fluctuating generator active power for generators No. 6 and No. 9 over extended periods were recorded (from several tens to hundreds of minutes). Statistical analysis of these data allowed to identify suboptimal operating ranges for these hydro generators. It was found that a stable oscillating mode occurs at a frequency of 1.14 Hz for both generators.

For generators No. 6, fluctuations in active power occur with an amplitude of 0.8-0.9 megawatts when the output power exceeds 105 megawatts. For generators No. 9, we can distinguish a zone of suboptimal operation between 80 and 105 MW, during which forced fluctuations in active power with an amplitude between 0.6 and 0.8 MW occur (Figure 10). It is worth noting that LFOs were also recorded outside of this range. However, their amplitude was not proportional to the output active power, and decreased to levels between 0.1 and 0.3 MW when going beyond both the upper and lower limits of the range.

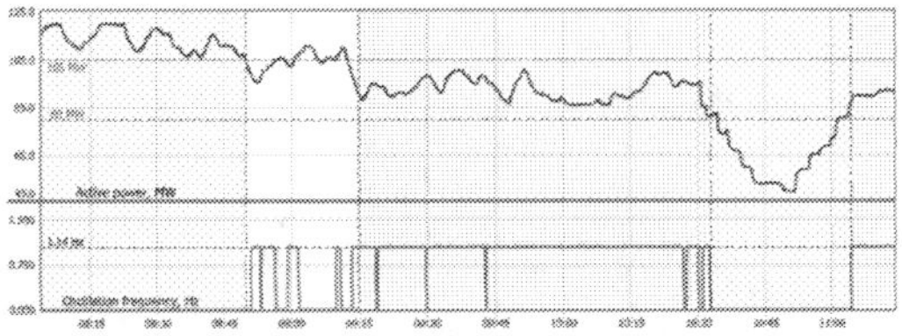

Fig. 10. Parameters of the electrical mode of the Volzhskaya power plant generator No. 9.

3 Conclusion

The development of WAMS technology and the continuous improvement of technical devices have significantly expanded the possibilities for using PMU data in practical monitoring and management tasks. One area of application for this technology, which has great potential for improving the reliability of equipment operation, is monitoring low-frequency oscillations of electrical parameters to identify sources of forced oscillations. Based on the practical experience, it can be concluded that:

1. A technique based on calculating integral values of dissipation energy flows is an effective way to identify sources of LFOs in an electric power system.
2. Under conditions of limited observability, in order to improve the efficiency of the process of finding the source of LFOs, it is necessary to systematically analyze the dissipation energy flows of both generators and power lines.
3. The components of *DEF* of generators - oscillating sources should be used as the main criteria in analyzing the causes of LFOs.
4. Statistical analysis of the results from continuous monitoring of amplitude and frequency of forced oscillations at electrical stations allows to identify a correlation between LFOs parameters and electrical mode parameters, thus revealing suboptimal operating conditions of generating equipment.

References

1. A.V. Zhukov, V.V. Dyachkov, E.I. Satsuk, D.M. Dubinin, Prospects of application of synchrophasor technology for the development of monitoring and control systems for future power system, CIGRE (2020)
2. The Order of the Ministry of Energy of the Russian Federation No. 548 dated 12/07/2018 "On Approval of Requirements for Ensuring the Reliability of Electric Power Systems, Reliability and Safety of Electric Power Facilities and Power Receiving Installations "Rules for Preventing the Development and Elimination of Violations of the Normal Operation of the Electrical Part of Power Systems and Electric Power Facilities"
3. A.V. Zhukov, D.M. Dubinin, O.L. Opalev, The results of identification of sources of low-frequency oscillations in the EYES of Russia according to the PMU data, EEGM (2016)
4. Order No. 141p of PJSC RusHydro, dated June 23, 2016, "Temporary Guidelines for Monitoring and Controlling the Vibration State of Hydraulic Units Equipped with Stationary Vibration Diagnostics Systems".
5. N. Hu, X. Li, Y. Yang, Research for space-time variational features of frequency for power system forced oscillations and disturbance source location, Journal of Sic. Un., **45(6)**, 132 (2013)
6. T. Geng, Z. Zhang, L. Xiang, A locating and splitting scheme for disturbance source of forced power oscillation based on the distribution characteristic, Pow. Sys. Prot. and Con., **43(6)**, 98 (2015)
7. Q. Dong, L. Zhang, X. Yan, X. Liu, Automatic locating source method of compelled resonance low frequency oscillation in power grid, Proc. of the Chin. Soc. of Elec. Eng., **32(28)**, 68 (2012)
8. Y. Li, Y. Huang, J. Liu, Power system oscillation source location based on damping torque analysis, Pow. Sys. Prot. and Con., **43(14)**, 84 (2015)
9. S. Maslennikov, E. Litvinov, ISO New England Experience in Locating the Source of Oscillations Online, IEEE Transactions, **36(1)**, 495 (2021)

© The Authors, published by EDP Sciences. This is an open access article distributed under the terms of the Creative Commons Attribution License 4.0 (https://creativecommons.org/licenses/by/4.0/).

Experience in implementing the algorithm for emergency control system of unloading in case of overload in terms of power based on synchrophasor angle

Mikhail Govorun[1*], *Evgeniy* Satsuk[1], *Yuriy* Luzhkovskiy[1], *Dmitriy* Dubinin[1], *Sergey* Titov[1] and *Aleksandr* Khokhrin[2]

[1]«System Operator of the United Power System», Joint-stock Company, Moscow, Russia
[2] «Prosoft-Systems», Yekaterinburg, Russia

Abstract. An algorithm for automatic unloading during a power overload is described. It is based on measurement of the phase angle between voltages on the Tyumen - Ural power transmission line, which is determined by information from phasor measurement units. The algorithm is implemented in a specialized controller K-100.

1 Introduction

Emergency control system of unloading in case of overload of active power (ECSU) is part of the emergency control system of stability disturbance prevention (ECSS). It is designed to protect the sections of the electrical network from increases in active power overflows resulting from disconnection of network and generating equipment, which can lead to instability and the occurrence of asynchronous mode. ECSU usually operates with such an imbalance of active power that is not fixed by other devices included in the ECSS. Traditionally, automation is performed with an active power trigger. If the active power exceeds the emergency permissible value along the controlled section of the electric network or along the power transmission line, control actions are implemented to turn off generation or to turn off the load of consumers. ECSU allows to increase the degree of utilization of network bandwidth by increasing the maximum allowable overflows of active power.

2 Features of the implementation of the ECSU in the Tyumen energy system

The Tyumen power system is redundant in terms of active capacity. Large thermal power plants are located here, including the most powerful thermal power plant in Russia – Surgutskaya-2 power plant (5,687 MW). The diagram of the power system is shown in Fig.1.

The existing ECSU is installed on the 500 kV Tyumen substation, controls the Tyumen – Ural section (2x500 kV Tyumen – Reftinskaya power plant and 500 kV Berkut - Kurgan transmission lines). This control system is designed to prevent violations of the stability of the Reftinskaya power plant - Surgutskaya power plant power transmission in case of active power surges that are not controlled by centralized emergency control system of the Tyumen power system [1]. This type of power surges on power transmission may occur due to the discharge of the active load of consumers in the north of the Tyumen power system, associated with voltage drawdowns as a result of long-term short circuits in 110 - 220 kV networks and the operation of technological protections on the equipment of oil and gas production enterprises. Active power surges up to 1000 MW were observed. There is an ECSU for unloading two generators of Surgutskaya-2 power plant with a nominal power of 800 MW (400 MW per unit). The action is carried out through pulse unloading of the generator with subsequent limitation of the output power.

The disadvantage of an ECSU with a start trigger by active power is a complicated setup, due to the strong dependence of the limit of transmitted power over a controlled cross-section on repairs of lines and reactors in a 500 kV network, on the composition of generating equipment in the Tyumen power system. The existing ECSU at the 500 kV Tyumen substation provides 11 power stages (from 1200 MW to 2700 MW), which are introduced promptly depending on the current state of the network circuit, and the number of controlled states of the network circuit is several hundred. The complexity of the setup leads to an increased likelihood of erroneous actions by the staff. So on 17.06.2024, a false action of the ECSU was recorded due to an incorrectly setpoint for active power.

In addition to the active power, the danger of violation of stability can also be assessed by the phase angle between voltages at various power transmission points. Based on the type of angular characteristics of the power transmission, the use of ECSU with phase angle control allows to exclude or limit as much as possible the control of the state of the network.

* Corresponding author: Govorun-MN@so-ups.ru

© The Authors, published by EDP Sciences. This is an open access article distributed under the terms of the Creative Commons Attribution License 4.0 (https://creativecommons.org/licenses/by/4.0/).

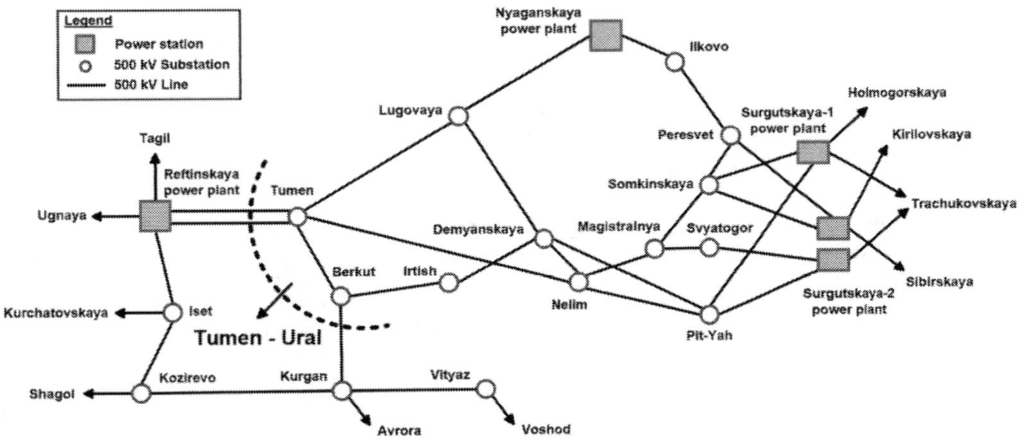

Fig. 1. Reftinskaya - Surgutskaya power transmission scheme

In various system network, when the limit of the transmitted power is changed, the angle corresponding to the value of the limit between the EMF remains close (although not always equal) to 90 degrees. Due to the peculiarities of the electrical circuit of the Tyumen power grid, the angle between the voltage vectors at Surgutskaya power plant and Reftinskaya power plant is close to the angle between the EMF vectors.

3 Description of the ECSU based on the synchrophasor angle

ECSU based on synchrophasor angle (ECSUa) is a function of local ECSU of power transmission Reftinskaya power station – Surgutskaya power station, using as a starting factor (trigger) the excess of the phase angle of a given setpoint between the voltage vectors at the electric power facilities of the Sverdlovsk and Tyumen power systems. The measurement of the phase angle is performed using a wide area measurement system (WAMS), which includes phasor measurement units (PMU). Synchronization allows you to measure the angles between vectors with high accuracy up to 1 degree. To organize ECSU, existing PMU devices are used, installed at all major stations of the Tyumen power system, as well as at the Reftinskaya power plant and transmitting data online.

The algorithm was implemented in the K-100 controller, developed by "Prosoft-Systems", and tests were conducted based on real PMU data using the controller. The PMU data is received and processed according to protocol C37.118 2005/2011. A validation algorithm was implemented for the input and processing of PMU data. The validation algorithm and the algorithm are implemented in the FBD (Function Block Diagram) visual programming language of the international standard IEC 61131-3 [2]. The IEC 60870-5-104 protocol is available in the controller to receive additional signals. The IEC 61850-8.1 protocol (GOOSE messages) is available to implement control actions and send control commands.

The device has a modular software architecture with a floating algorithm processing cycle. The configured software modules are executed sequentially, transmitting data to the next module using input and output signals. The device operates on the basis of generated data chains. A chain is a set of synchronized data generated by the device based on the data of the PMU and modules.

4 Description of the algorithm

Based on the conducted studies of static and dynamic stability, an algorithm for the operation of ECSU based on the synchrophasor angle has been developed. In the device, at each moment of time (continuously), the equivalent phase angle of the Reftinskaya – Surgutskaya power transmission is calculated using the formula:

$$\delta = \frac{\sum_{i=1}^{6} \delta_i \cdot \mathrm{P}_i}{\sum_{i=1}^{6} \mathrm{P}_i} - \frac{k_1\delta_7 + k_2\delta_8 + k_3\delta_9}{k_1 + k_2 + k_3} \quad (1)$$

where δ_1 – phase angle of the voltage vector on 500 kV bus system of Surgutskaya-1 power plant;
δ_2– phase angle of the voltage vector on 220 kV bus system of Surgutskaya-1 power plant;
δ_3– phase angle of the voltage vector at bus system №1 or №2 500 kV Surgutskaya-2 power plant;
δ_4 – phase angle of the voltage vector at bus system №3 or №4 500 kV Surgutskaya-2 power plant;
δ_5 – phase angle of the voltage vector on the 500 kV bus system of the Nyaganskaya power plant;
δ_6 – phase angle of the voltage vector on the 220 kV bus system of the Nyaganskaya power plant;
δ_7 – phase angle of the voltage vector on the 500 kV bus system of the Reftinskaya power plant;
δ_8 – phase angle of the voltage vector on the 500 kV busbar 500 kV Kurgan;
δ_9 – phase angle of the voltage vector on the 500 kV buses of the 500 kV Vityaz (control of this value will be required only after switching the 220 kV Vityaz – Irtysh overhead line to a voltage of 500 kV);
P_1 – total output active power of the Surgutskaya-1 power plant generators connected to 500 kV buses, these include TG-6 – TG-16;

© The Authors, published by EDP Sciences. This is an open access article distributed under the terms of the Creative Commons Attribution License 4.0 (https://creativecommons.org/licenses/by/4.0/).

P_2 – total output active power of the Surgutskaya-1 power plant generators connected to 220 kV buses, these include TG-1 – TG-5;

P_3 – total output active power of the Surgutskaya-2 power plant generators connected to 1 and 2 500 kV bus system, these include TG-4 – TG-8;

P_4 S-2 – total output active power of the Surgutskaya-2 power plant generators connected to 3 and 4 bus system 500 kV, these include TG-1 – TG-3;

P_5 – total output active power of Nyaganskaya power plant generators connected to 500 kV buses, these include TG-3;

P_6 – total output active power of the Nyaganskaya power plant generators connected to 220 kV buses, these include TG-1 and TG-2;

k_1, k_2, k_3 are the weighting coefficients, the values of which depend on the state of the network circuit.

Below is a simplified diagram of the connection of the PMU (Fig.2).

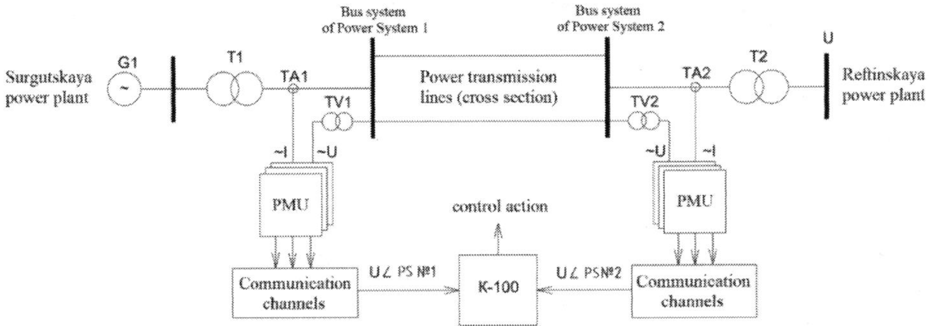

Fig. 2. Reftinskaya - Surgutskaya power transmission scheme

The fixation of an emergency overload of the power transmission is performed under several conditions. The algorithm provides two steps in terms of angle: sensitive and coarse. The sensitive angle step is designed to fix the fact of overload, and is determined by the lower limit of the range of angle values at the emergency permissible value of the active power overflow in the Tyumen-Ural section. The setpoints are set by constants, and are connected via discrete input signals (keys). The possibility of setting five groups of settings is involved. It is possible to manually set the settings. The rough angle step is designed to adjust the operation of the device in normal (quasi-steady-state) modes with overflows of active power at the level of the maximum permissible overflow, taking into account the operation of ECSUa, and is determined by the upper limit of the range of values of the phase angle in this mode. The setpoints are set by constants, and are connected via discrete input signals (keys). The possibility of setting five groups of settings is involved. It is possible to manually set the settings.

The algorithm of functioning provides for two steps in terms of the speed of increasing the angle of transmission. The first step in the rate of angle change is designed to determine high rates of angle change and is designed to take into account fast transients associated with the occurrence of short circuit in the network, synchronous swings and asynchronous modes. The second step in the rate of angle change is designed to determine the average (or small) rates of angle change and is designed to take into account normal (quasi-steady-state) modes. The speed is determined as the average value of the angle change rate in the interval of 50 ms. The speed measurement is performed continuously, and its comparison with the setpoints is carried out within 50-100 ms after the transmission angle exceeds one of the setpoints in terms of angle (sensitive or coarse).

The conditions for triggering the sensitive stage are as follows: exceeding the setpoint for the angle of the sensitive stage, the rate of angle change is less than the first stage, but more than the second stage in terms of the rate of angle change. The conditions for triggering the coarse stage are as follows: exceeding the setpoint for the angle of the sensitive and coarse, the rate of angle change is less than the first stage, but more than the second stage in terms of the rate of angle change.

The response time of the ECSUa stages consists of the time required to measure angles and power by the PMU data, the time of transmission of information from the PMU to the ECSUa, the time of operation of the ECSU to process the information received from the PMU and form the control action, the time of transmission of the command from the ECSUa to the actuator, the time of operation of the actuator and the time of implementation of the control action.

The algorithm of functioning of the ECSU provides for blocking its operation in the event of short circuits and synchronous oscillations. The lock is performed for the duration of the transient attenuation after the short circuit is eliminated. Additionally, blocking is provided for the presence and magnitude of the reverse sequence voltage.

The dosage of control actions is not provided, due to the fact that power surges on the power transmission are associated with motor load outages in the Tyumen power system (including low-voltage motors), which occur as a result of voltage drawdowns in the network. In this case, the disconnection of consumers does not occur at the same time, and the intensity and sequence of emergency shutdowns of the motor load may be different each time.

When the phase angle of the power transmission reaches the setpoint of the sensitive stage, the speed of

© The Authors, published by EDP Sciences. This is an open access article distributed under the terms of the Creative Commons Attribution License 4.0 (https://creativecommons.org/licenses/by/4.0/).

the angle is compared with the speed settings, while the following situations are possible:

1) The speed of the power transmission angle exceeded the first speed setpoint (occurrence of short circuit in the 500 – 220 kV network). The device is blocked for the duration of the short circuit elimination and attenuation of subsequent synchronous oscillations.

2) The speed of the power transmission angle did not reach the setpoint of the first stage in terms of speed, but reached or exceeded the setpoint of the second stage in terms of speed (emergency power surge on the power transmission). With the time delay necessary for detuning from synchronous swings, a command is formed to enter the control action. If, during the time delay, the value of the phase angle of the power transmission decreases below the setpoint of the sensitive stage, the device will return to its original state.

3) The speed of the transmission angle has not reached the setpoint of the second speed stage (normal (quasi-steady-state) modes in which the angle may increase due to changes in the balance situation in the power system, daily load fluctuations, etc.). Coarse stage is unblocked and the timer is started. If the sensitive stage does not return during the timer operation or the rough stage does not work, an emergency warning signal is generated to attract the attention of the operational dispatch personnel. In this case, the organ of the sensitive stage in terms of angle can remain in the actuated state for an indefinite period of time until one of the events occurs:

– the value of the transmission angle will decrease below the setpoint of the sensitive stage in terms of the angle. The device returns to its original state.

– the value of the transmission angle will increase and exceed the setpoint of the rough stage in terms of the angle. The speed will not exceed the setpoint value. The ECSUa device is triggered.

– the value of the transmission angle will increase and exceed the setpoint of the rough stage in terms of the angle. The speed will exceed the setpoint value. The device is blocked for the time necessary to eliminate the short circuit and attenuate subsequent synchronous oscillations.

5 Verifying the validity of PMU data

The PMU data input and processing unit is implemented in the ECSUa. It is based on an algorithm for verifying measurements of transducers or telemechanics devices using the IEC 60870-5-104 transmission protocol. The algorithm implements the following functions:

- quality control of PMU data;
- fixing the unreliability of measurements;
- exclusion of a cast in the data;
- comparison of measurement values obtained through two channels by an amount exceeding a setpoint during a set time;
- checking the increase of measurement values from a given technological range within a given time;
- verification of measurement quality codes (reliable/unreliable) received from telemechanic devices;
- switching the measurement from the main PMU to the backup one for cases when some PMUs are out of operation;
- generating a signal to block the ECSU.

6 The results of the tests of the controller and the algorithm

The tests were carried out at the System Operator using real PMU data. During the tests, algorithms were refined: PMU data validation, processing of PMU data skipping, processing of surge during data transmission. The main emphasis during the tests was placed on checking the reliability of the algorithms and the controller in conditions of a decrease in the quality of the PMU data stream for various reasons. The correct operation of the controller in conditions of working with data from PMU devices installed in a real power system is demonstrated.

The tests carried out have shown the efficiency of the algorithm and the correctness of calculating the mutual angle, including taking into account the rate of angle change. In case of malfunctions of PMU devices, false alarms of the algorithm did not occur.

7 Conclusion

The ECSUa algorithm was developed based on studies of static and dynamic stability, tested on a real device (controller) with real PMU data. The quality of PMU data, regulated by international and Russian documents, as well as the information and communication infrastructure created, make it possible to effectively use PMU data for emergency control system.

The K100 controller allows you to implement various algorithms using WAMS, while algorithms for receiving and processing data from PMU devices received via communication channels have been developed and tested. Effective algorithms for the reliability of PMU data have been tested, which increases the reliability of emergency control system.

References

1. E.I. Satsuk, A.N. Filinkov, A.V. Yudin Automation of preventing violations of dynamic stability in the Tyumen power system during overload of connections with the Urals. 5th Int sci and tech conf "Mod. Tr. in the Dev. of Rel. Prot. Syst. and Auto. of Pow. Sys." (2015)
2. The program "SoftConstructor2". Integrated development environment for algorithms and automation circuits. User's Guide. 62.01.29.000-102 (2022)

© The Authors, published by EDP Sciences. This is an open access article distributed under the terms of the Creative Commons Attribution License 4.0 (https://creativecommons.org/licenses/by/4.0/).

Fault Location with Increased Reliability for Overhead Transmission Lines

Vladimir Lachugin[1]*, *Andrey* Podshivalin[2], *Grigory* Ismukov[2], *Alexandr* Lvov[3]

[1]Joint Stock Company Scientific and Technical Centre of Federal Grid Company of the Unified Energy System, Kashirskoe shosse, 22, Moscow, Russia
[2]Relematica Limited Liability Company, pr. I.Yakovleva, 1, Cheboksary, Russia
[3] Branch of Public Joint Stock Company ROSSETI – IES of East, ul. Dzerzinskogo, 43, Khabarovsk, Russia

Abstract. The double-ended traveling wave (TW) method is a promising tool for fault location on transmission lines. It implies that high-frequency transients during fault inception are recorded in specific devices, which identify the precise time, when a front of an electromagnetic traveling wave reaches the corresponding end of the transmission line. The principles of operation, methods of calculations and test data of the developed traveling wave FL (TWFL) system are presented. The results of field tests and experience of operation of the TWFL system on the double-circuit 220 kV Tommot-Maya transmission line including short-circuit faults caused by lightning strikes are discussed.

1 Introduction

Accurate localization of a fault on a transmission line is may significantly reduce the time spent for restoration of a line after an emergency outage. There are various methods of fault location, but the traveling wave method based on measurements of the time of electromagnetic wave propagation along the overhead line from the fault location to each of the ends of the transmission line is known for the highest accuracy.

This method differs from the other methods by the fact that it uses information from the waves originating at the fault location on the transmission line, not yet distorted by the waves reflected from the node points of the network. The distance l from the fault location to one of the line ends (line length L) when using the TWFL method of double-ended measurements is determined by the expression [1]

$$l = \frac{L}{2} - \frac{t_2 - t_1}{2} v = \frac{L}{2} - \frac{\Delta t}{2} v ,$$

where t_1 and t_2 are the timestamps of arrival of the electromagnetic wave front to each end of the transmission line, respectively, v is the propagation velocity of the electromagnetic wave.

2 TWFL Principles, Calculation Methods and Laboratory Tests

In the proposed method, time-synchronized (using a satellite global navigation system) instantaneous values of transmission line voltages or currents in a system of independent modal channels are processed [2]. To fix the time of wave front arrival, the method of mathematical statistics is used, which allows calculating the coefficients of excess [3] for electromagnetic wave signals in interphase modal channels of a damaged transmission line by the expression

$$K_E = \frac{\mu_4}{\sigma^4} - 3 ,$$

where μ_4 and σ are the fourth and second central moments of the statistical distribution, respectively, defined as

$$\mu_4 = \frac{1}{n} \sum_{i=1}^{n} (x_i - \bar{x})^4 ,$$

$$\sigma^2 = \frac{1}{n} \sum_{i=1}^{n} (x_i - \bar{x})^2 ,$$

where x_i is the signal under study, and $\bar{x}$ is the arithmetic mean value, n is the number of signal

* Corresponding author: Lachugin_VF@ntc-power.ru

© The Authors, published by EDP Sciences. This is an open access article distributed under the terms of the Creative Commons Attribution License 4.0 (https://creativecommons.org/licenses/by/4.0/).

samples, chosen in accordance with [6] and actually accepted equal to 800.

In the absence of a transient (before the appearance of the wave front), the controlled signal (e.g. voltage drop of the incident wave, in relative units) is close to a normal distribution, so the value of the kurtosis coefficient will be at the level of zero. In the presence of a transient process, in particular caused by the occurrence of a short-circuit, the kurtosis increases sharply. This allows to determine quite accurately the fact of the wave front appearance in the points of its control [4-6] and to provide tuning of the TWFL system from interference [7]. Fig. 1 shows the dependencies characterising the change in the kurtosis coefficient for a single-phase earth fault that occurred on a 220 kV power line at a distance of 40 km from the control point.

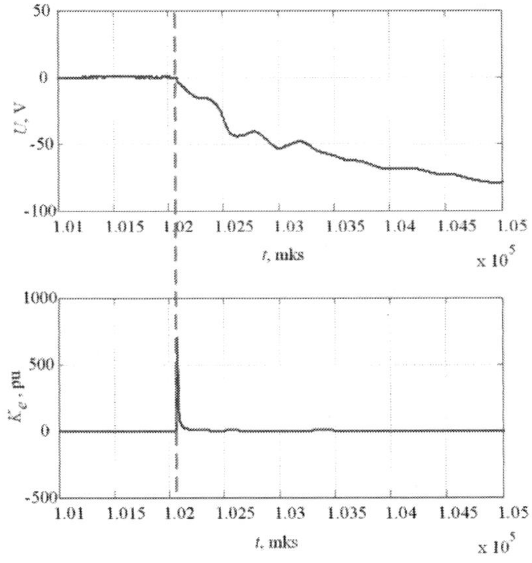

Fig. 1. Determining the timestamp of a wavefront using the kurtosis coefficient

On the basis of the proposed method, a TWFL system [8] was designed for FL on 110-750 kV overhead lines. It implements the transition from the devices used individually for each line, responding to the wave transient currents in each overhead line, to a system of devices in which the measurement of the wave voltage on the busbar system of the switchgear with a possibility to locate a fault on any transmission line of this busbar system [7]. In this regard, one TWFL device controls not one, but all lines connected to this busbar system, which leads to a significant reduction in the total number of devices used in the electrical network, and reduces the cost of transmission companies spent for installation and operation of a TWFL system.

In the developed algorithm of the TWFL system, the superimposed values of interphase voltage components on the switchgear busbars are identified, the time of arrival of the front of these components at each line end is synchronously recorded, and the process of calculating the distance to the fault location is started only on the damaged transmission line after the operation of its selective protection. The use of interphase wave voltage allows increasing the accuracy of the TWFL system, since the degree of deformation of the wave front in the inter-phase channel [2] when propagating from the location of a short-circuit to the point of installation of the TWFL device is significantly lower than that of the phase and zero components.

To study the efficiency of the TWFL system, a model of the electrical network was created, taking into account the presence of elements directly affecting fast transients during short-circuit (transmission lines, voltage transformers, coupling capacitors and high-frequency barriers) [9]. During the calculations, the distance to the fault location, the types of faults and the value of the transient resistance at the fault location were varied. The busbar systems with the largest number of overhead lines, where the wave voltage is minimal, were considered as the design ones.

For the laboratory tests of the TWFL system, a test bench [9] was developed, including a computer and a generator of arbitrary waveform signals, simulating the transient voltage during short-circuit faults on transmission lines. The computer was loaded with a database of files of calculated test signals containing arrays of instantaneous values of voltages at the locations of the TWFL devices for different types and locations of short-circuit faults. During the tests, the terminals of the TWFL system communicated within a local network and synchronized to a satellite navigation system independently. The type of fault, place of short-circuit, instantaneous values of the voltage preceding the short-circuit, as well as the values of transient resistances in the location of a short-circuit were varied. The tests showed that in the case of a fault on 330 kV overhead line, the error of the TWFL system did not exceed 0.3 km, being within the span length of the power line.

3 Results of Field Tests and Operating Experience

A two-circuit 220 kV Tommot-Maya transmission line located in the Republic of Sakha (Yakutia) was

© The Authors, published by EDP Sciences. This is an open access article distributed under the terms of the Creative Commons Attribution License 4.0 (https://creativecommons.org/licenses/by/4.0/).

selected as the object of field tests. The devices of the TWFL system were connected to inductive voltage transformer (VT) installed on the 220 kV busbar of Tommot and Maya substations. The length of the double-circuit 220 kV line Tommot - Maya [9] is 434.7 km. In the course of field tests, two experiments of single-phase earth faults were performed - through a metal support and in the middle of the span between the supports of the first circuit (through a specially cut down pine tree installed under the first circuit), while maintaining transit on the second circuit. In the first experiment the error of the TWFL system on the side of the Tommot substation was 570 m and on the side of the 220 kV Maya substation - 626 m, and in the second experiment - 268 m and 314 m, respectively. At the same time, the error of the TWFL devices operating on this transmission line, reacting to the parameters of the industrial frequency, reached several kilometers.

After completion of field tests, the TWFL system was put into pilot operation on both circuits of 220 kV Tommot-Maya transmission line. During operation, several short-circuits caused by lightning overlaps of the transmission line suspension insulators were recorded, which was confirmed by the data of the transmission line bypass. The high probability of lightning outages is consistent with the general statistics of outages in this region [10]. The short-circuit outages occurred without significant damage, with traces of overlaps visible only upon close inspection. Inspection of all insulators on 1287 poles of the Tommot-Maya 220 kV transmission line is very difficult, so an accurate solution of the TWFL problem is highly relevant.

Table 1 presents the performance of the TWFL system for short-circuit faults on the Tommot-Maya transmission line. They show a significantly higher accuracy of the TWFL system [11] compared to the phasor based devices [12] installed on the same transmission line.

According to the operational data [10], the time of fault location search on this transmission line after installation of the developed TWFL system was reduced from one week to one day. This indicates the significant contribution of the TWFL system in improving the reliability of power supply to consumers by reducing the time of restoration of the 220 kV power network in conditions of difficult climatic conditions and undeveloped transport infrastructure. All short-circuits presented in the table occurred during thunderstorms. Therefore, it was important to assess the noise immunity of the TWFL system in these modes.

In the No. 1 case there was no increase in the level of interference immediately before the short-circuit in the recording, as well as during the field tests, in contrast to the No. 2 - 4 cases. At the same time, the duration of interference preceding the short-circuit counted from several hundreds of microseconds to several milliseconds. In particular, on the basis of the analysis of the data in Fig. 2, corresponding to the No. 4 case, it was determined that the value of the excess coefficient at the beginning of the flashover effect is not more than 15.1, and then increases to 117.3 after the short circuit occured, which is quite sufficient to ensure reliable selftuning of the TWFL system from the interference of a lightning and ensure high sensitivity to the wave front caused by a short circuit. Similar results were obtained when analysing the disturbances corresponding to other cases of short-circuit faults under consideration.

Table 1. Operation results of a TWFL system for short-circuits on a transmission line

No	Date, time	Actual distance to fault, km	FL by phasors, km	TWFL, km	TWFL error, km	Interference
1	22.05.2022 17:50:06	61,8	57,9	61,0	0,8	No
2	03.07.2022 18:41:44	314,4	335,9	314,1	0,3	Yes
3	03.07.2022 21:41:35	210,9	212,1	208,2	2,7	Yes
4	03.08.2022 12:36:34	181,5	182,4	181,3	0,2	Yes

Consequently, the TWFL system retains its functional reliability under lightning disturbances, not caused by the fact of a short-circuit, providing the calculation of the distance to the fault location only after the formation of the short-circuit channel.

© The Authors, published by EDP Sciences. This is an open access article distributed under the terms of the Creative Commons Attribution License 4.0 (https://creativecommons.org/licenses/by/4.0/).

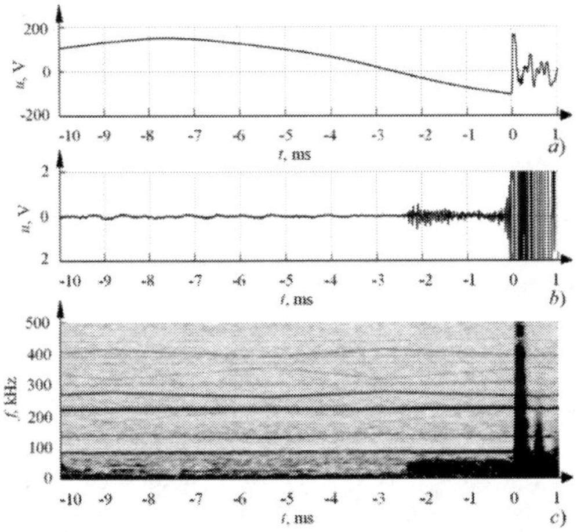

Fig 2. Development of the transient process before the short-circuit accompanied by lightning disturbances: *a)* - voltage signal measured by the VT; *b)* - voltage signal after filtering of the fundamental frequency component; *c)* – spectrogram of the voltage signal

4 Conclusions

1. The transition to a double-end TWFL system with terminals connected to the busbar VT reduces the total number of WFL devices in the electrical network and their maintenance costs.

2. The results of tests and operating experience on 220 kV Tommot-Maya transmission line, including thunderstorm disturbances, confirmed the high accuracy and reliability of the developed TWFL system. The algorithm determines the distance to the fault location on the basis of statistical method, adjusted for the impact of interference, not associated with the occurrence of short-circuit.

3.The developed TWFL system provides the increase of operational reliability of the high-voltage power network by reducing the mean time spent for the search of the actual fault location on long transmission lines from one week to one day.

References

1. G.M. Shalyt, Fault location in electric networks (in Russian), Energoizdat (1982)
2. M.V Kostenko., L.S. Perelman, Y.P. Shkarin, Wave processes and electrical disturbances in high voltage multi-conductor lines (in Russian), Energia (1973)
3. E.S. Ventzel, L.A. Ovcharov, Probability theory and its engineering applications (in Russian), Higher School (2010)
4. V.F Lachugin, D.I. Panfilov, A.N. Smirnov, S.A. Obraztsov, A.A. Ryvkin, A.O. Shimina, Multifunctional device for process registration, power quality control and fault location on the power transmission lines (in Russian), Electrical power stations, No. 8, P. 29 – 36 (2013)
5. V.F. Lachugin, D.I. Panfilov, A.N. Smirnov, Realisation of the wave method for the fault location on the power lines using the statistical methods of the data analysis (in Russian), Izvestiya RAN. Energetika, No.6, P. 137 – 146 (2013)
6. A.N. Podshivalin, Investigation of the properties of the kurtosis coefficient in the task of the wave determination of the fault location of the power lines (in Russian), Relay protection and automation, №1, P. 50-57 (2023).
7. A.N. Podshivalin, G.N. Ismukov, Statistical measuring organ of the wave RPA devices (in Russian), Izvestiya Akademiya Elektrotekhnicheskikh Nauk RF, Issue 23, July, P. 16-27 (2021).
8. V.F. Lachugin, P.S. Platonov, S.V. Ivanov, G.N. Ismukov, A.N. Podshivalin, V.G. Alekseev, M.V. Vaziulin, N.N. Mitrofanov, S.G. Popov, A.S. Shemetov, Method for determination of the fault location on the transmission line (in Russian), RU 2739433. Published 24.12.2020, Bulletin No. 36.
9. V.F. Lachugin, D.I. Panfilov, S.G. Popov, P.S. Platonov, V.G. Alekseev, N.G. Klushkin, A.N. Podshivalin, Development and application of devices for determining the fault location on the power transmission lines using wave methods (in Russian), Energy of the Unified Network, № 5-6 (60-61), P. 50-66 (2021)
10. A.Yu. Kiselev, A.P. Lvov, P.F. Vasiliev, P.S. Pinchukov, Investigation of the prerequisites of the reverse overlapping of the insulator garlands of the main power transmission lines and the determination of their damage location (in Russian), Energy of the united network. №3-4 (70), P. 20-24 (2023)
11. A.P. Lvov, V.F. Lachugin, P.S. Pinchukov, A.Yu. Kiselev, Application of the fault location devices on the 220 kV overhead power lines of the western and central districts of the Republic of Sakha (Yakutia) (in Russian), Energia of the united network, №5-6 (71), P. 24-28 (2023)
12. E.A. Arzhannikov, V.Yu. Lukoyanov, M.Sh. Misrikhanov, Determination of the short-circuit location on the high-voltage power lines (in Russian), Edited by V.A. Shuin, Energoatomizdat (2003)

© The Authors, published by EDP Sciences. This is an open access article distributed under the terms of the Creative Commons Attribution License 4.0 (https://creativecommons.org/licenses/by/4.0/).

Monitoring of medium voltage distribution networks by balance smart meters

Irina Golub[1,*], *Oleg* Voitov[1], *Evgeny* Boloev[1], and *Anatoly* Buchinsky[2]

[1]Energy Systems Institute of Siberian Branch of the Russian Academy of Sciences (ESI SB RAS), 664033, Irkutsk, Russia
[2]Regional State Unitary Energy Enterprise «Electric Grid Company for the operation of electric networks of Oblkommunenergo», 664009, Irkutsk, Russia

Abstract. This study addresses the issue of monitoring the unbalanced load flow of power distribution networks. To that end, instead of relying on medium voltage meters, we make use of balance smart meters installed on the low-voltage side of distribution transformers. We use a mathematical model of such transformers for real-time calculation of the powers and voltages on the medium voltage side.

1 Introduction

The load flow control of medium voltage (MV) distribution networks (DN) requires the operator to have accurate information about the state variables to ensure reliable, high quality, and economic operation of the system. Such applications that require current loading data rather than reference metering data or loading forecasts include reactive power and voltage regulation, network reconfiguration and reinforcement, and protective relaying and automatic controls.

Load unbalance in low voltage (LV) networks leads to unbalanced load flow of MV networks, thus requiring detailed three-phase modeling of the transmission lines and distribution transformers (DT) connecting the MV and LV networks. In order to estimate the parameter values of the state of the MV electrical network, one needs a software tool for load flow analysis of a three-phase unbalanced network based on measurement data and the availability of measurements in real time.

Load flow analysis of unbalanced networks can be performed by the SDO-7 software package, which introduces a single-line representation of three-phase lines, allowing the use of symmetrical load flow analysis programs for the unbalanced load flow analysis [1].

As far as direct measurements in LV networks are concerned, the load measurements available at individual nodes are not sufficient to ensure the observability of the DN. The substantial contribution of this study is that instead of measuring the loads and voltages on the primary side of the LV directly, we propose to calculate them through load monitoring by a balance smart meter (BSM) [2] on the secondary side of the LV DT. Such an approach proves effective due to the low cost of BSMs installed on the LV network as compared to devices installed on the MV network, and even more so when they are installed on both sides.

We performed a comprehensive analysis of the mathematical model of 6(10)/0.4 kV transformers used in Russian DTs that rely on the Y/Y_0 arrangement of windings (wye-grounded wye). This allowed us to develop a simple algorithm for calculating loads and voltages on the primary side of DT. The algorithm showed, unlike what was reported in [3], that the total power losses were independent of the value of DT zero-sequence impedance. Furthermore, unlike the algorithm reported in [4], it did not require one to take into account the singularity of the block partitions of the DT nodal admittance matrix.

The calculations reported in this paper were performed for measurements made in a real-world DN and were validated against a measurement experiment on a physical model.

2 Model of a three-phase distribution transformer with the winding connection scheme Y/Y_0

Impedance values of windings of a three-phase DT winding reduced to the LV side are specified as a square matrix of impedances under the phase frame of reference Z_{abc} as obtained by the diagonal matrix Z_{120} of impedances of positive z_1, negative z_2, and zero z_0 sequences and the matrix A of transition from symmetrical components to the phase frame of reference

$$Z_{120} = \begin{pmatrix} z_1 & & \\ & z_2 & \\ & & z_0 \end{pmatrix}; Z_{abc} = AZ_{120}A^{-1};$$

$$A = \begin{pmatrix} 1 & 1 & 1 \\ a^2 & a & 1 \\ a & a^2 & 1 \end{pmatrix}; a = e^{j\frac{2\pi}{3}}. \quad (1)$$

The impedance matrix Z_{abc} is used to arrive at the admittance matrix of the transformer windings $Y_t = Z_{abc}^{-1}$.

When modeling a DT, we assume that its primary winding is reduced to the secondary winding. The reduced currents I_1° on the primary side and currents I_2

* Corresponding author: golub@isem.irk.ru

© The Authors, published by EDP Sciences. This is an open access article distributed under the terms of the Creative Commons Attribution License 4.0 (https://creativecommons.org/licenses/by/4.0/).

on the secondary side of the DT, can be determined by the matrix of nodal admittances Y_{Y/Y_0}, the reduced voltage on the primary side U_1°, and the voltage on the secondary side U_2 as reported in [5,6]

$$\begin{pmatrix} I_1^\circ \\ I_2 \end{pmatrix} = \begin{pmatrix} Y_{11} & Y_{12} \\ Y_{21} & Y_{22} \end{pmatrix} \begin{pmatrix} U_1^\circ \\ U_2 \end{pmatrix} = Y_{Y/Y_0} \begin{pmatrix} U_1^\circ \\ U_2 \end{pmatrix}. \quad (2)$$

The structure of block partitions of the nodal admittance matrix is determined by the winding connection scheme Y/Y_0, for which, as shown in [5], block partitions Y_{11} and Y_{22} completely coincide, and block partitions Y_{12} and Y_{21} coincide with them while having the opposite signs, with all blocks being singular

$$Y_{11} = Y_{22} = \frac{1}{3}\begin{pmatrix} 2 & -1 & -1 \\ -1 & 2 & -1 \\ -1 & -1 & 2 \end{pmatrix} Y_t;$$

$$Y_{12} = Y_{21} = \frac{1}{3}\begin{pmatrix} -2 & 1 & 1 \\ 1 & -2 & 1 \\ 1 & 1 & -2 \end{pmatrix} Y_t. \quad (3)$$

3 Algorithm for calculating voltages and powers on the primary side of the DT by measurements coming from BSMs installed on the secondary side

We assume that active and reactive powers and voltages on the primary side of the DT are to be determined by snapshots [a] of measurements of average phase values of active and reactive powers and voltage magnitudes relative to the neutral point, as performed by a three-phase BSM installed on the secondary side of the DT.

The voltage vector of the secondary winding of the DT can be represented as follows, according to the measurements of voltage magnitudes in phases *a, b, c* for each of the measurement snapshot:

$$U_2 = \begin{pmatrix} U_2^a \cdot (1 + j0) \\ U_2^b \cdot \left(-0.5 \cdot -j\sqrt{3/2}\right) \\ U_2^c \cdot \left(-0.5 + j\sqrt{3/2}\right) \end{pmatrix}. \quad (4)$$

Currents in the windings of phases f on the secondary side of DT are calculated by measuring phase values of active P_2^f and reactive Q_2^f power and voltage values U_2^f

$$I_2^f = \left(P_2^f - jQ_2^f\right)/U_2^{f*}. \quad (5)$$

In the case of unbalanced loading, voltages and currents contain positive, negative, and zero-sequence components [7].

The reduced current I_1° on the primary side of the transformer, as per expression (2), can be determined as

$$I_1^\circ = Y_{11}U_1^\circ + Y_{12}U_2. \quad (6)$$

The lack of grounding of the neutral point of the transformer primary winding results in the absence of symmetrical components of the zero-sequence in I_1°, while the voltages U_1° on the primary side of the transformer have positive, negative, and zero sequences.

[a] Measurement snapshots mean a set of records of simultaneous readings of all SMs taken at a specified time interval separating individual records

The reduced voltage U_1° included in (6) can be found from the equation

$$U_1^\circ = (Y_{21})^{-1}(I_2 - Y_{22}U_2). \quad (7)$$

Since the matrix Y_{21} is singular, a regularization procedure [8] can be applied to obtain its inverse matrix, which is achieved by adding very small quantities to the diagonal entries of the matrix being inverted, thus simulating the introduction of shunts with small line-to-ground conductances in the ungrounded neutral point of the primary winding [4]. This approach can hardly be considered practical because of the large uncertainty of the resulting solution.

Another technique reported in [9] to overcome the singularity of the matrix Y_{21} is to determine the transformer U_1° primary side voltages omitting the zero-sequence component. Such a solution can be obtained by replacing the last row of the matrix Y_{21}, which is being inverted, with ones, and replacing the last row of the matrix Y_{22} and the last entry of the vector I_2 with zeros. Since the values of entries of the vector U_1° depend on the number of the row Y_{21} replaced by ones, in what follows we calculate their averages when adjusting each of the rows in (7), matrices Y_{21}, Y_{22}, and vector I_2 that correspond to phases *a, b, c*.

An alternative approach to obtaining the components of all three sequences for U_1° is to determine the voltage U_1° in the conventional way [7], i.e., by adding the voltage drop ΔU_{12} (defined as the product of the current I_2 and the matrix Z_{abc} of transformer winding impedances) to the voltage U_2.

$$U_1^\circ = U_2 + \Delta U_{12} = U_2 + Z_{abc} \cdot I_2. \quad (8)$$

An important result of comparing techniques (7) and (8) for determining the voltage U_1° is the complete coincidence of the positive- and negative-sequence components obtained therein. It should also be noted that the non-zero-sequence current I_1°, defined as (6), is independent of the way the voltage U_1° is calculated (i.e., without (7) or with (8) the zero-sequence component). Furthermore, the components of non-zero sequences of currents I_1° and I_2 coincide, which allows determining the current I_1° by removing the zero-sequence component from the current I_2 instead of using expression (6)

$$I_1^\circ = I_2 - \frac{1}{3}(I_2^a + I_2^b + I_2^c). \quad (9)$$

The currents and voltages obtained on the primary side of the transformer are used to determine the powers of the phase loads in the MV network

$$S_1^f = P_1^f + jQ_1^f = U_1^{\circ f} \cdot I_1^{\circ f^*}. \quad (10)$$

When referring to equation (10), we will use the notation "without ZS" and "with ZS" to differentiate between the cases when it uses the voltage without and with the zero sequence, respectively.

There is no doubt that real-time calculation of loads with calculation intervals matching those of BSM measurements can prove a game-changer when compared to the use of control measurements or forecasts used to the same effect. The new approach enables load flow analysis in MV networks and its use as a foundation for solving a number of problems requiring knowledge of current load flow parameters.

© The Authors, published by EDP Sciences. This is an open access article distributed under the terms of the Creative Commons Attribution License 4.0 (https://creativecommons.org/licenses/by/4.0/).

Our research has shown that in order to determine the phase loads S_1^f on the primary side of the transformer according to BSM measurements on the secondary side of DT with the winding connection scheme Y/Y_0, average values of active P_2^f and reactive Q_2^f load powers and voltage magnitudes U_2^f, it is necessary to perform calculations as per the expressions indicated in brackets $I_2^f \to (5) \Rightarrow U_1^{\circ f} \to (8) \Rightarrow I_1^{\circ f} \to (9) \Rightarrow S_1^f \to (10)$.

Due to the lack of power and voltage measurements on the primary side of the transformer, one way to validate the solutions obtained in this way can be to compare the total active and reactive power losses in transformers Y/Y_0 with the losses calculated by the simple technique recommended in [10] and, according to the authors, validated against the results of numerous experiments on a physical model.

Ref. [3, 10] calculated the total power losses in transformers with winding connection scheme Y/Y_0 by positive, negative, and zero-sequence currents i_1, i_2, i_0 and impedances $r_1 + jx_1, r_2 + jx_2, r_0 + jx_0$ as follows

$$\Delta S = \Delta P + j\Delta Q = \\ = 3 \sum_{k=1,2,0} i_k i_k^* r_k + j3 \sum_{k=1,2,0} i_k i_k^* x_k. \quad (11)$$

The symmetrical components of the currents are determined from the phase currents on the secondary side of the transformer (2) as

$$i_{120} = A^{-1} I_2. \quad (12)$$

Another way to determine the power losses coinciding with those obtained using expression (11) is to calculate the total power losses under the phase frame of reference

$$\Delta S = I_2^T Z_{abc} I_2\,, \quad (13)$$

and power losses in phases *a, b, c*

$$\Delta S_{f=abc} = \left(z_f \cdot I_2\right) \cdot I_2^*, \quad (14)$$

where z_f – row $f = abc$ of matrix Z_{abc}. The powers on the primary side of the transformer, taking into account (13), can be determined as follows

$$S_1^f = P_1^f + jQ_1^f = S_2^f + \Delta S_{f=abc}. \quad (15)$$

4 A case study of calculating MV network loads from BSM

We used 1,344 30-minute snapshots of measurements of active and reactive powers and voltage magnitudes to demonstrate how to calculate powers on the primary side of DT. The measurements were made by a three-phase BSM connected on the secondary side of DT 10/0.4. The DT had a nominal capacity of 160 kVA and served single-phase and three-phase loads of 39 residential buildings of a real-word LV electrical network.

Fig. 1 shows the graphs of average phase values of active and reactive powers measured on the secondary side of DT for two days (96 measurement snapshots) and calculated on the primary side of DT by technique (10) employing two ways of calculating voltages U_1° (with ZS and without ZS). Despite the discrepancy between the phase power values obtained for the two ways of calculation on the primary side of the DT, the total active and reactive power values for the two ways of calculation were equal to each other.

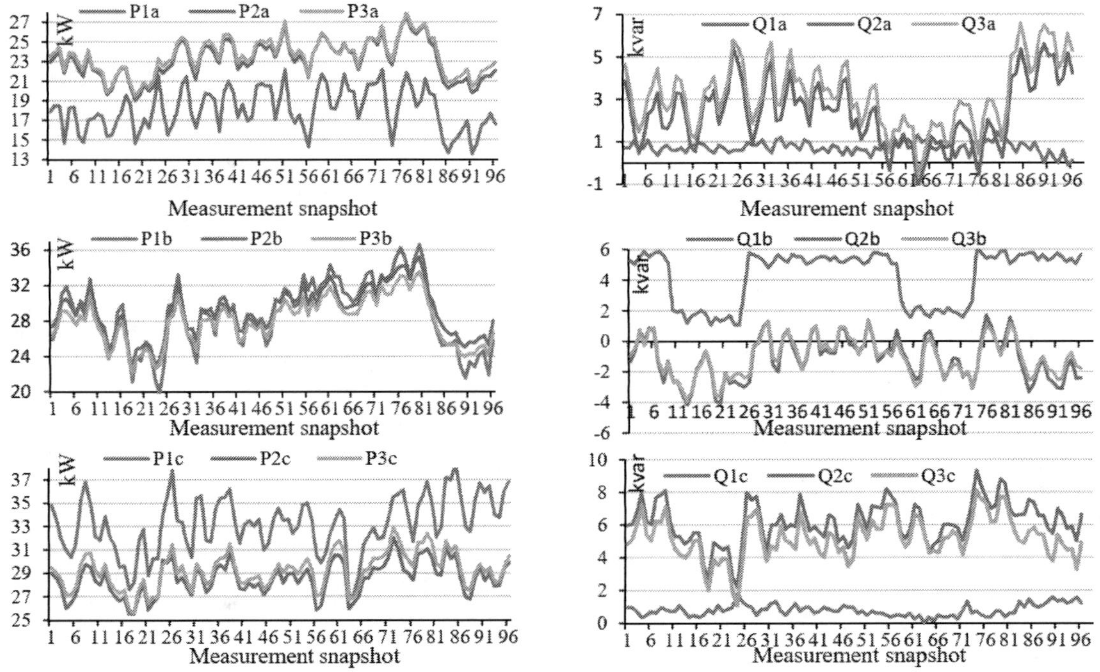

Fig. 1. Half-hourly average values of active and reactive powers in phases a, b, c: BSM -measured for two days - P1a, P1b, P1c, Q1a, Q1b, Q1c; as calculated on the primary side with ZS - P2a, P2b, P2c, Q2a, Q2b, Q2c and without ZS - P3a, P2b, P3c, Q3a, Q3b, Q3c.

© The Authors, published by EDP Sciences. This is an open access article distributed under the terms of the Creative Commons Attribution License 4.0 (https://creativecommons.org/licenses/by/4.0/).

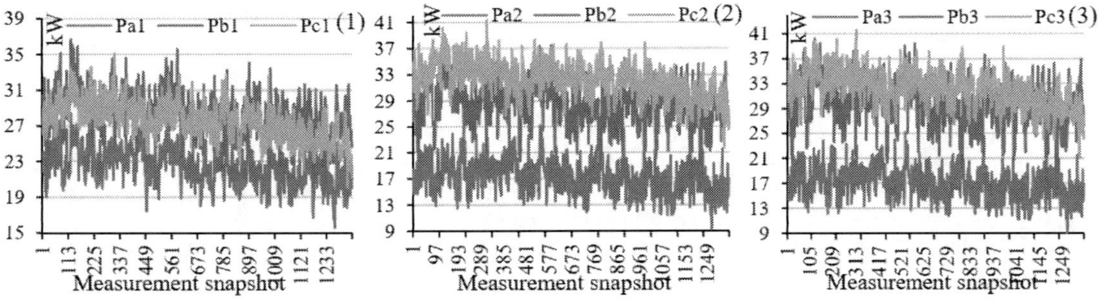

Fig. 2. Half-hourly average values of active powers in phases a, b, c,(1) - as calculated on the primary side of DT with ZS - Pa1, Pb1, Pc1,(2)- as measured on the secondary side of DT - Pa2, Pb2, Pc2, (3) as calculated on the primary side of DT as per (15) - Pa3, Pb3,Pc3.

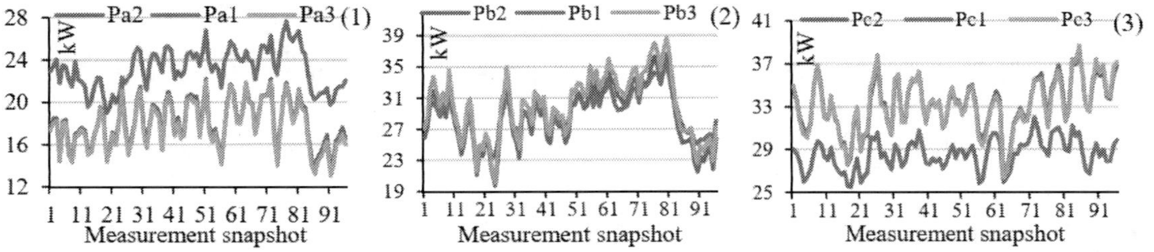

Fig. 3. Active half-hourly average values of powers in phases a, b, c: (1) as calculated on the primary side of DT with ZS - Pa1, Pb1, Pc1; (2) as measured on the secondary side of DT - Pa2, Pb2, Ps2; (3) as calculated on the primary side of DT as per (15) - Pa3, Pb3,Ps3.

To establish the relationship between the mismatched phase power values and the equality of their total values, Fig. 2 shows the graphs of active power in phases for 1,344 measurement snapshots during a month: 1) as calculated on the primary side with ZS; 2) as measured on the secondary side of the transformer; 3) as calculated on the primary side of the DT taking into account the power losses in phases as per (15).

Comparison of graphs 2.1 and 2.2 shows that redistribution of the total power of loads on the primary side of the DT with ZS led to their balancing, in which the power of the least loaded phase - *a* on the secondary side of the transformer increased, and the power of the most loaded phase - *c*, decreased. In the case of technique 2.3 for calculating the power on the primary side of DT with the use of power losses, there was no balancing of loads; on the contrary, the power of the least loaded phase *a* slightly decreased, and the power of the medium-loaded phase *b* increased.

Similar to Fig.2, Fig. 3 shows graphs of powers plotted for 96 first measurements out of 1,344 measurement snapshots, separately for each phase. They further confirmed the above conclusion that the calculation of power on the primary side of the transformer as per expression (10) led, unlike the technique (15), to balancing of power values in phases.

The ability of the investigated DT model to make balancing of phase loads on its primary side, as shown by calculations, was confirmed by the results of the experimental measurements, during which a stepwise varying load was connected to the phases on the low side of the three-phase transformer TZS-1.5/1-380/36. Two "UF2M Power Quality Analyzers" were connected to the windings on the primary and secondary sides of the transformer. Vector measurements of magnitudes and phases of currents and voltages, as well as active powers were performed. The measurements were GPS-synchronized, with an averaging interval of 200ms, or 5 measurements per second. A total of 4,691 measurement snapshots were made, from which 15 groups were selected that included 5 measurements per second for active powers in the transformer windings on its primary and secondary sides. Fig.4 shows that balancing was achieved by flattening the loads on the primary side of the transformer for 15 groups of 5 measurements including 75 snapshots.

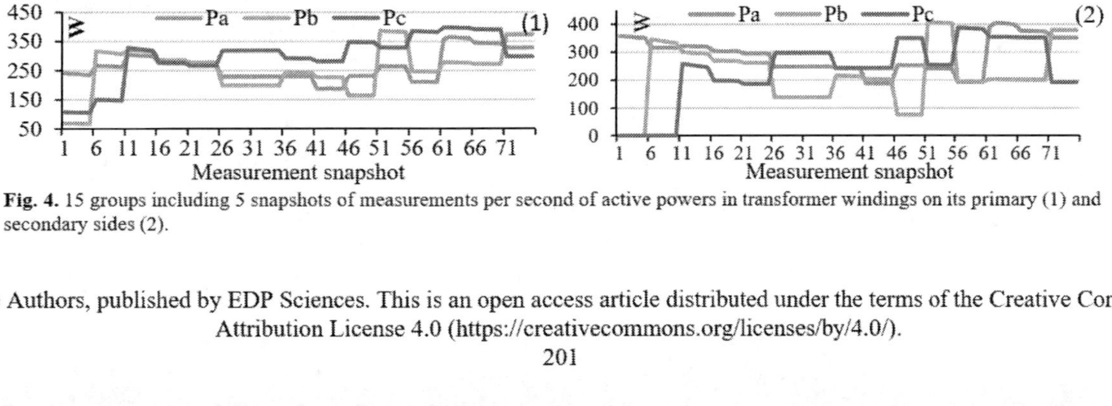

Fig. 4. 15 groups including 5 snapshots of measurements per second of active powers in transformer windings on its primary (1) and secondary sides (2).

© The Authors, published by EDP Sciences. This is an open access article distributed under the terms of the Creative Commons Attribution License 4.0 (https://creativecommons.org/licenses/by/4.0/).

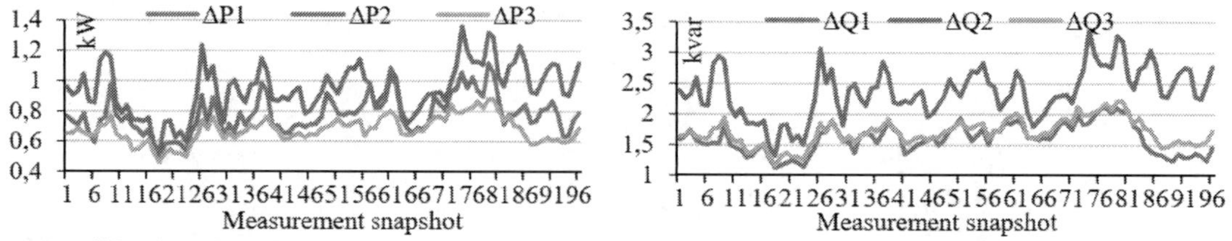

Fig. 5. Half-hourly total average values of active and reactive power losses for two days: ΔP1, ΔQ1 - losses obtained with the first technique; ΔP2, ΔQ2 - losses obtained as per expression (13) for nominal impedance values, and ΔP3, ΔQ3 - for zero-sequence impedance values reduced 10-fold.

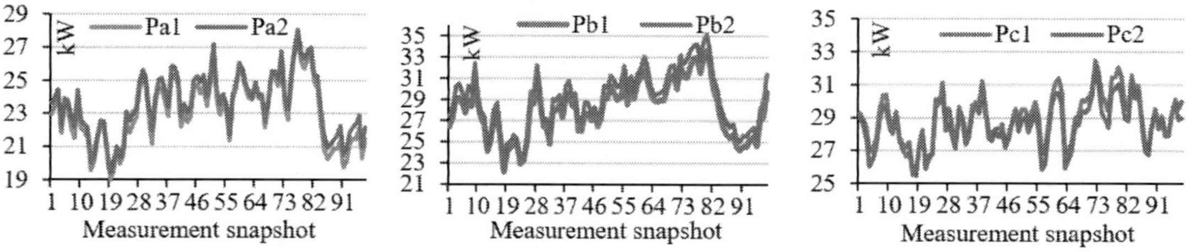

Fig. 6. Half-hourly average phase values of active powers with ZS, Pa1, Pb1, Pc1 corresponding to nominal values, and Pa2, Pb2, Pc2 with zero-sequence impedances increased 10-fold.

At the next stage of our study, we compared the total active and reactive power losses as calculated in two different ways: 1) as the difference of total power values calculated on the primary side of the DT and measured by BSM on its secondary side and 2) as total losses found as per expression (13) as recommended in [3,10].

In the case of applying the former technique, the total losses, similar to the total powers, did not change when the zero-sequence impedance was changed. The losses calculated as per the latter technique exceeded (in the case of standard zero-sequence impedance) or became less or equal (in the case of zero-sequence impedances reduced 10-fold) to the losses calculated as per the former technique, as shown in Fig. 5.

Fig. 6 shows how a change in zero sequence impedance, which did not affect the values of total power and total losses, led to a redistribution of total power between phases.

In what follows we show a load flow analysis performed for a real-world 10 kV MV DN fed from a 110/10 substation. The equivalent circuit of the DN, as presented in Fig. 7, contained 20 nodes. Eleven 10/04 DTs were connected to 9 of them and BSMs were installed on the LV side. The meters measured the load powers and voltage magnitudes at half-hourly intervals. The rated power of DT at node 15 was 400 kVA, nodes 19 and 20 were 250 kVA each, nodes 1 and 17 were 160 kVA each, nodes 5, 8, 16, and 18 were 100 kVA each, and nodes 4 and 9 were 63 kVA each.

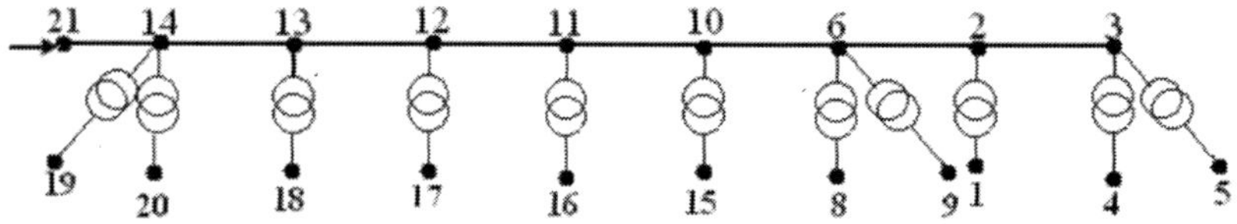

Fig. 7. Schematic of the MV DN with DT 10/0.4 at nodes 5, 4, 1, 9, 8, 15 – 20.

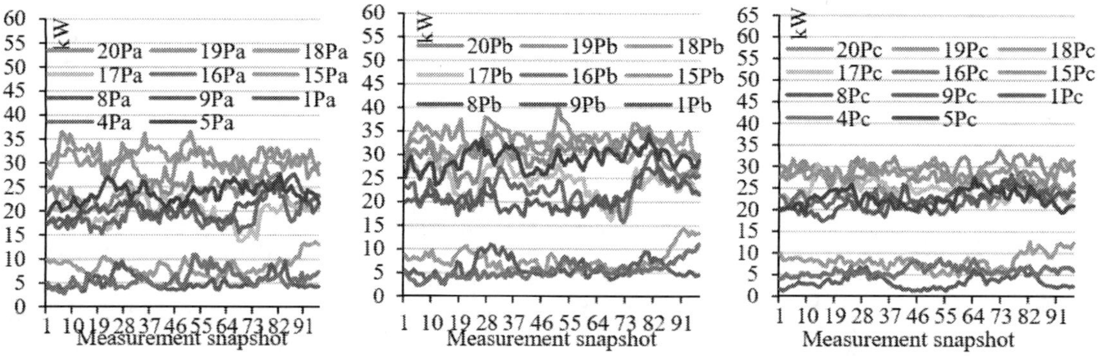

Fig. 8. Half-hourly active powers of loads in the MV network in phases a, b, c as averaged over two days and obtained with ZS.

© The Authors, published by EDP Sciences. This is an open access article distributed under the terms of the Creative Commons Attribution License 4.0 (https://creativecommons.org/licenses/by/4.0/).

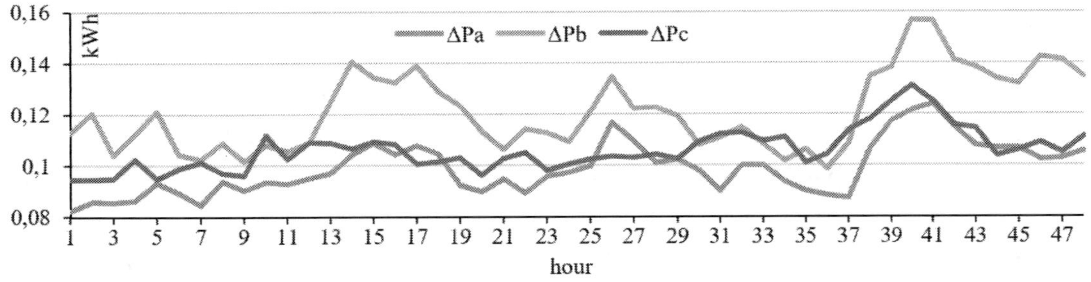

Fig. 9. Hourly power losses in the phases of the DN main wire for two days.

Fig. 8 shows the graphs of half-hourly average powers of active loads in the phases of the MV network, as calculated on the primary side of transformers. Similarly, the reactive powers of loads were obtained for each phase.

We relied on the SDO-7 [1] software package for load flow analysis in the unbalanced MV network (see Fig. 7) based on the values of active and reactive loads. To this end, Fig. 9 shows the values of average hourly active power losses in the phases of the main wire (which were equal to power losses per hour), as obtained from the load flow analysis for each hour for two days.

5 Conclusion

1. We proved experimentally the equality of non-zero components of voltage on the primary side of the DT for the cases when it was calculated both a) without considering the zero-sequence component and b) relying on the conventional technique of adding the voltage drop to the voltage on the secondary side of the DT, which allowed one to determine the zero-sequence component.

2. Our study showed that the total active and reactive powers on the primary side of the DT and the total power losses did not depend on either the zero-sequence component of the voltage or the value of the zero-sequence impedance of the DT.

3. The zero-sequence voltage component and the value of zero-sequence impedance affected the distribution of total active and reactive powers on the primary side between the phases of the DT.

4. Our calculations and an experimental measurement proved that calculation of loads on the primary side of the DT as the product of current and voltage led to balancing of phase loads as compared to loads on the secondary side of the DT.

5. We established that when the loads on the primary side were determined with the use of power losses calculated as the product of the matrix of self- and mutual impedances of the DT and the square of the current does it did not lead to the balancing of phase loads.

6. The efficacy of the contributed approach to calculating the loads on the primary side of the DT has been demonstrated through a case study of a real-world DN with 11 DTs, for which the load flow analysis was performed based on the loads on the MV side.

References

1. I. Golub, O. Voitov, E. Boloev, E3S Web of Conferences. **209**, 02013 (2020)
2. I. Golub, E. Boloev, International Conference on Electrical, Communication, and Computer Engineering (2021)
3. F. Kosoukhov, A. Epifanov, N. Vasiliev, N. Krishtopa, A. Gorbunov, A. Boroshnin, Electric power. Transmission and distribution. **6**, 81 (2023)
4. M.J. Gorman, J.J. Grainger, IEEE Transactions on Power Delivery. **7**, 2 (1992)
5. W. Dillon, M. Chen, IEEE PAS Summer Power Meeting (1972)
6. J. Arrillaga, D. Bradley, P. Bodger, *Power system harmonics* (John Willey &Sons, Chichester, 1972)
7. M. Kostenko, L. Piotrovsky *Electric machines. Part 2. AC machines. Transformers.* (Energiya, Leningrad, 1972)
8. M. Bazrafshan, N. Gatsis, IEEE Transactions on Power Systems. **33**, 2 (2018)
9. M. Wang, F. Chen, J. Li, IEEE Transactions on Power Systems. **19**, 4 (2004)
10. F. Kosoukhov, N. Vasiliev, A. Boroshnin, *Energy-saving in low-voltage electric networks under asymmetrical loading* (Lan', St. Petersburg, 2016)

© The Authors, published by EDP Sciences. This is an open access article distributed under the terms of the Creative Commons Attribution License 4.0 (https://creativecommons.org/licenses/by/4.0/).

E3S Web of Conferences **584**, 01042 (2024) https://doi.org/10.1051/e3sconf/202458401042
RSES 2024

Optimization of Data Processing Schemes for Locating Low-Frequency Oscillation Sources

Kirill Butin[1,2*], *Aleksandr* Popov[1,2], and *Andrey* Rodionov[1]

[1]Engineering Center «Energoservice», Russia
[2]Northern (Arctic) Federal University

Abstract. The organization of data processing for measurement data arrays is one of the key factors for the effective application of synchronized phasor measurement technology (PMU) within a Wide Area Monitoring System (WAMS). This encompasses not only ensuring high computational performance and scalability but also the openness of solutions to the continually evolving data analysis methodologies. In both the scientific community and current operational-dispatching practice, there is a growing interest in the issue of power system oscillatory stability. This interest is driven by the increasing complexity of modern power systems and enhanced observability of processes due to the deployment of WAMS. One of the emerging tasks is the real-time detection of low-frequency oscillation (LFO) sources using PMU data. The dissipating energy flow (DEF) method has shown high effectiveness in addressing this task by enabling the assessment of the magnitude, speed, and direction of oscillation energy propagation. It has been previously demonstrated that, while maintaining the general computational framework of the DEF method, the time required to reach a stable decision about the LFO source depends on the choice of the combination of subtask algorithms and their parameters. This study discusses a prototype computational platform for optimizing generalized data processing schemes based on PMU data. Using the DEF method and a dataset of actual LFOs in the Russian power system as an example, the study explores mechanisms for expanding the solution space and the measurement data base. The paper presents experimental results that provide an estimation of the expected time from the onset of oscillations to the identification of their source.

Introduction

In recent years, the Wide Area Monitoring System (WAMS) has been actively developed in Russia, made possible by the implementation of synchronized phasor measurement (PMU) technology and PMU-enabled devices [1]. This has led to an increased interest in the issue of forced low-frequency oscillations (LFOs) in power systems. As the complexity of power systems grows and their observability improves, there is a need to find effective methods for identifying or locating LFO sources in real-time with minimal delay after the oscillatory process begins.

There are numerous approaches to locating oscillation sources [2, 3], among which the Dissipating Energy Flow (DEF) method [4] stands out. This method has demonstrated high efficiency and has gained widespread adoption [5, 6]. The DEF method allows for the estimation of the magnitude, speed, and direction of energy propagation dissipated as a result of oscillations. The method employs a mathematical formula describing dissipating energy as a function of time. It involves modeling the signal as a sum of oscillatory components (modes) and harmonic signals with variable parameters.

The process of processing raw PMU signals involves several stages preceding the application of the DEF method's criterion: preliminary data processing for quality control and enhancement, identification of the LFO mode and its parameters, extraction of informative components from measurement signals, calculation of dynamic oscillation parameters, and assessment of the objects contribution to the development of the oscillation process. Each of these stages may involve different methods, algorithms, and parameters, giving the solution a combinatorial nature. The concept of generalized computational schemes (plans) allows for the structuring of this process [7, 8].

Previously, the authors demonstrated [9] using real LFO data from Russia that the time required to make a stable and accurate decision about the LFO source, while maintaining the general computational framework of the DEF method, depends on the combination of applied algorithms and their parameters.

This study takes a step toward creating a tool for optimizing computational schemes for LFO analysis based on an available set of data processing algorithms. The combinations of algorithms and parameters form a solution space that is tested on a dataset of measurement data.

The paper provides a brief overview of the DEF method and formalizes the concept of stable LFO source identification. It describes the generalized DEF computational scheme and the solution space. Approaches to automating the optimization of

* Corresponding author: k.butin@narfu.ru

© The Authors, published by EDP Sciences. This is an open access article distributed under the terms of the Creative Commons Attribution License 4.0 (https://creativecommons.org/licenses/by/4.0/).

computational schemes are considered. The results of computational experiments with real LFO data are analyzed.

Previously, the authors developed [10] and implemented [1] an approach for describing computations in the form of generalized schemes with the ability to vary the algorithms used. A computational scheme (plan) represents a set of interconnected computational stages (tasks). The plan defines the structure of the problem, and tasks correspond to subtasks. To execute the plan, each task must be assigned an executor – a specific data processing algorithm implementation.

1 Oscillation Source Identification and Solution Stability

Methods for identifying LFO sources based on oscillation energy assessment have shown high effectiveness. The most well-known of these methods is the Dissipative Energy Flow (DEF) method [4, 5]. In DEF, for each object, the quantity W_{ij}^D is calculated over signal segments ranging from 5 to 20 periods of LFO:

$$W_{ij}^D \approx \int 2\pi\Delta P_{ij}\Delta f_i dt + \Delta Q_{ij}\frac{d(\Delta V_i)}{\tilde{V}_i+\Delta V_i} \quad (1)$$

The symbol Δ here denotes the operation of extracting the oscillatory component from the measuring signal. P_{ij} – active power on line $i-j$, Q_{ij} – reactive power, f_i – frequency, V_i – voltage, $\tilde{V}_i$ – average voltage value at point i. The direction of the energy dissipation flow is determined by the slope of the linear regression:

$$W_{ij}^D(t) = DE_{ij} * t + b_{ij} \quad (2)$$

Positive values of DE_{ij} indicate the flow of oscillation energy from the object to the system, and negative values indicate the opposite direction of the flow. Thus, the largest value of DE_{ij} corresponds to the LFO source.

In online analysis, the decision regarding the LFO source should be made when the DEF results reach a certain stability [9]. Suppose the LFO source is known and the PMU data contain signals from this source. A function $s(t)$ is introduced:

$$s(t) = \begin{cases} 1, & \textit{if the source is correctly identified} \\ 0 & \textit{otherwise} \end{cases}$$

Consider also the function $r(t)$ – the time until the moment t, during which $s(t) = 1$ and $s'(t) = 0$, that is, the decision was correct and did not change.

In discrete form:

$$r(n) = \begin{cases} 0, & n = 0 \\ s(n)\big(r(n-1) + s(n)\big), & n > 0 \end{cases}$$

We emphasize that the output of the function $r(t)$ is time. We will compare $r(t)$ with some setting τ_r: if $r(t) > \tau_r$, then we will say that at time t there is a stable correct solution.

2 DEF template

Fig. 1 shows one of the possible calculation schemes for the DEF method. First, a set of quantities is prepared that are necessary for (1). Then the signals are packed into a special structure, the trend is removed, the modes are distinguished, the oscillation energy is calculated, and the rate of its change is calculated.

The * symbol means that the work is applied to all elements of the input data. For example, if filtering is used to extract a mode, then the same filter is applied to all values. The line thickness reflects the number of data elements in the stream.

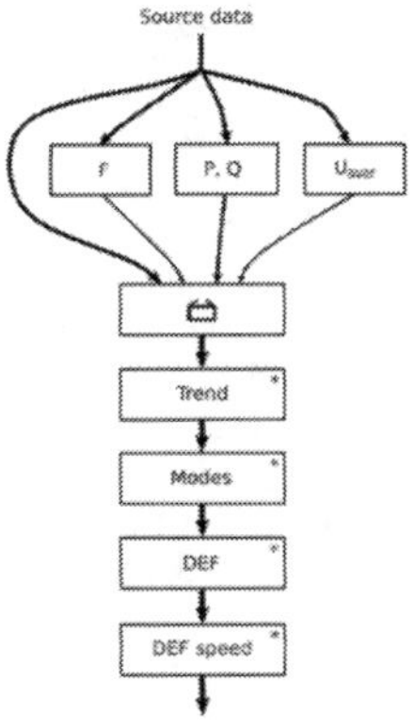

Fig. 1. DEF template.

Table 1 lists the possible performers for the selected jobs and the variable parameters.

Table 1. Calculation steps.

	Work	Workers	Parameters
1	f calculation	From data	
		Left shift	
2	**P** and **Q** calculation	From data	
		Function	
3	$\tilde{V}$ calculation		Number of averaged values
4	Removing a trend (detrending)	No	
		Averaging filter	Window function
			Win length
		Savitzky-Golay filter	Polynomial degree
			Length
5	Mode extraction	No	
		Finite impulse response (FIR) filter	Bandwidth
			Order
			Window function
		Infinite impulse response (IIR) filter	Order
			Bandwidth
6	DEF calculation	With Q	
		Without Q	
7	DEF slope calculation	Linear Regression	Window length

It is supposed to be possible to add any other performers that are compatible in terms of the interaction interface.

3 Utility for parameter selection in LFO source identification methods

The authors have developed a prototype computational platform for selecting the most suitable computational scheme for identifying LFO sources. As shown in Figure 1, sets of possible executors and their parameters

© The Authors, published by EDP Sciences. This is an open access article distributed under the terms of the Creative Commons Attribution License 4.0 (https://creativecommons.org/licenses/by/4.0/).

E3S Web of Conferences **584**, 01042 (2024) https://doi.org/10.1051/e3sconf/202458401042
RSES 2024

are defined. The filter and window lengths vary in the range from 5 to 1001 samples. A Butterworth filter of orders 2 and 3 is used as an IIR filter. For the Savitzky-Golay filter, polynomial degrees 5 and 6 are applied. The bandpass filter ranges cover the entire LFO interval: [0.1 Hz, 4.0 Hz], as well as intervals of 0.1, 0.2, 0.4, and 0.6 Hz, covering the mode. This parameter space is used by default but can be modified.

The input data for parameter selection are frequency, current, and voltage synchrophasors. The utility is adapted to read data from CSV files containing archived PMU data. All signals are processed in a streaming mode without introducing artificial delays.

The starting point t_0 is chosen to be a moment in time located 5-10 seconds before the onset of LFOs, with $\tau_r = 30$ seconds. Thus, starting from time t_0, the DEF method should consistently produce the same result for 30 seconds. This duration corresponds to approximately 25 oscillation waves, which is more than sufficient to confirm the detection of the source [4]. The moment when the DEF method starts producing the correct answer (which is known in advance) is considered the moment of source detection.

As shown in [11], identifying the LFO source itself requires high-performance computations. In this study, the number of different parameter combinations for the DEF method exceeds 10^{12}, making a full search infeasible. The Monte Carlo method [12] is applied for this investigation. At each step, one of the possible configurations is generated with equal probability, and the DEF calculation is launched for it. The configuration and the time of source detection are recorded for each run. This approach allows for exploring the entire parameter space and enables the distribution of the task across multiple computational nodes, where each node generates a report file. All files are then collected together to create summary tables and graphs.

4 Real LFO case processing

In [9], a case (Figure 2) was considered that occurred in January 2022 in the western part of Russia (Case 1).

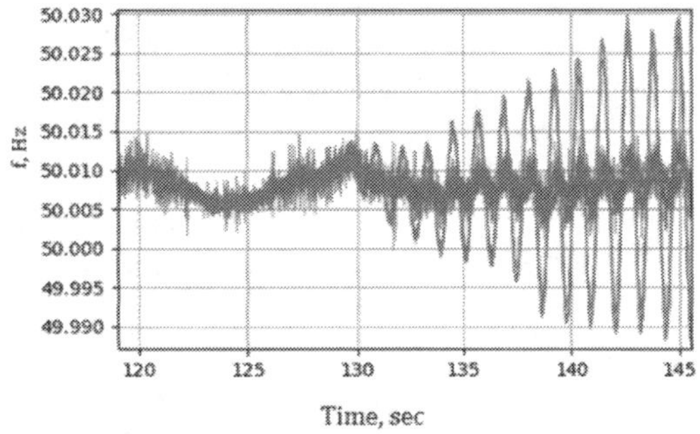

Fig. 2.a LFO at frequencies (Case 1).

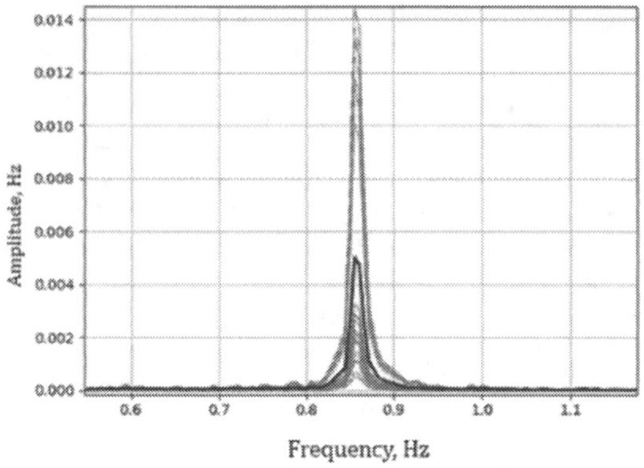

Fig. 2.b Amplitude spectrum (Case 1).

In Figure 2.a, the time segment marking the beginning of the LFO is shown. The raw frequency data for 21 input signals are presented with red lines. On the 2.b, the amplitude spectrum of frequencies is displayed. The blue line represents the average value across all spectra. A mode is observed around 0.85 Hz. This case is characterized by a stable mode and continuous dominance of the LFO source in the DEF method.

Now, consider Case 2 (Figure 3), which occurred in Siberia in 2024. Data from six generators with a reliably known LFO source are available. The recording duration is 30 minutes. The figure shows an unstable mode around 0.9 Hz. The oscillatory mode lasts about 16 minutes, with the amplitude of active power oscillations reaching 25 MW.

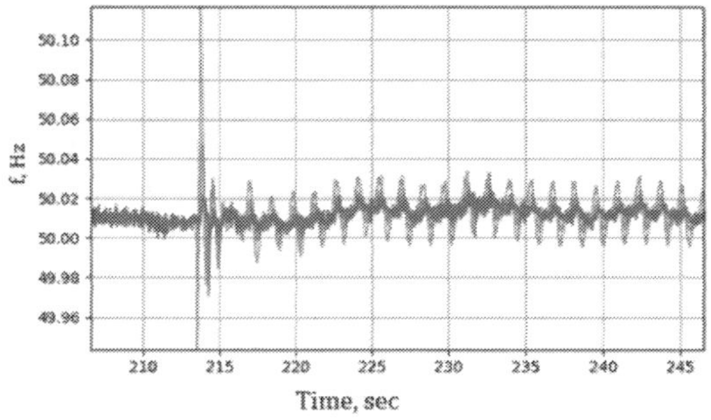

Fig. 3.a LFO at frequencies (Case 2).

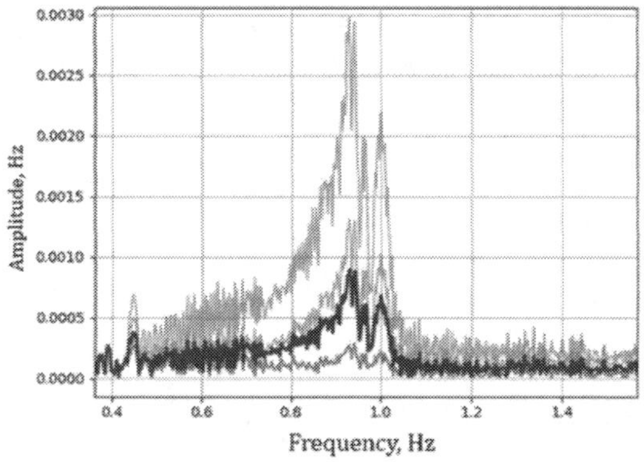

Fig. 3.b Amplitude spectrum (Case 2).

© The Authors, published by EDP Sciences. This is an open access article distributed under the terms of the Creative Commons Attribution License 4.0 (https://creativecommons.org/licenses/by/4.0/).

The second case is characterized by a strong dependence of the DEF results on the computational parameters. Within the first 30 seconds, the DEF method's answers alternate.

Here are the key characteristics of the calculations: 588,790 runs, 64.6 hours of machine time. The average processing time for a single configuration is 396 ms.

Figure 4 shows a histogram reflecting the number of correct detections (a total of 312,352) of the LFO source over time. The peak around 230 seconds corresponds to the most frequent source detection. This is primarily related to the cumulative delay of the methods. The worst results in terms of detection time are from methods with either the maximum cumulative delay or an excessively short window for linear regression coefficient estimation. This occurs because the DEF integral oscillates around the trend, causing the regression coefficient to constantly change its sign.

In the combinations yielding the best results, the calculation of P and Q is derived from the data, the DEF window length is about 120 samples, and a second-order IIR filter is applied. In Case 1, the best-performing methods involved detrending without mode filtering. The detrending subtask is one of the most critical aspects. Depending on how it is performed, in some cases, opposite-sign values of the regression coefficient are obtained. Detrending can be done using bandpass filtering, which is managed by an IIR filter. However, the IIR filter has drawbacks such as accumulated computational errors and a nonlinear phase characteristic.

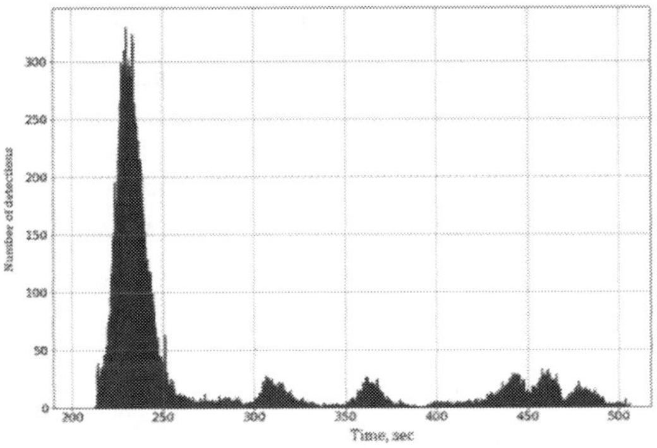

Fig. 4. Source detection frequency diagram.

Figure 5 illustrates the impact of window lengths (in samples) on the time required to detect the LFO source. Each blue dot represents a successful experiment result, and the red lines show the linear regression graph. The longer the window, the more delay it introduces in obtaining the answer. The slope indicates that the window length used for linear regression in the DEF method has the greatest influence.

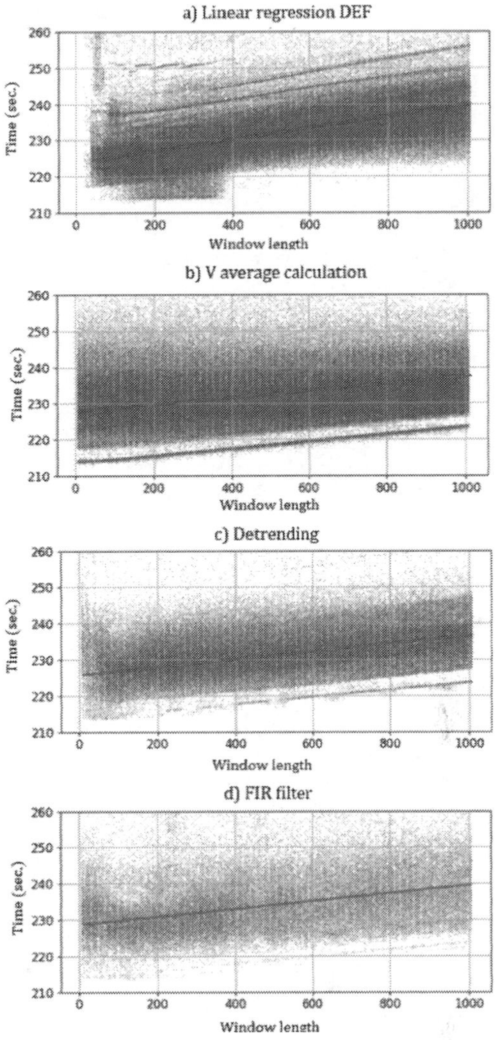

Fig. 5. The influence of window lengths on the result.

Table 2 shows the results grouped by parameters, presenting the average LFO source detection time for each group in both cases.

Table 2. Average values by parameter groups.

Parameter group	Average detection time from LFO onset (sec.)	
	Case 1	Case 2
All groups	21.2	59.0
f from data	21.7	63.3
f calculated	20.8	54.9
P from data	21.3	59.2
P calculated	21.2	58.8
Without detrending	19.0	62.4
With detrending	21.9	58.1
Without mode filtering	22.4	34.3
FIR filter	22.0	77.8
IIR filter	20.1	59.6
DEF with Q	21.4	74.8
DEF without Q	21.1	47.3

© The Authors, published by EDP Sciences. This is an open access article distributed under the terms of the Creative Commons Attribution License 4.0 (https://creativecommons.org/licenses/by/4.0/).

In Case 2, it took significantly longer on average to obtain a stable answer. For both cases, the IIR filter shows a clear advantage over the FIR filter. Additionally, better results are obtained with calculated f and P values. More experiments on a larger number of cases are needed for a more accurate assessment.

Conclusion

Currently, real-time monitoring of LFOs using WAMS data is highly sought after in dispatch centers, with the critical task being the identification of LFO sources. Numerous methods for source detection exist, with the Dissipative Energy Flow (DEF) demonstrating the highest efficiency in practical applications. The software implementation of the DEF method involves several computation steps with variable algorithms and parameters. A set of possible executors and their parameters has been defined. The authors have developed a utility for optimizing the selection of parameters and executors and have conducted computational experiments on real LFO cases using high-performance computing to find configurations that reliably determine the LFO source within an acceptable timeframe.

Dependencies of the final results on the computational parameters were identified, and the feasibility of online application of the method was experimentally verified. Examples were provided of computation combinations where the LFO source is correctly identified at the very onset of the process, allowing the source location to be known as soon as the LFO is detected. Future work will focus on expanding the test set and the parameter space. Given only two LFO cases, drawing conclusions about a universally optimal configuration is challenging. The question of the strong dependence of the source detection time on the inclusion of reactive power in the DEF calculation remains unexplored. Machine learning methods could be employed to accelerate the configuration optimization process.

References

1. A. Popov et al. *Framework for synchrophasor measurements data processing and the case studies of the low-frequency oscillations* (https://www.naspi.org/sites/default/files/2021-04/D3S7_01_butin_energoservice_naspi_20210415.pdf)
2. L.G. Meegahapola et al. *Review on Oscillatory Stability in Power Grids with Renewable Energy Sources: Monitoring, Analysis, and Control using Synchrophasor Technology* in IEEE Transactions on Industrial Electronics, **68**, pp. 519 – 531 (2020)
3. B. Wang, K. Sun. *Location methods of oscillation sources in power systems: a survey* in Journal of modern power systems and clean energy **5**, pp. 151 – 159 (2017)
4. S. Maslennikov et al. *Dissipating energy flow method for locating the source of sustained oscillations* in International Journal of Electrical Power & Energy Systems, **88,** pp. 55 – 62 (2017)
5. B. Wang, S. Maslennikov. *IEEE-NASPI Oscillation Source Location Contest-Case Development and Results* in Tech. rep. National Renewable Energy Lab (2021)
6. S. Maslennikov, E. Litvinov. *ISO New England Experience in Locating the Source of Oscillations Online* in IEEE Transactions on Power Systems (2020)
7. A. Popov. et al. *Examples of processing low-frequency oscillations in Russia and ways to improve the analysis* // 2022 International Conference on Smart Grid Synchronized Measurements and Analytics (SGSMA), pp. 1 – 6 (2022)
8. A. Popov et al. *Combined method for identifying a source of low-frequency oscillations in power system* in AIP Conference Proceedings **2552, 060004**, pp. 125 – 132 (2023)
9. K. Butin K. et al. *Investigation the influence of the parameters of the computational scheme on detecting the source of low-frequency oscillations* //E3S Web of Conferences. – EDP Sciences (2023)
10. A. Rodionov et al. *Practical Issues of Processing Synchrophasor Measurement Data in the Analysis of Low-Frequency Oscillations in Power System* in Proceedings of 2nd International Youth Scientific and Technical Conference on Relay Protection and Automation, RPA, pp. 1 – 16 (2019)
11. K. Butin. *Monitoring of Low-Frequency Oscillations in Power Systems from the Perspective of Computational Complexity*. Energy of the Arctic, pp. 20 – 24 (2023).
12. E. Okun et al. *Application of Monte Carlo Numerical Methods in Simulation Modeling*. Young Researcher of the Don, pp. 8 – 11 (2021)

© The Authors, published by EDP Sciences. This is an open access article distributed under the terms of the Creative Commons Attribution License 4.0 (https://creativecommons.org/licenses/by/4.0/).

E3S Web of Conferences **584**, 01043 (2024)
RSES 2024
https://doi.org/10.1051/e3sconf/202458401043

Neural networks and damage pattern recognition in power transformer diagnostics

Alexander Khrennikov[1*], *Nikolay* Aleksandrov[2], *Konstantin* Mikhailov[3], *Sergey* Mikhailov[4]

[1]JSC "S&TC FGC UES" Rosseti, Moscow, Russia
[2]NPP "Dynamics", Cheboksary, Russia
[3]Branch of JSC "SO UES" ODU Middle Volga, Samara, Russia
[4]National Research University MPEI, Moscow, Russia

Abstract. The results of detecting deformations and damage of power transformer windings using the transformer Frequency Response Analysis (SFRA) are presented taking into account RG CIGRE A2.26, Standard IEC 60076-18, Standard IEEE C57.149. The technology of pattern recognition by signal images for diagnostics of winding damage, type of defect, its localization is implemented. Neural networks are used - where in a multilayer perceptron with backpropagation of error - the work of neurons in a hierarchical network is imitated. Parametric methods - such as the Naive Bayes classifier - a probabilistic classifier based on the Bayes formula with the assumption of independence of features among themselves for a given class, which greatly simplifies the classification task due to the assessment of one-dimensional probability densities instead of one multidimensional. An algorithm for recognizing patterns of defects and damage to power transformers using the results of diagnostics by the FRA method has been developed.

1 Mechanical deformations of transformer windings

Mechanical deformations of transformer windings lead to their damage and, accordingly, cause interruptions in the power supply to consumers. Mechanical deformations are mainly a consequence of electromagnetic forces acting on the transformer windings during short circuits [1,2]. In the case of significant deformations, subsequent internal electrical damage in the form of a turn-to-turn short circuit is possible. Repairing such a transformer requires significant time and effort, as well as large financial costs due to the time of decommissioning and the cost of the repair itself. Damage to power transformers can also have a serious negative effect on the environment, such as a fire or a spill of transformer oil.

An example of the effective use of the FRA method can be the detection of residual deformations in the LV winding of the TRDN-32000/110 power transformer, which was used for powering the auxiliary needs of a power plant. After a short circuit occurred on the 6 kV side, due to the insulation breakdown of the cables supplying the equipment of the power plant, significant amplitude-frequency changes were detected in the FRA spectra of the LV winding compared to those taken a year before the short circuit (Fig. 1).

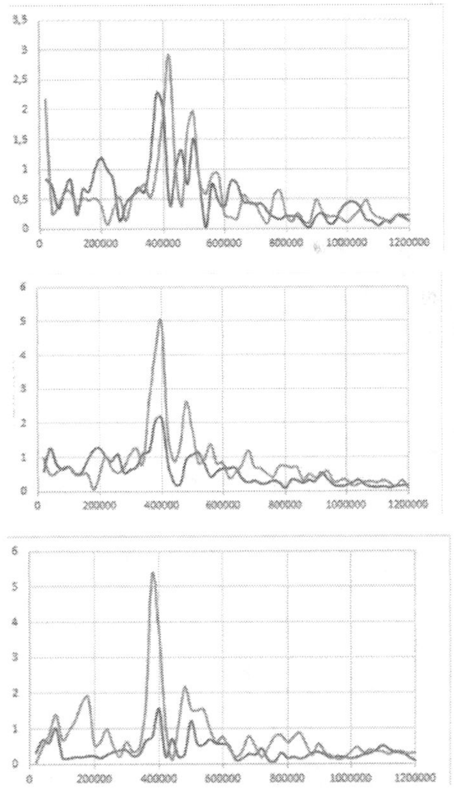

Fig. 1. Frequency characteristics of the TRDN-32000/110 transformer winding in the configurations a1-b1, b1-c1, a1-c1

* Corresponding author: Hrennikov_AY@ntc-power.ru

© The Authors, published by EDP Sciences. This is an open access article distributed under the terms of the Creative Commons Attribution License 4.0 (https://creativecommons.org/licenses/by/4.0/).

 https://doi.org/10.1051/e3sconf/202458401043

The greatest changes occurred in the frequency characteristics where the differences in signals a-c and b-c were recorded, i.e., involving phase C (Fig. 1).

Table 1. Calculation of power transformer TRDN-32000/110 winding parameters after a short circuit.

Phase	A	B	C
CCF	0.512	0.653	0.628
ASLE	0.029	0.03	0.031
CSD	0.0051	0.0057	0.005

Thus, taking into account that according to the technical brochure of the CIGRE working group A2.26, the IEC 60076-18 Standard, the IEEE C57.149 Standard and the statistical analysis methods for comparing responses, the limit values for the numerical indicators of the cross correlation factor (CCF) should not be less than 0.9974; for the absolute sum of the logarithmic error module, the tolerance limit is more than 0.125, we have significant deviations from the permissible values. The CSD indicator is a type of standard deviation and is calculated using the formula

$$CSD = \frac{\sqrt{\sum_{i=0}^{N}((x_i-\bar{x})-(y_i-\bar{y}))^2}}{N-1} \quad (1)$$

This confirms the presence of significant residual deformations in the windings of the TRDN-32000/110 transformer [1, 2]. According to these recommendations, the transformer was replaced with a new one of higher power.

Causes of internal damage to the windings of power transformers: damage to high-voltage bushings, breakdown of turn insulation as a result of insulation destruction, deformation of the windings during a short circuit (Fig. 2) [1, 2].

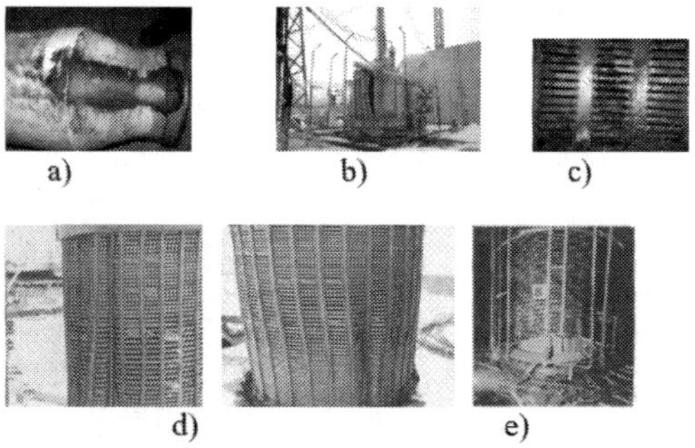

Fig. 2. Damage to high-voltage bushings, breakdown of turn insulation, deformation of windings due to the electrodynamic effect of short-circuit currents: a) carbonized traces of discharge on the lower porcelain cover of the 220 kV bushing at the 500 kV substation; b) rupture of the wall of the AT 167 MVA/500 kV-220 kV tank due to damage to the 500 kV bushing; c) turn-to-turn short circuit in the LV winding of 35 kV AT 167/500 kV-220 kV at the 500 kV «Srednyaya Volga» substation; d) deformation of the 220 kV windings of the 125 MVA autotransformer; e) loss of axial stability of the windings.

Fig. 3 shows the oscillograms of the LVI and FRA, graphs of the change in resistance (voltage) of the short circuit Zk (ΔZk) for diagnostics of the winding's mechanical condition windings.

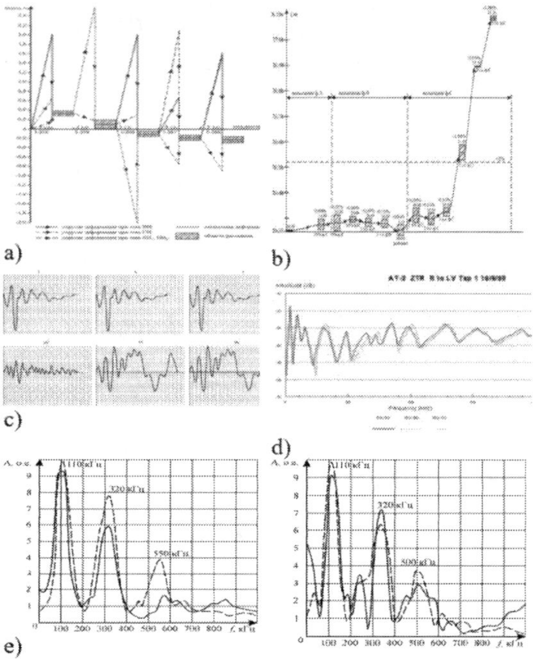

Fig. 3. Methods of diagnostics of mechanical condition of windings:a) change in short-circuit resistance Zk during tests of transformer ORDTNZ-25000/110, b) change in ΔZk value during short-circuit in LV winding TDC-250000/220, c) Oscillograms of NVI of LV winding 6 kV AT-1 "Kostroma-2", d) spectra of FRA signals of AT-2 windings illustrating residual deformations, e) FRA spectra of LV winding TDC-250000/220 after short-circuit.

2 Nonparametric methods. Neural networks.

Neural networks are a large class of systems whose architecture is similar to the construction of nervous tissue from neurons. In a multilayer perceptron with backpropagation of errors, the work of neurons is simulated as part of a hierarchical network, where each neuron of a higher level is connected by its inputs to the outputs of neurons of the underlying layer (Fig. 4).

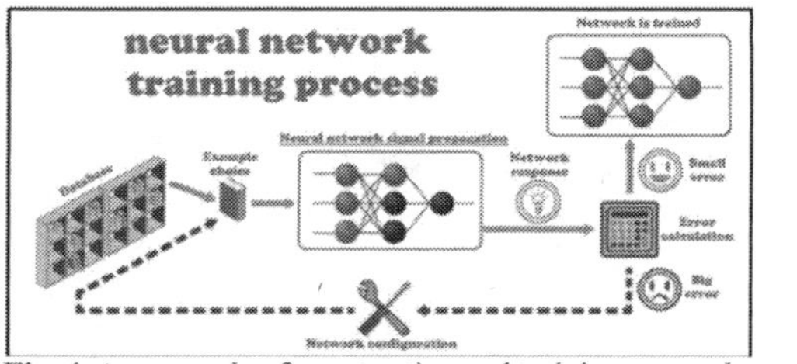

Fig. 4. An example of constructing and training a neural network [4, 5].

© The Authors, published by EDP Sciences. This is an open access article distributed under the terms of the Creative Commons Attribution License 4.0 (https://creativecommons.org/licenses/by/4.0/).

3 Parametric methods.

Naive Bayes classifier is a probabilistic classifier based on Bayes' formula with a strict (naive) assumption about the independence of features among themselves for a given class, which greatly simplifies the classification task due to the evaluation of one-dimensional probability densities instead of one multidimensional one. Bayes' equation:

$$P(A|B) = \frac{P(B|A)P(A)}{P(B)} \tag{2}$$

where: P(A|B) is the posterior probability of event A given that event B occurs; P(B|A) is the conditional probability of event B given that event A occurs; P(A) and P(B) are the prior probabilities of events A and B.

In the context of filtering transformer defects, the naive Bayes classifier is based on the frequency of occurrence of defect features in the matrices for damage and non-damage, and maximizing the product of their probabilities. Naivety will consist in the assumption that the features in the matrices are independent of order and context. Then Bayes' equation takes the following form:

$$P(C|M) \propto P(C)\prod_{i=1}^{n} P(w_i|C), \quad w_i \in M \tag{3}$$

where: C is the class of damage or not damage; M is the feature; w_i is the i-th diagnostic feature in the message M; ∝ is the proportionality sign.

The probabilities of a defect being damage or not damage are calculated, and the final forecast will be the class with the maximum probability.

$$\begin{aligned} P(C|M) = {} & P(C)\cdot P('Hi'|C)\cdot P('you'|C)\cdot P('won'|C)\cdot P('a'|C)\cdot \\ & \cdot P('discount'|C)\cdot P('and'|C)\cdot P('you'|C)\cdot P('can'|C)\cdot P('get'|C)\cdot \\ & \cdot P('the'|C)\cdot P('prize'|C)\cdot P('this'|C)\cdot P('evening'|C) \end{aligned} \tag{4}$$

4 Application of pattern recognition technology for diagnostics of damage to power transformer windings.

Defects and damage to transformers and reactors corresponding to the FRA spectra in Fig. 1, 3 are shown in Fig. 2. Let us set the task of recognizing patterns of defects and damage to power transformer windings according to the algorithm for recognizing patterns of defects and damage (Fig. 5).

The first is the "aligner" network. Stage No. 1. The program cuts out the necessary sections of diagnostic signals. It takes images of frequency response (FRA) of the transformer winding, which are fed to the input from sensors on electrical equipment, detects, "cuts out" and "aligns" them according to the signal sweep, frequency range, etc.

Stage No. 2. The program recognizes key points on the FRA characteristics of the transformer winding. The first is the high-frequency region (about or more than 2 MHz), does not carry diagnostic information - interference, interference, circuit influence. The second is the mid-frequency region, where the influence of capacitive couplings is present, but the influence of inductive couplings is already evident. The third is the low-frequency region (kHz), the winding inductance during deformations or turn-to-turn short circuits.

Stage #3. The program brings the signal images to a standard form. Rotates and adjusts the size so that the high-frequency region, the mid-frequency and low-frequency regions are in certain places convenient for mathematical processing.

The second is the "recognizer" network. Stage #1. The program accepts aligned images transmitted by the first neural network at the input, and outputs a vector (or matrix) of the diagnostic signal - a set of numbers of a fixed length.

Stage #2. The program, using the vector (or matrix) of the diagnostic signal, determines the type of defect or damage (residual deformations, turn-to-turn short circuits, insulation breakdown, compression, input damage).

Stage #3. During the training of the neural network, the program correlates the type of defect or damage to electrical equipment with the photographs of defects and damages available in the database in Fig. 2, and provides recommendations for localizing the location of the defect and its size (winding, shape, length of breakdown tracks) [2]. The algorithm for recognizing images of defects and damages to power transformers using the results of diagnostics using the FRA method is presented in Fig. 5.

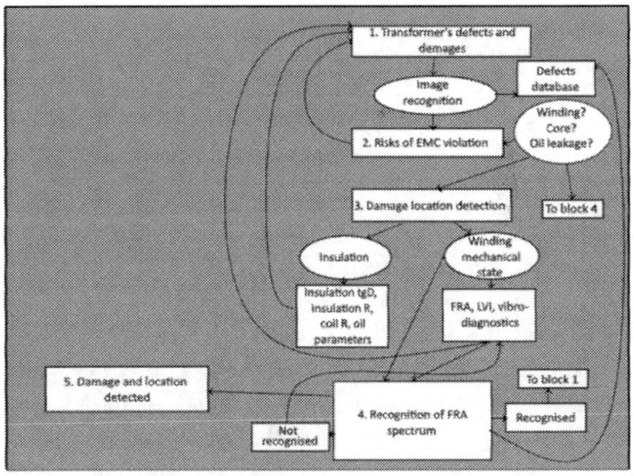

Fig. 5. Algorithm for recognizing patterns of defects and damages of power transformers using the results of diagnostics by the FRA method.

During the training process of the neural network, the program correlates the type of defect or damage to electrical equipment with photographs of defects and damage available in the database.

© The Authors, published by EDP Sciences. This is an open access article distributed under the terms of the Creative Commons Attribution License 4.0 (https://creativecommons.org/licenses/by/4.0/).

Conclusions

1. The presented work provides an example of using and training neural networks to analyze the results of diagnostics of power transformers and reactors in order to detect deformations and damage to transformer windings using the frequency response analysis (SFRA) method and other diagnostic methods, taking into account the results of the work of the CIGRE WG A2.26, IEC 60076-18 Standard, IEEE C57.149 Standard.

2. Since the diagnostic results are presented in graphical form, the technology of pattern recognition from signal images is implemented to diagnose damage to windings, the type of defect, and its localization.

3. An algorithm for recognizing patterns of defects and damage to power transformers has been developed using the results of diagnostics using the frequency response analysis (SFRA) method using a Bayesian classifier (Naive Bayes classifier) based on the Bayes formula.

4. The algorithm for recognizing patterns of defects and damage to power transformers consists of several subroutine networks - the "aligner" network, which in turn consists of 3 stages, and the "recognizer" network.

5. The "aligner" network at the 1st stage "cuts out" signal sections, detects, and "aligns" them according to the signal sweep, frequency range; at the 2nd - recognizes key points; at the 3rd - brings signal images to a standard form; the "recognizer" network at the 1st stage produces a vector (matrix) of the signal at the output; at stage #2 it determines the type of defect or damage; at stage #3, in the process of training the neural network, the type of defect (damage) is compared with the database of defects and damages, recommendations are issued for localizing the location of the defect and its size.

References

1. Khrennikov A.Yu. High-voltage equipment in electrical systems: diagnostics, defects, damage, monitoring. tutorial /A.Yu. Khrennikov. - M .: INFRA-M, 2019. - 186 p.
2. Khrennikov A.Yu., Vakhnina V.V., Kuvshinov A.A., Aleksandrov N.M. Power transformers in energy facilities: testing, diagnostics, defects, damage, monitoring. Tutorial. Direct-Media -M, 2021. 335 p.
3. Pattern recognition methods: from simple to complex. [Electronic resource] URL: https://www.simbirsoft.com/blog/metody-raspoznavaniya-obrazov-ot-prostykh-do-slozhnykh/(date accessed: 02.05.24)
4. Mazurov V.D. Mathematical methods of pattern recognition. Tutorial. Ekaterinburg, 2010, 101 p.
5. Vikramkumar, B. Vijaykumar, Trilochan. Bayes and Naive-Bayes Classifier, Rajiv Gandhi, Andhra Pradesh. Computer Science, Mathematics. 2014.
6. Khrennikov A.Yu., Klyushkin N.G., Aleksandrov N.M. Analysis of emergency events in electrical networks: Tutorial // Moscow; Direct-Media, 2023. - 149 p.
7. Khrennikov A.Yu., Aleksandrov N.M. Checking and adjusting electrical equipment. Tutorial// Moscow: KNORUS, 2024. — 360 p.
8. Aleksandrov N.M., Kashin M.A. Maintenance of substations. Tutorial// Moscow: KNORUS, 2024. — 245 p.
9. Khrennikov A.Yu., Lyubarsky Yu.Ya. Principles of constructing artificial intelligence systems and their application in the electric power industry // Energetik, 2022, No. 3. – P. 22-30. DOI: http://dx.doi.org/10.34831/EP.2022.82.70.005
10. Khrennikov A.Yu., Aleksandrov N.M., Mikhailov S.A. Automation and IT in diagnostics: Interpretation of the results of the study using the frequency response method (SFRA), the model of deformation of power transformer windings. Part 1 // Energy of the Unified Grid, 2023, No. 5-6. - P. 36-47.
11. Khrennikov A.Yu. Diagnostics of Electrical Equipment Faults and Power Overhead Transmission Line Condition by Monitoring Systems (Smart Grid): Short-Circuit Testing Of Power Transformers / Nova publishers, New York, 2016, USA, p. 165 . ISBN: 978-1-63484-159-7
12. Khrennikov A.Yu. Providing Reliable Operation of Electric Networks / Nova science publishers, New York, 2019, USA, p. 308, ISBN: 978-1-53615-422-1

© The Authors, published by EDP Sciences. This is an open access article distributed under the terms of the Creative Commons Attribution License 4.0 (https://creativecommons.org/licenses/by/4.0/).

E3S Web of Conferences **584**, 01044 (2024) https://doi.org/10.1051/e3sconf/202458401044
RSES 2024

Principles of controlling intersystem power flows in the conditions of the electricity market

Nurali Yusifbayli, *Rana* Alizade*

Azerbaijan Technical University, AZ1073 Baku, Azerbaijan

Abstract. The principles of electricity market control in relation to intersystem communications is analyzed in the paper. Electricity market can be both intersystem communications and legal entities of transmission and distribution networks of power systems. As an example, parallel operation of two power systems through the intersystem communication with different operating costs was considered.
Key words: power system, intersystem communications, electricity, electricity market

1 Introduction

In accordance with the current legislation of many countries, including the Republic of Azerbaijan, radical structural changes have taken place in the energy sector. The state monopoly in the power sector has been eliminated, and some sectors have completely changed their form of ownership [1, 2].

Under these conditions, economic principles are added to the main technological principles of control. Moreover, both of these principles are not independent, but closely interrelated and interdependent. Their joint realization works as a whole for the integral indicator-efficiency. This is associated with the need to maximize the satisfaction of both the economic interests of energy producers (electric power industry) and the economic interests of its consumers while complying with numerous restrictions of technological and economic nature, as well as the interests of the state and society in the ensuring energy security of the country [3].

2 National Dispatch Center

In a market economy, the National Dispatch Center (NDC) operates in a multi-criteria environment. NDC becomes a guarantor of effective functioning not only of the electric power industry, but also of the entire economy.

The experience of North American and European UPSs (UCTE, NORDEL, CENTREL) shows that the NDC implements almost independent control over the operation of their power systems, ensures their reliable and economical operation, regulates technological and economic relations within the energy association [4].

Economic aspects within the framework of dispatch services tasks were previously expressed mainly in the problem of conduct of optimal mode with observance of the criterion of minimum costs under the existing constraints. This is the problem of optimal distribution of powers [5].

In market conditions, a much wider and more serious range of problems is revealed [2], one of which is reflected in this paper in the form of recommendation criteria.

Intersystem communications are a market where acts of purchase and sale of electricity are realized. In this case, it is determined how much can be sold/bought and at what price. Subjects on the market in this case are power systems of power systems of energy association.

Similar processes take place in power systems, where as a market are substations of connection of transmission and distribution networks, at the ends of which are concentrated, respectively, suppliers (power plants) and buyers (consumers-loads) [6-8].

In both cases, the regulators of these processes are dispatch centres that perform the combination of technical and economic criteria. The activity of dispatch centres under these conditions should be based on two main provisions:

1. In the electric power system (EPS), the optimal capacity allocation should be ensured according to the criterion of minimum costs and consideration of constraints [5, 6].

* Corresponding author: rena_alizade@mail.ru

© The Authors, published by EDP Sciences. This is an open access article distributed under the terms of the Creative Commons Attribution License 4.0 (https://creativecommons.org/licenses/by/4.0/).

2. The action of NDC should be subject to the basic laws of market economy - the laws of Supply and Demand [4].

Moreover, both of these provisions are inseparably linked. The first provision is well known and is successfully applied in the practice of power system regime management at all hierarchical levels of ASDC. Therefore, in our paper we will focus on the second provision.

3 Laws of Supply and Demand

The Laws of Demand and Supply are represented as the respective curves shown in Figure 1 [4, 9, 10]. The directions of slope of the curves reflect the logic of the laws on the market:

The higher is the price of a product, the less it is bought (Demand curve - S) and vice versa.

The higher is the price on the market, the more goods the manufacturer is ready to offer to the market (Supply curve -D) and vice versa.

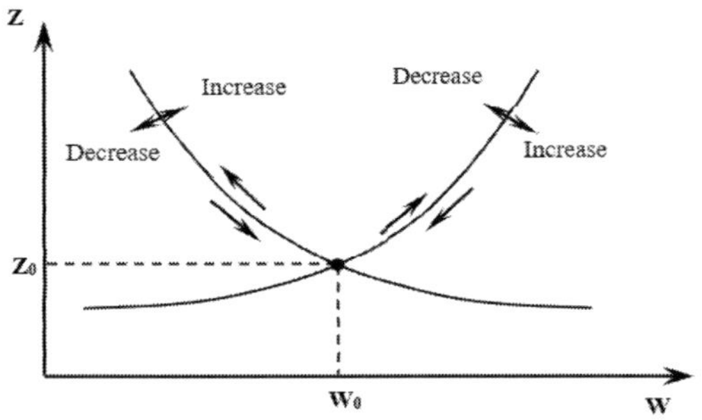

Fig. 1. Demand and Supply curves. Z - costs, W - quantity of product

The issue of the price that would satisfy both subjects is solved in the market at the point of intersection of the curves, where the so-called equilibrium price is established. Subjects in the interconnection market are power systems, each of which, depending on the load schedule, can be either a seller or a buyer of electricity, i.e. the power flow through the intersystem communication can be reversible.

In the electricity market, the Demand and Supply curves can be represented in the form of dependencies of the relative increment of costs (fuel or cost) on the transferred active power. Such characteristics are available in power systems. The peculiarity is that they undergo changes under the influence of a number of factors. These factors can be divided into two categories. The first category of factors leads to an upward or downward shift of the Demand and Supply curves, while the second category requires a transition from one "operating point" to another while keeping the curve positions unchanged (Fig. 1).

The first category includes such factors as:

- forecast changes in loads;
- changes in fuel prices;
- change in equipment prices;
- seasonal factor;

The effect of these factors should be taken into account by dispatch centres mainly at the level of long-term and partially short-term planning.

Such a factor as a change in the number of consumers or producers of load fluctuations lead to the movement of the "operating" point along the corresponding curves, which should be taken into account at the levels of short-term planning and operational management.

Demand and Supply curves for EES are represented as dependencies of relative growth of costs on the increase in generated capacity. Some publications [3] introduce the concept of commercial cost - the value of the system relative increase in costs (fuel or cost) when EES operates in the optimal mode. In other words, it is the cost per additional 1 MWh of electricity produced.

When solving the issue of sale-purchase of electric power, the operating personnel should be guided by the ratio of the existing tariff for electric power on the wholesale market and its commercial cost within the power system.

4 Parallel operation of power systems

As an example, let's consider parallel operation of two power systems through intersystem communication (Fig. 2). Both power systems have a power reserve. This means that each of the EPSs is ready to sell or buy free capacity. However, the EPSs have different operating costs.

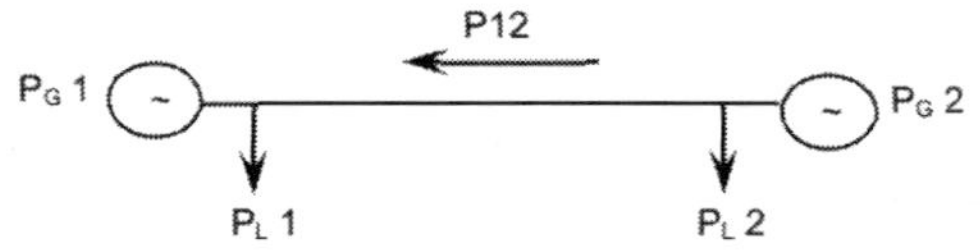

Fig. 2. Parallel operation of 2 power systems via inter-system communication

If we assume that within the time of short-term planning and operational management the tariff for electric power on the wholesale market is stable, then the dependence Topt (P) is linear and parallel to the abscissa axis. The degree of benefit (or damage) from the purchase of electricity on the wholesale market can be estimated by combining this characteristic with a

© The Authors, published by EDP Sciences. This is an open access article distributed under the terms of the Creative Commons Attribution License 4.0 (https://creativecommons.org/licenses/by/4.0/).

similar characteristic of the power system T(P). In the initial optimal mode, the load power is equal to P0 at costs $Z_0=T_0\,P_0$

When the load of the power system increases by ΔP, consumers can be supplied either by additional loading of their power plants, but the costs will increase to $Z_1=T_0(P_0+\Delta P)$, or by purchasing from the wholesale market, and in this case the costs will be equal to $Z_{ws}=T_{ws}(P_0+\Delta P)$.

Let us show the practical realization with a concrete example. The deficit "Absheron" node (Fig. 2) along with its own generating sources (Shimal PP, Baku PP, Sumgait PP) is fed by capacities from two sides: from the Azerbaijan TPP through the "Sharg" section and from JSC UES of Russia through the intersystem communication of the 330 kV overhead line "Yashma - Derbend". In case of load increase in the "Absheron" node, make-up can be carried out either from the Azerbaijan TPP - at the internal tariff (3.88 cent/kWh), or through the intersystem communication - at the wholesale tariff (3.0 cent/kWh) [11].

In the process of operational regulation the dispatcher will decide the question: "Until when it is profitable to load the units of Azerbaijan TPP, and when the purchase from the wholesale market gives economic benefit (and from Azerbaijan TPP - damage)?".

Below, Table 1 summarizes the baseline and calculation information for one 300 MW unit of the Azerbaijan TPP.

Table 1. 300 MW unit of Azerbaijan TPP

P, MW	**E gr/kWh**	***Z₁*, ц.**	***C₁*, cent/kWh**	***Z*, million dollars.$**
150	18,41	0,1105	3,99	23,94
200	24,46	0,1468	4,03	32,21
250	30,51	0,1832	4,06	40,63
300	36,56	0,2195	4,10	49,19

In it:

E – is the relative increase in fuel consumption.

E = 0,26+0,000121P

Z_1 – cost increment in US dollars

C_2 - specific cost of thousand m^3 of gas ($69).

K - conversion coefficient of fuel equivalent in m^3 of gas (0.87).

C_1 - electricity tariff taking into account the incremental cost for additional kWh

$C_1 = C_0 + Z_1$ (C_0 = 3.88 cent/kWh)

Z - costs,

$Z = PTC_1$

If the interconnection capacity limit is 400 MW, the number of hours of its use is 1960 and the wholesale tariff is 3.0 cent/kWh, the costs are as follows:

$Z = 400{\cdot}10^3$ kW$\cdot$ 2750 h$\cdot$ 0.03$/kWh = 33 mln. $

Figure 3 shows the cost characteristics of domestic and external wholesale tariff.

In the initial mode the unit of Azerbaijan TPP operates with 15-16% reserve, i.e. the capacity of the unit is 250 MW.

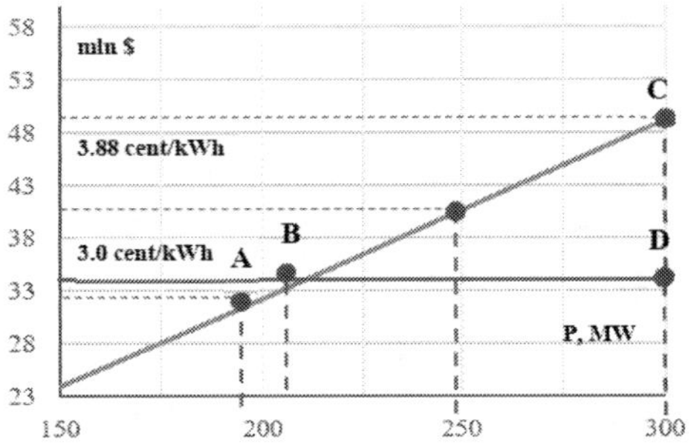

Fig. 3. To determine the benefit (loss) of purchasing electricity on the wholesale market

When the load in the system increases, the dispatcher can increase the capacity of the *AB* unit to 205 MW. Further satisfaction of consumers is beneficial due to the flow through the intersystem communication (*BC*).

According to the internal tariff with benefit, the dispatcher has the *AB* range on the graph.

Benefit or damage is determined from the ratio of the areas under the wholesale characteristic and the EPS characteristic. Point *B* is the equilibrium point: at this point, mutually beneficial energy exchange is ensured and they are ready to conclude contractual agreements, because after it $S_{BC}300\ 281 > S_{BD}300\ 281$, the energy system benefits from purchasing electricity from the wholesale market at the equilibrium tariff (T_0) for each megawatt hour.

Thus, both provisions apply: optimal distribution of capacities under the condition of ensuring minimum costs and the Laws of Supply and Demand.

A similar result can be obtained by equating the costs (cent):

$З_1 = PTC_1$ – production of electric energy (1)

and $З2 = PTqC_2$ – for used fuel (2)

where q is the specific consumption per 1 kWh

In this case, C_1 refers to the fuel component of the tariff, which is about 80%.

From equality (1) and (2) follows $C_1 = qC_2$ (3)

The calculation results and their comparison with the first approach are given in Table 2,

(q = 320 g/kW, C_2 = 69 dollars/thousand m^3).

© The Authors, published by EDP Sciences. This is an open access article distributed under the terms of the Creative Commons Attribution License 4.0 (https://creativecommons.org/licenses/by/4.0/).

 https://doi.org/10.1051/e3sconf/202458401044

Table 2. Comparison of calculation approaches

P MW	Costs in million dollars USA		Difference%
	Method I	Method II	
150	23,94	24,26	1,3
200	32,21	32,35	0,4
250	40,63	40,44	0,5
300	49,19	48,52	1,4

As we can see, the difference in the results is insignificant.

A similar method can be applied to a power plant or the system as a whole. Moreover, in each case, method II is preferable. Thus, according to the readings of JSC Azerenerjy for 2023, the tariff can be determined.

$C_1 = qC_2 = 3.1$ cents/kWh.

An increase in the cost of 1000 m^3 of gas to the selling prices of natural gas abroad (by 2.7 times) leads to an increase in the tariff to 8.37 cents/kWh, which indicates an underestimated level of the current tariff of 3.88 cents/kWh, corresponding to q = 320 g/kWh.

The special point *B* on the characteristic moves in the operational cycle of power system load control in accordance with the dispatcher load schedule and is controlled by the dispatcher, as well as when changing the specific fuel consumption, the fuel component of the electricity tariff (60-65%), the fuel tariff. Transferring these characteristics to analytics and introducing them into the computer system will allow for the operational management of power flows while observing technical and economic criteria.

Conclusions

1. It is necessary to work out methods for obtaining economic characteristics from station and EPS units in real time within the framework of the APCS of stations and the EPS ASDC.

2. It is necessary to work out methods for obtaining functions of the proposal for economic characteristics within the framework of the EPS ASDC. It is important that these functions are obtained by the seller and the consumer using a single method.

3. The NDC must have a method for continuously monitoring these functions in order to make timely corrections.

4. There must be open and accessible information communication in the system and between the systems.

References

1. N.Yusifbayli, V.Nasibov, A.Huseynov, R.Alizade, K.Suleymanov Rudenko International Conference “Methodological Problems in Reliability Study of Large Energy Systems” (2023)
2. Guseinov A.M., Yusifov N.A. Problems of Energy, **No. 2**, (2000)
3. McConnell K.R., Brew S.L., Flin S.M. Economics, principles, problems and policy, (2011)
4. Osak A.B., Domyshev A.V., Sorokin I.V. System research in energy. (2005)
5. Venikov V.A., Zhuravlev V.G., Filipova T.A. Optimization of power plant modes of energy systems // Textbook for universities of energy specialization. Universities, (1981)
6. Yusifbeyli N.A., Nasibov V.Kh., Alizade R.R. “Methodological issues of studying the reliability of large energy systems”, “Reliability of energy systems in the context of energy transition” (2022)
7. Letun V.M., Gluz I.S. Electric stations., **№3**, (2003)
8. Volkova I.O., Shuvalova D.G., Salnikova E.A. Active consumer in the intelligent energy system: possibilities and prospects, https://publications.hse.ru/mirror/pubs/share/folder/xnto5eu7bd/direct/71899800
9. Pazouki, S., Mohsenzadeh, A., Haghifam, M. R. IEEE Pes General Meeting Conference & Exposition, National Harbor, MD, USA: 27-31 July, (2014)
10. Zhang, X. et al. IEEE Trans. Power Syst. **31(1)** (2015)
11. N.A. Yusifbayli, A.M. Huseynov, H.B. Guliyev, A.F. Aliyev, O.B. Azadkhanov Methodological issues in the study of the reliability of large power systems. Issue 68, Irkutsk (2017)

© The Authors, published by EDP Sciences. This is an open access article distributed under the terms of the Creative Commons Attribution License 4.0 (https://creativecommons.org/licenses/by/4.0/).

Operation modes of the power system of Azerbaijan under conditions of penetration of large volumes of RES

Rana Alizade*, *Valeh* Nasibov

Azerbaijan Scientific-Research and Design-Prospecting Power Engineering Institute, AZ1012, Baku, Azerbaijan

Abstract. The nature of changes in the operating modes of the power system when introducing large volumes of renewable energy sources (RES), mainly wind and solar energy, is considered in the paper. Studies were conducted for two modes: 1) at parallel operation with the Russian power system - the Russian power system carries out normalized primary frequency regulation; 2) at isolated operation - the Azerbaijani power system itself carries out frequency regulation, while the Georgian power system operates in parallel with the Azerbaijani power system. Static and dynamic models of the power system were developed.
Key words: power system, sun energy, wind energy, power consumption, load schedule, power system frequency

1 Introduction

One of the most important problems of the modern world is the problem of climate change. Despite an increase in power consumption in recent decades, global carbon emissions associated with energy consumption have remained stable. This stabilization is partly due to the increased use of renewable energy sources, which has accelerated significantly in recent years.

By increasing the extraction of fossil resources, Azerbaijan has also accelerated the development of the renewable energy sources (RES) sector, and has set a target to reduce greenhouse gas emissions by 35% by 2030 and 40% by 2050. So far, Azerbaijan's share in global emissions is about one thousandth, at that energy sector accounts for about 70% of emissions. In this regard, the country is witnessing a rapid transition from traditional hydrocarbon energy sources to renewable energy sources [1-3].

2 Integration of RES into the power system

So far, thermal power plants account for more than 80% of the installed capacity in the Azerbaijan power system and produce about 94% of electricity. Taking into account the capacity of large HPPs, the share of RES capacity in the total installed capacity does not exceed 20% and they produce about 6% of electricity. Solar power plant was commissioned in 2023. Two more RES-based power plants (240 MW- solar power plant (SPP) and 240 MW-wind power plant (WPP)) are planned to be built by 2025. Thus, by 2025, the installed capacity of large solar and wind power plants will reach 710 MW. By 2027, it is planned to increase the installed capacity of RES-based power plants (SPP and WPP) to 1850 MW, and by 2030 - more than 2500 MW (according to other estimates, the capacity of RES taking into account the capacity of large HPPs will reach 5000 MW).

Under these conditions, it is necessary to study the possible impact of large volumes of renewable energy sources on the operation modes of the power system [4, 5].

It should be noted that the impact of different volumes of renewable energy sources on the operation modes of the power system is different. Six phases of integrated RES volumes into the power system are distinguished in the references [6, 7, 8].

The four phases of variable renewable energy sources (VRES) penetration and the measures recommended by the International Energy Agency for long-term power system planning and safe conduct of power system modes are discussed below.

The first phase (3-5%) of VRES penetration is not accompanied by any problem, and during the conduct of the power system mode the VRES presence is hardly felt.

In the second phase (5-15%) of VRES penetration, it is possible to introduce the specified volumes of VRES into the power system by carrying out measures.

In the third phase (15-20%), the issue of provision of power system flexibility becomes topical. At this phase of VRES volume penetration in the power system, there is a high degree of uncertainty in ensuring the power balance, since the variability of the VRES generation is significant.

The fourth phase (20-60%) of VRES penetration is characterized by the key problem, the threat of instability of power system operation modes, in a short time the power system should have the ability to resist various disturbances in the power system.

In the next few years Azerbaijan power system will be in the fourth phase of VRES penetration, where

* Corresponding author: rena_alizade@mail.ru

© The Authors, published by EDP Sciences. This is an open access article distributed under the terms of the Creative Commons Attribution License 4.0 (https://creativecommons.org/licenses/by/4.0/).

special attention will be paid to the stability of the power system.

The impact of VRES integration is being analyzed comprehensively: for the power grids, for the generation system and for the relay protection and automation system.

First, the transmission system capabilities and the connection of RES substations to the existing transmission system are investigated. Second, the impact of RES production on the operating modes of other stations is investigated. Then, the impacts of VRES (VRES power fluctuation) on such power system parameters as active power reserve for frequency regulation and intersystem power flow security are studied.

3 Analysis of power system operation at RES integration

To connect the specified volumes of VRES, a comparative analysis of many schemes for distributing PRES power to the power system was carried out in terms of reliability and cost-effectiveness of the compared options, taking into account the prospects for the development of transmission networks and large power consumption nodes. A preferred scheme is selected and, based on it, calculations of typical normal modes and calculations of all possible modes with disconnection of power system elements are made, including disconnection of the largest unit of the power system (a unit at a power plant with a capacity of 400 MW).

In the general case, the stability of the power system is analysed for different time periods, where the main parameters are the stability of the rotor angle, frequency stability and voltage stability for different time periods.

It was found that the static stability of the operating modes of the power system is ensured (the power line is not overloaded and the voltage at the substations is within the permissible limits during any shutdowns) [9-11].

The stability of the operating modes of the power system is analysed taking into account the criteria shown in Figure 1.

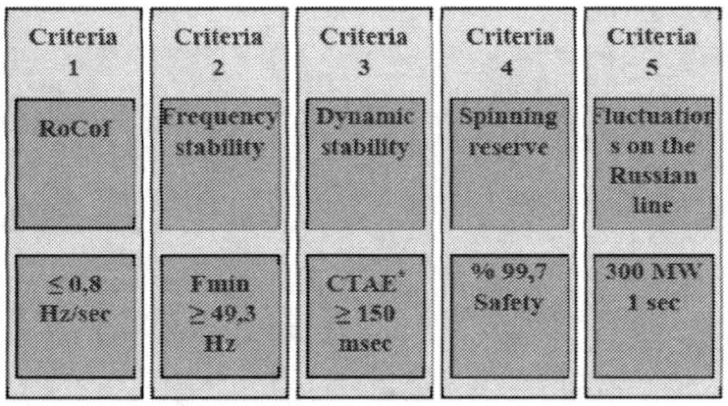

*CTAE- critical time of accident elimination

Fig. 1. Stability of power system operation modes

The possibility of supporting the frequency change rate parameter (RoCoF) with a large volume of PRES penetration and the impact of VRES on the level of rotating reserve is mainly analysed in this paper.

To determine a unit reference event that will create the maximum imbalance, three options for switching off/on of any power system element are compared:

– Shutdown of the largest generator (one unit of 400 MW Shimal PP);

– On/off the highest load of 250 MW (load of Mushvig substation)

– Shutdown of 330 kV Derbend-Khachmaz international interconnection line shutdown, through which 300-350 MW of capacity is transferred and through which the normalized primary frequency regulation is carried out while the Azerbaijan power system works in parallel with the Russian power system.

Analysis and comparison of the mentioned events as a reference system event was selected, the shutdown of one power unit in Shimal PP at full load 380 MW.

Due to the penetration of large volumes of VRES, the number of synchronous machines with high inertia is reduced, which can lead to exceeding the allowable rate of frequency change (RoCoF) at emergency situations. To keep the RoCoF parameter within the allowed limits, it is necessary to define the minimum composition of synchronous generators of power system being in operation.

As is known, the actual power imbalance and frequency change are bound with the following equation:

$$\frac{P_m(t)-P_e(t)}{P_{load}} = 2\cdot\frac{\sum_i^N H_i\cdot S_i}{P_{load}}\cdot\frac{df_{pu}}{dt} \qquad (1)$$

where Hi and Si- inertia and the nominal full generator power, and P_{load} represents the total system load.

Since the worst event is defined as the disconnection of one unit at the Shimal PP, and the RoCoF limit is assumed to be 0.8 Hz/sec [12], the equation takes the form:

$$\int_{t=0}^{t=0,5}\frac{380MW}{P_{load}}dt = 2\cdot\frac{\sum_i^N H_i S_i}{P_{load}}\frac{0.4Hz}{50Hz} \qquad (2)$$

As is obvious from the equation above, the P_{load} parameter is balanced in both parts of the equation. This means that the frequency change rate is almost not related to the total load of the system, but mainly depends on the inertia of rotating masses of synchronous machines.

The equation permutation results in:

$$\sum_i^N H_i S_i = \frac{380\cdot t\Big|_{t=0}^{t=0.5}}{0.008\cdot 2} = \frac{380\cdot 0.5}{0.008\cdot 2} = 11.875MWs \quad (3)$$

According to the above calculations:

– Taking into account the maximum disconnected capacity of the 380 MW unit at Shimal PP, the total estimated required inertia of the power system is 11.875 MW.

– It should be noted that the calculation assumes the disconnection of one unit at Shimal PP. This means that

© The Authors, published by EDP Sciences. This is an open access article distributed under the terms of the Creative Commons Attribution License 4.0 (https://creativecommons.org/licenses/by/4.0/).

the calculated inertia corresponds to the generator being in operation. However, to calculate the inertia of the power system before disconnecting one unit, it is necessary to add the inertia of the disconnected unit equal to 6.65 MW to the calculation. c/MVA* 560 MVA = 3.724 MWs.

– Summing these two values, the minimum inertia of the power system is calculated as 15.599 MWs. Thus, the total inertia of the power system should always be maintained at the level above 15.6 GWs so that RoCoF does not exceed 0.8 Hz/sec.

To create this level of inertia in the power system, synchronous machines must always provide about 1,500 MW of capacity. To achieve this goal, a minimum set of generators is installed, providing the minimum necessary inertia of the power system.

RoCoF calculations have been performed in different scenarios of change of level of synchronous generation when one generator of Shimal PP is switched off, the results are shown in Figure 2.

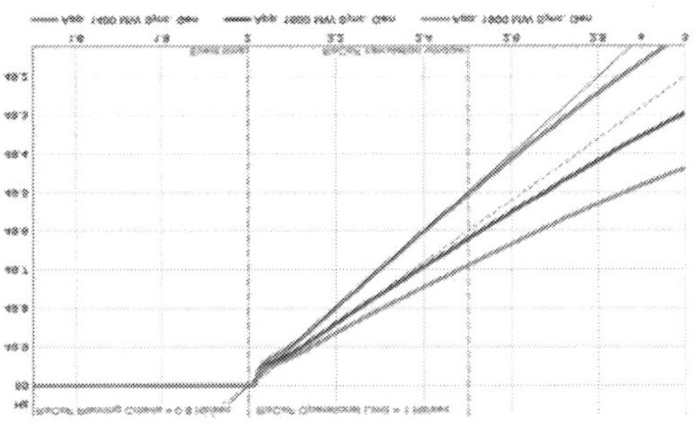

Fig. 2. Changes in the level of synchronous generation when one generator of Shimal PP is disconnected

As shown in Figure 2, to maintain RoCoF at the required level, the total synchronous generation capacity must be at least 1500 MW.

4 Frequency and voltage stability

To assess the frequency and voltage stability of the power system, during simulation the faults are studied under different operating conditions in order to evaluate the fluctuations of these critical parameters. Limits used for frequency and voltage:

– Frequency sweeping: working limit +/- 0.8 Hz, planning criteria +/- 0.7 Hz.

– Voltage oscillations on high-voltage buses: working limit +/- 10%, planning criteria +/- 8%.

The analysis takes into account the most severe regime [7, 12, 13].

4.1 Complex mode scenario -1

The power system operates in island mode, i.e. isolated from the Russian and Georgian power systems with minimal inertia and reactive power support. The total installed capacity of RES is 710 MW (240 MW condensing power plant (CPP) and 470 MW SPP), the total load of the power system is about 1700 MW (minimum load at night), the total capacity of synchronous machines is about 1500 MW. In this case, the shutdown of power unit of Shimal PP with a capacity of 380 MW occurrs.

4.2 Complex mode scenario-2

The power system operates in island mode, i.e. isolated from the Russian and Georgian power systems with minimal inertia and reactive capacity support. Total installed capacity of RES is 710 MW (240 MW CPP and 470 MW SPP), total system load is about 2300 MW (with low load before noon and high output of SPP), total synchronous machine capacity is about 1500 MW. In this case, the shutdown of power unit of Shimal PP with a capacity of 380 MW occurrs.

As shown above, the frequency control for the two scenarios is almost the same (same composition of synchronous machines), but there is a slight difference in voltage control. In terms of voltage regulation, scenario-1 can be considered as the worst case, considering the additional voltage support that solar power plants can provide after sunrise.

Accordingly, the simulation is carried out for scenario-1, the frequency and voltage swing results are shown in Figure 3 and Figure 4, respectively.

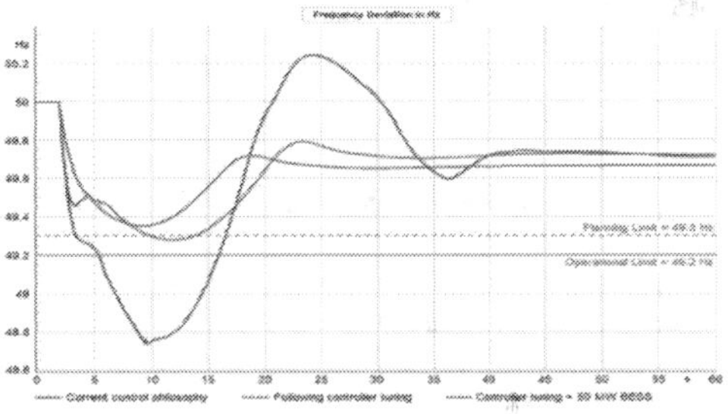

Fig. 3. Scenario-1, frequency swing results.

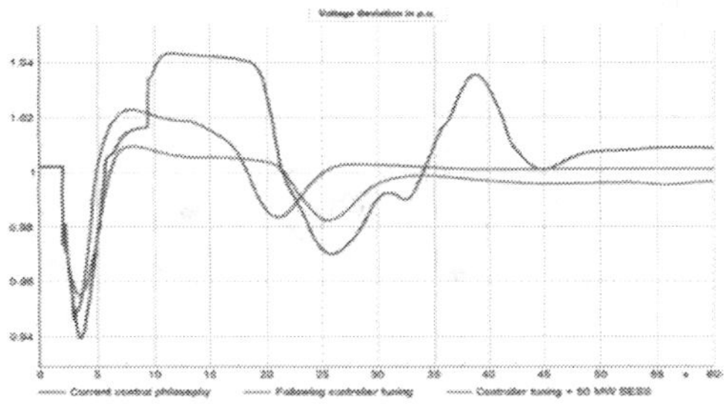

Fig. 4. Scenario-1, voltage swing results

As can be seen in Figures 3 and 4, three conditions are investigated in the modeling:

– Current control methodology when some power plant regulators are insensitive or deactivated.

– The second simulation is performed with the condition that the power plant regulators are activated and participate in frequency control. Some unit

© The Authors, published by EDP Sciences. This is an open access article distributed under the terms of the Creative Commons Attribution License 4.0 (https://creativecommons.org/licenses/by/4.0/).

regulators with power plant frequency control capability are inactive and the second simulation involves modifying the logic and tuning of the regulators for proper operation.

– The third simulation is performed using a battery pack that is actively involved in the frequency support process, with very fast control capability (response speed < 1 second).

As can be seen in Figure 3, the simulation results show the following:

– As RES levels increase, the current control methodology does not reliably control frequency fluctuations after the event in reference system. Power system frequency can only be maintained after activation of load shedding from several levels of frequency drops.

– Adjustments to the controllers (including minor software changes to the generator / power plant control systems) significantly improve the controllability and stability of the power system. Due to these adjustments, the power system frequency can be kept very close to 49.3 Hz (planning criterion) even under the most severe modes and load levels in the power system.

– In addition, increased flexibility due to the 50 MW battery pack improves RoCoF and minimum frequency, allowing the power system to maintain system frequency above 49.3 Hz even under the most complicated conditions.

As can be seen in Figure 4, synchronous operation of all units is ensured without disturbing the operating limits of voltage levels.

4.3 Power system load fluctuations

Rapid inter-system active power overflow fluctuations along the 330 kV Khachmaz-Derbend interconnection line are taken from the SCADA system, as shown in Figure 5. Considering the long-term power flow data, the standard system load fluctuation for 1-minute and 10-minute variations is considered to be 50 MW.

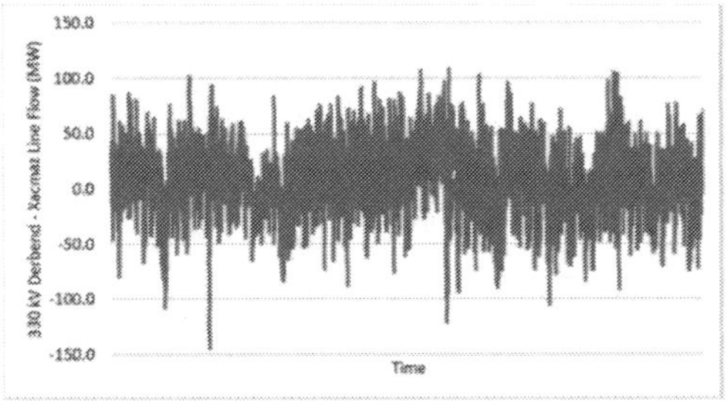

Fig. 5. Active power flow of 330 kV Derbent - Khachmaz line

The study also provides probabilistic data on the one-second power changes of the 230 MW SPP, as shown in Figure 6.

Fig. 6. Change in active power of 230 MW SES during one day (second measurements)

4.4 Requirements for the rotating power reserve

For safe and reliable operation of power systems, balance must be maintained at all times. For this purpose, it is necessary to install some rotating power reserve available in all modes of operation of the power system. The main objective of the study is to have a reserve capable of covering 99.7% of the fluctuations in the system according to the ENTSO-E reserve policy.

Based on the installed capacity of contracted and planned RES facilities and their active power change probabilities for 10-minute windows, the mandatory rotating reserves are calculated as follows:

Table 1. Font styles for a reference.

Installed RES capacity, MW	Total rotating reserve, MW
≤ 300	480
≤ 900	600
≤ 1200	650
≤ 1600	700
≤ 2000	750
≤ 2600	800

As can be seen from the table, the reserve demand increases with the installed capacity of VRES and starts to congest above a certain level of VRES. However, this reserve requirement creates another problem for power system operation, since the reserve availability condition depends on the level of synchronous generators being on. Thus, taking into account the running generators, the minimum levels of rotating reserves for stable operation of the power system are determined, as shown in Figure 7.

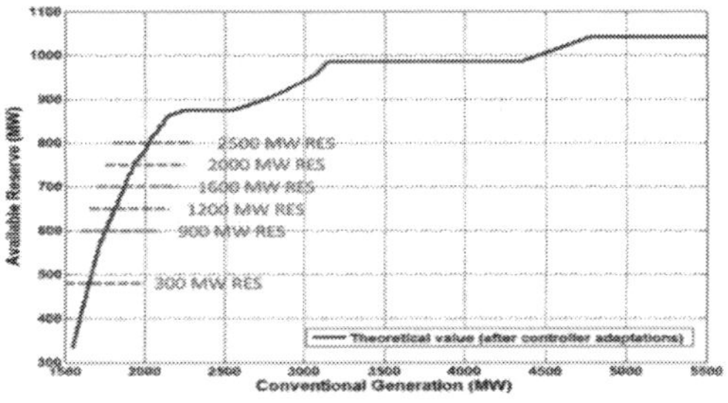

Fig. 7. Minimum levels of rotating reserves

In order to provide the required level of rotating reserve without building new TPPs, it is proposed to

© The Authors, published by EDP Sciences. This is an open access article distributed under the terms of the Creative Commons Attribution License 4.0 (https://creativecommons.org/licenses/by/4.0/).

install batteries according to the level of VRES to be penetrated.

Table 2. Rotating reserve level using battery packs

RES installed capacity, MW	BSS, MW
710	50
1200	100-200
1850	250
2500	300

Conclusions

1. In the next few years, the Azerbaijan power system will be in the fourth phase of penetration of variable RES, which is characterized by a key problem, the threat of instability of power system operation modes, in a short time the system must have the capacity to withstand various disturbances in the power system.
2. In order to maintain the parameter of frequency change rate within the permissible limits in the power system, all synchronous machine regulators must fulfill their frequency control functions and synchronous generators of at least 1500 MW capacity must be put into operation.
3. Large volumes of VRES penetration in the power system require maintaining a rotating power reserve greater than the capacity of the largest unit, the larger the volume of VRES, the larger the rotating reserve.

Part of the rotating power reserve should be covered by a battery put into service quickly.

References

1. Senderov S. M., Yusifbeyli N. A., Rabchuk V. I., et al., GESG 2018, **69**, (2018).
2. I. Prokofiev World Energy, **№2**, (2008).
3. N.N. Normamatov, A.E. Shanazarov et al. "Electric drive, electric technologies and electrical equipment of enterprises" Collection of scientific papers of the V International Scientific and Technical Conference (2020)
4. Yusifbayli N.A, Nasibov V.Kh., SOCAR Proceedings, Renewable energy, Issue **No. 1** (2022)
5. Yusifbayli N.A, Nasibov V.Kh., Alizade R.R. "Machine-building and Energy: New Concepts and Technologies" International Scientific-practical Conference, December 2-3, (2021)
6. Lenzi, V., Ulbig, A., Andersson, G. In: Proceedings of the 2013 IEEE Grenoble Conference Power Tech, POWERTECH (2013).
7. Dreidy, M., Mokhlis, H., Mekhilef, S. Renewable and Sustainable Energy Reviews, **69**, (2017).
8. Bayindir, R., Demirbas, S., Irmak, E., et al. IEEE International Power Electronics and Motion Control Conference, (2016)
9. Tambunan, H. B., Hakam, D. F., Prahastono, I., et al. Case Study of Java-Bali Power System. Energies, **13(22)**, (2020).
10. Ackermann, T., Prevost, T., Vittal, V., et al. IEEE Power and Energy Magazine, **15(6)**, (2017)
11. Kundur, P., Balu, N. J., Lauby, M. G. Power system stability and control. New York, NY, USA: McGraw-Hill, (1994).
12. EPRA Annual Report (2023) https://www2.deloitte.com/content/dam/Deloitte/de/Documents/real-estate/EPRA-Annual-Report-Survey-Sept-2023.pdf
13. Tielens, P., Van Hertem, D. Proceedings of the Young Researchers Symposium in Electrical Power Engineering, Delft, (2012, April).

© The Authors, published by EDP Sciences. This is an open access article distributed under the terms of the Creative Commons Attribution License 4.0 (https://creativecommons.org/licenses/by/4.0/).

Method of identification and operational control of unauthorized electricity withdrawal in distribution networks

Turatbek Omorov[1]*, *Kubnychbek* Zakiriaev[2], *Beishenaly* Takyrbashev[1], and *Tolkunbek* Zholdoshov[3]

[1]National Academy of Sciences, Institute of Mechanical Engineering and Automation, 265a Chuy, Bishkek, Kyrgyz Republic
[2]Issyk-Kul State University named after K. Tynystanov, Department of Information Systems and Technologies, 103 Abdrakhmanov, Karakol, Kyrgyz Republic
[3] Osh State University, Department of Applied Mathematics, 331 Lenin, Osh, Kyrgyz Republic

Abstract. This article addresses the problem of operational identification and control of unauthorized electricity withdrawals (UEW) in distribution electric networks (DEN) under the operation of the Automated System for Commercial Metering of Electricity (ASCME). The issue involves significant uncertainty regarding the current state of the network and external disturbances, represented by unauthorized electricity consumers. To reduce uncertainty and obtain necessary information, the concept of a virtual DEN model is introduced, representing the ideal state of the network without unauthorized consumers. A new method based on this virtual model is proposed for solving the problem. Conditions for identifying the current state of the DEN are established. Vectors of actual current and voltage values in the virtual network are identified. Additionally, a criterion for localizing UEW and a computational algorithm for automated control are formulated. The results improve the reliability of DEN operations and can be used to develop specialized software for the UEW localization subsystem within the ASCME framework.

Introduction

One of the most critical indicators of the efficiency of distribution electric networks (DEN) is the level of electricity losses. These losses are caused not only by internal factors, such as current and voltage imbalances [1,2], nonlinear properties of individual loads [3,4], but also by external factors, particularly the presence of unauthorized electricity withdrawals (UEW) [5,6]. Therefore, when developing new automation systems for DEN, it is essential to design new methods and technologies that minimize electricity losses. In this regard, the development of an information system aimed at managing, monitoring, and controlling electricity losses in real-time is seen as a more promising approach [6-11]. This system would include a subsystem for promptly detecting and localizing UEW sites, allowing for a reduction in electricity losses within the network. Analysis shows that the development of such a system is best implemented as part of the ASCME hardware and software complexes [12-14]. For this purpose, the ideas and approaches for identifying and locating non-technical electricity losses, as outlined in [5,15-19], can be utilized. This article proposes a new method for solving the problem, designed for use within ASCME.

Problem statement

We are considering a four-wire DEN with a voltage of 0.4 kV, the design diagram of which is shown in fig.1, where k and ν are index variables indicating, respectively, the numbers of phases A, B, C $(k = \overline{1,3})$ and electrical circuits of the network $(\nu = \overline{1,n})$; $\widetilde{E_k}$ – instantaneous EMF power supply for the k-th phase; $\tilde{I}'_{k\nu}$, $\tilde{U}'_{k\nu}$, $\dot{Z}_{k\nu}$ – instantaneous current, voltage and complex resistance of the subscriber load with coordinate (k,ν), respectively; $\tilde{\imath}'_{k\nu}$, $\dot{z}_{k\nu}$ – current and complex resistance of the ν-th inter-subscriber section of the k-th phase; $\tilde{J}'_{\nu}$, $\dot{z}_{\nu}$ – instantaneous current and complex resistance of the ν-th section of the neutral wire; $\tilde{I}'_k$, $\tilde{U}'_k$, $\dot{Z}_k$ – instantaneous currents, voltages and complex resistance at the inputs of the corresponding phases.

We will assume that the following conditions are met:

1. There are UEWs in the network, the locations of which are unknown.
2. The phase and neutral wires of the network have the same cross-sections, i.e. inter-subscriber resistances $\dot{z}_{k\nu} = \dot{z}_k$.
3. The modules of resistances Z_ν of inter-subscriber sections are determined in advance and are stored in the ASCME database.
4. At a discrete moment in time $t = t_0$, using the ASCME electricity meters, synchronous measurements of the effective values of currents I'_k, $I'_{k\nu}$ and voltages U'_k, $U'_{k\nu}$ respectively at the phase inputs and at the network loads, as well as the power factors $c_k = \cos\varphi'_k$ and $c\_c_{k\nu} = \cos\varphi'_{k\nu}$ between them are

* Corresponding author: omorovtt@mail.ru

© The Authors, published by EDP Sciences. This is an open access article distributed under the terms of the Creative Commons Attribution License 4.0 (https://creativecommons.org/licenses/by/4.0/).

performed. The received information is recorded in the ASCME database and current and voltage vectors are formed on their basis:

$$\bar{I}'_k = [I'_{k1}, I'_{k2}, \dots, I'_{kn}],$$
$$\bar{U}'_k = [U'_{k1}, U'_{k2}, \dots, U'_{kn}],$$
$$C'_k = [c_{k1}, c_{k2}, \dots, c_{kn}],\ k = \overline{1,3}.$$
$$I' = [I'_1, I'_2, I'_3],\ U' = [U'_1, U'_2, U'_3],$$
$$C' = [c_1, c_2, c_3].$$

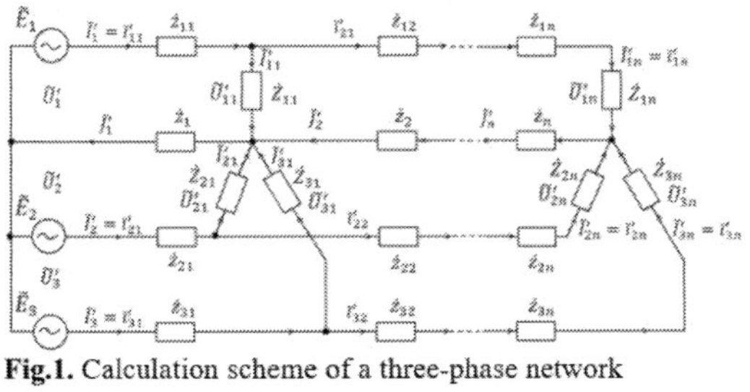

Fig.1. Calculation scheme of a three-phase network

The task is to ensure operational control of unauthorized power withdrawals in the DEN based on the identification of their current coordinates.

The solution to the formulated problem includes the following main stages:

1. Construction of a virtual model of the DEN.
2. Identification of the current state of the real DEN.
3. Formation of criterial functions.
4. Formation of a general scheme for identifying the locations of unauthorized power withdrawals.
5. Evaluation of inter-user currents of the virtual network.
6. Evaluation of the resistances of inter-user sections of the network.
7. Construction of an algorithm for identifying unauthorized power withdrawals.

Construction of the Virtual DEN Model

Under conditions of uncertainty of the current state of the DEN, in order to solve the problem formulated above, it is necessary to form additional information about the object in addition to the primary initial data obtained from the ASCME meters. For this purpose, we will introduce the concept of a virtual model of the DEN, which characterizes the desired state of a real three-phase network in the absence of external disturbances in the form of unauthorized consumers, on whose loads complex currents $\dot{I}^x_k$, $k = \overline{1,3}$ flow. Fig.2 shows the calculation scheme of the k-th phase of such a virtual network.

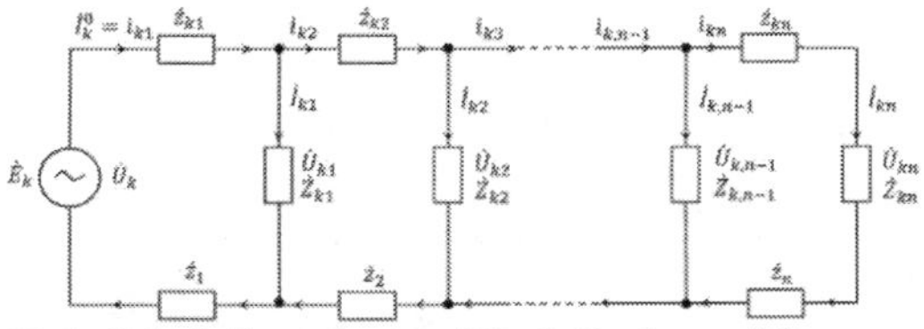

Fig.2. Calculation scheme of the k-th phase of the virtual network

Here $\dot{E}_k$, $\dot{I}^0_k$ are the complex EMF and current at the input of the k-th phase; $\dot{I}_{kv}$, $\dot{U}_{kv}$, $\dot{Z}_{kv}$ are the complex current, voltage and load resistance of the virtual subscriber with coordinate (k, v); $\dot{i}_{kv}$ is the complex current of the corresponding inter-subscriber section, which is determined by the following sum:

$$\dot{i}_{kv} = \sum_{\xi=v+1}^{n} \dot{I}'_{k\xi} \qquad v = \overline{1,n}, \tag{1}$$

where $\dot{I}'_{k\xi}$ are the components of the original vector $\bar{I}'_k$.

Input phase currents $\dot{I}^0_k = \dot{i}_{k1}, k = \overline{1,3}$, which we will further call desired currents, as they characterize (describe) the state of the network in the absence of UEW. We will assume that the following functional dependencies exist between the currents $\dot{I}'_k$, $\dot{I}^0_k$ and $\dot{I}^x_k$:

$$\dot{I}'_k = \dot{I}^0_k + \dot{I}^x_k, \qquad k = \overline{1,3}. \tag{2}$$

The values of currents $\dot{I}_{kv}, \dot{i}_{kv}$ and voltages $\dot{U}_{kv}$, describing the state of the network under consideration, differ from their corresponding values characterizing the state of the original - real DEN (Fig.1), i.e. $\dot{I}_{kv} \neq \dot{I}'_{kv}$, $\dot{i}_{kv} \neq \dot{i}'_{kv}$, $\dot{U}_{kv} \neq \dot{U}'_{kv}$. In this case, the values of resistances $\dot{Z}_{kv}$, $\dot{z}_{kv}$ and $\dot{z}_v$of subscriber loads and inter-subscriber sections of the real and virtual network have the same values. In what follows, the complex currents $I^0_k$and i_{kv} will be presented in the following complex form [14]:

$$\dot{I}^0_k = I^0_k e^{j(\beta_k + \alpha^0_k)}, \tag{3}$$

$$\dot{i}_{kv} = l_{kv}\, e^{j(\beta_k + \tilde{\alpha}_{kv})},$$

where I^0_k, l_{kv} are the moduli of the corresponding complex quantities that make up the vectors $\bar{I}^0_k = [I^0_1, I^0_2, I^0_3]$ and $\bar{l}_k = [l_{k1}, l_{k2}, \dots, l_{kn}]$; $\alpha^0_k, \tilde{\alpha}_{kv}$ are the deviations of the phase shifts from their base values β_k, determined by the formulas $\beta_k = (2(k-1)\pi)/3$. The methodology for estimating the components of the vectors $\bar{I}^0_k$ and $\bar{l}_k$ will be described below.

Identifying the current state of the network

According to the conditions of the problem under consideration, the current values of phase currents $I'_k = I'_k(t_0)$ at the input of the initial real distribution network (Fig.1) at the time $t = t_0$ are measured by the main three-phase meter, which are contained in the ASCME database. The analysis shows that for identifying the current state of the distribution network, it is advisable to use the following conditions:

$$|I'_k - I^0_k| \leq \Delta I, \qquad k = \overline{1,3}, \tag{4}$$

where ΔI is the maximum permissible error in measuring currents in the automated metering system.

It is obvious that when relations (4) are satisfied, there are no UEWs in the network, and if at least one

© The Authors, published by EDP Sciences. This is an open access article distributed under the terms of the Creative Commons Attribution License 4.0 (https://creativecommons.org/licenses/by/4.0/).

 https://doi.org/10.1051/e3sconf/202458401046

of the conditions (4) is not satisfied, then there are unauthorized consumers in the network. Thus, relations (4) can be used as a criterion condition for identifying the current state of the DEN.

Formation of criteria functions

Let us consider the k-th phase of a real DEN (Fig. 1), the states of the loads of which are determined by the vector $\bar{U}'_k = [U'_{k1}, U'_{k2}, ..., U'_{kn}]$. Based on its components, we determine the following differences:

$$\Delta U'_{kv} = U'_{kv} - U'_{k,v+1}, \quad v = \overline{1, n-1}. \tag{5}$$

The values $\Delta U'_{kv}$ characterize the voltage losses in the inter-subscriber sections of the k-th phase, having coordinates (k, v).

Now, let us assume that unauthorized power withdrawals in the network are carried out in the vicinity of the nodes of a real DEN with coordinates (k, m_k), which is conventionally shown in fig. 3, where $v = m_k$, and $m_k < n$. Here $\dot{I}_k^x$, $\dot{Z}_k^x$ are the current and load resistance of the unauthorized consumer, respectively. In this case, each phase of the three-phase network relative to these nodes can be conventionally divided into two parts. Obviously, throughout the first (initial) part of the k-th phase of the DEN, an additional current $\dot{I}_k^x > 0$ flows through its inter-user sections, i.e. the values of the inter-user currents I'_{kv} of the real network, taking into account (2), are determined by the expressions

$$I'_{kv} = I_{kv} + I_k^x, \quad v = \overline{1, m_k}.$$

In this case, the voltage losses $\Delta U'_{kv}$ at the initial inter-subscriber sections, taking into account (2), are determined by the expressions

$$\Delta U'_{kv} = I'_{kv} Z_{kv} = (I_{kv} + I_k^x) Z_{kv}, \tag{6}$$

where I_{kv} is the effective value of the desired current on the inter-subscriber section with coordinate (k, v) provided that there are no UEWs in the network.

In the second part of the phase under consideration, the current I_k^x of the unauthorized consumer is absent, and the voltage losses ΔU_{kv} on the corresponding inter-subscriber sections can be estimated with a certain accuracy using the formulas:

$$\Delta U_{kv} = I_{kv} Z_{kv}, \quad k = \overline{1,3}, \ v = \overline{m_k + 1, n}. \tag{7}$$

Now we introduce the following relative discrete quantities into consideration:

$$G_{kv} = \frac{\Delta U'_{kv}}{\Delta U_{kv}}, \quad v = \overline{1, n}.$$

For the initial sections of the network, taking into account (6) and (7), the functions G_{kv} can be written in the form

$$G_{kv} = \frac{\Delta U'_{kv}}{\Delta U_{kv}} = \frac{I_{kv} + I_k^x}{I_{kv}}, \quad k = \overline{1,3}, \quad v = \overline{1, n}.$$

From this it is evident that the relative values of G_{kv} at the initial sections of the network ($v = \overline{1, m_k}$) differ significantly from the values of the differences $\Delta U_{k\xi}$, $\xi = \overline{m_{k+1}, n}$, related to the sections where there are no UEFs ($I_k^x = 0$). With an increase in leakage currents I_k^x, the values of G_{kv} for the initial sections of the network are significantly greater than $\Delta U_{k\xi}$ calculated for the second part of the network, i.e.

$$G_{kv} \gg G_{k\xi}, \ v = \overline{1, m_k}, \ \xi = \overline{m_{k+1}, n}. \tag{8}$$

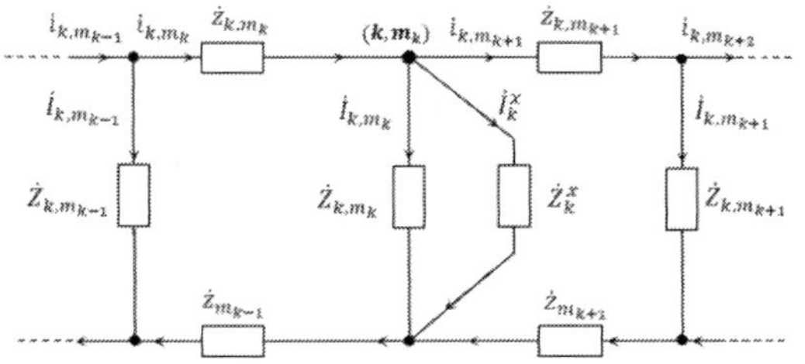

Fig.3. Conditional fragment of the network where the UEW is carried out

This property (position), defined by relation (8), can be used to construct a constructive algorithm for searching for coordinates (locations) of the UEW. For this purpose, we write the discrete functions G_{kv} taking into account (7) in the form:

$$G_{kv} = \frac{\Delta U'_{kv}}{\Delta U_{kv}} = \frac{\Delta U'_{kv}}{I_{kv} Z_{kv}}, \qquad k = \overline{1,3}, \ v = \overline{1, n}. \tag{9}$$

An analysis of the structure of the lattice functions G_{kv} shows that in order to localize the coordinates of the UEW, it is advisable to introduce the following discrete functions:

$$F_{kv} = |G_{kv} - G_{k,v+1}|, \ k = \overline{1,3}, \ v = \overline{1, n-1}. \tag{10}$$

The set of values F_{kv} for a given k is the vector $F_k = [F_{k1}, F_{k2}, ..., F_{k,n-1}]$. Analysis shows that, taking into account relations (8) and the condition $m_k < n$, the graphical representation of the function F_{kv} is presented in the form shown in fig. 4.

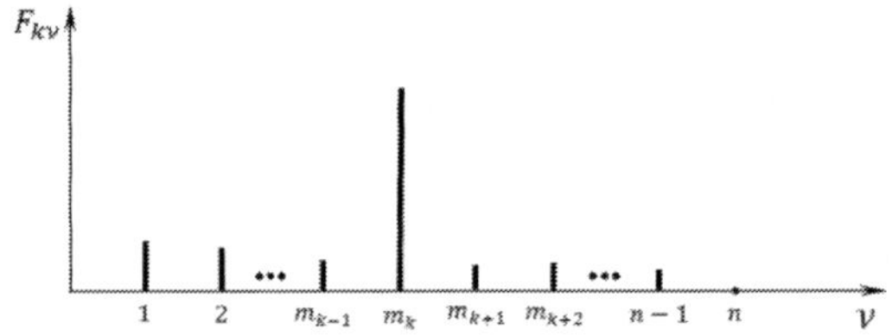

Fig.4. Graphical representation of the function F_{kv}

As can be seen from the figure, the coordinate of the UEW in the k-th phase of the network is determined based on the solution of the following extremum problem:

© The Authors, published by EDP Sciences. This is an open access article distributed under the terms of the Creative Commons Attribution License 4.0 (https://creativecommons.org/licenses/by/4.0/).

$$\max_{v \in V} F_{kv} = F_{k,m_k}, \quad k = \overline{1,3}, \qquad (11)$$

where m_k is the number of the node of the k-th phase, in the vicinity of which unauthorized power extraction is observed; V is a discrete subset consisting of the numbers of network subscribers, i.e. $V = \{1, 2, ..., n\}$. As a result, the extremum problem (11) is reduced to finding the maximum element of the vector $F_k = [F_{k1}, F_{k2}, ..., F_{k,n-1}]$, $k = \overline{1,3}$, which is solved by fairly simple means.

Thus, the introduction of the concept of a virtual model of the distribution network and the above-proposed procedure for constructing discrete functions F_{kv} make it possible to identify locations of unauthorized electricity withdrawals in the distribution network. In this case, the system of functions (9) and (10) can be used to construct a criterion for identifying unauthorized electricity withdrawals, which can be formulated as follows: the coordinates of unauthorized consumers in the distribution network are determined based on an analysis of the structure of the functions F_{kv}, $v = \overline{1,n}$, $k = \overline{1,3}$, by solving the extremal problem (11).

Estimation of inter-subscriber currents of a virtual network

These currents in complex form are described by expressions (1). Let us set the task of estimating the components of the vectors $\bar{I}_k^0 = [I_1^0, I_2^0, I_3^0]$ and $\bar{I}_k = [I_{k1}, I_{k2}, ..., I_{kn}]$. Let us consider a virtual network, the diagram of which is shown in fig. 2. For this purpose, it is necessary to first determine the complex resistances of the network subscriber loads:

$$\dot{Z}_{kv} = Z_{kv}\, e^{j\varphi_{kv}}, \quad k = \overline{1,3}, \; v = \overline{1,n}, \qquad (12)$$

where Z_{kv}, φ_{kv} are the module and phase of $\dot{Z}_{kv}$ respectively, calculated based on the readings (data) of subscriber electricity meters using the formulas:

$$Z_{kv} = \frac{U'_{kv}}{I'_{kv}}, \qquad \varphi_{kv} = \arccos(c_{kv}). \qquad (13)$$

Corresponding conductivity:

$$\dot{Y}_{kv} = \frac{1}{\dot{Z}_{kv}} = Y_{kv}e^{-j\varphi_{kv}}, \; k = \overline{1,3}, \; v = \overline{1,n},$$

where $Y_{kv} = \frac{1}{Z_{kv}}$. Then the powers $\dot{S}_{kv}$ consumed by network subscribers:

$$\dot{S}_{kv} = (I'_{kv})^2\dot{Z}_{kv}, \quad k = \overline{1,3}, \; v = \overline{1,n}. \qquad (14)$$

Now for the initial nodes of the k-th phase of the virtual network with coordinate (k, v) we introduce the concept of the total power $\dot{S}_k^a$, consumed by its subscribers. It is assumed that the latter are connected to the k-th phase beyond the specified point, i.e. having coordinates $(k, v + 1), (k, v + 2), ..., (k, n)$.

At the same time

$$\dot{S}_{kv}^a = \sum_{\xi=v}^{n} \dot{S}_{k\xi} = S_{kv}^a e^{j\varphi_{kv}^a}, \quad k = \overline{1,3}, \; v = \overline{1,n}, \qquad (15)$$

where S_{kv}^a, φ_{kv}^a are the modulus and phase of the total complex power $\dot{S}_k^a$, respectively.

Then, for the considered node of the k-th phase, the following functional relationships can be written:

$$\dot{S}_{kv}^a = \ell_{kv}{}^2\dot{Z}_{kv}^a, \; \overline{k = 1,3}, \; v = \overline{1,n}, \qquad (16)$$

where ℓ_{kv} is the desired effective current; $\dot{Z}_{kv}^a$ is the total resistance of the sections of the k-th phase located behind the point $(k, 1)$, which are calculated as the reciprocal of the sum of the corresponding conductivities $\dot{Y}_{kv}$:

$$\dot{Z}_{kv}^a = \frac{1}{\dot{Y}_{kv}^a} = \frac{1}{\sum_{v=1}^{n} \dot{Y}_{kv}} = Z_{kv}^a e^{j\varphi_{kv}^a}. \qquad (17)$$

Relations (16), taking into account (15) and (17), can be represented as

$$S_{kv}^a e^{j\varphi_{kv}^a} = \ell_{kv}{}^2 Z_{kv}^a e^{j\varphi_{kv}^a}, \quad \overline{k = 1,3}, \qquad v = \overline{1,n}.$$

To ensure the latter conditions, the following equalities must be satisfied:

$$S_{kv}^a = \ell_{kv}{}^2 Z_{kv}^a, \quad k = \overline{1,3}, \; v = \overline{1,n}.$$

From here we can calculate the values of the required input (desired) phase currents I_k^0 of the virtual network, since S_{kv}^a and Z_{kv}^a are known values

$$l_{kv} = \sqrt{\frac{S_{kv}^a}{Z_{kv}^a}} = \sqrt{S_{kv}^a Y_{kv}^a}, \quad k = \overline{1,3}, \; v = \overline{1,n}. \qquad (18)$$

In this case, the desired input phase currents of the virtual network are

$$I_k^0 = l_{kv}, \qquad k = \overline{1,3}. \qquad (19)$$

Evaluation of resistances of inter-subscriber network sections

Note that the resistances of subscriber loads $\dot{Z}_{kv}$ are pre-calculated based on subscriber electricity meters, and inter-subscriber resistances $\dot{z}_{kv}$ and $\dot{z}_v$ can be pre-identified and recorded in the ASCME database. It is advisable to estimate these resistances directly during the implementation of the identification procedure of the UEW in real time. In this case, in particular, the methods described in [10] can be used.

© The Authors, published by EDP Sciences. This is an open access article distributed under the terms of the Creative Commons Attribution License 4.0 (https://creativecommons.org/licenses/by/4.0/).

Construction of the algorithm for identification of the UEW

The results obtained above allow us to formulate the following algorithm for identification and control of the UEW in the three-phase distribution network.

1. Cyclic polling of the main three-phase and subscriber electricity meters of the ASCME at a given discrete point in time.
2. Recording the information received from the meters in the ASCME database and forming vectors $\overline{U}'_k = [I'_k, C'_k, I', U', C']$, $k = \overline{1,3}$.
3. Formation of a virtual model of the distribution network (Fig. 2.).
4. Determination of the input phase currents of the virtual network I^o_k and I_{kv}, $= \overline{1,3}$. According to formulas (18) and (19), respectively.
5. Identification of the current state of the distribution network by checking conditions (4).
6. When relations (4) are fulfilled, there are no UEW in the network. Go to point 1. Otherwise, go to point 7.
7. Calculation of voltage losses $\Delta U'_{kv}$ on inter-subscriber network sections using formulas (5).
8. Estimation of discrete functions G_{kv} using formulas (9).
9. Formation of criterial functions F_{kv}, $k = \overline{1,3}$, $v = \overline{1,n}$, using formulas (10) and criterial vectors $F_k = [F_{k1}, F_{k2}, \ldots, F_{k,n-1}]$.
10. Solution of extremal problem (11) and determination of coordinates of UEW in a real RES.
11. Completion of the process of identifying and localizing UEW locations in a DEN.

Conclusion

The practice of operating 0.4 kV distributed electric networks (DEN) shows that significant power losses due to unauthorized power consumption may be observed in them during certain periods of time. The hardware and software systems of the automated metering system (ASCME), which are implemented in low-voltage networks for the purposes of automation and informatization of the processes of control and accounting of electric power, do not include technical and software tools designed to identify and control unauthorized power consumption in the DEN. A new method and algorithm for solving this problem are proposed based on the concept of a virtual model of a three-phase network. The latter is introduced to describe the desired state of a real network, to the phases of which unauthorized consumers of electric power are connected. The computational procedure of the developed algorithm, in contrast to existing methods, is more adapted for use in the ASCME in real time. The obtained results can be used to create a subsystem for automated control of UEW as part of the ASCME.

References

1. F. D. Kosoukhob, N. V. Vasilev, A. O. Filipov, Reducing losses from current unbalance and improving electric-energy quality in 0.38-kV networks with municipal loads, Russ. Electr. Eng. 85, 6 (2014) 350-353.
2. T. T. Omorov, B. K. Takyrbashev, R. Ch. Osmonova, To the problem of mathematical modelling of the three-phase asymmetrical distribution network, Izv. Vyssh. Uchebnykh Zavedeniy. Problemy Energetiki 22, 1 (2020) 93-102.
3. A. G. Harutyunyan, On the calculation of additional power losses in three-phase four-wire networks, Electricity 10 (2015) 55.
4. Yu. S. Zhelezko, Losses of Electricity. Reactive Power. Power Quality (ENAS, Moscow, 2009) 456 p.
5. V. O. Samoylenko, A. M. Verkhozin, V. A. Tashchilin, A. V. Pazderin, N. D. Mukhlynin, Analysis of the effectiveness of modern information methods for detecting non-technical losses of electricity, Electr. Transm. Distrib. 4(79) (2023) 64-74.
6. A. A. Sapronov, S. L. Kuzhekov, V. G. Tynyansky, Operational detection of uncontrolled electricity consumption in electrical networks up to 1 kV, Izv. Vuzov. Electromechanics 1 (2004) 55-58.
7. T. T. Omorov, B. K. Takyrbashev, T. D. Koibagarov, Management of power losses in distribution networks as part of automated control and metering systems, Mechatronics, Automation, Control 22, 4 (2021) 191-199.
8. T.T. Omorov, B. Takyrbashev, K. Zakiriaev, T. Koibagarov, Digital Control of Electric Power Flows in Unbalanced Distribution Networks as Part of the Automated Metering and Control System, Energy Syst. Res., 4 (1), 38-46 (2021).
9. T. Omorov, K. Zakiriaev, B. Takyrbashev, Zh. Imanakunova, Automated control of unauthorized power take-offs in a distributed electrical network, Mechatronics, Automation, Control 24, 1 (2023) 24-32.
10. T. Omorov, B. Takyrbashev, T. Koibagarov, T. Zhanybaev, Identification and Monitoring of Power Losses in Distribution Networks as a Composition of AMR, AIP Conference Proceedings 2552 (2022) 020002.
11. T. Omorov, T. Koibagarov, B. Takyrbashev, A. Boronin, T. Zhanybaev, For the construction of subsystem for power line diagnostic states of distribution networks as a part of AMRCS, E3S Web Conf. 2020, 01042.
12. http://www.sigmatelas.lt/
13. http://www.mir-omsk.ru/stuff/career/vacancies
14. http://www.energomera.ru/

© The Authors, published by EDP Sciences. This is an open access article distributed under the terms of the Creative Commons Attribution License 4.0 (https://creativecommons.org/licenses/by/4.0/).

15. G. M. Messinis, N. D. Hatziargyriou, Review of non-technical loss detection methods, Electr. Power Syst. Res. 158 (2018) 250-266.
16. M. G. Chuwa, F. Wang, A review of non-technical loss attack models and detection methods in the smart grid, Electr. Power Syst. Res. 199 (2021) 18 p.
17. T. S. D. Ferreira, F. C. L. Trindade, J. C. M. Vieira, Load Flow-Based Method for Nontechnical Electrical Loss Detection and Location in Distribution Systems Using Smart Meters, IEEE Trans. Power Syst. 35, 5 (2020) 3671-3681.
18. V. A. Stennikov, I. I. Golub, E. V. Boloev, Use of smart meter measurements to identify those responsible for unauthorized electricity consumption, Energetik 4 (2023) 3.
19. T. T. Omorov, On the problem of localization of unauthorized electricity withdrawals in distribution networks as part of ASCME, Devices and Systems. Control, Diagnostics 7 (2017) 27-32.

© The Authors, published by EDP Sciences. This is an open access article distributed under the terms of the Creative Commons Attribution License 4.0 (https://creativecommons.org/licenses/by/4.0/).

Automated experimental complex on the platform of modern ASCME

Turatbek Omorov[1]*, *Abai* Asiev[2], *Rima* Osmonova[2], *Beishenaly* Takyrbashev[1], *Zgenishkul* Imanakunova[2], and *Manas* Suerkulov[2]

[1]National Academy of Sciences, Institute of Mechanical Engineering and Automation, 265a Chuy, Bishkek, Kyrgyz Republic
[2]Kyrgyz State Technical University named after I. Razzakov, 66a Aitmatov, Bishkek, Kyrgyz Republic

Abstract. Currently, in order to automate and digitalize processes in electrical distribution networks (DEN), ASCME software and hardware systems are widely used, which are mainly designed to solve problems of commercial electricity metering. At the same time, these automated systems do not solve such urgent problems as optimizing the operating modes of distribution networks, operational monitoring of technical and commercial losses of electricity, as well as diagnosing network conditions, including localizing the coordinates of unauthorized withdrawals of electricity (UWE). To date, the National Academy of Sciences of the Kyrgyz Republic and a number of foreign scientific centers have developed the scientific foundations and technologies for solving the above problems in real time. In order to test and study the effectiveness of these new results, it is advisable to create an automated experimental complex (AEC), which is essentially a physical model of ASCME. This is due to the fact that the use of automated information systems currently used in distribution networks for this purpose is practically impossible. This article discusses the general structure, functional tasks and main directions of research using AEC. The complex includes a physical model of the distribution network and the main functional elements of modern automated control and accounting systems for electricity (ASCAE), such as a data concentrator, a computer server, single-phase and three-phase electricity meters and a telecommunications module for transmitting and receiving data. The basis of the special software of the experimental complex is made up of new methods and technologies for solving the above functional problems, which are not included in modern AMR systems. AEC is planned to be used not only for conducting scientific research and experiments, but also for the educational process in technical universities with areas of training in the electrical power industry.

Introduction

Currently, ASCME software and hardware systems are being actively implemented in electrical distribution networks (DEN) [1-3], which are created using Smart Grid technology [4, 5]. They belong to the class of information-measuring systems, the composition of which functions is aimed at automating commercial electricity metering by remotely collecting data from electricity meters installed at network subscribers. At the same time, modern (traditional) ASCME lacks the means to solve a number of important functional problems, the solution of which makes it possible to significantly increase the efficiency of using automated systems and the economic indicators of distribution companies, as well as improve the quality of electricity. Such tasks, in particular, include the tasks of optimizing the operating modes of distribution zones by balancing a three-phase network [6-10], operational monitoring of electricity losses [11-13] and localizing places of unauthorized electricity withdrawals (UEW) [14-17]. As is known, the solution to these functional problems is directly related to the problem of reducing electricity losses in networks, which currently have a fairly high level. In this regard, in order to minimize technical and commercial losses of electricity in distribution zones, the National Academy of Sciences of the Kyrgyz Republic is conducting theoretical and applied research aimed at improving traditional ASCME, which is currently being actively implemented in distribution networks. As a result of these studies, new methods and digital technologies have been developed [7, 10, 13, 15, 18-20], oriented for use as part of traditional ASCME. It should be noted that in order to study and test the effectiveness of these technologies, it is advisable to have the necessary set of data from real objects operating under the operating conditions of AMR. However, obtaining such data with given characteristics in practice presents great difficulties in organizational and technical plans. In this regard, in order to study and develop new technical and software solutions in the field of automation and digitalization of processes in distribution networks, an automated experimental complex (AEC) has been developed, which is essentially a physical model of ASCME in a simplified version.

* Corresponding author: omorovtt@mail.ru

© The Authors, published by EDP Sciences. This is an open access article distributed under the terms of the Creative Commons Attribution License 4.0 (https://creativecommons.org/licenses/by/4.0/).

Functional structure of the AEC

Traditional ASCME belong to the class of information-measuring systems. As is known, their composition mainly solves the following problems:

- commercial metering of electricity consumed by distribution network subscribers;
- automated data collection from the system's electricity meters and transmission of the necessary data to the upper management level;
- monitoring the status of electricity meters and other technical means.

Since these automated systems do not solve the problems of optimizing operating modes and minimizing electricity losses in distribution zones, in order to increase the efficiency of traditional ASCME it is proposed to introduce additional information subsystems designed to solve the following problems:

- Optimization of the operating mode of distribution networks based on balancing of phase loads.
- Identification and monitoring of technical and commercial power losses in real time.
- Identification and localization of places of unauthorized withdrawals (thefts) of electricity (UWE).
- Prompt identification of the electrical state of the DEN.

The structure of AEC is shown in Fig. 1. It includes new functional information subsystems that are not included in traditional ASCME. These are the subsystems: optimization of the network operating mode (NOM); monitoring of power losses (MPL); localization of places of unauthorized electricity withdrawals (LUEW); identification of the electrical state of the network (ESN).

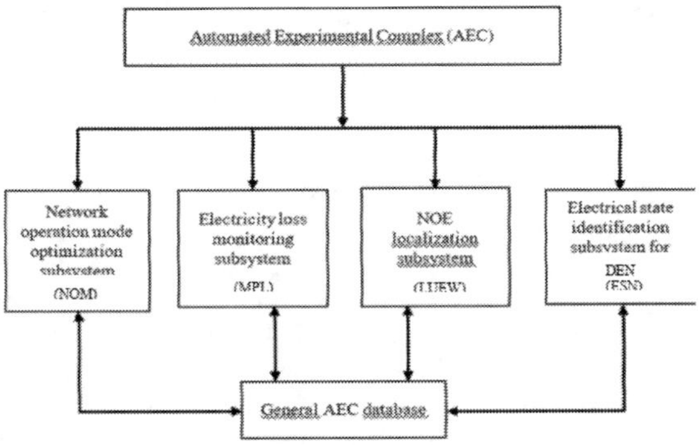

Fig.1. Functional structure of the AEC

The NOM subsystem is technically implemented through a digital controller (DCR) [7, 10]. The DCR block, based on a special algorithm (control law), generates control actions u^* to the object. In this case, the control signal u^* is a command digital code, which is formed in the form of a vector $u^* = [\Phi_1, \Phi_2, \beta]$, где Φ_1, Φ_2- numbers (names) of phase pairs in which it is necessary to switch network subscribers from a more loaded phase (Φ_1) to a less loaded (Φ_2); β – a vector composed of the coordinates (addresses) of phase subscribers to be switched. This control signal u^* via a communication channel (CC) it is transmitted to the executive elements of the system, which are used as phase current switches (PCS), designed to carry out the required switching of loads of network subscribers from one phase to another [7, 10]. The phase current switch (PCS) is based on a separate microcontroller. Methods, algorithms and technologies for constructing an NOM subsystem are proposed in [7, 10].

The main functions of the "Electricity Loss Monitoring" (ELM) subsystem are the identification and continuous monitoring of technical and commercial electricity losses in distribution zones. In the case when there are unauthorized withdrawals (thefts) of electricity from the network, the balance of complex capacities is determined by the following relationships:

$$\dot{S}_k(\xi) = \dot{S}_k^a(\xi) + \dot{S}_k^T(\xi) + \dot{S}_k^x(\xi), \quad k = \overline{1,3},$$

where k - is index variable indicating the number of the corresponding phase (A, B, C), $k = \overline{1,3}$, $\dot{S}_k$ - complex power, consumption k – phase at a discrete point in time $t = t_\xi$; $\dot{S}_k^a$– total complex power, consumed by all subscribers k – phases; $\dot{S}_k^T$– technical power losses k – phases; $\dot{S}_k^x$- uncontrolled power loss (NOE) k – network phase. At the same time, the power $\dot{S}_k(\xi)$ и $\dot{S}_k^a(\xi)$ are known quantities. It should be noted that in existing ASCME technical $\dot{S}_k^T$ and commercial $\dot{S}_k^x$ losses are not determined, but only total power is assessed $\dot{S}_k$, consumed by network phases and electricity consumers $\dot{S}_k^a$. The main task of the MPE subsystem is to identify quantities $\dot{S}_k^T$ and $\dot{S}_k^x$ and on their basis, carry out continuous monitoring of uncontrolled electricity losses in the distribution network. At the same time, in order to effectively solve the problems of the MPE subsystem, models of virtual RES are introduced into consideration, describing the desired states of real networks in the absence of unauthorized consumers in them. Methods and algorithms for solving problems of the subsystem under consideration were proposed in [13, 19].

The LUEW subsystem solves the problem of identifying and localizing places of unauthorized electricity withdrawals in the distribution zone [15].

The IES subsystem is designed to identify parameters and variables of the network state that are inaccessible for measurement [13, 18, 19], which are the complex resistances of the inter-subscriber sections of the network and the voltage and current vectors at given points of the DEN, as well as the internal resistance of the transformer power supply of the network.

The general AEC database includes regulatory, reference and operational data necessary for the functioning of the complex. These include, in particular: the structure and parameters of the physical model of the DEN; measurement data obtained from electricity meters; results of functioning of AEC information subsystems.

© The Authors, published by EDP Sciences. This is an open access article distributed under the terms of the Creative Commons Attribution License 4.0 (https://creativecommons.org/licenses/by/4.0/).

 https://doi.org/10.1051/e3sconf/202458401047

From the outputs of each electricity meter at a given time $t = t_0$ the following basic quantities are read and written to the common database: effective current values I_{kv} and stresses U_{kv} on DEN loads; power factors $c_{kv} = \cos\varphi_{kv}$, where $k = 1,3$, $v = \overline{1, n_k}$. At the same time k - phase number (A, B, C); n_k - number of subscribers connected to phase number k.

The composition of the functional elements of the AEC is shown in Fig. 2. It can be noted that the developed AEC is essentially a mini model of the ASCME software and hardware complex, since it includes almost all of its main functional elements and the physical model of the distribution network in a simplified version.

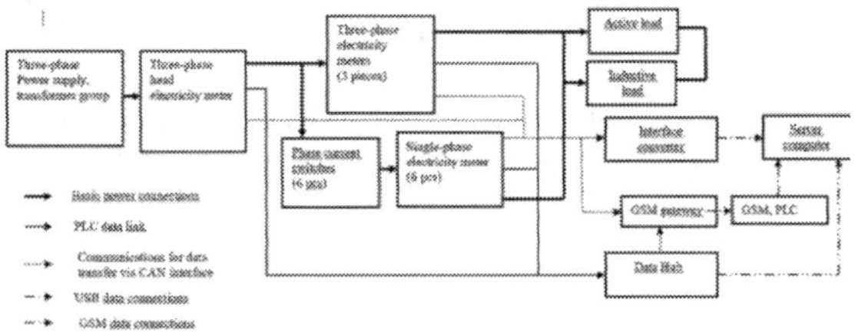

Fig. 2. Block diagram of AEC blocks

Composition of AEC technical means

The main functional elements of AEC are:

- Physical model of the distribution network.
- Transformer power supply.
- Computer – server.
- Data concentrator (DC).
- Single-phase and three-phase electricity meters.
- Phase current switch (PCS).
- Communication channel.

The computer server implements the algorithmic and special software of the AEC functional subsystems (NOM, MPL, LUEW, ESN), which was developed in the C software environment. The central block of the NOM subsystem is a digital controller (DCR), which forms an algorithm for controlling objects, which are subscriber electricity meters. The synthesis of the control algorithm is carried out on the basis of the principles and methods of automatic control [21-23]. The initial data necessary for the functioning of the latter and included in the structure of the common database comes from the hub database (DC). The latter is part of traditional ASCME and performs the following main functions:

- Cyclic collection of data on the energy consumption of subscribers and network parameters in specified nodes in an automatic or automated mode.
- Pre-processing of information coming from electricity meters.
- Transfer of necessary data on energy consumption and the status of metering devices.
- Data exchange between DC and the general system database.
- The phase current switch (PCT) performs the following functions:
- Implementation of control actions on the object by controlling the state of subscriber meters that are subject to switching from one phase to another in accordance with the command signal from the DCR.
- Generating control signals about the state of subscriber meters.

The technology for receiving data from ASCME electricity meters, developed by SigmaTelas based on GSM, is used as a communication channel (CC), and PLC technology, as well as a radio channel, can be used to transmit control command signals to the control object (electricity meters).

Electricity meters provide the following functions [2, 3]:

- Measurement of electrical parameters (currents, voltages, active and reactive powers, etc.) characterizing the state of network loads.
- Monitoring the status of functional units.
- Reception and transmission of control and monitoring signals, as well as measurement data.

In the complex, it is possible to simulate various modes of operation of the distribution network, for example, symmetrical and asymmetrical modes of network operation; broken power line; connect unauthorized consumers to the network.

Main directions of research based on AEC

The main idea of creating AEC is to conduct experimental research and develop new methods and digital technologies in order to improve traditional ASCME. In this regard, modeling of the joint functioning of the developed AEC information subsystems (NOM, MPL, LUEW, ESN) and the physical model of the DEN of a given structure is carried out.

The results of measurements of the parameters of the physical network model and the functioning of the AEC subsystems at appropriate times are recorded in a common database. Next, a comparative analysis of the obtained calculation and measurement data is carried out. Based on this analysis, the effectiveness of the proposed information subsystems for use as part of traditional ASCME is determined. A generalized scheme of experimental studies based on AEC is shown in Fig. 3.

© The Authors, published by EDP Sciences. This is an open access article distributed under the terms of the Creative Commons Attribution License 4.0 (https://creativecommons.org/licenses/by/4.0/).

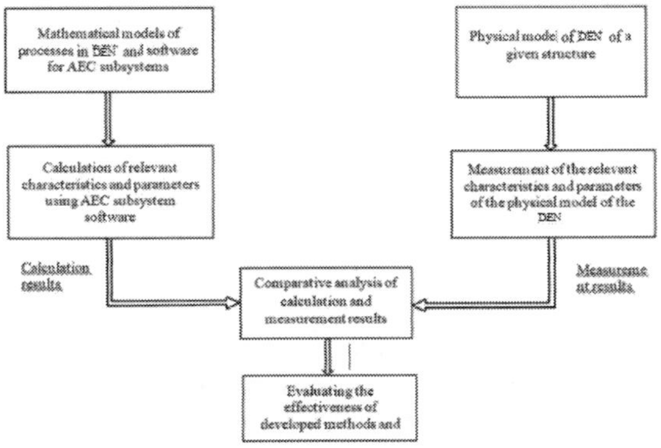

Fig. 3. Generalized scheme of experimental studies based on AEC

It can be noted that as a result of the functioning of AEC information subsystems, physical quantities are formed that characterize the current state of the RES and system efficiency indicators. In particular, the following quantities are identified:

- assessment of technical power losses in the network;
- assessment of commercial power losses in distribution zones;
- effective values of currents and voltages that are not available for measurement;
- complex resistances of inter-subscriber sections of the distribution network, which are not controlled as part of traditional ASCME;
- assessment of the power values consumed by the network phases in the presence of uncontrolled current leaks in the network;
- assessment of neutral wire currents in inter-subscriber sections of the network.

At the same time, using appropriate measuring instruments, it is possible to obtain data on the values of the above variables and system efficiency indicators at appropriate points in time.

The results of AEC functioning allow us to determine:

- characteristics of the accuracy of the assessment of identified variables (voltages, currents, power, etc.) and three-phase network parameters that are not available for measurement;
- efficiency indicators of functioning AEC subsystems, which, in particular, can be used as the share of technical and commercial losses of electricity in the total volume of electricity supplied to the input of the distribution network.

Thus, based on a comparison of data obtained simultaneously by the AEC information subsystems and with the help of measuring instruments, it is possible to evaluate the effectiveness of the developed methods and digital technologies aimed at improving traditional ASCME.

Conclusions

Improving traditional ASCME is the most important direction in the automation and informatization of processes in distribution networks. One of the tasks is to develop new functional subsystems in their composition based on scientifically based methods and digital technologies, aimed at reducing electricity losses in the network in real time. In order to study and test these new developments, an automated experimental complex (AEC) has been created, the structure of which includes all the main elements of ASCME and developed functional subsystems, implementing new methods and algorithms for solving problems of optimizing operating modes of distribution networks and monitoring technical and commercial losses of electricity. The results of experimental studies make it possible to evaluate the effectiveness of the developed AEC information subsystems and the possibility of their practical use as part of traditional ASCME. The developed complex can also be used in the educational process when preparing students, undergraduates and graduate students studying in areas related to automation and digitalization of management and control processes in electrical networks.

References

1. http://www.sigmatelas.lt/
2. http://www.mir-omsk.ru/stuff/career/vacancies
3. http://www.energomera.ru/
4. El-Hawary, M.E. The Smart Grid—State-of-the-art and Future Trends. Electr. Power Compon. Syst. 2014, 42, pp239–250.
5. Kazmi S.A., Shahzad M.K., Khan A.Z., Shin D.R. Smart Distribution Networks: A Review of Modern // Energies, 2017.
6. Kiselev M.G., Lepanov M.G. Balancing currents in power supply networks with a power electric regulator of inactive power // Electrical engineering. 2018. No. 11. P.63-70.
7. Omorov T.T. Takyrbashev B.K., Koibagarov T.Dzh. Management of power losses in distribution networks as part of automated power control and metering systems // Mechatronics, automation, control. V. 22. №4. 2021. page 192 - 199.
8. Patent No. 2490768 (RF). I.V. Naumov, D.A.Ivanov, S.V. Podyachikh, Gantulga Damdinsuren. Balancing device for three-phase networks with a neutral wire // Bulletin. No. 23. 08.20.2013.
9. Patent No. 2548656 (RF). Samokish V.V. Method for balancing phase currents of a three-phase four-wire line and a device for its implementation // Bulletin. No. 11. 12/27/2013.
10. Omorov, T. T., Takyrbashev B. K. On the problem of optimizing asymmetric operating modes of distribution networks // Devices and systems. Management, control, diagnostics. 2016. No. 6. pp. 11–15.
11. Averbukh M.A., Zhilin E.V. On electricity losses in power supply systems of individual housing construction // Energetik. 2016. No. 6. – pp. 54-56.

© The Authors, published by EDP Sciences. This is an open access article distributed under the terms of the Creative Commons Attribution License 4.0 (https://creativecommons.org/licenses/by/4.0/).

12. A. V. Ded, S. V. Biryukov, A. V. Parshukova. Estimation of additional power losses in 0.38 kV electrical networks based on experimental data // Advances in modern science – 2014. – No. 11. P. 64-67.
13. Omorov T., Takyrbashev B., Koibagarov T., ZhanybaevT. Identification and Monitoring of Power Losses in Distribution Networks as Composition of AMR // AIP Conference Proceedings, 2022, 2552, 020002.
14. Stennikov V.A., Golub I.I., Boloev E.V. Using smart meter measurements to identify those responsible for unaccounted for electricity consumption // Energetik. No. 4. pp. 3-13.
15. Omorov T.T., Takyrbashev B.K., Osmonova R.Ch., Koibagarov T.Zh. Identification of the coordinates of unauthorized withdrawal of electricity in the distribution network as part of the ASCME // Control. Diagnostics. 2019. No. 1. P. 50-55.
16. Kirankumar, T. Power theft detection using probabilistic neural network classifier / T. Kirankumar, G.N. Sri Madhu // International Research Journal of Engineering and Technology. – 2018. – Vol. 5, no. 8. – P. 834–838.
17. Jeyaranjani, J. Machine learning algorithm for efficient power theft detection using smart meter data / J. Jeyaranjani, D. Devaraj // International Journal of Engineering & Technology. – 2018. – Vol. 7, no. 3. – P. 900–904.
18. Omorov, T.T. Methodology for identifying parameters of the main line of a distribution network according to ASCAE data / T.T. Omorov, R.Ch. Osmonova, etc. // Bulletin of the Kazan State Energy University. -2021. -T. 13. No. 3 (51). - WITH. 168-177.
19. Omorov T., Zakiriaev K., Takyrbashev B., Imanakunova Zh. Automated control of unauthorized power take-offs in a distributed electrical network//Mechatronics, Automation, Control.2023.V.24.№1.C.24-32.
20. Omorov T., Koibagarov T., Takyrbashev B., Boronin A., Zhanybaev T. For the construction of subsystem for power line diagnostic states of distribution networks as a part of AMRCS // В сборнике: E3S Web of Conferences. 2020. с. 01042.
21. Methods of classical and modern theory of automatic control: in 5 volumes / ed. K.A. Pupkova, N.D. Egupova. – M.: Publishing house of MSTU im. N.E. Bauman, 2004.
22. T.T. Omorov, B.O. Dzholdoshov. A brief overview of methods for analysis and synthesis of nonlinear automatic control systems // News of the Kyrgyz State Technical University named after. I. Razzakova, Bishkek, 2012, No. 26, -S. 28-36.
23. Omorov, T.T., Kozhekova G.A. Synthesis of a control system for a synchronous generator // Devices and systems. Management, control, diagnostics. 2011. No. 1. P. 5-9.

© The Authors, published by EDP Sciences. This is an open access article distributed under the terms of the Creative Commons Attribution License 4.0 (https://creativecommons.org/licenses/by/4.0/).

https://doi.org/10.1051/e3sconf/202458401048

The use of information and computing technologies in solving problems of monitoring and managing the condition of cable equipment to ensure their reliability and efficiency

V. Ivanova *[1], *V.* Tsipkina [1], *U.* Mamadalieva[1]

[1] Tashkent state technical University, Republic of Uzbekistan

Abstract. The article discusses the issues of integrating data on the condition of cable equipment through information and computing monitoring and management procedures. Possibility of using information and computing technologies in order to solve management of production and technological tasks of manufacturing cable products which are produced in cable factories of the Republic of Uzbekistan has been studied. The results of the work of the developed Integrated Technological Platform as a system combining heterogeneous information and computing resources of cable equipment units and involved production systems are presented.

1 Introduction

Ensuring the technological stability of the production process for the manufacture of cable conductor products (CCP) should allow the engineering service of the cable plant to maintain the reliability and stability of the cable product manufacturing technology at the required level by optimizing production, time, technological and human resources. The reliability and stability of the production process can be achieved through the introduction of a specialized IT structure with a transparent technological stack, which allows to ensure the continuity of production, primarily in case of failure of technological equipment.

Equipping cable production facilities with modern technological equipment makes it possible to achieve a higher level of efficiency and reliability of the technological process (TP) for the manufacture of CCP. However, all cable plants can not afford to completely re-equip the production of the enterprise with new cable equipment (CE). The assessment of the technical condition of cable machines (CM) installed at the production sites of cable plants in Uzbekistan showed that more than 50% of the total equipment has a service life of more than 10 years, which indicates a large moral and physical wear of technological machines (Table 1).

Table 1. The age composition of cable processing equipment installed at cable plants in Uzbekistan

Name of the cable equipment	Service life, year		
	5	10	20 and more
Drawing- machine	15%	30%	55%
Extrusion line	10%	45%	45%
Twisting machines	15%	50%	35%
Rewinding equipment	10%	20%	70%

According to the results of the analysis of the equipment (Table 1), it was found that CM in operation have uneven physical wear (deterioration), the degree of which is determined by: the level of production load; terms and duration of operation (three-shift operation); operating conditions; qualifications of employees and maintenance personnel; regularity and quality of maintenance. Consequently, the deterioration of the technical characteristics of the CM, compared with the passport data, occurs in proportion to the increase in the service life of the cable equipment.

Ensuring the reliability and stability of the production process of the CCP can be achieved by creating a unified information system (UIS) that comprehensively combines all the technologies operating at the enterprise for monitoring, managing and optimizing production processes that ensure the reliability and safety of cable production (CP) by intellectualizing and digitalizing the control system for both a single TP and and the full production cycle of the gearbox, in order to predict the parameters of reliability and safety in the operation of the entire production system.

When analyzing the reliability status of a CM considered as an object of research (RO), it was found that the reliability characteristics (durability, working life, probability of trouble-free operation, probability of failure, etc.) and machine parameters (expected durability, compliance with technical equipment with specified requirements, changes in the composition of system elements, the forecast of reliability indicators, the probability of an emergency failure of individual parts, assemblies and mechanisms). At the same time, the RO analysis considers it as the operation of a complex multidimensional system in which the failure

* Corresponding author: v.p.ivanova.tsipkina@gmail.com

© The Authors, published by EDP Sciences. This is an open access article distributed under the terms of the Creative Commons Attribution License 4.0 (https://creativecommons.org/licenses/by/4.0/).

of one node leads to the shutdown of the entire machine and, accordingly, violates the technology where the CM in question is involved. Ensuring conditions for reducing the number of probable possibilities of an emergency CM failure determines an increase in the operability of equipment, reliability and stability of the entire technology as a whole.

The probability of deterioration of the work of a single CM may adversely affect the fulfillment of a production order (delay or lag) due to increased equipment downtime; increased production and technological cable waste; overspending of raw materials and materials; defective CCP. Shutdown of one machine and the resulting rupture and continuity of technology may cause the absence of transitional (interoperative, incomplete) products along the entire technological route of manufacture (including technological transitions). An indicator of the reliability of the CM operation, which allows a comprehensive assessment of both the reliability of the technological process and the functionality of the EMS of each CM, is the integral reliability indicator (IRI) EMS [12], which includes 5-7 indicators (Table 2) characterizing various reliability properties of the maintenance.

Table 2. Reliability indicators of the twisting machine «RS 1+6+12+24+30/630»

The name of the indicator	Designations	Value
Average time to failure	T_0	at least 8000 hours
Average life before major repairs	$T_{ркр}$	at least 20000 hours
Assigned full rotor life	$T_{рр}$	at least 4000 hours
Assigned full service life	$T_{сл\ н}$	at least 10 year
The coefficient of technical use	$K_{ти}$	0,8
Average retention period	$T_{с\ ср}$	at least 20 months

The process of integration into the production process of the UIS should take into account: the technical condition of all equipments involved in the technology; compliance with technological regimes and regulations; processing of system indicators and quality control of the finished product.

Technical audit of the twisting machine of the brand "RS 1+6+12+24+30/630" the production of "Jiangsu Jiacheng Technology Co., Ltd." (China) conducted in 2014 at the plant of the joint venture "YUQORICHIRCHIQ ENERGY SYSTEMS LLC" (Uzbekistan) showed that the real values of the production capacity and speed parameters of this line are only they correspond by 50% to their passport data for the type of CE in question (Table 3).

Table 3. Technical condition of the torsion machine of the brand «RS 1+6+12+24+30/630»

Work unit	Quantity, pcs.	Total maximum power, kW		Operating speed, m/s	
		passport	fact	passport	fact
Working wheel	4	300	350	120	90
Receiving device	1	15	20	25	10
Wheel type pulling device	1	110	100	25	10
Total		**425**	**470**	**20**	**10**

The results of the assessment of the technical condition of the twisting machine (Table 3) confirmed a significant decrease in its technical indicators and parameters compared with the passport data (service life of 10 years). During the period of operation, maintenance of this twisting equipment included: scheduled preventive and emergency repair and restoration work aimed at eliminating breakdowns, failures and equipment failure, followed by gradual replacement of a significant part of the working mechanisms, assemblies and parts (only 10-20% of the CM configuration remained with basic (native) parts, assemblies and the element base). The assessment of the efficiency and reliability of the twisting machine carried out as part of the technical audit was carried out on the basis of real performance indicators (Tables 2, 4).

Table 4. Reliability indicators of the options for the twisting machine "RS1+6+12+24+30/630"

Alterna-tive	Reliability indicators					
	T_0	$T_{ркр}$	$T_{рр}$	$T_{сл\ н}$	$K_{ти}$	$T_{с\ ср}$
1	8,5	20	43	10	0,9	24
2	11	23	43	8	0,87	24
3	9	21,5	41	12	0,89	36

The process of calculating the IRI EMS for CM (Fig. 1) is carried out on the basis of these options for the reliability of maintenance (Table 4).

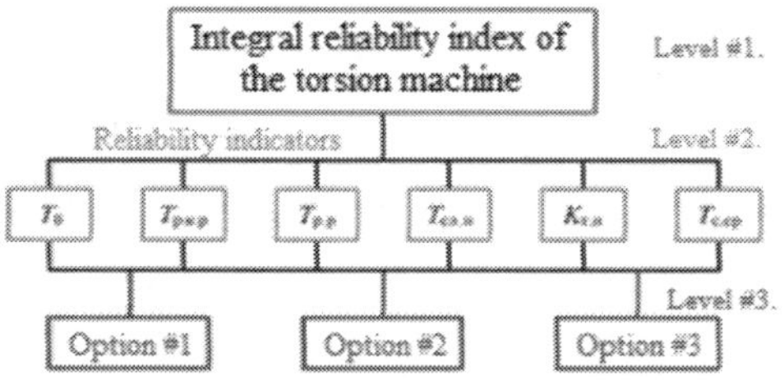

Fig. 1. The scheme for determining the IRI CM

Cable equipment with a service life of less than 10 years is able to ensure the reliability of the technology performed within the framework of its functionality. Thus, CM is a complex, multifunctional system (technological cable line) equipped with an integrated control system (CS) with the function of processing large flows of production, technological and technical data (big data).

© The Authors, published by EDP Sciences. This is an open access article distributed under the terms of the Creative Commons Attribution License 4.0 (https://creativecommons.org/licenses/by/4.0/).

Unfortunately, not all CE can be considered as RO within the framework of the development of the unified control system (UCS), since many have a service life of more than 10 years, and the process of adapting their CS with subsequent integration into the automated process control system is impossible without appropriate modernization. Thus, the efficiency of the automated process control system, within the framework of an operating enterprise, is estimated by the number of CM involved in the TP. Nowdays, there are about 58 cable plants operating in Uzbekistan, which 15 are large enterprises that meet Uzbekistan's need for a checkpoint by 95%. The annalization of the level of digital technology implementation at the cable enterprises of the Republic of Uzbekistan has shown that there is a very significant lag in the field of digitalization of production and technology in general.

2. Research methods

The solution to the issue of increasing the reliability of cable technology is reduced to the introduction of a system capable of monitoring and controlling the technological parameters of the CM in the technological process, by developing a special automated management and control system (AMaCS) with an additional information level that provides monitoring and management of the state of the cable management system while maintaining the required level of reliability and efficiency of all systems.

The evaluation of the results of the reliability analysis depends primarily on the level of digital maturity of the enterprise. Various IT methods are used in foreign practice: the Center for Digital Business at the Massachusetts Institute of Technology and Capgemini Consulting, the Digital Maturity Model, the Digital Transformation Index, Digital Business Capabilities, DBA, the Global Center for Digital Business Transformation, Acatech [2, 3, 4].

The solution to the issue of introducing modern automated process control systems with advanced functions for cable production is based on increasing the level of informatization of TP/CE ensure the maximum increase in the efficiency/stability of the production process of manufacturing the CCP, assessed by the level of quality and operational reliability of the finished product. The use of information and computing technologies for both a single CM and for the entire enterprise is an urgent solution, and the introduction of a network of intelligent (SMART) measuring sensors that monitor technology and control IT will solve many production and technological problems.

Thus, the generalized CM has the necessary level of reliability, which can be represented as a complex automated control system that ensures the smooth functioning of working units and mechanisms by monitoring the established modes and technical parameters in the event of the probabilistic occurrence of various kinds of production and technological uncertainties. CM control system integrated into the automated control system should ensure synchronization in the operation of all equipment units without exception (according to the technological route) during the execution of a production order, provides quality control (input, interoperative and output) of finished products, forecast of CM maintenance and repair times, etc.

Cable production management is a technically difficult task that the production service performs in a real factory. The question of the correctness and adequacy of the decisions made by this service largely depends on the qualifications of the employees of the production service, which means there is a high probability of delays in decision-making or erroneous actions that are committed due to lack of work experience or the presence of a human factor [5].

3. Results

Modern CM is a complex, multi-level structure, which includes interconnected subsystems, the joint functioning of which ensures compliance with a large number of requirements for product quality and reliability of both technology and equipment. A decrease in reliability and performance indicators for a single CE may also be associated with uneven production load, distortions in technological modes that occur during operation and conditions of interaction with the environment (Fig. 2).

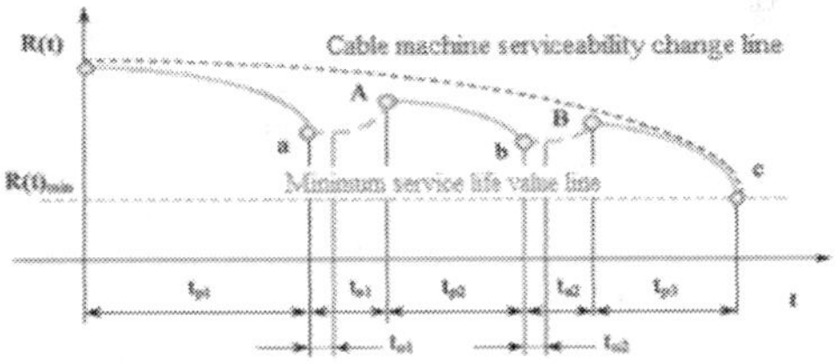

Fig.2. Changes in CM performance during operation

The occurrence of various kinds of abnormal shutdowns in the CM operation leads to failures of the ACS and the technology as a whole, which negatively effects the entire system, because it leads to various violations due to the lack of integrity of incoming signals (input, output, control and disturbing), as well as software failure of all equipment. The developed system of an additional level of control can maintain the required level of reliability and quality (Fig. 3), the function of which is to solve the problem of complex control of engineering, technological and production parameters of the technology being performed; formation of solutions with confirmation of their acceptance (adequacy); determination of the possibility and expediency of their implementation. ACS CM with advanced functions should be interconnected at the subsystem level and be able to integrate it into a Single Process Control System (ESU TP) for the manufacture of a CCP.

© The Authors, published by EDP Sciences. This is an open access article distributed under the terms of the Creative Commons Attribution License 4.0 (https://creativecommons.org/licenses/by/4.0/).

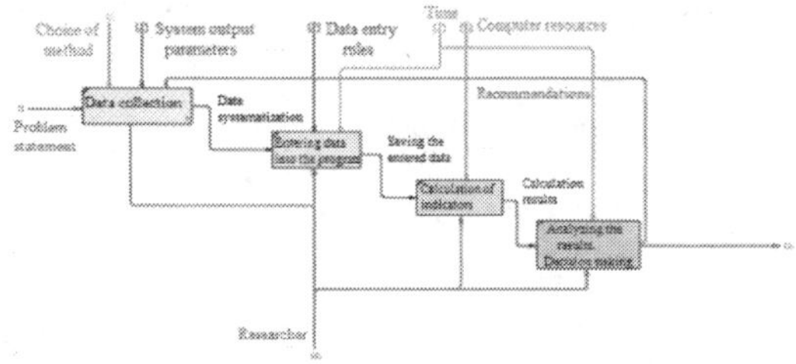

Fig. 3. Functional diagram of computer control of the main processes by data flows

UCS TP of the manufacture of the checkpoint — this is a centralized system consisting of many subsystems (ACS) that ensures the stability, reliability and efficiency of the production cycle within the framework of the technology: the production route, main and auxiliary equipment, transitional (interoperative) processes [5, 6], which is an SCS (smart control system) with functionality built on strict control of the design parameters of the CP (cable products), operating modes of CM, deadlines for the execution of a production order, as well as ensuring high reliability and production efficiency in conditions of various kinds of uncertainties, ensuring the performance, stability and operability of maintenance in various dynamic conditions when performing the technology [7]. The development of the UCS TP was carried out on the basis of the current functional scheme of computer control (Fig.3) in the Embarcadero RAD Studio XE3 environment [4] for a real cable enterprise.

The structure of the SCS is shown in Fig. 4, in which X1, X2 are input signals determined by the parameters and properties of raw materials and materials, according to the design of the cable product; F - disturbances and interference; Y - output signal (qualitative parameters of the finished product, construction length).

The procedure for creating an SCS is to develop a numerical, multi-valued (fuzzy) algorithm/a symbolic form, which is basic for subsequent mathematical modeling [5, 6, 7], which allows for rapid processing of information flows of data obtained on the basis of the results of the corresponding numerical calculations.

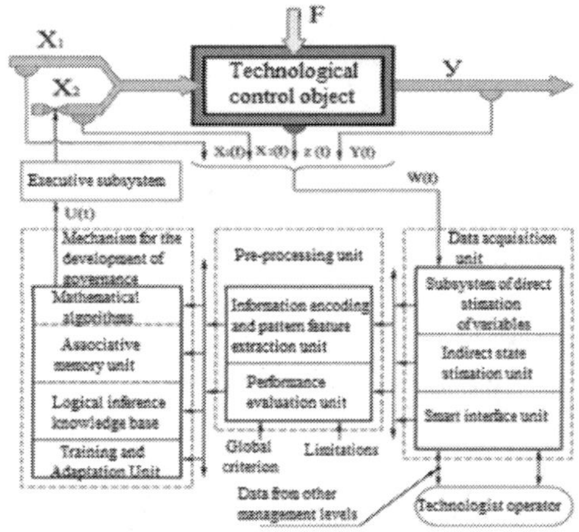

Fig. 4. SCS block structure: X1, X2 – input signals; F - disturbances and interference; Y is the output signal.

The developed mathematical model describes the interaction of a complex of "Systems" and "Subsystems" as the operation of a complex, multicomponent intelligent cable production system:

$$X(t) = \frac{X(t) - X^{min}}{X^{max} - X^{min}} \quad (1)$$

$$Y = \left[1 + exp\left(\sum_{m=0}^{n-1} W_{n-1}(t)\left[1 + exp\left(\sum_{m=0}^{n} W_n(t) \ldots \left[1 + exp\left[\sum_{m=0}^{1} W_1(t)\right]\right]\right)\right]\right)\right] \quad (2)$$

where, X1, X2 – input signals, determined parameters and properties of raw materials and materials according to the design of the cable product; W – generalized system signal; Y –output signal, determining the quality of the exhaust CCP and its construction length.

By means of a mathematical model, variants of information movement from "Data transmission Channels" are worked out by analyzing system elements (accumulated knowledge, generalized and analyzed data), according to the developed algorithms for controlling and evaluating numerical values of dynamic parameters obtained from the "Database" of production and technological basic values [8, 9, 10, 11].

The development of the functional structure of the SCS TP is based on the interaction of the multilayer structure of an artificial neural network (INS) (Fig.5).

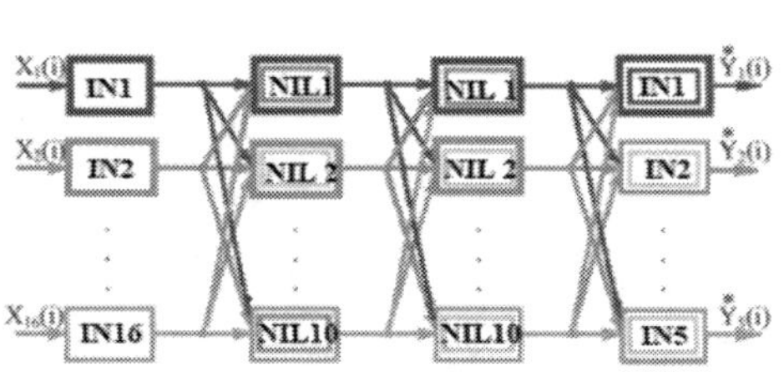

Fig.5. Functional structure of the automated control system: IE1, ... IE16 – input elements (normalizing and distributing converters); NIL1, ... NEL10- neurons of the intermediate layer; ON1, ... ON5 – output neurons

Conclusions

Mathematical modeling of technology with SCS TP is a multi-complex task solution that considers two options for the operation of the control unit within the technological process of manufacturing the CCP (Fig. 6), both a dynamic structure fully connected (a) and an optimized complex (b). The input data selected random values of the technological process of production of the cable wire product that brand is LSOH PVC 4x16, according to technological route: drawing (fig.6, blue), twisting of the conductive core (Fig.6, green), insulation (Fig. 6, yellow color), core twisting (Fig.6, blue color), flanging (Fig.6, purple color), rewinding (Fig.6, red color) with a production time cycle of 20 hours.

© The Authors, published by EDP Sciences. This is an open access article distributed under the terms of the Creative Commons Attribution License 4.0 (https://creativecommons.org/licenses/by/4.0/).

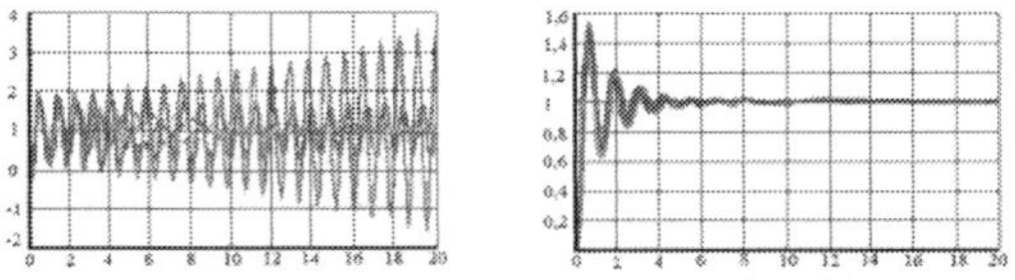

a) b)

Fig.6. The results of mathematical modeling of the SCS TP of the LSOH PVC 4x16 cable production: dynamic structure (a), optimized structure (b).

The simulation results are shown in Fig.6, which are based on a special indicator – the reproducibility index, which is a numerical coefficient of statistical reproducibility (stability) - Cpk [12] and the results obtained by the analytical study method:

• the dynamic structure (Fig.6, a) is unstable, which is based on the isolated operation of separate non-coordinated subsystems - CM, dynamically adjusted to the requirements of the production process, taking into account the redistribution of the production capacities involved;

• the optimized system (fig.6, b) is considered as flexible and multilevel, which operates within the production process and technology. This is the result of optimizing the technological route, formed taking into account the production workload of the CM and its technical condition. This production system becomes stable, it stabilizes over time due to the emergence of a synergistic effect due to the deep integration of subsystems of the Integrated Platform operating within the unified information space of the cable enterprise.

The considered solution to the problem ensures the reliability of the control unit and the efficiency of the technological process of manufacturing CCP, which is achieved by monitoring data obtained from the operation of information and computing technologies, which allows you to obtain the required stability of the technological process (Fig.6, b) and reducing the number of unplanned failures in the operation of equipment, optimizing the route of technological operations and CM operation. The application of the obtained results made it possible to solve many problems of evaluating and analyzing reliability and stability indicators both for the operation of an individual CM, and taking into account an integrated approach – the technological process as a whole for a cable enterprise and the construction of forecast (planned) technological routes for fulfilling production orders for the manufacture of checkpoints for the existing CCP.

References

1. Javadov G.A. Management of scientific and technical progress. - Moscow: Knowledge, 2012 – 64p.

2. Gileva T.A. Digital maturity of the enterprise: methods of assessment and management, Economic sciences, Economics and management in industries and spheres of activity. Bulletin USPTU. Science, education, economy. Series economy. № 1 (27), 2019. file:///G:/tt/tsifrovaya-zrelost-predpriyatiya-metody-otsenki-i-upravleniya.pdf

3. Digital Transformation — How to Become Digital Leader. Study 2015 Results. Available at: http://www.adlittle.com/sites/ default/files/viewpoints/ADL_HowtoBecomeDigitalLeader_02.pdf (accessed 13.07.2024).

4. Are You Ready for Digital Transformation? Measuring Your Digital Business Aptitude. URL: https://assets.kpmg/content/dam/kpmg/ pdf/2016/04/ measuring-digital-businessaptitude.pdf (дата обращения: 13.07.2024).

5. Pupkov K.A. Intelligent systems / K.A. Pupkov, V.G. Konkov - M.: Publishing house of Bauman Moscow State Technical University, 2003. - 348 p.

6. Hangos K.M. Intelligent control systems: an introduction with examples / K.M. Hangos, R. Lakner, M. Gerzson - New York: Kluwer academic publishers, 2001. - 301 p.

7. Gavrilova T.A. Knowledge bases of intelligent systems / T.A. Gavrilova, V.F. Khoroshevsky - St. Petersburg: Piter, 2001. - 384 p.

8. Modeling of a resource-saving method of drawing Victoria Tsypkina, Veronica Ivanova E3S Web Conf. 139 01073 (2019) DOI: 10.1051/e3sconf/ 201913901073.

9. Improving the reliability of cable lines operation in hot climates Daniyar Bakhtiyarovich Madrakhimov, Vera Pavlovna Ivanova, Victoria Vyacheslavovna Tsypkina E3S Web of Conf. 216 01151 (2020) DOI: 10.1051/e3sconf/202021601151

10. Improving the reliability of cable and wire products by improving the insulation application technology // Universum: technical sciences: electronic. scientific journal. Ivanova V.P. [et al.]. 2020. 11(80). URL: https://7universum.com/ru/ tech/archive/ item/10970.

11. Digital transformation of the process of assessing the technical condition of technological equipment in cable production // Universum: technical sciences: electronic scientific journal. Pirmatov N.B. [et al.]. 2024. 5(122). URL: https://7universum.com/ru/tech/ archive/item/17636.

12. GOST R 50779.11-2000.

© The Authors, published by EDP Sciences. This is an open access article distributed under the terms of the Creative Commons Attribution License 4.0 (https://creativecommons.org/licenses/by/4.0/).

Ensuring the reliability of energy systems through the application of a new method for introducing soil-replacement mixtures during the installation of deep grounding devices

Ivan Pavlovich[1]*, *Siarhei* Baraishuk[1] *Muzaffar* Murodov[2] , *Khurshed* Abdulkhayev[2]

[1]Belarusian State Agrarian Technical University, Department of practical training of students, Minsk, Belarus
[2] Namangan Engineering-Construction Institute, Department of Energy, Namangan, Uzbekistan

Abstract. This paper discusses a method for using mixtures to optimize the electrical parameters of grounding devices during the installation of vertical composite ground electrodes. The authors propose a design for the coupling and tip for vertical composite ground electrodes. Various techniques for artificially reducing the resistance of the grounding circuit are examined. Results from vertical electrode soundings of the soil at grounding locations are presented. The proposed method allows for the introduction of a mixture simultaneously with the vertical composite grounding electrode, including designs for the coupling, tip, and auxiliary device. Experimental studies of the proposed design have been conducted, and results measuring the resistance to current spread of such grounding devices are presented for both standard and proposed couplings.

Keywords: resistance, grounding circuit, hydrogel, graphite, deep grounding device, near-electrode space, seasonality coefficient, deep grounding device, connecting coupling

Introduction

To ensure the safety of maintenance personnel and the reliability of energy equipment, it is essential that the resistance of the grounding device remains within established norms [1]. Due to changes in weather and climatic conditions, fluctuations in the soil's resistivity occur, which, in turn, leads to instability in the resistance values of the grounding circuit [2]. A number of methods are being developed [3-6] to reduce the grounding resistance in such conditions, each of which has its own advantages and disadvantages.

The objective of this work is to develop a method for introducing soil-replacement mixtures to optimize the electrical parameters of the soil during the installation of composite deep grounding devices without using vertical drilling technology.

1 Methods for Reducing the Seasonality Coefficient

When designing protective grounding for electrical installations exceeding 1000 V, it is essential to adhere to the requirements for their resistance, structural execution, and voltage limitation on the grounding device. The voltage on the grounding device during a ground fault should not exceed 10 kV. Any voltage above this level can be applied to protective devices without transmitting potential beyond the boundaries of buildings and external enclosures of electrical installations.

If the voltage higher 5000 volts, protective measures should be in place regarding the insulation of outgoing communication and telemechanic cables to prevent the transmission of dangerous potentials beyond the electrical installation. The grounding device used must have a resistance no greater than 0.5 ohms at all times throughout the year (including the resistance of natural grounding devices).

Reducing the seasonality of soil resistance allows for increased stability in the operation of grounding devices, ensures uniform electrical resistance throughout the year, and guarantees more reliable functioning of electrical systems. This is particularly important for systems requiring stable electrical conductivity, such as power engineering, telecommunications, industry, and agriculture.

It is also known that to avoid fire hazards in energy systems, appropriate electrical protection devices and appliances are employed. However, fires typically occur when protection fails to perform its functions. Regarding electrical station and substation equipment, it should be noted that the status of electrical protection for power supply, signalling, and control circuits is directly linked to fire safety issues, which are determined by the condition of protection devices and grounding circuits during the operation of energy facilities.

When designing artificial grounding systems in areas with high soil resistivity, the following measures are recommended:

- The use of vertical ground electrodes of increased length in cases where deeper soil layers exhibit a gradual decrease in resistivity, and natural deep

* Corresponding author: mrChesel20@gmail.com

© The Authors, published by EDP Sciences. This is an open access article distributed under the terms of the Creative Commons Attribution License 4.0 (https://creativecommons.org/licenses/by/4.0/).

grounding devices (such as wells with metallic casings) are absent;

- The application of remote grounding devices in cases where areas with lower soil resistivity are located nearby (up to 2 km) from the electrical installation;
- Laying horizontal grounding electrodes in trenches, followed by compaction and backfilling with gravel to the top of the trench in rock formations of wet clay soil.
- Artificial treatment of the soil to reduce its resistivity when other methods cannot be applied or do not yield the desired effect. Previous studies have shown that the resistivity of the soil at a depth of 0.3 m increases tenfold when the temperature drops from 0°C to -10°C, and at a depth of 0.5 m, it increases threefold [7]. The work in [8] discusses how an increase in depth, represented in an arithmetic progression, leads to a decrease in temperature amplitude in a geometric progression.

To mitigate the influence of temperature, vertical deep grounding electrodes longer than 10 meters are typically used. A similar solution involves using electrodes immersed in deep wells filled with conductive soil (such as coke dust, concrete, etc.). Deep grounding is preferred when the soil at significant depths exhibits good conductivity, especially when reaching a saturated layer. The simplest and most common solution is to create such a grounding system in the form of homogeneous long rods connected by couplings. Montage can be accomplished using various methods such as screwing, driving, drilling, etc. This solution generally provides more reliable grounding and improves protection against surges and leakage currents.

When designing grounding systems, the structure of the soil is often not considered, assuming it to be uniform, and the influence of groundwater is also overlooked. In the study [4], the authors conducted measurements of the grounding resistance depending on its length, concluding that as the vertical grounding electrode increases in height, the resistance decreases.

As shown previously [4-6, 5, 6, 9, 10], to reduce the resistivity of the soil at the location of the grounding electrodes, various soil-replacement mixtures can be used. These mixtures help retain moisture more reliably in the electrode space using hydrogels; moreover, containing graphite and clay, they help lower the overall soil resistance.

However, the application of such mixtures in conjunction with deep grounding electrodes usually requires prior drilling of a hole for the grounding electrode and mixture [11, 12]. This method allows the distribution of the compositional mixture along the entire length of the deep grounding electrode, which reduces the overall resistance of the grounding system and maintains its stable value throughout its operational life. This leads to increased construction costs. To reduce these expenses, we propose a method for introducing these mixtures without the need for prior drilling of a hole for the electrode.

2 Experimental Studies

During the field experiment, a control deep ground electrode was installed, made of standard composite rod electrodes and factory couplings included in the kit. A similar ground electrode was made using the mixture, and the experimental ground electrode was assembled using the couplings and tip developed by us, as shown in Figure 1 below. To optimize the electrophysical parameters of the grounding devices, a previously developed mixture was used [11, 13 14, 15].

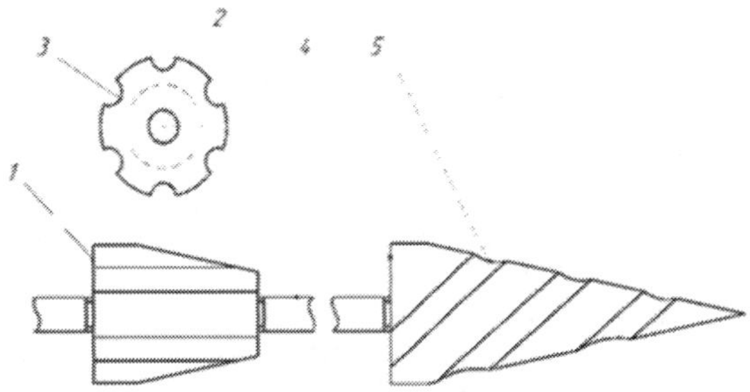

Fig. 1. Appearance of the tip and coupling: 1 – coupling, 2 – composite grounding electrode, 3 – grooves on the coupling, 4 – tip, 5 – grooves on the tip.

Before the installation of deep grounding electrodes, we conducted measurements of the soil's specific electrical resistance at the installation sites using the vertical electrical soundings (VES) method. The results of the VES are presented in Figure 2. The VES data obtained allow us to conclude that the optimal length of a deep ground electrode should be 11-15 meters.

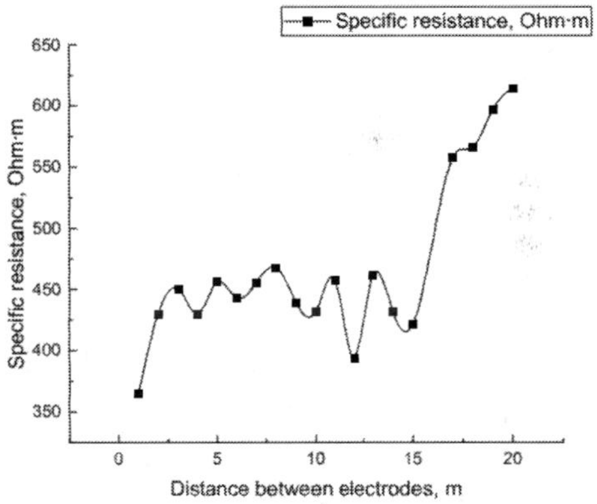

Fig. 2. Results of the VES

After the installation of these grounding systems, resistance measurements were conducted, which were repeated every few weeks over the course of a year. The graphs are presented in Figure 3.

© The Authors, published by EDP Sciences. This is an open access article distributed under the terms of the Creative Commons Attribution License 4.0 (https://creativecommons.org/licenses/by/4.0/).

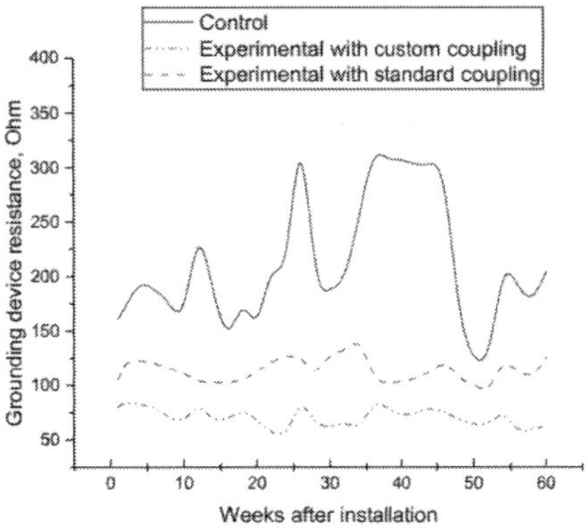

Fig. 3. Annual variations in the resistance of the control, experimental with a non-standard coupling, and experimental with a standard coupling of deep electrodes.

From the obtained dependencies, it is evident that the resistance of the experimental grounding systems (GS) is significantly lower than that of the control grounding electrode. Furthermore, for the developed couplings, this resistance is noticeably lower than that of standard couplings, which can be attributed to the larger volume of the mixture surrounding the electrode body. Additionally, the graph indicates that for standard couplings, the seasonality effect reduces by approximately 1.64 times, while for the developed couplings, the reduction in seasonality is 2.1 times. This can be explained by the much better distribution of the mixture composition along the entire length of the grounding system. Such distribution leads to a reduction in resistance, as due to the capillary effect of the hydrogel (transport functions) [16], moisture is pulled up from the depths and distributed along the entire length of the grounding electrode.

Using the reverse modelling method, the calculation of the "apparent" diameter of the grounding electrode was conducted when using a soil-replacement mixture. The calculation demonstrates the effectiveness of applying this mixture in conjunction with the grounding electrode.

The resistance of the current dispersion of the deep grounding electrode was calculated without taking into account the mixture, using the seasonality coefficient recommended in the regulatory documentation for the Minsk region. According to the expression presented in [17], we obtained:

$$R_V = \frac{\varphi \cdot \rho}{2 \cdot \pi \cdot l_v} \cdot \left(\ln \frac{2 \cdot l_v}{d} + \frac{1}{2} \cdot \ln \frac{4 \cdot t + l_v}{4 \cdot t - l_v} \right),\ Ohm \quad (1)$$

where: t – depth of the vertical electrode installation (measured from the ground surface to the midpoint of the electrode, in meters);

l_v – length of the vertical deep grounding electrode, in meters;

d – diameter of the grounding electrode, in millimeters;

ρ – specific resistance of the soil at the installation site, in ohm-meters;

φ – seasonality coefficient [17]

According to measurements, the specific resistance of the soil is 526 Ohm·m, so by substituting these values into the expression, we obtain:

$$R_V = \frac{2,4 \cdot 526}{2 \cdot \pi \cdot 11} \cdot \left(\ln \frac{2 \cdot 11}{0,012} + \frac{1}{2} \cdot \ln \frac{4 \cdot 6 + 11}{4 \cdot 6 - 11} \right) = 146,4\ Ohm$$

This value is close to the actual resistance of the grounding system obtained experimentally, the average value of which is 156.4 Ohms. Using the expressions from work [18], a calculation of the resistance of the deep grounding electrode was performed when using a soil-replacing mixture to optimize the electrical parameters of the grounding device:

$$R_V = \varphi \cdot \rho_{eq} \frac{1}{2\pi l_v} \left(\ln \frac{2 l_v}{d} + \frac{1}{2} \ln \frac{4t + l_v}{4t - l_v} \right), \quad (2)$$

where φ – experimental seasonality coefficient for the mixture;

ρ_{eq} – specific resistance of the soil at the installation site according to the formula, Ohm·m.

$$\rho_{экв} = \rho_g \cdot 0,98^{\frac{d_{mixtures}}{d_s}} \cdot \left(\frac{\rho_{mixtures}}{\rho_g} \right)^{\frac{1}{3\pi}}, \quad (3)$$

ρ_{eq} – the desired calculated specific resistance, ohm·m;

ρ_g – specific resistance of the soil, ohm·m;

$d_{mixtures}$ – equivalent diameter of the grounding electrode and mixture ($d_{mixtures} = \sqrt{\frac{V_{mixtures}}{l}}$), m;

d_g – diameter of the grounding electrode, m;

$\rho_{mixtures}$ – specific resistance of the mixture, ohm·m.

$$d_{mixtures} = \sqrt{\frac{0,02}{3}} + 0,012 = 0,082 + 0,012 = 0,094\ m$$

$$\rho_{eq} = 526 \cdot 0,98^{\frac{0,1}{0,012}} \cdot \left(\frac{50}{526} \right)^{\left(\frac{1}{3\pi}\right)} = 349,75\ Ohm \cdot m$$

$$R_V = 1,64 \cdot 349,75 \frac{1}{2\pi \cdot 11} \left(\ln \frac{2 \cdot 11}{0,012} + \frac{1}{2} \ln \frac{4 \cdot 6 + 11}{46 - 11} \right) = 66,5\ Ohm$$

The obtained value is close to the experimental data, which is equal to 70.7 ohms. Using the method of reverse modelling, the calculation of the "imaginary" diameter of the grounding electrode when using a soil-replacing mixture was carried out. The calculation demonstrates the effectiveness of using this mixture in conjunction with the grounding electrode. To do this, let's express the value of d from formula (1):

$$d = \frac{2 \cdot l_в}{e^{\left(\frac{R_V - \frac{\varphi \cdot \rho}{2 \cdot \pi \cdot l_v} \cdot \frac{1}{2} \cdot \ln \frac{4 \cdot t + l_v}{4 \cdot t - l_v}}{\frac{\varphi \cdot \rho}{2 \cdot \pi \cdot l_v}} \right)}} \quad (4)$$

The reverse calculation of the imaginary diameter of the grounding electrode showed that when using this mixture in conjunction with a deep grounding electrode, the imaginary diameter of the electrode increased. For standard couplings, it is 0.07 m, while

© The Authors, published by EDP Sciences. This is an open access article distributed under the terms of the Creative Commons Attribution License 4.0 (https://creativecommons.org/licenses/by/4.0/).

for the embedded type, it is 0.94 m. The use of these devices allows for a twofold reduction in the costs of installing the grounding loop. The demonstrated decrease in resistance of the grounding device and the overall increase in the stability of its value throughout the year, regardless of seasonal changes in climatic conditions, completely offsets the increase in the production costs of the specified devices. The maximum increase in the production costs of couplings and tips will not exceed 10-15%.

Conclusion

The use of a special coupling allows for moisture movement between individual segments of the grounding device and achieves lower resistance values for the loop than when using standard couplings. Additionally, it significantly reduces seasonal variations in resistance to current dispersion in vertical composite grounding electrodes. The results of the calculations indicate that the use of a hydrogel-based mixture has led to a tenfold increase in the imaginary diameter of the grounding electrode, positively impacting the resistance of the grounding device and the stability of its properties. This makes the proposed method promising for enhancing the reliability of power systems.

References

1. PUE-6. Rules for Electrical Installations, 6th ed., Energoservice (2002), 342 p. (Introduced 01.06.1985).
2. V.I. Glushko, O.E. Yamny, E.P. Kovalev, Energy, Reports of Higher Educational Institutions and Energy Associations of the CIS, No. 2, 60-68 (2001).
3. Al-Ammar, E., Khan, Y., Malik, N., & Wani, N. (2010, December). Development of low resistivity material for grounding resistance reduction. In 2010 IEEE International Energy Conference (pp. 700-703). IEEE.
4. Khan, Y., Malik, N. H., Al-Arainy, A. A., Qureshi, M. I., & Pazheri, F. R. (2010, November). Efficient use of low resistivity material for grounding resistance reduction in high soil resistivity areas. In *TENCON 2010-2010 IEEE Region 10 Conference* (pp. 620-624). IEEE.
5. Al-Arainy, A. A., Malik, N. H., Qureshi, M. I., & Khan, Y. (2011). Grounding pit optimization using low resistivity materials for applications in high resistivity soils. *International Journal of Emerging Electric Power Systems*, *12*(1).
6. Harahap, P. H. P., & Al-Ani, W. K. A. (2021). The Effect of Charcoal on the Improvement of Grounding Resistance as a Soil Treatment in Reducing Grounding Resistance. Journal of Renewable Energy, Electrical, and Computer Engineering, 1(1), 12-15.
7. L.M. Vedeneyeva, A.V. Chudinov, Bulletin of Perm National Research Polytechnic University, Geology, Oil and Gas and Mining, 16, 1, 89-100 (2017).
8. Coelho, V. L., Piantini, A., Almaguer, H. A., Coelho, R. A., Boaventura, W. D. C., & Paulino, J. O. S. (2015). The influence of seasonal soil moisture on the behavior of soil resistivity and power distribution grounding systems. Electric power systems research, 118, 76-82.
9. N.M. Zaitseva, B.B. Isabekova, M.Ya. Kletsel, Electricity, No. 7, 19-24 (2011).
10. Meng, Q., He, J., Dawalibi, F. P., & Ma, J. (1999). A new method to decrease ground resistances of substation grounding systems in high resistivity regions. IEEE Transactions on Power Delivery, 14(3), 911-916.
11. S. Baraishuk et all.. The use of hydrogels in mixtures to reduce the transient resistance of the soil - grounding device 5th International Conference on Energetics, Civil and Agricultural Engineering/ E3S Web Conf., 497 (2024) 03001/ ICECAE 2024. DOI: https://doi.org/10.1051/e3sconf/202449703001
12. M.A. Drako, S.M. Barayshuk, I.A. Pavlovich, News of Higher Educational Institutions, Problems of Energy, 23, 1, 80-92 (2023).
13. S.M. Barayshuk et al., Agropanorama, 5, 28-33 (2021).
14. I.A. Pavlovich, S.M. Barayshuk, Energy, Reports of Higher Educational Institutions and Energy Associations of the CIS, 66, 4, 322-332 (2023).
15. S.M. Barayshuk, I.A. Pavlovich, Mixture for Reducing Electrode-Soil Transition Resistance, Patent BY 24181 (Published 28.02.2024).
16. GOST R 58882-2020, Grounding Devices. Technical Requirements.
17. A.V. Smagin, N.B. Sadovnikova, E.A. Belyaeva et al., Soil Science, 55, 9, 1150-1164 (2021).
18. I.A. Pavlovich, S.M. Barayshuk, V.V. Bogdanovich, Bulletin of FFI, 4(106), 146-157 (2023).

© The Authors, published by EDP Sciences. This is an open access article distributed under the terms of the Creative Commons Attribution License 4.0 (https://creativecommons.org/licenses/by/4.0/).

Efficiency Evaluation of Evacuated Solar Tube Collectors in Sakha Republic (Yakutia)

Dmitry V. Prokhorov[*] and *Semen S.* Vasilev

Larionov Institute of Physical and Technical Problems of the North SB RAS, Energy Problems Department, 677980 Yakutsk, Russia

Abstract. Calculations of solar radiation incident on an inclined surface were performed considering the installation location, local time, azimuth angle, tilt angle, atmospheric transparency coefficient and surface albedo. The calculations were based on long-term meteorological observation archives. The accuracy of the solar radiation calculations for the inclined surface was verified. A comparative analysis was conducted between the calculation results carried out for the installation location, tilt angle, and azimuth angle of the experimental setup and pyranometer measurements. Thermal energy production by evacuated solar tube collectors was also calculated. These calculations were performed on an hourly basis, taking into account the total solar radiation falling on the inclined surface and the atmospheric air temperature. The calculated values were compared with experimental data obtained from an ultrasonic heat meter. The payback period of solar heating systems for various locations was used as an efficiency indicator for the solar collectors. The article estimated the payback period for Verkhoyansk town in the Sakha Republic (Yakutia). Economic efficiency strongly depends on the installation location of the solar collector and the thermal energy tariff at the installation site.

1 Introduction

The regions of the Far North and those equivalent to them encompass a vast territory of more than 7,500,000 square kilometers in the northeast of the Russian Federation. These areas are primarily characterized by harsh Arctic climate conditions, low annual air temperatures, and a significant need for heat supply [1]. The territory of the Sakha Republic (Yakutia) falls within the Far North regions. The average length of the heating season in Yakutia is 8-9 months, and in some settlements in the Arctic zone, it lasts almost all year round. Due to the influence of the climate, the amount of energy required for heating 1 m² of housing is twice as much as the national average in the Russian Federation. As a result, economically justified heat tariffs for utilities are significantly higher than the average for the country [2].

The severe climatic conditions, remoteness of settlements, and complex transport infrastructure create significant challenges for the development of the Sakha Republic (Yakutia), particularly in terms of energy supply. The current situation in the heat supply sector is marked by serious issues, such as outdated technology, low economic efficiency of heating facilities, and high fuel costs. The increase in fuel supply costs is driven by rising purchase prices at shipping points and the growing expense of transporting fuel to remote areas. Consequently, the final cost of energy resources far exceeds their purchase price.

In the Republic of Sakha (Yakutia), lost income for heating companies due to state-regulated tariffs is compensated through the regional budget. Regional legislation also provides targeted subsidies to help cover part of the heating costs for the population. However, since more than 90% of the republic's budget expenditures are mandatory, further increases in subsidies (since 2018) have become difficult. Between 2011 and 2021, the share of energy supply subsidies relative to regional budget revenues (excluding receipts to local budgets) ranged from 5.6% to 11.6%. In terms of the total volume of subsidies at current prices, a stagnation plateau at around 25 billion rubles has been observed from 2017 to the present [3-4].

Given the high cost of heating in the populated areas of the republic, there are significant opportunities for the implementation of energy-saving technologies. Their adoption could be facilitated through energy service contracts or other state programs aimed at reducing budget expenditures.

The authors consider the use of a hybrid solar heating system with evacuated solar tube collectors to be promising in Yakutia due to the region's high solar energy potential. The Sakha Republic has a substantial solar resource potential, with the gross potential estimated at 392 billion tons of oil equivalent [5].

The total cost of such an auxiliary heating system, including installation, delivery, piping systems, and other expenses, starts at around 400,000 rubles for a single building. The production of thermal energy by solar collectors is expected to grow and will increasingly replace heat production from traditional sources as the number of installed systems increases. The optimal solution is specific to each building and requires detailed investigation. Energy savings in buildings can range from 1% to 50%, reducing the heating period by up to 66

* Corresponding author:prokhorovdv@gmail.com

© The Authors, published by EDP Sciences. This is an open access article distributed under the terms of the Creative Commons Attribution License 4.0 (https://creativecommons.org/licenses/by/4.0/).

days in certain regions. For a one-story building measuring 8x10 meters, the additional thermal energy consumption from traditional heating systems is estimated to range from 258,514 MJ to 125,280 MJ.

The economic success of installing solar heating systems can be ensured by combining two main factors: the high costs of heat supply and the significant amount of solar energy available in the region. Based on these considerations, the following areas of the Sakha Republic (Yakutia) are most suitable for the implementation of hybrid solar heating systems: Abyisky, Verkhoyansky, Verkhnekolymsky, Momsky, Oymyakonsky, and Srednekolymsky districts.

2 Test facility and experimental data

Since the beginning of 2022, the performance of a solar heating system based on an evacuated solar tube collector has been investigated under natural conditions in Yakutsk, at a specialized test facility developed by the Department of Energy Problems of the IPTPN SB RAS. The experimental setup allows for continuous monitoring of the operating parameters of the solar heating system and the solar radiation incident on the collector's surface. Measuring the useful thermal energy produced with an ultrasonic heat meter enables the assessment of the solar collector's performance under varying climatic conditions.

The thermal energy source is the evacuated solar tube collector SILA R5-15, which has 15 tubes. The total collector area is 2.41 m², with an aperture area of 1.64 m². The collector is installed at a height of 22 meters above ground level. The azimuth angle, aligned with the building's roof orientation, is 217°, the tilt angle relative to the horizon is 65°, and the roof slope is 10°. A detailed description of the experimental setup and the equipment used is provided in [6].

To assess the efficiency of the solar collector, data on the total solar radiation incident on the collector's surface, as well as the generated thermal energy for 2023, were analyzed. Solar radiation was measured at five-minute intervals, and hourly data were obtained by calculating the arithmetic mean of the measurements over one hour. A large amount of data was processed using specially developed software for measuring and analyzing the data. Figures 1 and 2 show the monthly total solar radiation incident on the collector's surface and the monthly thermal energy generated by the evacuated solar tube collector, respectively.

Generated heat by solar collector in Yakutsk since March until August is quite large and much more than in the autumn and winter.

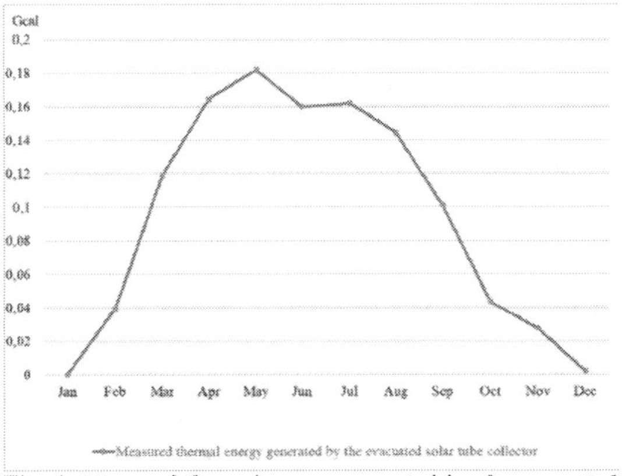

Fig. 1. Measured thermal energy generated by the evacuated solar tube collector in 2023.

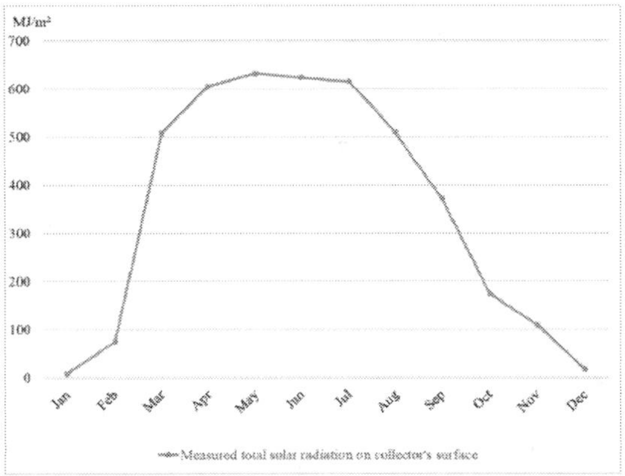

Fig. 2. Measured monthly total solar radiation on the surface of the collector in 2023.

3 Solar radiation on a tilted surface

The developed method is designed to recalculate solar radiation incident on a horizontal plane onto any inclined plane in locations between 58.6° and 68.5° north latitude and 112.5° and 153.8° east longitude. This is achieved using empirical coefficients and the hour angle equation, which are specifically applicable to this region.

The methodology for evaluating total solar radiation incident on a tilted surface is described in numerous studies [7-9]. The main difference between the approaches of various authors lies in the lack of a unified method, as well as the difficulty of determining certain parameters, such as the hour angle, cloud cover, ground albedo, and diffuse radiation. In this article, we assume that diffuse solar radiation and ground albedo values are available from reference sources and hydrometeorological service data, and therefore, we do not perform their calculations.

The hour angle is determined using the following expression [10]:

$$\omega = 15(t - t_{noon}) + E + \psi - \psi_{zone}, \quad (1)$$

where t – current local time, h; t_{noon} - local time of solar noon in the time zone of the receiver, h; E - equation of time; ψ – longitude, rad; ψ_{zone} - longitude of the

© The Authors, published by EDP Sciences. This is an open access article distributed under the terms of the Creative Commons Attribution License 4.0 (https://creativecommons.org/licenses/by/4.0/).

https://doi.org/10.1051/e3sconf/202458401050

meridional plane in which noon coincides with true solar noon, rad.

Equation of time can be defined by [7]:

$$E = 229{,}2 \cdot (0{,}000075 + 0{,}001868 \cdot \cos B - 0{,}032077 \cdot \sin B - -0{,}014615 \cdot \cos 2B - 0{,}04089 \cdot \sin 2B), \quad (2)$$

where B – time coefficient, rad.

Time coefficient can be defined by [7]:

$$B = \frac{(n-1)\cdot 360}{365}, \quad (3)$$

where n – day of the year.

Estimates of total solar radiation taking into account total cloudiness require adjustment for lower cloudiness, which increases the accuracy of the calculations [11].

In accordance with the methodology of P.P. Kuzmin [11-12]:

$$H_{Tcl} = H_T \cdot (1 - A \cdot l - C \cdot (n - l)), \quad (4)$$

where H_{Tcl} - total solar radiation on a tilted surface per hour considering cloudiness; H_T - total solar radiation on a tilted surface per hour; l - lower cloudiness expressed as a fraction of a unit; A,C – empirical coefficients characterizing the attenuation of total radiation by clouds of the lower and lower with middle tiers.

According to research [12] for arctic latitudes A = 0,62; C = 0,24.

Using this method, the calculated total solar radiation on a tilted surface was estimated and compared with the experimental data of the stand.

The comparison results for Yakutsk can be seen in the Figure 3.

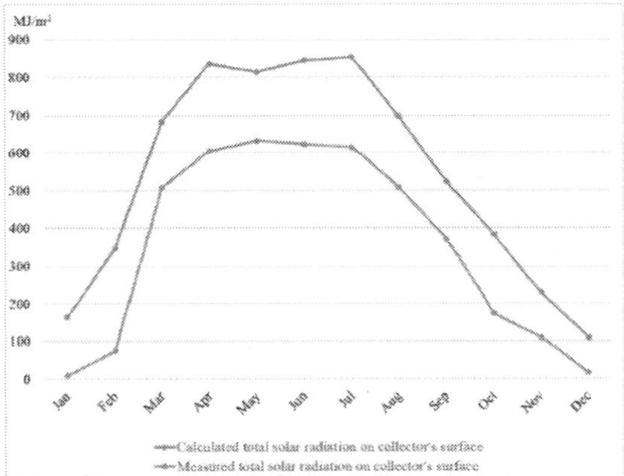

Fig. 3. Measured and calculated monthly total solar radiation on the collector's surface in 2023.

The results of the comparative analysis show that the measured and calculated total solar radiation on the collector's surface are generally comparable. Relative discrepancies in the monthly values can be attributed to annual variations in solar radiation, deviations in climatic conditions from reference data, and the influence of various natural factors during the measurement process (such as dust, precipitation, frost formation, etc.). The discrepancy between the measured and calculated annual total solar radiation on the collector's surface was 53%, which can be explained by the overestimated values produced by existing solar radiation calculation models.

4 Thermal energy generation and heat losses

Previously calculated solar insolation on a tilted surface is required to determine the production of thermal energy by solar collectors. The following is the methodology used by the authors for calculating the production of thermal energy by solar collectors and determining thermal energy losses.

Production of thermal energy by solar collectors can be determined as:

$$Q_C = H_{Tcl} \cdot N \cdot A_C \cdot \eta_C, \quad (5)$$

where N – number of collectors; A_C - aperture area of one solar collector, m^2 (for the considered collectors 1.87 m^2); η_C – thermal efficiency of collector.

Thermal efficiency of evacuated solar tube collector is given by this equation [13-14]:

$$\eta_C = \eta_{opt} - u_{L1} \cdot \frac{t_c - t_{amb}}{G} - u_{L2} \cdot \frac{(t_c - t_{amb})^2}{G}, \quad (6)$$

where η_{opt} – optical efficiency of collector (for the used collector is 84%); u_{L1}, u_{L2} - coefficients of heat loss due to conduction and convection, respectively (in our case u_{L1}= 2,118, u_{L2} =0,008; t_C - temperature of the heat-transfer fluid in the solar collector, °C; t_{amb} - ambient air temperature, °C; aperture area of one solar collector, m^2 (for the considered collectors 1.87 m^2); G – solar irradiance, W/m^2.

Let us assume that the heat generated by the collector is completely transferred to the heat-transfer fluid in the storage tank, excluding losses in the pipeline.

Heat energy losses in the external pipeline are determined as:

$$Q_{loss} = \frac{\pi \cdot (t_C - t_{amb}) \cdot l_{p1}}{\frac{1}{\alpha_1 \cdot d_{in}} + \frac{1}{2 \cdot \lambda_p} \cdot \ln\frac{d_{ext}}{d_{in}} + \frac{1}{2 \cdot \lambda_{isol}} \cdot \ln\frac{d_{ext} + 2t}{d_{ext}} + \frac{1}{\alpha_2 \cdot (d_{ext} + 2t)}}, \quad (7)$$

where α_1 – heat transfer coefficient from the heat-transfer fluid to the pipe surface, $W/(m^2K)$; α_2 – heat transfer coefficient from the outer surface of the insulation to the air, $W/(m^2\ K)$; λ_p - thermal conductivity coefficient of the pipe material, W/(mK); λ_{isol} - thermal conductivity coefficient of the insulation, W/(mK); l_{p1} – length of pipe outside, m; d_{in}- inner diameter of the pipe , m; d_{out} - outer diameter of the pipe, m; t - insulation thickness, m.

Heat transferred to the storage tank defined as:

$$Q_{ST} = Q_C - Q_{loss}. \quad (8)$$

The total monthly output of the solar collector was 932 MJ according to the calculation results and 589 MJ according to the heat meter. The relative discrepancy for the month was 58%, which occurs due to the discrepancy in the total solar radiation of up to 53%.

© The Authors, published by EDP Sciences. This is an open access article distributed under the terms of the Creative Commons Attribution License 4.0 (https://creativecommons.org/licenses/by/4.0/).

5 Efficiency of hybrid solar heating system

To evaluate the efficiency of the hybrid solar heating system, we will consider the heat supply for a two-story office building with an area of 200 m² and a ceiling height of 2.75 meters, located in Verkhoyansk, a remote area of the Sakha Republic (Yakutia). Based on these inputs, the building's volume is 1,100 m³. As an additional heat source, we will consider 10 evacuated solar tube collectors, each with 20 tubes, identical to those used in the test facility. The number of collectors is variable, which will affect the project's economics. The storage tank volume is assumed to be 1 m³, based on the standard hot water consumption in the building. The total cost of the installation, based on the vacuum solar collectors, is estimated at 2 million rubles.

Thermal energy from the solar collectors is used for heating during the heating season, replacing part of the thermal energy from boiler houses, and for hot water supply during the non-heating period, reducing electricity consumption for electric water heaters. The total costs of producing thermal energy from the solar heating system include the costs of building and installing generating equipment, production costs, and ongoing operating expenses.

Figure 4 shows the thermal energy consumption of the building in Verkhoyansk and the thermal energy produced by the solar collectors. The thermal energy production by the solar collectors was estimated using the methodology described earlier in this work. Data on the total solar radiation on a horizontal surface were obtained from the meteorological station in Verkhoyansk.

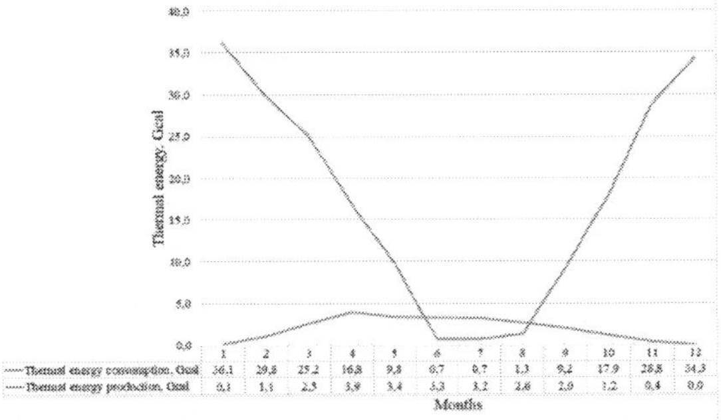

Fig. 4. Nature of thermal energy consumption and production.

The money savings by replacing thermal energy for heating and hot water production with solar collectors are shown in the Figure 5.

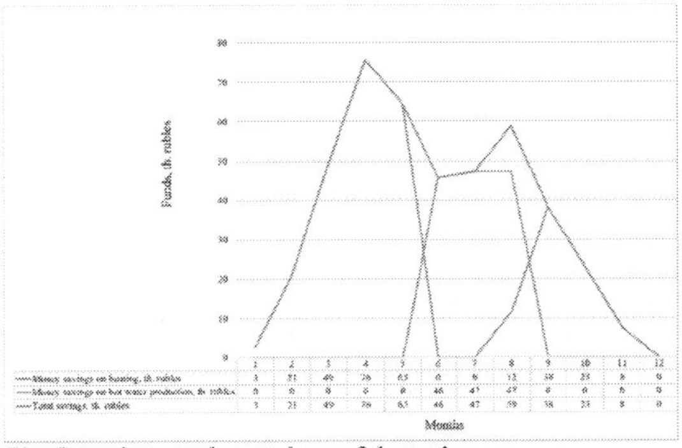

Fig. 5. Savings on the purchase of thermal energy.

Economic characteristics for Verkhoyansk can be seen in Table 1.

Table 1. Economic characteristics of the project.

№	Parameter	Value
1	Payback period, years	5.9
2	Discounted payback period, years	8
3	Operating costs, th. rubles	93.6
4	Investments, th. rubles	2000
5	Revenue, th. rubles/year	433.6
6	Gross profit, th. rubles/year	340.1
7	Economically justified tariff for thermal energy, rubles/Gcal	19195.9
8	Economically justified tariff for electricity, rubles/Gcal	62.6

Due to high tariffs for heat and electricity, the payback period for hybrid solar heating systems is quite favorable. This value may vary with different input data.

6 Conclusion

The work involved calculating the total solar radiation on an inclined surface using long-term archive data of total solar radiation on a horizontal surface obtained from meteorological stations. The recalculation method for total solar radiation on inclined surfaces was validated using experimental data from the test facility. Calculations of heat production by evacuated solar tube collectors were also performed, based on available data on incoming solar radiation. The payback period of solar heating systems in various locations was used as an efficiency measure for the solar collectors.

The results of the comparative analysis show that the results from the developed calculation methods and experimental data are generally comparable. Relative discrepancies in the monthly values can be explained by annual variations in solar radiation, deviations in climatic conditions from reference data, and the influence of various natural factors during the measurement process (e.g., dust, precipitation, frost formation, etc.). The discrepancy between the monthly values of total solar radiation was up to 53%, which can be attributed to the overestimated values produced by existing solar radiation calculation models.

A comparison of the calculated and experimental data on thermal energy production by the evacuated solar tube collector revealed that the discrepancy over the

© The Authors, published by EDP Sciences. This is an open access article distributed under the terms of the Creative Commons Attribution License 4.0 (https://creativecommons.org/licenses/by/4.0/).

month was up to 58%, which is comparable to the discrepancies observed in the solar radiation values.

Thus, for solar heating systems in Yakutsk, where the energy cost for thermal energy is approximately 2,800 rubles/Gcal, the payback period of the simulated system is around 29 years. In remote settlements, such as the town of Verkhoyansk, where the energy cost for thermal energy is more than 19,000 rubles/Gcal, the payback period is 5.9 years. Economic efficiency strongly depends on the installation location of the solar collector and the heating tariff.

Acknowledgments

This work was carried out within the framework of the State Assignment project No. FWRS-2024-0031 of the fundamental research program of the Russian Federation for 2021-2030.

References

1. Ed. by N.A. Petrov, B.G. Saneev, G.F. Alekseev, et al., Energy strategy of the Republic of Sakha (Yakutia) until 2030 (Yakutia Media Group, Yakutsk, 2010) (in Russian)
2. Decree of the Head of the Republic of Sakha (Yakutia) (Dec. 13, 2019), No. 897. The State Program of the Republic of Sakha (Yakutia) "Providing quality housing and improving the quality of housing and public services for 2020-2024," 132 p. URL: https://www.consultant.ru/regbase/cgi/online.cgi?req=doc;base=RLAW249;n=77806#EcCwOQUWs79FgPw (in Russian)
3. A.E. Ivanova, V.E. Zakharov, T.N. Petrova, Analysis of subsidizing thermal energy industry in the Arctic zone of the Sakha Republic (Yakutia). Bulletin of Irkutsk State Technical University. **24(1)**, pp. 123-134 (2020). https://doi.org/10.21285/1814-3520-2020-1-123-134 (in Russian)
4. V.E. Zakharov, T.N. Petrova, Analysis of budget costs for subsidizing electricity supply in the Sakha Republic (Yakutia) in 2011-2021, MCITO **64(5)**, September 11-15 (2023) (in Russian)
5. Handbook of the climate of the USSR. Issue 24. Yakut ASSR. Part 1. Solar radiation, radiation balance and insolation (Gidrometeoizdat, Leningrad, 1967) (in Russian)
6. D.V. Prokhorov, V.E. Zakharov, S.S. Vasilev, Evacuated solar tube collectors for heating in Central Yakutia. Energy Systems Research. **4(24)**, pp. 67-71 (2023) doi: 10.25729/esr.2023.04.0007
7. J.A. Duffie, W.A. Beckman, Solar engineering of thermal processes. 4th Edition (John Wiley & Sons, New York, 2013)
8. M. Jamil Ahmad, G.N. Tiwari, Optimization of tilt angle for solar collector to receive maximum radiation. The Open Renewable Energy Journal. **2**, pp. 19-24 (2009). doi: 10.2174/1876387100902010019
9. T.O. Kaddoura, M.A.M. Ramli, Y.A. Al-Turki, On the estimation of the optimum tilt angle of PV panel in Saudi Arabia. Renewable and Sustainable Energy Reviews. **65**, pp. 626-634 (2016). https://doi.org/10.1016/j.rser.2016.07.032
10. V.A. Shakirov, A technique for estimating global solar radiation on inclined surfaces using multi-year archives of meteorological data. Systems. Methods. Technologies. **4(36)**, pp. 115-121 (2017). https://doi.org/10.18324/2077-5415-2017-4-115-121 (in Russian)
11. S.I. Sivkov, Methods for calculating solar radiation characteristics (Gidrometeoizdat, Leningrad, 1968)
12. B.M. Galperin, T.A. Chentsova, On the calculation of monthly solar radiation amounts based on total and lower cloudiness. Proceedings of the LGMI. **48**, pp. 119-124 (1972) (in Russian)
13. S.A. Klein, W.A. Beckman, J.A. Duffie, A design procedure for solar air heating systems. Solar Energy. **19(5)**, pp. 509-512 (1977). https://doi.org/10.1016/0038-092X(77)90106-2
14. Methods of testing to determine the performance of solar collectors, (ASHRAE standard, Atlanta, 2003).

© The Authors, published by EDP Sciences. This is an open access article distributed under the terms of the Creative Commons Attribution License 4.0 (https://creativecommons.org/licenses/by/4.0/).

AUTHOR INDEX

Abdulkhayev Khurshed 238
Akhmetbaev Arman 64, 69
Akhmetbaev Dauren 64, 69
Aleksandrov Nikolay 209
Alizade Rana 213, 217
Alzakkar Ahmad 121
Asiev Abai 228
Askarov Alisher 75, 80
Babayeva Aytek 89
Baraishuk Siarhei 131, 238
Batukhtin Andrey 121
Boldyrev Ilya 161
Boloev Evgeny 198
Boyarkin Denis 116
Buchinsky Anatoly 198
Butin Kirill 181, 204
Bystrova Svetlana 64
Chukreyev Iurii 1, 5
Chukreyev Mikhail 1, 9
Danilov Gleb 19
Darian L. A. 50
Drako Mikhail 131, 136
Dubinin Dmitriy 186, 190
Dzhandigulov Abdygali 64
Edelev Alexey 24
Emelyanov Alexander 126
Gabdushev Dmitry 101
Galiaskarov Irek 85
Golub Irina 198
Govorun Mikhail 186, 190
Grabchak Evgeny 55
Grigoriev Vladimir 55, 59
Guliyev Huseyngulu 89, 95
Hashimov Arif 89
Hvijuzov Mikhail 126
Iakubovskii Dmitrii 116
Ilyushin Pavel 36
Imanakunova Zgenishkul 228
Ismukov Grigory 194
Ivanitckii Maxim 161
Ivanova V. 233
Karimov Nikita 186
Kashyn Mikhail 41
Khamisov Oleg 19
Khokhrin Aleksandr 190
Khrennikov Alexander 209
Khromov Sergey 121
Kievets Anton 75
Klassen Viktor 166
Kolik Viachaslau 136
Korotchenko Evgeniy 85
Krupenev Dmitry 112, 116
Lachugin Vladimir 194
Loginov Evgeny 55, 59
Logunov V. A. 50
Luzhkovskiy Yuriy 190
Lvov Alexandr 194
Maiseyenkà Àleh 136
Maksimchenko A. V. 50
Mamadalieva U. 233
Mammadov Zaur 95
Mashalov Eugene 166
Menshikov Pavel 161
Mihajlenko Andrej 166
Mikhailov Konstantin 209
Mikhailov Sergey 209
Misrikhanov Misrikhan 85
Mokeev Alexey 151, 156
Molochko Andrey 41
Motovilov Alexey 126
Murodov Muzaffar 238
Nasibov Valeh 217
Navolochny Alexander 141
Neuymin Vladimir 166
Novikov Alexander 41, 171
Novikov Nikolay 41, 171
Obraztsov R. M. 50
Omorov Turatbek 222, 228
Onisova Olga 141
Osmonova Rima 228
Pavlovich Ivan 131, 238
Piskunov Sergey 151, 156
Plakidin Roman 177
Podshivalin Andrey 194
Polishchuk V. P. 50
Popov Aleksandr 181, 204
Prokhorov Dmitry V. 242
Rahmanov Nariman 95
Rodionov Andrey 181, 204
Romanova Victoria 121
Rudnik Vladimir 80
Ryabchenko Vladimir 85
Sabirov Irek 146
Salmin Alexander 141
Satsuk Evgeniy 186, 190
Senderov Sergey 14, 19
Shavlovskiy Sergey 36
Sheveleva Galina I. 107
Shkuta Alexander 59
Shuntov Andrey 85
Smirnova Elena 28
Snegirev Denis 166
Solovyev Ivan 126
Sorokin Dmitry 45
Suerkulov Manas 228
Sultanov Mahsud 161
Suprunov Igor 171
Suvorov Aleksey 75, 80

AUTHOR INDEX

Takyrbashev Beishenaly 222, 228
Teplukhin Daniil.. 166
Titov Sergey... 190
Tsipkina V. 233
Tulsky Vladimir 101
Ufa Ruslan 75
Ulyanov Dmitriy.. 177
Ulyanov Dmitry... 156
Vanin Artem 101
Vankov Yuri... 146
Vasilev Semen S.. 242
Verakso D. E. 50
Voitov Oleg 198
Vorobev Sergey.. 19, 24
Ýbtahimov Famil 95
Yusifbayli Nurali 213
Zakiriaev Kubnychbek 222
Zholdoshov Tolkunbek 222
Zhukov Andrey 186
Zvonareva Julia.. 146

EDP Sciences – Web of Conferences
17, Avenue du Hoggar
Parc d'Activité de Courtabouf
BP 112
F-91944 Les Ulis Cedex A
France

ISBN 979-8-3313-0871-1

CURRAN ASSOCIATES INC.
proceedings.com